Fifth Edition

HUMAN HEREDITY

Principles and Issues

Michael R. Cummings
University of Illinois at Chicago

Brooks/Cole
Thomson Learning ™

Pacific Grove • Albany • Belmont • Boston • Cincinnati • Johannesburg • London
Madrid • Melbourne • Mexico City • New York • Scottsdale • Singapore • Tokyo • Toronto

Sponsoring Editor: *Nina Horne*
Project Development Editor: *Marie Carigma-Sambilay*
Marketing Team: *Tami Cueny, Kelly Fielding,*
  and Dena Donnelly
Editorial Assistant: *John-Paul Ramin*
Production Coordinator: *Mary Anne Shahidi*
Production Service: *Carlisle Publishers Services*
Permissions Editor: *Mary Kay Hancharick*
Interior Design: *Laurie Janssen*

Cover Design: *Irene Morris*
Cover Photos: *Charles C. Benton*
Interior Illustration: *Veronica Burnett*
Photo Researcher: *Sue C. Howard*
Print Buyer: *Vena Dyer*
Typesetting: *Carlisle Communications*
Cover Printing: *Phoenix*
Printing and Binding: *R. R. Donnelley—*
  Roanoke

For more information, contact:
BROOKS/COLE
511 Forest Lodge Road
Pacific Grove, CA 93950 USA
www.brookscole.com

For permission to use material from this work, contact us by
Web: www.thomsonrights.com
fax: 1-800-730-2215
phone: 1-800-730-2214

Printed in United States of America

10 9 8 7 6 5 4 3 2 1

Library of Congress Cataloging-in-Publication Data

Cummings, Michael R. [date]
   Human heredity : principles and issues / Michael R. Cummings.—
5th ed.
      p.   cm.
   Includes bibliographical references and index.
   ISBN 0-534-52372-2
   1. Human genetics.   2. Heredity, Human.   I. Title.
QH431.C897 2000
599.93'5—dc21
                              99-37549
                        CIP

To those who mean the most,

Lee Ann,

Brendan and Shelly,

Kerry, Terry, Colin, and Maggie.

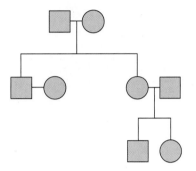

ABOUT THE AUTHOR

MICHAEL R. CUMMINGS received his Ph.D. in Biological Sciences from Northwestern University in 1968. His doctoral work, conducted in the laboratory of Dr. R. C. King, centered on ovarian development in *Drosophila melanogaster*. After a year on the faculty at Northwestern, he moved to a teaching and research position at the University of Illinois at Chicago. Here, he established a research program on the developmental genetics of *Drosophila* and began teaching courses in genetics, developmental genetics, and evolution. Currently an associate professor in the Department of Biological Sciences and in the Department of Genetics, he has also taught at Florida State University.

About ten years ago, Dr. Cummings developed a strong interest in scientific literacy. In addition to teaching genetics to biology majors, he organized and currently teaches a course in human genetics for non-majors, and participates in teaching general biology. He is now working to integrate the use of electronic resources such as the Internet and World Wide Web into the undergraduate teaching of genetics and general biology. His current research interests involve the role of the short arm/centromere region of human chromosome 21 in chromosomal aberrations. His laboratory is engaged in a collaborative effort to construct a physical map of this region of chromosome 21 to explore molecular mechanisms of chromosome interactions.

Dr. Cummings is the author and co-author of a number of widely used college textbooks, including *Biology: Science and Life, Concepts of Genetics,* and *Essentials of Genetics.* He has also written sections on genetics for the *McGraw-Hill Encyclopedia of Science and Technology,* and has published a newsletter on advances in human genetics for instructors and students.

He and his wife, Lee Ann, are parents of two adult children, Brendan and Kerry, and have two grandchildren, Colin and Maggie. He is an avid sailor, enjoys reading and collecting books (biography, history), music (baroque, opera, and urban electric blues), eating the fine cuisine at Al's Beef, and is a long-suffering Cubs fan.

Contents

CHAPTER 3

Transmission of Genes from Generation to Generation   46

CHAPTER 6

Cytogenetics   140

CHAPTER 7

Development and Sex Determination   174

CHAPTER 8

DNA Structure and Chromosomal Organization   200

CHAPTER 9

Gene Expression: How Proteins Are Made   219

CHAPTER 10

From Proteins to Phenotypes   240

CHAPTER 11

Mutation: The Source of Genetic Variation   267

CHAPTER 12

An Introduction to Cloning and Recombinant DNA 290

CHAPTER 13

Applications of Recombinant DNA Technology   313

CHAPTER 14

Genes and Cancer   340

CHAPTER 15

Genetics of the Immune System 367

CHAPTER 16

Genetics of Behavior   394

CHAPTER 17

Genes in Populations   421

CHAPTER 18

Human Diversity and Evolution   441

CHAPTER 19

Genetic Screening and Genetic Counseling   468

Preface

*W*hen the first edition of *Human Heredity* was published in 1988, molecular biology and recombinant DNA technologies were just beginning to transform the field of human genetics. At that time, biotechnology was a fledgling industry, and the scientific disciplines of genomics and bioinformatics did not yet exist. The Human Genome Project was in the early stages of planning, and its inception was two years away. In the twelve years since then, the biotechnology industry has grown to become a twenty billion dollar component of the economy, thousands of human genes have been identified, and the sequence of most of the human genome will be completed while this edition is being used in the classroom.

Keeping in mind how genetic research, technology, and knowledge are transforming research, industry, agriculture, and our everyday lives, I have written this book for a one-term introductory human genetics course for the non-major. It assumes that the student has little or no background in biology, chemistry, or mathematics, but has an interest in learning something about human genetics. Some descriptive chemistry is used after an appropriate introduction and definition of terms. In the same vein, math is used in several places, but no advanced math skills are required to calculate elementary probabilities or to calculate genotype and allele frequencies.

To emphasize the need for non-science students to acquire an understanding of basic genetic concepts rather than a collection of facts, each chapter is organized around one or two central ideas. For example, Chapter 3 deals with Mendelian principles, and uses crosses in pea plants to explain basic concepts. The use of an experimental organism (in this case, peas) to illustrate the principles that govern the transmission of traits from generation to generation makes it easy to define each concept and provide examples that are clear cut and unambiguous. With a firm grounding in these principles, students will be able to apply them to the transmission of traits in humans, where the methods are indirect and observational rather than experimental.

The speed at which genetics is advancing and transferring knowledge into many other areas including medicine, law, agriculture, and the pharmaceutical industry makes it clear that difficult, informed decisions need to be made at many levels, from the personal to the societal. The public, elected officials, and policy makers outside the biological sciences community need to have a working knowledge of genetic principles in order to support and help shape the course of research and its applications in our society. This text, based on a hierarchical approach that links a basic set of genetic concepts, has been written to transmit these principles without unnecessary jargon, detail, or the use of anecdotal stories.

This book is written by an author who works with undergraduates on a daily basis, teaching courses in a biology curriculum to both majors and non-majors. Over the last decade, students have offered suggestions about how to introduce a topic, have identified the most effective examples and analogies in explaining concepts, and more importantly, been forthright in clarifying what does not work in the classroom. As in the past, the organization and content for this edition have been developed by incorporating classroom-tested ideas refined by student feedback. With this background, this text has been written to achieve several well-defined goals, and this edition reinforces and extends these goals:

1. Present the principles of human genetics in a clear, concise manner that gives students a working knowledge of genetics. The premise behind this

approach is that the use of a limited number of clearly presented, interlinked concepts is the best way to learn a complex subject such as genetics.

2. Communicate an understanding of the origin and amount of genetic diversity present in the human population, and how this diversity has been shaped by natural selection.

3. Examine the social, cultural, and ethical implications associated with the use of genetic technology.

4. Begin the discussion of concepts at a level that students can understand, and provide relevant examples that students can apply to themselves, their families, and their work environment.

To achieve these goals, emphasis has been placed on clear writing with the use of accompanying photographs and artwork that teach rather than merely illustrate the ideas under discussion. In addition, the book features up-to-date coverage and flexible organization. A conscious effort has been made to pare down unnecessary terms and jargon and to present the material in a straightforward and engaging manner.

In general, the text consists of three sections: Chapters 1–7 cover cell division, transmission of traits from generation to generation and development. Chapters 8–13 emphasize molecular genetics, recombinant DNA and biotechnology. These chapters cover gene action, mutation, cloning, the applications of genetic technology, the Human Genome Project, and the social, legal and ethical issues related to genetics. Chapters 14–19 consider specialized topics, including cancer, the immune system, population genetics, as well as the social aspects of genetics including behavior, genetic screening, and genetic counseling.

Because courses in human genetics have a wide range of formats, the book is organized so it will be easy to use, no matter what order of topics an instructor chooses. After the section on transmission genetics, the chapters can be used in any order. Within each chapter, the outline lets the instructor and students easily identify central ideas.

FEATURES OF THE FIFTH EDITION

New Chapter

Chapter 1 has been rewritten to serve as an introduction to the science of genetics, and provides an overview of the structure and scope of the field, and a timeline of major developments in human genetics. The social impact of human genetics on law and social policy is emphasized, and the need for informed decisions in the future is outlined. This chapter provides a perspective for what follows in subsequent chapters.

New Organization

The order of chapters is the same as in the last edition. The reorganization and reordering of chapters that took place in the last edition has proven successful, and has been continued here. Mendelian inheritance and quantitative inheritance are in one section, and the chapter on mutation precedes the chapters on recombinant DNA technology and its applications. In this edition, emphasis has been placed on ensuring that each chapter focuses on one or two central ideas. As a result, material has been moved, rewritten to shift emphasis, and new sections have been added as needed.

A few examples of these changes will serve to illustrate how this focus was developed. In Chapter 2, the sections on the human chromosome set and the box on making a human karyotype have been moved to Chapter 6, leaving the central ideas of cell structure, mitosis, and meiosis intact. In addition, the essay on cell membranes has been moved from Chapter 4 to Chapter 2, and the essay from Chapter 2 on aging has been moved to Chapter 4. In Chapter 7, the section on human reproduction has been eliminated, and the chapter now begins with human development from fertilization to birth. This places the sections on sex determination and differentiation into a broader context. In Chapter 8, the section on viruses has been eliminated, and the chapter begins with a discussion of the evidence that DNA is a carrier of genetic information. Once this has been established, a discussion of DNA as a part of human chromosomes follows, along with sections on the structure and replication of DNA. The result is a chapter that deals with a few central ideas, using clear-cut examples for each idea.

Expanded Questions and Problems

Recognizing that many students have difficulty solving genetics problems, the end-of-chapter questions and problems have been revised and expanded. The revisions and additions have been contributed by adopters of the text who have used their experience and student input to redesign, rewrite, and add to the problem sets. The questions and problems use both an objective question and problem format, and are arranged by level of difficulty. Because some quantitative skills are necessary in human genetics, almost all chapters include some problems that require the students to organize the concepts in the chapter and use these concepts in reasoning to a conclusion. Answers to selected problems are provided in Appendix B.

New Case Histories

To make issues in human genetics relevant to situations that students may encounter outside the classroom, a new section on case histories has been added at the end of each chapter. This section contains scenarios and examples of genetic issues related to health, reproduction, personal decision making, public health, and ethics. Many of these can be used as the basis for classroom discussions, student presentations, and role playing.

New Topics

Many new topics have been introduced, or have received greater emphasis. It is impossible to list them all, but included among them are pharmacogenetics; background radiation exposure, radiation as a source of mutation, and DNA repair mechanisms; cloning of mammals from somatic cells; mapping genes by positional cloning, preimplantation genetic testing, the use of DNA chips in genetics, biopharming, and the ethical issues surrounding the Human Genome Project; DNA repair mutations in cancer, the role of gatekeeper and caretaker genes in cancer, genomic instability and cancer, and behavior and cancer; the search for genes controlling manic depression and schizophrenia; and genetic evidence for the spread of humans across the world.

Guest Essays

This edition features a series of essays written by distinguished scientists, describing how they became interested in science, what they have chosen to study, and how their research relates to the larger context of human genetics. These essays are not just *about* scientists, they are written *by* scientists, giving non-majors some insight into the lives, thoughts, and motives of biologists and geneticists.

Genetic Databases

To foster awareness of the vast array of databases dealing with genetics, the genetic disorders mentioned in the book are referenced with the indexing number assigned to them in the comprehensive catalog assembled by Victor McKusick and his colleagues.

This catalog is available in book form as *Mendelian Inheritance in Man: Catalog of Human Genes and Genetic Disorders*. It is also available at several World Wide Web sites as *Online Mendelian Inheritance in Man (OMIM)*. The online version (with daily updates) contains text, pictures, and videos along with references to the literature and links to other databases, including those related to the Human Genome Project. Students and an informed public need to be aware of the existence and relevance of such databases, and OMIM is used here in part to promote such awareness.

Students wishing to learn more about a particular genetic disorder can use OMIM to obtain detailed information about the disorder, its mode of inheritance, phenotype and clinical symptoms, mapping information, biochemical properties, the molecular nature of the defect, and availability of cloned sequences for the gene in question. In the classroom, OMIM and its links to other databases are a valuable resource for student projects and presentations.

New Internet Activities

The World Wide Web (WWW) is an important and valuable resource in teaching human genetics, and the exercises included in this edition can be used to expand on concepts covered in the text, and provide detailed information about specific genetic disorders. They can also be used to introduce the social, legal, and ethical aspects of human genetics into the classroom and serve as a point of contact with support groups and testing services.

This edition contains updated and expanded end-of-chapter Internet activities for students. These activities use WWW resources to enhance the topics covered in the chapter, and are designed to generate interaction and thought rather than passive observation. Sites for these activities can be reached through the book's home page.

PEDAGOGICAL FEATURES

The basic organization within chapters, which has been successful as a teaching resource, has been continued in this edition. Many of these features have been updated and revised to reflect current topics, engage student interest, and improve the pedagogy.

Opening Vignettes

Each chapter begins with a short prologue directly related to the main ideas of the chapter, often drawn from real life. Topics include genetic discoveries, such as the chromosomal basis of Down syndrome, the development of genetic technologies such as *in vitro* fertilization, and the birth of Louise Brown, the first IVF baby. These vignettes are designed to engage student interest in the topics covered in the chapter and to demonstrate that laboratory research often has a direct impact on everyday life.

Chapter Outlines

At the beginning of each chapter, an outline provides an overview of the main concepts, secondary ideas, and examples. To help students grasp the central points, many of the headings have been rewritten as narratives or summaries of the ideas that follow. These outlines also serve as convenient starting points for students to review the material in the chapter.

Concepts and Controversies

Within most chapters, students will find boxes that present ideas and applications related to the central concepts in the chapter. Some of these present interesting but tangential examples that should be of interest to the student, while others examine controversies that arise as genetic knowledge is transferred into technology and services.

Guest Essays

Scattered throughout the book are essays written by prominent scientists. The essays introduce the human side of scientists, and summarize how they became interested in science, and how their work relates to larger issues in society.

Margin Glossary

A glossary in the page margins gives students immediate access to definitions of terms as they are introduced in the text. This format also allows definitions to be identified when students are studying or preparing for examinations. These definitions have been gathered into an alphabetical glossary at the back of the book. Because an understanding of the concepts of genetics depends on understanding the relevant terms, more than 350 terms are included in the glossary.

Sidebars

Throughout the book, sidebars are used to highlight applications of concepts, present the latest findings, and point out controversial ideas without interrupting the flow of the text.

END-OF-CHAPTER–FEATURES

The end-of-chapter features have been revised and updated for this edition, and a new section on case histories has been added.

Case Studies

As described earlier, this new section contains case histories of individuals and families using various genetic services, large-scale issues such as radioactive pollution, and the impact of the Human Genome Project. Many of these can be used as the basis for classroom discussions, student presentations, and role playing.

Summary

Each chapter ends with a numbered summary that restates the major ideas covered in the chapter. Beginning each chapter with an outline and ending with a summary of the major concepts and their applications helps focus the students' attention on the conceptual framework and minimizes the chance that they will attempt to learn by rote memorization of facts.

Questions and Problems

The questions and problems at the end of each chapter are designed to test students' knowledge of the facts and their ability to reason from the facts to conclusions. This section has been revised and expanded by contributions from adopters of previous editions, by professionals in genetic research and health care services, and by students, who suggested many of the revisions and new problems.

Internet Activities

Activities at the end of each chapter use Web sites to engage the student in activities related to the concepts discussed in the text. This section has been revised and expanded in this edition.

For Further Reading

A list of readings is presented at the end of each chapter. These include reviews and general articles that are accessible to the non-scientist, as well as the key papers that describe discoveries covered in the chapter. The references have been updated just before publication to provide the most current coverage of the literature.

ANCILLARY MATERIALS

The expanded array of ancillary materials that accompany this edition are designed to assist the instructor in preparing lectures and examinations, and to help keep instructors abreast of the latest developments in the field.

Instructor's Manual

An expanded and updated instructor's manual is available to help instructors in preparing class materials. It contains chapter outlines, chapter summaries, teaching/learning objectives, key terms, additional test questions, and discussion questions. It also contains answers to end-of-chapter questions and problems that are not given in the text.

Test Bank

A computerized test bank, available on CD-ROM, can be customized and printed according to the instructor's preferences and needs. It contains approximately 800 test items.

Transparencies

A set of 100 color transparencies featuring key figures—including drawings, charts, and diagrams from the text—is available to adopters.

Study Guide

A student study guide has been prepared by Nancy Shontz of Grand Valley State University. It is intended to enhance understanding of the text and course material. It includes chapter objectives and summaries, lists of terms, case worksheets (based on case studies in the text), discussion problems and questions, and other practice test items in multiple-choice, fill-in-the-blank, and modified true/false formats.

Gene Link 2.0

A lecture/presentation tool for instructors, which allows them to select from a large database of images from this and other related Brooks/Cole texts as well as animations that can be arranged and edited to create dynamic, motivating lectures. The lectures created in this format (including animations, images, notes and URLs), can be posted on the Web.

Online Genetics Newsletter

An electronic newsletter (on the Brooks/Cole Biology Resource Center) that features articles and essays on current topics and discoveries in the field of human and general genetics with links to related sites.

Current Perspectives in Genetics

A reader prepared by Shelly Cummings of the University of Chicago features approximately 40 articles in molecular, classical, and human genetics. Each article begins with a brief introduction and ends with critical thinking or discussion questions. Answers are provided in the back of the book.

Genetics on the Web: A Brief Guide to the Internet

A handy reference for students who need an introduction to research on the Internet. It includes a list of relevant URLs listed by genetics topics.

Instructor's Edition

A special edition of the text for instructors that features extra frontmatter, including a visual preface.

CONTACTING THE AUTHOR

I welcome questions and comments from faculty and students about the book or about human genetics. Please contact me at:

cummings@uic.edu

ACKNOWLEDGMENTS

A text is originally shaped by the vision and teaching philosophy of its author. By the time it reaches the fifth edition, it has been reshaped by changes in the field, and more importantly, by reviewers, editors, instructors, and students. Hopefully, with each of these changes, it becomes a more useful teaching tool that represents the distilled wisdom of those who have contributed to it. Over its lifetime, this text has had many reviewers, but three stand out for the time and effort they have expended in improving the presentation of topics, sharpening the language, and often, teaching or re-teaching me things about genetics and pedagogy. Through their patience and efforts, they have helped me become a better author and a better teacher. I owe them a great debt of gratitude, and I thank them for their contributions to this book and to the teaching of human genetics. They are: H. Eldon Sutton of the University of Texas, George Hudock of Indiana University, and Werner Heim of Colorado College.

To all the reviewers who helped in the preparation of this edition, I extend my thanks and gratitude for their efforts and many suggestions. Their efforts have enhanced the focus and presentation of the material.

Thomas Breen
Southern Illinois University

James Brennan
Bridgewater State College

Richard Cosby
Southern University, New Orleans

Joe Dickinson
University of Utah

Carl Frankel
Pennsylvania State University, Hazelton

Meredith Hamilton
Oklahoma State University

John Haynie
SUNY College, Genesco

George Hudock
Indiana University

Cran Lucas
Louisiana State University, Shreveport

Wendell McKenzie
North Carolina State University

Charlotte Omoto
Washington State University

Douglas K. Walton
College of St. Scholastica

Frank Potter
Fort Hays State University

Once again, Michelle Murphy of Notre Dame University undertook the task of revising and adding to the end-of-chapter questions and problems. Writing good questions is one of the most difficult tasks in teaching, and Michelle's efforts show through in this expanded and pedagogically improved feature of the text. Shelly Cummings of the University of Chicago prepared the case histories at the end of each chapter. Her hard work and cheerful optimism have produced a new and valuable feature for this edition. The Internet activities were revised and expanded by Kim Finer of Kent State University, who is recognized as one of the pioneers in using electronic media in the classroom. She was assisted by Peter Follette.

The book has had a warm and welcoming reception in its new home at Brooks/Cole Publishing. The transition from West Publishing, via Wadsworth Publishing was guided by Gary Carlson, Executive Editor, who ensured that the transfer took place with a minimum of problems. He also worked to develop the approach and philosophy of this edition. My editor, Nina Horne, with her experience in publishing and knowledge of the marketplace, has helped me to emphasize the human aspects of human genetics, and guided the preparation of this edition. The heart and soul of this edition has been Marie Carigma-Sambilay, my project development editor. She worked tirelessly to coordinate all aspects of this project, and in spite of the inevitable delays, setbacks, and dilemmas, was always cheerful, upbeat, and optimistic. Editorial assistants Larisa Lieberman and John-Paul Ramin efficiently handled the review focus. Mary Anne Shahidi, the production coordinator, oversaw the book's production process. Her positive attitude was always a steadying influence. Chad Thomas and Terry Routley at Carlisle Communications smoothly guided the production of this edition, including the development of the design, art, and photo program. Roy Neuhaus was the creative force behind the cover design.

The concern and care expressed by the staff at Wadsworth and Brooks/Cole during my illness was heartwarming. I deeply appreciate their cards, flowers, and gifts, and I will always remember their kindness. I thank them and appreciate their efforts in overcoming the ensuing delays.

Once again, I wish to express special thanks to my colleague, Suzanne McCutcheon, who gave her time to help supervise the research of graduate students in my lab while this edition was in preparation. I also thank my students, Holly Dimitropoulos and David Wolff for their patience.

Michael R. Cummings

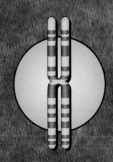

Genetics as a Human Endeavor

WELCOME TO GENETICS

In mid-December, 1998, the media reported that the Icelandic Parliament (*Althingi*) passed a controversial bill that has generated international debate on several key issues, including privacy, medical ethics, the mapping of human genes, and the role of biotechnology in health care. The bill allows a company, deCODE Genetics, to establish and operate a central database that includes the health records of all Icelanders. The company's database will also contain the health records of individuals dating back to 1915, when the National Health Service began. In addition, deCODE is compiling a database of the genealogies of the approximately 800,000 Icelanders who have lived, since the colonization of the island in the ninth and tenth centuries. In combination with blood and tissue samples (for DNA extraction) provided by patients, the database is a powerful tool in the hunt for disease-causing genes. The new law grants the company the right to sell this information (and the DNA samples) to third parties, including the research labs of pharmaceutical companies with the hope that once disease genes are identified, diagnostic tests and therapies will soon follow.

Why establish such a database in Iceland? The answer is in the history of the island and its people. The population of Iceland was established by a small number of individuals in the ninth and tenth centuries, and until fifty years ago, was almost completely isolated from outside immigration. In addition, bubonic plague (in the 1400s) and a volcanic eruption (in the 1700s) decimated the population, further reducing genetic variation. As a result, the 270,000 inhabitants of Iceland share a remarkably similar set of genes, providing fertile ground for gene hunters seeking to identify disease genes.

To illustrate the power of the approach, using its database and DNA samples, deCODE researchers took only 2 1/2 months to identify and map a gene for a disorder called familial essential tremor. Most gene searches take years or in some cases, decades to complete.

Why the controversy? Those opposed to the pooling and selling of Icelanders' genetic information point out that because of the size and relatedness of the population, the privacy provisions of the law are inadequate and may violate the ethical principle that health records must be kept confidential. Abuses and misunderstandings may affect employment, insurance, and even marriage. In addition, critics question whether a single company should have exclusive rights to medical information, and that the Icelandic population will derive no health benefits from this arrangement. The argument is far from settled, and may involve action in the European Court (in Strasbourg) as well as censure by international medical societies.

Almost every day the media carry a story about genetics. Often the stories are straightforward accounts reporting that a gene for a genetic disorder has been cloned, or that a new understanding of how a gene works has been achieved. On other occasions, as in the case of the Icelandic law, papers and newsmagazines devote several pages to a story, along with interviews of geneticists, industry spokesmen, and consumer groups.

With stories like those about Iceland becoming more common in the media, as we begin this book, we might pause and remember that genetics is more than a laboratory science; unlike some other areas of science, genetics and biotechnology have a direct impact on society.

■ **Genetics** The scientific study of heredity.

As a first step in studying human genetics, we should ask what *is* genetics? **Genetics** is the science of heredity. It encompasses the study of cells, individuals, as well as populations and species of organisms. As we will see in a later section, geneticists study the transmission of traits from generation to generation, the molecular nature of genes and gene products, how genes are expressed and regulated, and the dynamics of gene frequencies in populations. Other geneticists work to develop products for industries as diverse as agriculture and pharmaceutical firms by applying knowledge and methods first discovered in genetic research. This area of applied research, called biotechnology, has grown in recent years to become a multi-billion dollar component of our economy.

Genetic Disorders in Culture and Art

*I*t is difficult to pinpoint when the inheritance of specific traits in humans was first recognized. Descriptions of heritable disorders often appear in myths and legends of many different cultures. In some ancient cultures assigned social roles—from prophets and priests to kings and queens—were hereditary. The belief that certain traits were heritable helped shape the development of many cultures and social customs.

In some ancient societies, the birth of a deformed child was regarded as a sign of impending war or famine. Clay tablets excavated from Babylonian ruins record more than 60 types of birth defects, along with the dire consequences thought to accompany such births. Later societies, ranging from the Romans to 18th-century Europe, regarded malformed individuals (such as dwarfs) as curiosities rather than figures of impending doom, and they were highly prized by royalty as courtiers and entertainers.

Whether motivated by fear, curiosity, or an urge to record the many variations of the human form, artists have portrayed both famous and anonymous individuals with genetic disorders in paintings, sculptures, and other forms of the visual arts. These portrayals are often detailed, highly accurate, and easily recognizable today. In fact, across time, culture, and artistic medium, affected individuals in these portraits often resemble each other more closely than they do their siblings, peers, or family members. In some cases, the representations allow the disorder to be clearly diagnosed at a distance of several thousand years.

Throughout the book you will find fine-art representations of individuals with genetic disorders. These portraits represent the long-standing link between science and the arts in many cultures. They are not intended as a gallery of freaks or monsters, but as a reminder that being human encompasses a wide range of conditions. A more thorough discussion of genetic disorders in art is in *Genetics and malformations in art,* by J. Kunze and I. Nippert, published by Grosse Verläg, Berlin, 1986.

In a sense then, genetics is the key to all of biology, because genes control the structure and function of all cells, as well as large-scale processes such as development and reproduction. An understanding of genetic mechanisms is essential in deciphering how living systems are organized and how they function.

In the chapters that follow, we will examine how genes are transmitted from generation to generation, the nature and location of genes, how they are expressed as gene products called proteins, and how proteins result in traits that we can observe and study. Since this book is about human genetics, we will use human genetic disorders as examples of inherited traits (see Concepts and Controversies: Genetic Disorders in Culture and Art). In addition, we will go beyond a discussion of genetics as a science to examine the social, political, legal and ethical implications of genetics and genetic technology generated by advances in molecular genetics and biotechnology. In the rest of this chapter, we will preview some of the basic concepts of human genetics and introduce some of the social issues inspired by genetic research. These concepts and issues will be explored in more detail in the chapters that follow.

WHAT ARE GENES AND HOW DO THEY WORK?

Simply put, a gene is the functional unit of heredity. At the molecular level, a gene is a linear sequence of chemical building blocks called nucleotides in a **DNA** molecule (DNA is shorthand for deoxyribonucleic acid) (❯ Figure 1.1). The sequence of nucleotides in a DNA molecule stores information in the form of a **genetic code.** The "letters" of the code specify the type and order of chemical subunits (amino acids) that make up gene products (polypeptides). After the stored information is decoded in the cell, it is translated into a polypeptide, a molecule which folds into a three-dimensional structure that makes the molecule functional. These functional polypeptides are known as proteins (❯ Figure 1.2). The action of proteins produces

■ **DNA** A helical molecule consisting of two strands of nucleotides that is the primary carrier of genetic information.

■ **Genetic code** The sequence of nucleotides that encodes the information for amino acids in a polypeptide chain.

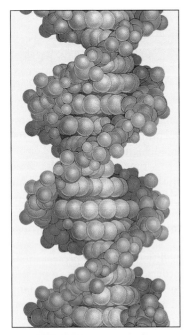

▶ **FIGURE 1.1** The structure of DNA, discovered by James Watson and Francis Crick in 1952, marked the beginning of molecular genetics. Fairbanks/Andersen, 1999.

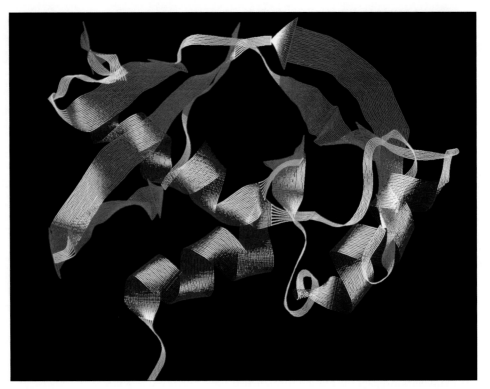

▶ **FIGURE 1.2** A diagram showing how a polypeptide chain becomes folded into the three dimensional structure of a protein.

observable traits that are studied for their pattern of transmission from generation to generation.

Genes can also be thought of in terms of their properties. Genes can undergo replication (they are copied), mutation (they can undergo change), expression (they can be turned on and off), and recombination (they can move from one chromosome to another).

HOW ARE GENES TRANSMITTED FROM PARENTS TO OFFSPRING?

Thanks to the work of Gregor Mendel (▶ Figure 1.3), an Augustinian monk, we understand how genes are passed from parents to offspring. In the late nineteenth century, Mendel conducted breeding experiments using pea plants. In his experiments, the parent plants each had a different, distinguishing characteristic called a **trait.** For example, he crossbred tall pea plants with short plants, or yellow-seeded plants with green-seeded ones, and kept careful records of the number of progeny with each trait. He followed the inheritance of traits through several generations, and based on his results, developed ideas about how traits are inherited. Mendel showed that traits are passed from generation to generation through the inheritance of "factors." We now know that these factors are genes. He reasoned that each parent carries a pair of genes for a given trait, and that each parent contributes one gene to the traits shown in the offspring.

Mendel concluded that pairs of genes separate from each other during the formation of egg and sperm. When the egg and sperm fuse during fertilization to form a zygote, the genes from the mother and father become members of a new gene pair in the offspring. Later, researchers discovered that genes are located on chromosomes, which are linear DNA molecules complexed with proteins. Chromosomes

■ **Trait** Any observable property of an organism.

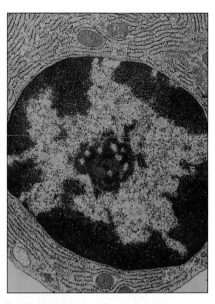

▶ **FIGURE 1.3** Gregor Mendel, the Augustinian monk whose work on pea plants provided the foundation for genetics as a scientific discipline.

▶ **FIGURE 1.4** The nucleus of a eukaryotic cell as seen in the electron microscope. The uncoiled chromosomes are represented by the dark clumps of chromatin throughout the nucleus.

are found in the nucleus of eukaryotic cells (▶ Figure 1.4). The basis of the separation of genes during the formation of sperm and egg and the reunion of genes at fertilization is explained by the behavior of chromosomes in a form of cell division called meiosis.

When Mendel published his work on the inheritance of traits in pea plants, ideas about inheritance that had held sway for centuries were replaced with new theories based on specific and fundamental biological mechanisms. Mendel began the scientific study of heredity, which has expanded in many directions in the past hundred years.

HOW DO SCIENTISTS STUDY GENES?

Ideas that form the foundation of genetics have been discovered by studying a wide range of organisms including viruses, bacteria, fungi, insects, plants, as well as mammals, including humans. Because the principles of genetics are universal across living systems, discoveries made using one organism can be applied to other species, including humans. In spite of the fact that geneticists study many different species, they use a small number of basic approaches in their work.

Approaches to the Study of Genetics

The most basic approach in genetics, called **transmission genetics,** studies the pattern of inheritance of traits from generation to generation. In work with experimental organisms, geneticists carry out mating experiments to analyze the transmission of traits (such as height, eye color, hemophilia, and so on) from parents to offspring. These experimental matings are analyzed through several generations to establish the mode of inheritance of a trait. Gregor Mendel did the first significant work in this area in the mid-nineteenth century, using pea plants as his experimental organism. The methods he developed form the foundation of transmission genetics. In humans, however, experimental matings are not possible for ethical reasons, and another method, called pedigree analysis is used. **Pedigree analysis** uses family history to

■ **Transmission genetics** The branch of genetics concerned with the mechanisms by which genes are transferred from parent to offspring.

■ **Pedigree analysis** The construction of family trees and their use to follow the transmission of genetic traits in families. It is the basic method of studying the inheritance of traits in humans.

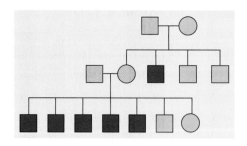

FIGURE 1.5 A pedigree is a representation of the inheritance of a trait through several generations of a family. In this pedigree, males are symbolized by squares, females by circles. Filled symbols indicate those affected by the trait.

■ **Cytogenetics** The branch of genetics that studies the organization and arrangement of genes and chromosomes using the techniques of microscopy.

■ **Karyotype** A complete set of chromosomes from a cell that has been photographed during cell division and arranged in a standard sequence.

■ **Molecular genetics** The study of genetic events at the biochemical level.

■ **Population genetics** The branch of genetics that studies inherited variation in populations of individuals and the forces that alter gene frequency.

follow a trait through several generations of a family and determine its pattern of inheritance (▶ Figure 1.5). Pedigree analysis is basic to the study of human heredity, and is the foundation for all further studies of a trait.

Cytogenetics is an approach that uses microscopy to study the nature, number and organization of chromosomes. It was used to make one of the most seminal discoveries in genetics, the chromosome theory of inheritance, the idea that genes are carried on chromosomes. One of the most important investigative approaches used in human genetics, cytogenetics is used to investigate abnormalities of chromosome structure and number. A basic method in human cytogenetics is the preparation of **karyotypes,** in which photographs of chromosomes are displayed in a standard arrangement to reveal abnormalities (▶ Figure 1.6). Cytogenetics is also used extensively in mapping genes to specific chromosomes and specific regions of chromosomes.

A third approach, **molecular genetics,** has had the greatest impact on human genetics over the last four or five decades. Molecular genetics uses recombinant DNA technology to identify, isolate, clone (produce multiple copies), and analyze genes. Cloned genes can be used to investigate the fundamental organization and function of genes; they also provide the basis for gene transfer between species, as well as gene therapy to treat human genetic disorders. Other applications of molecular genetics include the prenatal diagnosis of genetic disorders and DNA fingerprinting. Advances in molecular genetics have generated much of the debate about the social, legal and ethical aspects of genetics and biotechnology.

A fourth investigative approach involves the study of genes in populations, and the forces that change gene frequencies and drive the evolutionary process. Investigations in **population genetics** have defined how much genetic variation exists in populations, and how forces such as migration, population size and natural selection change this variation. The coupling of population genetics with recombinant DNA

FIGURE 1.6 A karyotype arranges the chromosomes in a standard format so they can be analyzed for abnormalities. This karyotype is that of a normal male.

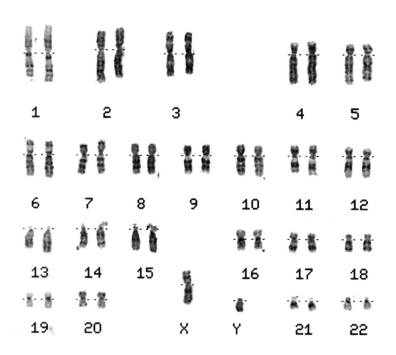

technology has helped us understand the evolutionary history of our species, and the patterns of migrations that distributed humans across the earth.

Basic and Applied Research

Because its principles have widespread uses, genetics is a discipline that crosses and recrosses the line between basic and applied research. Basic research is done by scientists in laboratory and field settings with the goal of understanding how something works, or why it works the way it does. In basic research, there is no goal of solving a practical problem, or making a commercial product; the knowledge itself is the goal. In turn, the results of basic research generate new ideas about how other systems work, and stimulate more basic research. In this way, we are able to gain detailed information about the internal working of organisms. Among other things, basic research in genetics has provided us with precise knowledge about genes, how they work, and what happens when they fail to work properly.

Applied research is usually done with the goal of solving a problem or turning a discovery into a commercial product. Applied research uses classical methods of genetics including experimental matings, as well as the latest methods of biotechnology. In agriculture, applied genetic research has helped increase crop yields, lower the fat content of pork by selective breeding, and make farm plants more resistant to disease. In medicine, new diagnostic tests, the synthesis of customized proteins for treating diseases, and the production of vaccines are just a few examples of applied genetic research. However, some advances are controversial, and generate debates about the merits and risks of applied research. Occasionally, the public rejects some products. For example, there are serious controversies about the environmental release of genetically modified organisms, the sale of food that has been modified by recombinant DNA technology (▶ Figure 1.7), the use of recombinant DNA-derived growth hormone in milk production, and the irradiation of food.

An understanding of the principles of genetics will help everyone make informed decisions about the applications of genetic research in our lives, ranging from the food we eat, the diagnostic tests we elect to have, and even the breeding of our pets. Hopefully, this course will provide you with knowledge of the basic concepts that can be used as a framework to make these informed decisions.

▶ **FIGURE 1.7** Transgenic tomatoes, genetically modified by recombinant DNA techniques to slow softening were withdrawn from the marketplace because of consumer resistance to genetically modified food.

WHAT IS THE HISTORICAL IMPACT OF GENETICS ON SOCIETY?

Genetics and biotechnology not only affect our personal lives, but also raise larger questions about ethics, social policy, and law. You may be surprised to learn that genetics has had a significant impact on society in the past. Knowledge about this era may help us avoid mistakes and pitfalls as we consider new issues raised by genetics.

Genetics and Social Policy

■ **Eugenics** The attempt to improve the human species by selective breeding.

After the publication of *The Origin of Species* by Charles Darwin, his cousin Francis Galton (▶ Figure 1.8) thought that natural selection could be used to improve the human species. He founded **eugenics,** a method he thought could improve the intellectual, economic, and social level of our species. Galton proposed that those having desirable traits such as leadership or musical ability be encouraged to have large families, while those with undesirable traits should be discouraged from reproducing. The error of Galton's reasoning was that he believed that traits were handed down without any environmental influence. The idea that all human traits are genetically determined is known as **hereditarianism.**

■ **Hereditarianism** The idea that human traits are determined solely by genetic inheritance, ignoring the contribution of the environment.

Eugenics took hold in the United States, and eugenicists worked to promote selective breeding in the human population (▶ Figure 1.9) and to control reproduction of those labeled as genetically defective. From about 1905 through 1933, eugenics was a powerful and influential force in passing laws regulating social policy.

Eugenics and Immigration Laws

In the early decades of this century, the high levels of unemployment, poverty and crime among immigrants from southern and eastern Europe were taken as evidence that these people were genetically inferior. Based on testimony by eugenics experts, Congress enacted the Immigration Act of 1924, which was signed into law by President Coolidge. This law, based on faulty genetic assumptions, effectively closed the door to America for millions of people in southern and eastern Europe by setting low entry quotas for immigrants from that region. The immigration of Asians was previously restricted by the Chinese Exclusion Acts of 1882 and 1902, and a 1907 agreement between the U.S. government and the Japanese government to exclude Japanese immigrants. At the time, there was little immigration from Africa, and lawmakers saw little need to regulate entry from this continent. This was not corrected until 1965, when a new immigration law was passed, admitting immigrants by their order of application rather than country of origin.

Eugenics and Reproductive Rights

In addition to influencing immigration policy, the eugenics movement in the United States became closely associated with laws that required sterilization of those labeled as genetically inferior. A committee of eugenicists concluded that up to 10% of the U.S. population should be isolated from the gene pool by being institutionalized or sterilized. Laws requiring sterilization for genetic defectives and those convicted of certain crimes were passed in many states. It may surprise you to learn that the U.S. Supreme Court in 1927 (*Buck v. Bell*) upheld the right of the state of Virginia (and all other states) to use forced sterilization for eugenic reasons. This ruling, which has never been overturned, includes the following statement:

> It is better for all the world, if instead of waiting to execute degenerate offspring for crime, or to let them starve for their imbecility, society can prevent those who are manifestly unfit from continuing their kind. The principle that sustains compulsory vaccination is broad enough to cover cutting the fallopian tubes.

▶ **FIGURE 1.8** Sir Francis Galton, cousin of Charles Darwin and the founder of the eugenics movement.

Soon after this decision, Carrie Buck, the plaintiff in the suit brought to the Supreme Court, was sterilized. By about 1930, some twenty-four states passed sterilization laws, and by the mid-1930s, about twenty thousand sterilizations had been carried out. In the 1960s, some states began to repeal these laws, but they are still on the books in more than a dozen states, although federal regulations restrict their use.

Eugenics and the Nazi Movement

In Germany, eugenics (known as *Rassenhygiene*) fused with genetics and the political philosophy of the Nazi movement (see Concepts and Controversies: Genetics, Eugenics, and Nazi Germany). This relationship evolved into the systematic killing of those defined as socially defective, such as the physically deformed, the retarded, and the mentally ill. Later this rationale was expanded in an attempt to eradicate entire ethnic groups such as the Gypsies and Jews. The association of eugenics with the government of Nazi Germany led to the decline of the eugenics movement in the United States by the mid-1930s.

With advances in genetics and biotechnology, we are once again facing a period when new social policies and laws are being formulated, based on what we know (or think we know) about genetics. As we contemplate the possibility of cloning humans, or modifying our genetic makeup by gene therapy, we would do well to remember the mistakes of the past, admit that our knowledge is incomplete, and formulate policy and laws wisely.

WHEN DID HUMAN GENETICS GET STARTED?

In the first part of the twentieth century, most geneticists avoided human genetics because of its association with eugenics. Immediately after World War II, however, serious research in human genetics began to focus on the identification of Mendelian

Concepts and Controversies

Genetics, Eugenics, and Nazi Germany

*I*n the first decades of this century, eugenics advocates in Germany were concerned with preservation of racial "purity," as were their colleagues in other countries, including the United States and England. By 1927, many states in the United States had enacted laws that prohibited marriage by "social misfits," and made sterilization compulsory for the "genetically unfit" and for certain crimes. In Germany, the laws of the Weimar government prohibited sterilization, and there were no laws restricting marriage on eugenic grounds. As a result, several leading eugenicists became associated with the National Socialist Party (Nazis), which advocated forced sterilization and other eugenic measures to preserve the purity of the Aryan "race."

Adolf Hitler and the Nazi party came to power in January 1933. By July of that year a sterilization law was in effect. Under the law, those regarded as having lives not worth living, including the feebleminded, epileptics, the deformed, those having hereditary forms of blindness or deafness, and alcoholics, were to be sterilized.

By the end of 1933, the law was amended to include mercy killing (*Gnadentod*) of newborns who were incurably ill with hereditary disorders or birth defects. This program was gradually expanded to include children up to 3 or 4 years of age, then adolescents, and finally, all institutionalized children, including juvenile delinquents and Jewish children. More than two dozen institutions in Germany, Austria, and Poland were assigned to carry out this program. Children were usually killed by poison or starvation.

In 1939, the program was extended to include mentally retarded and mentally defective adults, and adults with certain genetic disorders. This program began by killing adults in psychiatric hospitals. As increasing numbers were marked for death, gas chambers were installed at several institutions to kill people more efficiently and crematoria were used to dispose of the bodies. This practice spread from mental hospitals to include defective individuals in concentration camps, and then to whole groups of people in concentration camps, most of whom were Jews, Gypsies, Communists, homosexuals, or political opponents of the government.

traits and the use of mathematical formulas to study genes in different human populations. During this period, human genetics emerged as a separate branch of genetics.

In 1949, a group of researchers led by Linus Pauling at Cal Tech discovered that a genetic disorder known as sickle cell anemia was caused by a defective hemoglobin molecule ( ) Figure 1.10), giving birth to the field of human molecular genetics. Human cytogenetics began in 1956 when J. H. Tijo and A. Levan determined that humans carry 46 chromosomes, and in 1959 the chromosomal basis of Down syndrome was identified.

The revolution in molecular genetics that began with the discovery of the structure of DNA in 1953 had an immediate impact on human genetics. In the years following these discoveries, the molecular basis of a number of human genetic disorders was explained. This allowed the use of dietary restrictions to treat genetic disorders such as phenylketonuria and galactosemia.

With the development of recombinant DNA technology in the 1970s, human genetics has made rapid strides. In the span of about 20 years, we have learned how to predict the sex of unborn children, to diagnose many genetic disorders prenatally, and to manufacture gene products to prevent the deleterious effects of some genetic diseases. The Human Genome Project, started in 1991, is an international effort to map all the genetic information carried by humans at its most elemental level: the 3 billion nucleotides in our DNA. As of this writing, the project is ahead of its timetable, and has produced detailed genetic maps of all human chromosomes. Genetic technology has made it possible to produce human embryos by fusion of sperm and eggs in a laboratory dish ( ) Figure 1.11) and transfer the developing embryo to the womb of a surrogate mother. Embryos can also be frozen for transfer to a womb at a later time. In a small number of cases, genetic defects are now corrected by inserting normal genes to act in place of mutant genes, a technique called gene therapy. We can even insert human genes into animals, creating new types of organisms,

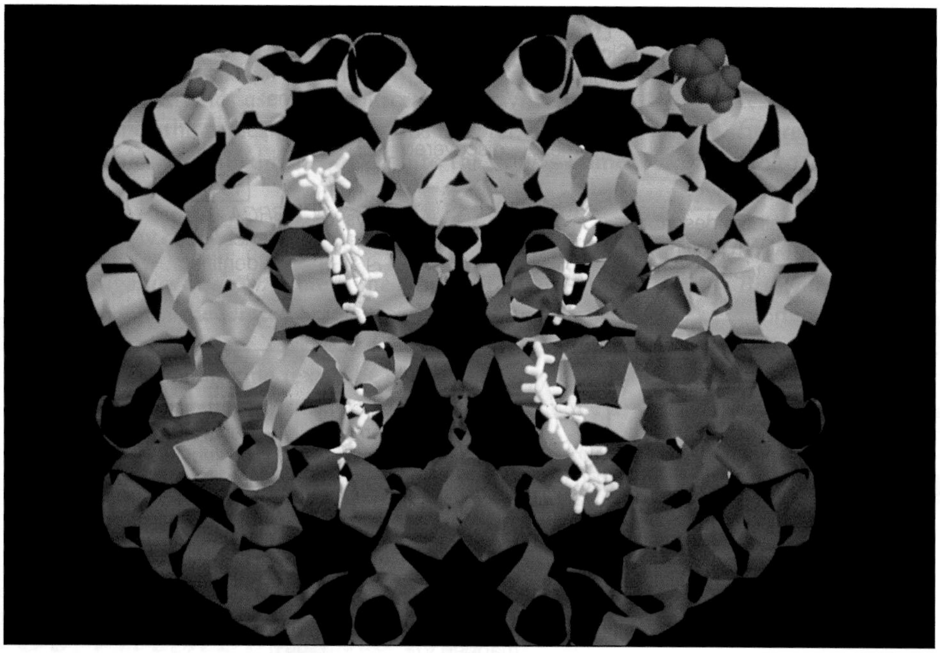

▶ **FIGURE 1.10** Hemoglobin is an oxygen-transporting protein found in red blood cells.

▶ **FIGURE 1.11** Human embryo, shortly after fertilization in the laboratory. Embryos at this stage of development can be analyzed for genetic disorders before implantation into the uterus of the egg donor, or that of another, surrogate mother.

which produce human proteins. These proteins are purified and used to treat human diseases such as emphysema.

WHERE IS HUMAN GENETICS GOING IN THE FUTURE?

Interest in heredity and the transmission of traits from parents to offspring can be traced back thousands of years. Evidence indicates that animals and plants were domesticated by preliterate cultures between 10,000 and 12,000 years ago. The heritable traits of domesticated species were manipulated to produce strains tailored to human needs, a process known as selective breeding. Although it is effective, selective

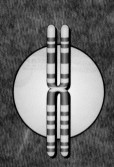

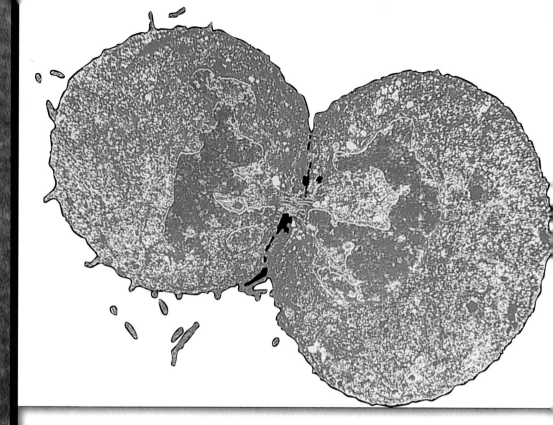

Cells, Chromosomes, and Cell Division

Chapter Outline

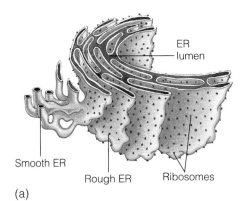

ER
lumen

Smooth ER

Rough ER   Ribosomes

(a)

▶ **FIGURE 2.4** (a) Three-dimensional representation of the endoplasmic reticulum (ER), showing the relationship between the smooth and rough ER. (b) An electron micrograph of ribosome-studded rough ER.

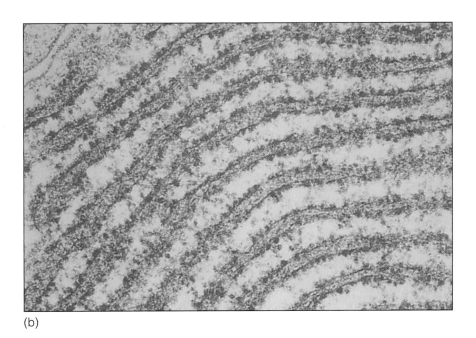

(b)

Endoplasmic Reticulum The **endoplasmic reticulum (ER)** is a network of membranous channels within the cytoplasm. The ER predominates in cells that make large amounts of protein (▶ Figure 2.4). The ER is the cellular site of protein synthesis. The inside of the ER, called the lumen, is a separate compartment within the cell, where proteins are modified and prepared for transport to other locations within the cell and for secretion out of the cell. In addition to being the site of protein synthesis, the ER is the source of most of the new membranes produced for maintenance and growth in the cell.

The outer surface of the ER is often studded with **ribosomes,** another cytoplasmic component (Figure 2.4). Ribosomes are the most numerous cellular structures and can be found free in the cytoplasm, attached to the outer surface of the ER or the nuclear membrane. Under an electron microscope, ribosomes appear as spherical structures but are actually composed of two subunits that function to assemble the proteins necessary for cellular growth and metabolism. The process of protein synthesis is covered in detail in Chapter 9.

Golgi Apparatus Animal cells contain clusters of flattened membrane sacs, called the **Golgi apparatus** (▶ Figure 2.5). The Golgi chemically modify and distribute proteins synthesized by the ER and are a source of membranes for other organelles, including lysosomes. **Lysosomes** are membrane-enclosed vesicles that store a collection of digestive enzymes. The enzymes stored here originate in the ER and are transported

◼ **Endoplasmic reticulum (ER)** A system of cytoplasmic membranes arranged into sheets and channels that functions in synthesizing and transporting gene products.

◼ **Ribosomes** Cytoplasmic particles composed of two subunits that are the site of protein synthesis.

◼ **Golgi apparatus** Membranous organelles composed of a series of flattened sacs. They sort, modify, and package proteins synthesized in the ER.

◼ **Lysosomes** Membrane-enclosed organelles that contain digestive enzymes.

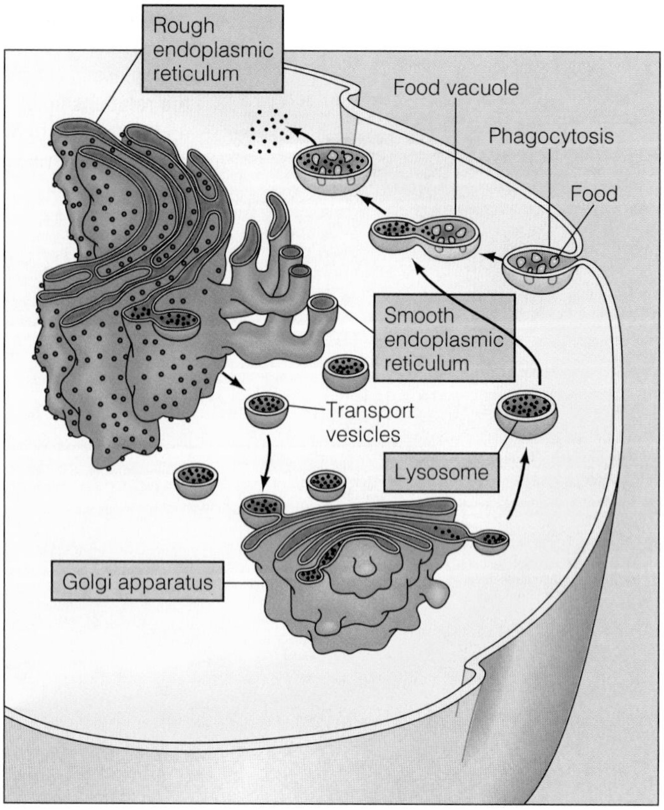

▶ **FIGURE 2.5** The relationship between the Golgi complex and lysosomes. Digestive enzymes are synthesized in the ER and move to the Golgi in transport vesicles. In the Golgi, the enzymes are modified and packaged. Lysosomes pinch off the end of the Golgi membrane. In the cytoplasm, lysosomes fuse with and digest the contents of vesicles that are internalized from the plasma membrane.

■ **Mitochondria (singular: mitochondrion)** Membrane-bound organelles present in the cytoplasm of all eukaryotic cells that are the sites of energy production within cells.

■ **Nucleus** The membrane-bounded organelle in eukaryotic cells that contains the chromosomes.

■ **Nucleolus (plural: nucleoli)** A nuclear region that functions in the synthesis of ribosomes.

■ **Chromatin** The component material of chromosomes, visible as clumps or threads in nuclei under a microscope.

■ **Chromosomes** The threadlike structures in the nucleus that carry genetic information.

to the Golgi, where they are packaged into vesicles. The vesicles bud off the Golgi to form the lysosomes (Figure 2.5).

Lysosomes The lysosomes degrade a wide range of materials, including proteins, fats, carbohydrates, and viruses that might enter the cell. The importance of these organelles in cellular maintenance is underscored by several genetic disorders that disrupt or halt lysosome function. These disorders, including **Tay–Sachs disease** (MIM/OMIM 272800)and **Pompe's disease** (MIM/OMIM 232300), act at the level of organelles within the cell, but their outcomes can include severe mental retardation, blindness, and death by the age of 3 or 4 years. These disorders serve to reinforce the point made earlier that the functioning of the organism can be explained by events that occur within its cells.

Mitochondria Energy transformation takes place in mitochondria (▶ Figure 2.6). They are somewhat variable in shape but are always enclosed by two double-layered membranes. Mitochondria carry their own genetic information in the form of circular DNA molecules. A typical liver cell contains about 1000 mitochondria. Mutations in mitochondrial DNA can cause a number of genetic disorders, and some of these are discussed in Chapter 4.

Nucleus The largest and most prominent cellular organelle is the nucleus (▶ Figure 2.7). This structure's double membrane is known as the nuclear envelope. The envelope is studded with pores that allow direct communication between the nucleus and cytoplasm (Figure 2.7). Within the nucleus, one or more dense regions known as **nucleoli** (singular: **nucleolus**) function in synthesizing cytoplasmic organelles called ribosomes. Under an electron microscope, thin strands and clumps of a substance called **chromatin** are seen throughout the nucleus (Figure 2.7). As the cells prepare to divide, the chromatin condenses and coils to form the **chromosomes**.

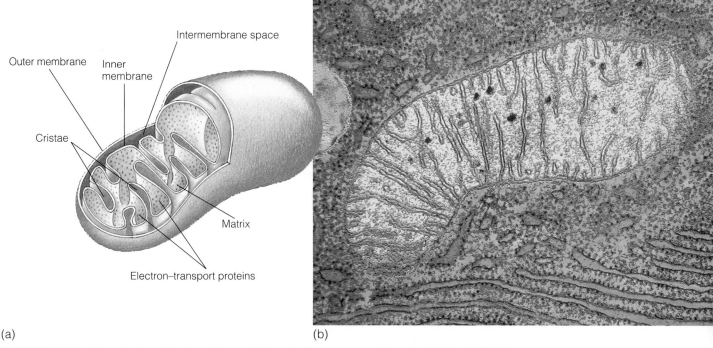

(a)　　　　　　　　　　　　　　　　　　　　(b)

▶ **FIGURE 2.6**　The mitochondrion is a cell organelle involved in energy transformation. (a) The infolded inner membrane forms two compartments where chemical reactions transfer energy from one form to another. (b) A transmission electron micrograph of a mitochondrion.

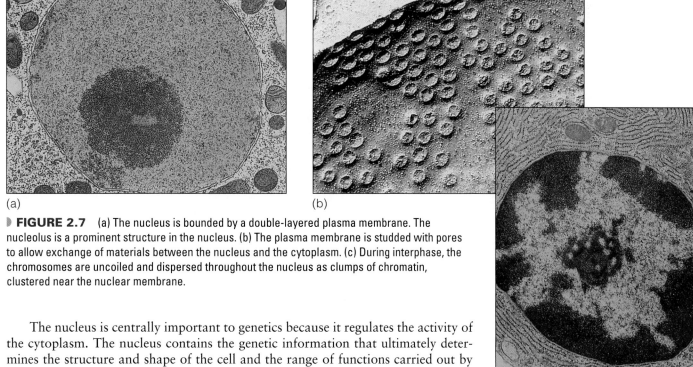

(a)　　　　　　　　　　　　　　　　　　(b)

▶ **FIGURE 2.7**　(a) The nucleus is bounded by a double-layered plasma membrane. The nucleolus is a prominent structure in the nucleus. (b) The plasma membrane is studded with pores to allow exchange of materials between the nucleus and the cytoplasm. (c) During interphase, the chromosomes are uncoiled and dispersed throughout the nucleus as clumps of chromatin, clustered near the nuclear membrane.

(c)

■ **Gene**　The fundamental unit of heredity.

The nucleus is centrally important to genetics because it regulates the activity of the cytoplasm. The nucleus contains the genetic information that ultimately determines the structure and shape of the cell and the range of functions carried out by the cell. In most cells, genetic information is composed of DNA, organized into units called **genes.** DNA and its associated proteins are organized into chromosomes, which are carried in the nucleus.

Because chromosomes carry genetic information and are the vehicles for transmitting genetic information from generation to generation, they occupy a central position in human genetics. The correct number of chromosomes in humans (46) was not determined until 1956. In the 1970s, developments in chromosome staining and refined methods of preparing chromosomes for examination led to advances in prenatal diagnosis, genetic screening, and mapping genes to specific chromosomal regions.

Exploring Membranes

Anne Walter

*W*hile growing up, it seemed to me that a biologist could be only a physician or a quintessential naturalist with sensible shoes, binoculars, and field notebook. Yet even though I loved biology, neither of these futures felt right. Fortunately, two wonderful teachers, Miss Hill and Miss Strosneider, taught biology in the Washington, D.C., public schools. With a paper chromatography experiment to extract plant pigments and a lab on enzymes that I remember to this day, these two remarkable teachers were the first to show me that the basis for much of biology was the precise and intricate interactions of specialized molecules. This was amazing! My interest must have showed, because I was encouraged to compete for an American Heart Association research opportunity that resulted in a summer at George Washington University working on lipid metabolism. Little did I expect lipids to be in my future.

My experiences in college helped me decide that I was interested in physiology and I started graduate studies with Maryanne Hughes, who wanted to understand how seagulls can drink seawater. These birds have several special adaptations, including a gland that secretes an incredibly salty solution after they've had a salty drink. I had learned that because water permeability across cell membranes is high, cells equilibrate rapidly with their external medium. So why didn't the seagull salt glands rapidly lose water and shrink? I concluded that there must be something unusual about their membrane lipids and went on to study at Duke University in a department that specializes in membranes. My research problem was to define the permeability properties of the lipid bilayer in the absence of protein. My results confirmed that hydrophobic solutes like CO_2, ethanol, and aspirin all penetrate the membrane rapidly, whereas hydrophilic molecules, such as glucose or amino acids, penetrate very slowly. Part of my research was to test our methods thoroughly to ensure that our values reflected the true permeabilities, allowing us to be very confident in our conclusion that the membrane behaves like a hydrocarbon that is very constrained . . . like an oil, but

not like an oil. One potential application of this research may be a phospholipid 'sponge' for cleaning up hydrocarbon pollutants. If lipid bilayers can be a sponge to soak up other molecules they can also release molecules. This idea is being used by the cosmetics and pharmaceutical industries to develop phospholipid dispersions as safe, slow-release systems.

My current research asks whether the behavior of transmembrane proteins is affected by their environment. With Neal Rote's research group at Wright State University, I have helped explore the possibility that the autoimmune disease called 'antiphospholipid antibody syndrome' might be caused by antibodies against phosphatidylserine that react when this lipid is exposed on the outside of the cell during platelet activation and possibly during placenta formation. I never would have guessed that expertise with membranes and lipids would be important in trying to figure out a disease process whose main symptoms are blood clotting disorders and poor placental development.

Lipid bilayers are essential to all living cells. In fact, it has been suggested that the 'primordial soup' contained lipids that spontaneously formed closed vesicles and bilayer surfaces that both protected and concentrated the protoenzymes as one of the first steps in the origin of living cells. Discovering the molecular basis for these properties is a puzzle that is turning out to be quite exciting to put together.

ANNE WALTER *is an associate professor in the Biology Department of St. Olaf College in Northfield, Minnesota. She received a B.A. in biology in 1973 from Grinnell College in Iowa and a Ph.D. in physiology and pharmacology in 1981 from Duke University, North Carolina.*

CHROMOSOMES HAVE A CHARACTERISTIC STRUCTURE

Chromosomes, though present, are not usually visible in the nuclei of nondividing cells but are uncoiled into threads of chromatin distributed throughout the nucleus. As a cell prepares to divide, the thin strands of chromatin condense into the coiled structures recognizable as chromosomes. As they become visible, certain structural features allow the chromosomes to be distinguished from one another. In addition, the number of chromosomes in the nucleus is characteristic for a given species: the fruit fly *Drosophila melanogaster* has 8, corn plants have 20, and humans contain 46 chromosomes. The chromosome numbers for several species of plants and animals are given in Table 2.3. The chromosomes of humans and most other eukaryotic organisms occur in pairs. One member of each chromosome pair is derived from the female parent and the other from the male parent. Members of a chromosome pair are known as **homologues.**

Cells that contain pairs of homologous chromosomes are known as **diploid** cells, and the number of chromosomes carried in such cells is known as the diploid, or $2n$, number of chromosomes. In humans, the diploid number of chromosomes is 46. Certain cells, such as eggs and sperm (gametes), that contain only one copy of each chromosome are called **haploid** cells. The chromosome number in these cells is known as the haploid, or n, number of chromosomes. In humans, the haploid number of chromosomes is 23. At fertilization the fusion of haploid gametes and their nuclei produces a cell, known as a **zygote,** that carries the diploid number of chromosomes.

Each chromosome contains a specialized region known as the **centromere.** The position of the centromere divides the chromosome into two arms, and its location is characteristic for a given chromosome (❱ Figure 2.8). Chromosomes with centromeres at or near the middle have arms of equal length and are known as **metacentric** chromosomes (❱ Figure 2.9). If the centromere is not centrally located and the arms are unequal in length, the chromosome is called **submetacentric.** If the centromere is located very close to one end, the chromosome is called an **acrocentric** chromosome.

In studying chromosomal structure, cytologists discovered that human males and females (and other animal species) have one pair of chromosomes that are not completely homologous. Members of this pair are involved in sex determination and are known as **sex chromosomes.** There are two types of sex chromosomes, X and Y. Females have two homologous X chromosomes, and males have a nonhomologous pair, consisting of one X and one Y chromosome. Chromosomes other than sex chromosomes are called **autosomes.**

◼ **Homologues** Members of a chromosomal pair.

◼ **Diploid** The condition in which each chromosome is represented twice as a member of a homologous pair.

◼ **Haploid** The condition in which each chromosome is represented once in an unpaired condition.

◼ **Zygote** The diploid cell resulting from the union of a male haploid gamete and a female haploid gamete.

◼ **Centromere** A region of a chromosome to which fibers attach during cell division. The location of a centromere gives a chromosome its characteristic shape.

◼ **Metacentric chromosome** A chromosome that has a centrally placed centromere.

◼ **Submetacentric chromosome** A chromosome whose centromere is placed closer to one end than the other.

◼ **Acrocentric chromosome** A chromosome whose centromere is placed very close to, but not at, one end.

◼ **Sex chromosome** In humans, the X and Y chromosomes that are involved in sex determination.

◼ **Autosomes** Chromosomes other than the sex chromosomes.

Table 2.3 Chromosome Number in Selected Organisms		
Organism	**Diploid Number ($2n$)**	**Haploid Number (n)**
Human (*Homo sapiens*)	46	23
Chimpanzee (*Pan troglodytes*)	48	24
Gorilla (*Gorilla gorilla*)	48	24
Dog (*Canis familiaris*)	78	39
Chicken (*Gallus domesticus*)	78	39
Frog (*Rana pipiens*)	26	13
Housefly (*Musca domestica*)	12	6
Onion (*Allium cepa*)	16	8
Corn (*Zea mays*)	20	10
Tobacco (*Nicotiana tobacum*)	48	24
House mouse (*Mus musculus*)	40	20
Fruit fly (*Drosophila melanogaster*)	8	4
Nematode (*Caenorhabditis elegans*)	12	6

▶ **FIGURE 2.8** Human chromosomes as seen at the metaphase of mitosis in the scanning electron microscope. The replicated chromosomes appear as double structures, consisting of sister chromatids joined by a single centromere.

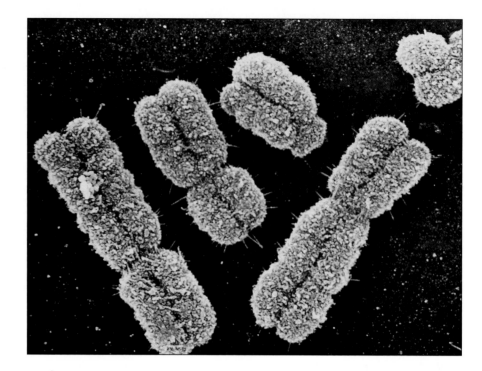

▶ **FIGURE 2.9** Human metaphase chromosomes are identified by size, centromere location, and banding pattern. The relative size, centromere locations, and banding patterns for three representative human chromosomes are shown. Chromosome 3 is one of the largest human chromosomes, and because the centromere is centrally located, is a metacentric chromosome. Chromosome 17 is a submetacentric chromosome because the centromere divides the chromosome into two arms of unequal size. Chromosome 21 has a centromere placed very close to one end and is called an acrocentric chromosome. In humans, the short arm of each chromosome is called the p arm, and the long arm is called the q arm.

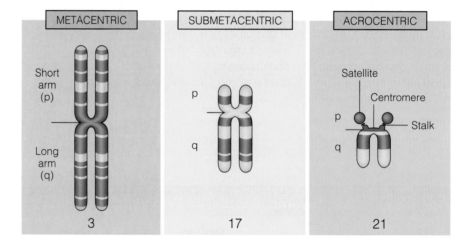

▪ **Cell cycle** The sequence of events that takes place between successive mitotic divisions.

▪ **Interphase** The period of time in the cell cycle between mitotic divisions.

▪ **Mitosis** Form of cell division that produces two cells, each of which has the same complement of chromosomes as the parent cell.

▪ **Cytokinesis** The process of cytoplasmic division that accompanies cell division.

THE CELL CYCLE DESCRIBES THE LIFE HISTORY OF A CELL

Many cells in the body alternate between states of division and nondivision. The interval between divisions can vary from minutes in embryonic cells to months or even years in some cells of adults. The sequence of events from one division to another is called the **cell cycle**, which consists of three phases, **interphase, mitosis**, and **cytokinesis** (▶ Figure 2.10). The period between cell divisions, known as interphase, is the first major part of the cell cycle. The period of division itself consists of two phases, mitosis and cytokinesis. Mitosis is the division of the chromosomes, and cytokinesis is the division of the cytoplasm.

Interphase Has Three Stages

A good place to begin a discussion of the cell cycle is with a cell that has just finished division. After a cell divides, the resulting daughter cells are about one-half the size of the parental cell. Before they can divide again, they must undergo a period of growth and synthesis. These events take place during the three stages of the interphase: G1, S, and G2.

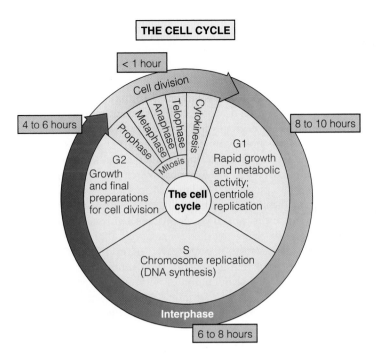

THE CELL CYCLE

< 1 hour

Cell division

Cytokinesis
Telophase
Anaphase
Metaphase
Prophase
Mitosis

4 to 6 hours

8 to 10 hours

G2
Growth and final preparations for cell division

G1
Rapid growth and metabolic activity; centriole replication

The cell cycle

S
Chromosome replication (DNA synthesis)

Interphase

6 to 8 hours

▶ **FIGURE 2.10** The cell cycle has three stages: interphase, mitosis, and cytokinesis. Interphase has three components: G1, S, and G2. Times shown for the stages are representative for cells grown in the laboratory.

The G1 stage, which begins immediately after division, is a period when many cytoplasmic components, including organelles, membranes, and ribosomes, are synthesized. The synthetic activity in G1 almost doubles the cell size and replaces components lost in the previous division. G1 is followed by the S (synthesis) phase, during which a duplicate copy of each chromosome is made. A second period of cellular growth, known as the G2 phase, takes place before the cell is ready to begin a new round of division. By the end of G2, the cell is ready to divide. The time spent in the three stages of interphase (G1, S, and G2) varies from 18 to 24 hours in animal cells grown under laboratory conditions. Because the events of mitosis usually take less than one hour, cells spend most of their time in interphase. Table 2.4 is a summary of the phases of the cell cycle.

Table 2.4 **Phases of the Cell Cycle**

Phase	Characteristics
Interphase	
G1 (gap 1)	Stage begins immediately after mitosis.
	RNA, protein, and other molecules are synthesized.
S (synthesis)	DNA is replicated.
	Chromosomes become double stranded.
G2 (gap 2)	Mitochondria divide. Precursors of spindle fibers are synthesized.
Mitosis	
Prophase	Chromosomes condense.
	Nuclear envelope disappears.
	Centrioles divide and migrate to opposite poles of the dividing cell.
	Spindle fibers form and attach to chromosomes.
Metaphase	Chromosomes line up on the midline of the dividing cell.
Anaphase	Chromosomes begin to separate.
Telophase	Chromosomes migrate or are pulled to opposite poles.
	New nuclear envelope forms.
	Chromosomes uncoil.
Cytokinesis	Cleavage furrow forms and deepens.
	Cytoplasm divides.

Sea Urchins, Cyclins, and Cancer

*A*dvances in human genetics and cancer research sometimes come from unexpected directions. One such story has its beginning at the Marine Biological Laboratories at Woods Hole, Massachusetts in 1982. There, a group of young scientists led by Tim Hunt gathered for the summer to study biochemical changes that take place after fertilization in sea urchin eggs. They fertilized a batch of sea urchin eggs and at ten-minute intervals, analyzed the newly made proteins during the first 2 to 3 hours of development. The fertilized egg first divides at about 1 hour and again about 2 hours after fertilization, resulting in a four-cell embryo.

Several new proteins appeared almost immediately after fertilization, including one that was continuously synthesized, but then destroyed just before each round of cell division. Because of its cyclic behavior, this protein was called cyclin. Work with newly fertilized clam eggs revealed that this species also has cyclins that disappear just before mitosis. Because of their pattern of synthesis and destruction, Hunt and his colleagues concluded that cyclins might be involved in controlling cell division.

Subsequent work showed that cyclins are present in the cells of many organisms and act as important switches in controlling cell division. Sea urchins have only one cyclin, but humans and other mammals have as many as eight to twelve different cyclins, each of which controls one or more steps in cell division. What does all this have to do with cancer? It turns out that some nondividing cells are arrested in the G1 phase. The mechanism that determines whether cells move through the cycle operates in G1. A critical switch point commits a cell to enter the S phase, G2, and mitosis, or causes the cell to leave the cycle and become nondividing. The nature of this switch point, one of the central regulatory mechanisms in all of biology, is slowly being revealed by research in genetics and cell biology. The synthesis and action of cyclins generate the chemical signals that are part of this switch point. At the G1 control point, a cyclin combines with another protein, causing a cascade of events that moves the cell from G1 into S.

Cancer cells have disabled this signal and can divide continuously. Mutations in genes that control the synthesis or action of cyclins are important in the transition of a normal cell into a cancer cell. This important discovery is built on a foundation of work done on sea urchin embryos. Because eukaryotic cells share many properties, work done on yeast, sea urchin eggs, or clam embryos can be used to understand and predict events in normal human cells and in cells that have undergone mutations and become cancerous.

The life history of cells and their relationship to the cell cycle vary for different cell types. Some cells, like those in bone marrow that give rise to red blood cells, pass through the cell cycle continuously and divide regularly. At the other extreme, some cell types become permanently arrested in G1 and never divide. In between are cell types that are arrested in G1 or G2 but can divide under certain circumstances.

When cells escape from the controls that are part of the cell cycle, they can become cancerous (see Concepts and Controversies: Sea Urchins, Cyclins, and Cancer).

Cell Division by Mitosis Occurs in Four Stages

When the cell reaches the end of the G2 stage, it is ready to undergo mitosis, the second major part of the cell cycle. During this period, two important steps are completed. A complete set of chromosomes is distributed to each daughter cell (mitosis), and the cytoplasm is distributed more or less equally to the two daughter cells (cytokinesis). The division of the cytoplasm is accomplished by splitting the cell into two parts, each of which receives a supply of organelles and plasma membrane to form the new cells. Although cytoplasmic division can be somewhat imprecise and still be operational, the division and distribution of the chromosomes must be accurate and unerring for the cell to function.

The chromosomes and the genetic information they contain are precisely replicated during the S stage of the interphase, and a complete set of chromosomes is distributed to each of the daughter cells during mitosis. The net result is two daughter cells. In humans, each daughter cell contains 46 chromosomes derived from a single parental cell that has 46 chromosomes. Although the distribution of chromosomes in cell division should be precise, errors in this process do occur. These mistakes often have serious genetic consequences and are discussed in detail in Chapter 6.

Although mitosis is a continuous process, for the sake of discussion it has been divided into four phases: prophase, metaphase, anaphase, and telophase (▶ Figure 2.11). Throughout the cell cycle, the chromosomes alternately coil and uncoil. When they are uncoiled (interphase), the chromosomes are dispersed throughout the nucleus and are visible as dark-staining clumps called chromatin. In mitosis, the replicated chromosomes become coiled, thicker and shorter and are visible under the microscope. At the end of mitosis, the chromosomes uncoil and become dispersed.

Prophase   At the beginning of **prophase**, the chromosomes coil, shorten and become recognizable as distinct structures under a microscope. At first, the chromosomes appear as long, thin, intertwined threads. As prophase continues, the chromosomes become shorter and thicker. In human cells, 46 such structures can be seen. The condensation of chromosomes at the beginning of prophase serves an important function. In a shortened and contracted form, the chromosomes untangle from each other and move freely during mitosis. Near the end of prophase, each chromosome consists of two longitudinal strands known as **chromatids**. The chromatids are separate structures, held together at the centromere. Two chromatids joined by a common centromere are known as **sister chromatids** (▶ Figure 2.12). Near the end of prophase, the nucleolus disappears, and the nuclear membrane breaks down. A collection of specialized tubules known as spindle fibers begins to form in the cytoplasm. When fully organized, the fibers stretch from centriole to centriole forming an axis along which mitosis will occur. (▶ Figure 2.13 illustrates the events at this stage of mitosis.

Metaphase   Metaphase begins when the nuclear membrane disappears completely, and the chromosomes are free in the cytoplasm. At this stage, each chromosome is made up of two chromatids attached to a single centromere, giving chromosomes an X-shaped appearance. The chromosomes move to the midline, or equator, of the cell, where spindle fibers become attached to the centromeres (Figure 2.13). At this stage there are 46 centromeres, each attached to two sister chromatids.

Anaphase   At the beginning of **anaphase**, the centromeres divide, converting each sister chromatid into a chromosome (Figure 2.11). The two chromosomes derived from sister chromatids are genetically and structurally identical. Spindle fibers attached to the centromeres begin to shorten, and the chromosomes migrate toward opposite sides of the cell. As the spindle fibers shorten, the centromere moves first, and the chromosome arms trail behind, making the chromosomes appear V-shaped or J-shaped. By the end of anaphase, a complete set of chromosomes is present at each pole of the cell. Although anaphase is the briefest stage of mitosis, it is essential to ensuring that each daughter cell receives a complete and identical set of 46 chromosomes. An abnormality of centromeric function during prenatal development is responsible for Roberts syndrome (MIM/OMIM 268300, (▶ Figure 2.14)

Telophase   Two events take place during **telophase**, the final stage of mitosis. These events are the formation of nuclei and division of the cytoplasm. As the chromosomes reach opposite poles of the cell, the spindle fibers break down into subunits, which are stored in the cytoplasm, or become incorporated into the cytoskeleton. Membrane buds from the ER form a new nuclear membrane. Inside the new nucleus, the chromosomes uncoil and become dispersed as chromatin. The cycle of chromosome uncoiling extends from the telophase through G1, and the cycle of condensation runs from G2 to anaphase. It is likely that genetic information in an uncoiled chromosome is more accessible and able to direct the synthesis of gene products during the interphase.

Cytokinesis, the division of the cytoplasm, begins with the formation of the **cell furrow**, a constriction of the cell membrane that forms at the equator of the cell

■ **Prophase**   A stage in mitosis during which the chromosomes become visible and split longitudinally except at the centromere.

■ **Chromatid**   One of the strands of a duplicated chromosome, joined by a single centromere to its sister chromatid.

■ **Sister chromatids**   Two chromatids joined by a common centromere. Each chromatid carries identical genetic information.

■ **Metaphase**   A stage in mitosis during which the chromosomes move and become arranged near the middle of the cell.

■ **Anaphase**   A stage in mitosis during which the centromeres split and the daughter chromosomes begin to separate.

■ **Telophase**   The last stage of mitosis, during which division of the cytoplasm occurs, the chromosomes of the daughter cells disperse, and the nucleus re-forms.

■ **Cell furrow**   A constriction of the cell membrane that forms at the same point of cytoplasmic cleavage during cell division.

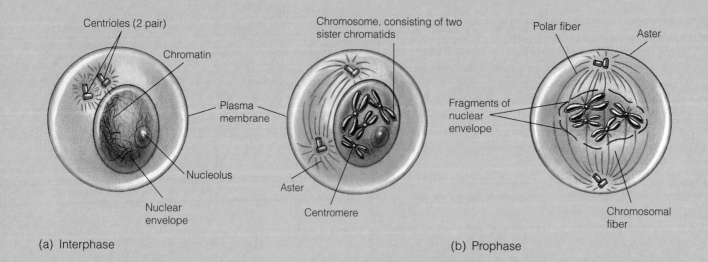

(a) Interphase

Centrioles (2 pair)
Chromatin
Plasma membrane
Nucleolus
Nuclear envelope

(b) Prophase

Chromosome, consisting of two sister chromatids
Aster
Centromere
Polar fiber
Aster
Fragments of nuclear envelope
Chromosomal fiber

Interphase.

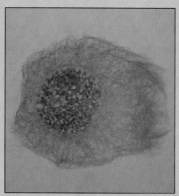

Early prophase.

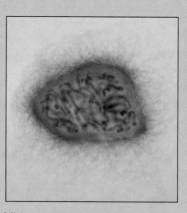

Midprophase.

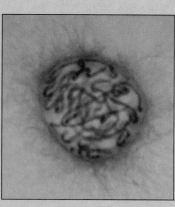

Late prophase.

▎**FIGURE 2.11**   Stages of mitosis. During the interphase (a) replication of chromosomes takes place. (b) In the prophase, the chromosomes coil and become visible as threadlike structures. In late prophase, they are visible as a double structure, consisting of sister chromatids joined by a single centromere. At the end of prophase, the nuclear membrane breaks down. In metaphase (c), chromosomes become aligned at the equator of the cell. In anaphase (d), the centromeres divide, converting the sister chromatids into chromosomes, which move toward opposite sides of the cell. At telophase (e) the chromosomes uncoil, the nuclear membrane reforms, and the cytoplasm divides.

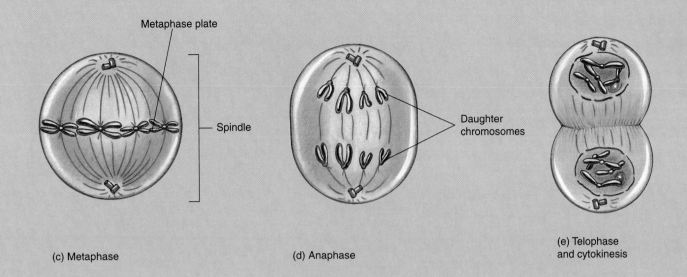

Metaphase plate

Spindle

(c) Metaphase

Daughter chromosomes

(d) Anaphase

(e) Telophase and cytokinesis

Early metaphase.

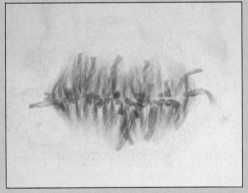

Metaphase.

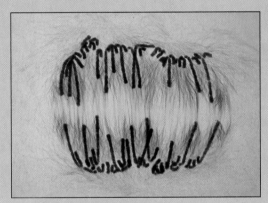

Mid-anaphase.

Late anaphase.

Telophase.

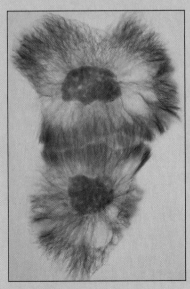

Interphase.

One chromosome (unreplicated)

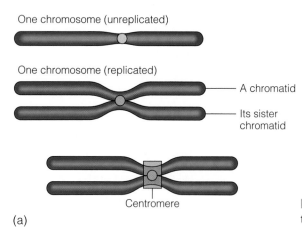

One chromosome (replicated)

A chromatid

Its sister
chromatid

Centromere

(a)

(b)

▌**FIGURE 2.12**   Chromosomes replicate during the S phase. While attached to the centromere, the replicated chromosomes are called sister chromatids.

▌**FIGURE 2.13**   Microtubules form the mitotic spindle during prophase. Some of the fibers connect to the centromeres and help move the chromosomes to opposite sides of the cell during anaphase.

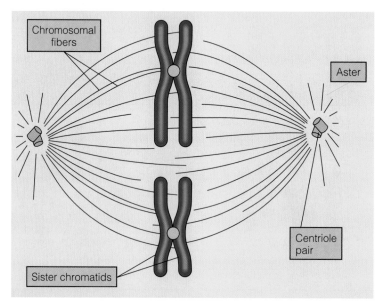

Chromosomal
fibers

Aster

Centriole
pair

Sister chromatids

▌**FIGURE 2.14**   Roberts syndrome is a genetic disorder caused by malfunction of centromeres during mitosis. In this painting by Goya (1746–1828), the child on the mother's lap lacks limb development that is characteristic of this syndrome.

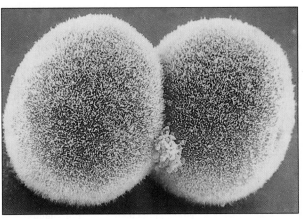
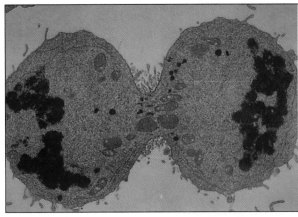

(a)   (b)

▶ **FIGURE 2.15**   Cytokinesis. (a) A scanning electron micrograph of cleavage as seen from the outside of the cell. (b) A transmission electron micrograph of cytokinesis in a cross section of a dividing cell.

Table 2.5 **Summary of Mitosis**	
Stage	**Characteristics**
Interphase	Replication of chromosomes takes place.
Prophase	Chromosomes become visible as threadlike structures. As they continue to condense, they are seen as double structures, with sister chromatids joined at a single centromere.
Metaphase	Chromosomes become aligned at equator of cell.
Anaphase	Centromeres divide, and chromosomes move toward opposite poles.
Telophase	Chromosomes uncoil, nuclear membrane forms, and cytoplasm divides.

(▶ Figure 2.15). The constriction gradually tightens and divides the cell in two. The major features of mitosis are summarized in Table 2.5.

MITOSIS IS ESSENTIAL FOR GROWTH AND CELL REPLACEMENT

Mitosis is an essential process in humans and all multicellular organisms. In humans—throughout adult life—some cells retain their capacity to divide, whereas others do not divide after adulthood is reached. Cells in the bone marrow continually move through the cell cycle and produce about 2 million red blood cells each second. Skin cells divide to replace dead cells that are continually sloughed off the surface of the body. Skin cells also divide during wound healing to repair tissue damaged in an injury. By contrast, most cells in the nervous system remain permanently in G1 and cannot divide. As a result, when nerves are damaged or destroyed, they cannot be replaced. For this reason many injuries to the spinal cord result in permanent paralysis.

The mechanism that determines whether cells are cycling or noncycling operates in the G1 phase of the cell cycle. The general features of this regulation are known and are discussed in Chapter 14, "The Genetics of Cancer."

Cells grown in the laboratory undergo a characteristic number of mitotic divisions. Once this number, known as the Hayflick limit, is reached, the cells die. Cells cultured from human embryos have a limit of about 50 doublings, a capacity that

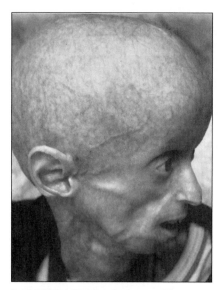

includes all divisions necessary to produce a human adult and cell replacement during a lifetime. Cells from adults and elderly individuals can divide only about 10 to 30 times in tissue culture before dying. Human cells and those of other multicellular organisms contain an internal mechanism that determines the maximum number of divisions, which in turn determines the ultimate life span of the individual. This process is under genetic control because several mutations that affect the aging process are known. One of these is a rare heritable condition known as **progeria** (MIM/OMIM 176670), in which affected individuals age rapidly. Seven- or 8-year-old children who have this disease physically and mentally resemble individuals of 70 or 80 years of age (▶ Figure 2.16). Affected individuals usually die of old age by the age of 14. **Werner syndrome** (MIM/OMIM 277700) is another genetic condition associated with premature aging. In this case, the disease process begins between the ages of 15 and 20 years, and affected individuals die of age-related problems by 45 to 50 years of age.

▶ **FIGURE 2.16** A 13-year-old girl who has progeria.

■ **Progeria** A genetic trait in humans associated with premature aging and early death.

■ **Werner syndrome** A genetic trait in humans that causes aging to accelerate in adolescence, leading to death by about age 50.

■ **Meiosis** The process of cell division during which one cycle of chromosomal replication is followed by two successive cell divisions to produce four haploid cells.

■ **Gamete** A haploid reproductive cell, such as the sperm or egg.

■ **Synapsis** The pairing of homologous chromosomes during prophase I of meiosis.

■ **Chiasmata** The crossing of nonsister chromatid strands seen in the first meiotic prophase. Chiasmata represent the structural evidence for crossing-over.

■ **Crossing-over** The process of exchanging parts between homologous chromosomes during meiosis, which produces new combinations of genetic information.

CELL DIVISION BY MEIOSIS: THE BASIS OF SEX

The genetic information we inherit is contained in two cells, the sperm and the egg. These cells are produced by a form of cell division known as **meiosis** (▶ Figure 2.17). Recall that in mitosis, each daughter cell receives 46 chromosomes. In meiosis, however, members of a chromosome pair are separated from each other to produce haploid cells, or **gametes**, each with 23 chromosomes. Union of the two gametes in fertilization restores the chromosome number to 46 and provides a full set of genetic information to the fertilized egg (zygote).

The distribution of chromosomes in meiosis is an exact process. Each gamete must contain one member of each chromosomal pair, not a random selection of 23 of the 46 chromosomes. How the precise reduction in the chromosome number is accomplished is centrally important in human genetics.

Cells in the testis and ovary that give rise to gametes are diploid and divide by mitosis. Some of the progeny of these germ cells (but no other cells in the body) undergo meiosis. In meiosis, diploid (*2n*) cells undergo one round of chromosomal replication followed by two divisions to produce four cells. Each cell has the haploid (*n*) number of chromosomes. The two divisions are referred to as meiosis I and II, respectively.

Meiosis I Reduces the Chromosome Number

Before cells enter meiosis, the chromosomes are replicated during interphase (Figure 2.17a). In the first part of prophase I, the chromosomes coil, condense and become visible under a microscope (Figure 2.17b). As the chromosomes coil, the nucleoli and nuclear membrane disappear, and the spindle becomes organized. Each chromosome physically associates with its homologue, and the two chromosomes line up side by side in a process known as **synapsis.** Chromosomal pairing usually begins at one or more points along the chromosome and proceeds until the chromosomes are aligned.

Next, the sister chromatids of each chromosome become visible (Figure 2.17b). At this stage, each chromosome consists of two sister chromatids joined by a single centromere. Also during this stage, the chromatids of homologous chromosomes form cross-shaped or X-shaped regions known as **chiasmata** (singular: chiasma) (Figure 2.17b). These structures are the points at which physical exchange of chromosome material between chromatids can take place. This event, known as **crossing-over,** is of great genetic significance and is discussed later.

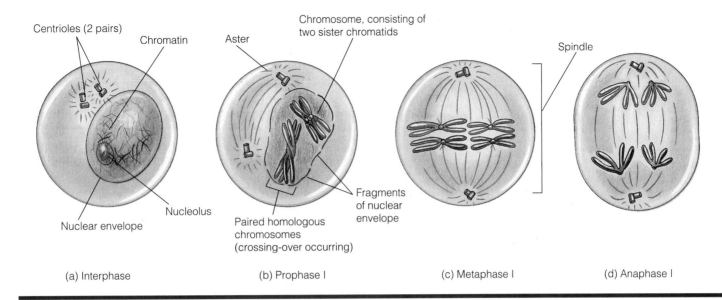

(a) Interphase (b) Prophase I (c) Metaphase I (d) Anaphase I

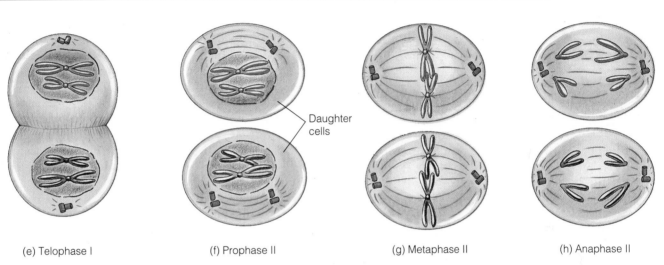

Daughter cells

(e) Telophase I (f) Prophase II (g) Metaphase II (h) Anaphase II

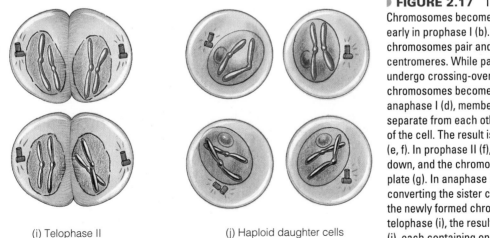

(i) Telophase II (j) Haploid daughter cells

▶ **FIGURE 2.17** The stages of meiosis. Chromosomes become visible as threadlike structures early in prophase I (b). Once visible, homologous chromosomes pair and split longitudinally except at centromeres. While paired, homologous chromosomes undergo crossing-over. In metaphase I (c), the chromosomes become arranged at the cell's equator. In anaphase I (d), members of all chromosome pairs separate from each other and migrate to opposite sides of the cell. The result is the formation of two new cells (e, f). In prophase II (f), the nuclear membrane breaks down, and the chromosomes align at the metaphase plate (g). In anaphase II, the centromeres split, converting the sister chromatids into chromosomes, and the newly formed chromosomes move apart. After telophase (i), the result is four haploid daughter cells (j), each containing one copy of each chromosome.

In metaphase I (Figure 2.17c), the members of a chromosomal pair line up at the equator of the cell. An important feature of meiosis should be noted here. The orientation of each chromosomal pair is random at metaphase. Remember that one chromosome in each pair is paternal, and the other is maternal. These chromosomes have no fixed pattern of arrangement. The genetic consequences of this random arrangement at metaphase I is discussed in a later section.

In anaphase I, members of each chromosomal pair separate from each other and move toward opposite poles of the dividing cell (Figure 2.17d). Each daughter cell receives one member of each chromosomal pair. The chromosomes begin to unwind slightly in telophase I, and cytokinesis occurs, producing two haploid cells (Figure 2.17e). Two major events take place during meiosis I: *reduction,* the random separation of maternal and paternal chromatids, and *crossing-over,* the physical exchange of segments between homologous chromosomes.

Meiosis II Begins with Haploid Cells

In prophase II, the chromosomes coil and condense, and a spindle forms. The chromosomes cannot synapse because there are no homologues with which to pair (Figure 2.17f). As the chromosomes become more condensed, they move toward the equator of the cell. At metaphase II (Figure 2.17g), the 23 chromosomes, each consisting of a centromere and two sister chromatids, line up on the **metaphase plate** with spindle fibers attached to the centromeres. At the beginning of anaphase II (Figure 2.17h), the centromere of each chromosome divides for the first time, and the 46 chromatids are converted to chromosomes and move to opposite poles of the cell.

In telophase II, the chromosomes uncoil and become diffuse, the nuclear membrane re-forms, and division of the cytoplasm takes place (Figure 2.17i). Now, the process of meiosis is complete. One diploid cell that contains 46 chromosomes has undergone one round of chromosomal replication and two rounds of division to produce four haploid cells, each of which contains one copy of each chromosome (Figure 2.17j).

The movement of chromosomes is summarized in ▶ Figure 2.18, and the events of meiosis are presented in Table 2.6. ▶ Figure 2.19 compares the events of mitosis and meiosis.

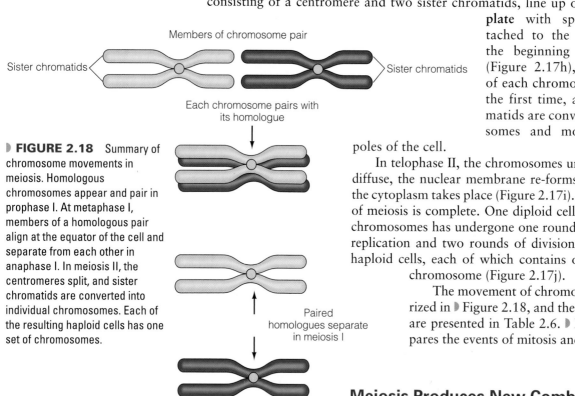

Members of chromosome pair

Sister chromatids ◁   ▷ Sister chromatids

Each chromosome pairs with its homologue

▶ **FIGURE 2.18** Summary of chromosome movements in meiosis. Homologous chromosomes appear and pair in prophase I. At metaphase I, members of a homologous pair align at the equator of the cell and separate from each other in anaphase I. In meiosis II, the centromeres split, and sister chromatids are converted into individual chromosomes. Each of the resulting haploid cells has one set of chromosomes.

Paired homologues separate in meiosis I

Meiosis Produces New Combinations of Genes in Two Ways

Meiosis produces new combinations of genes in two ways: by random **assortment**, if the genes are located on different chromosomes, and by **recombination**, if the genes are located on the same chromosome.

As discussed previously, each chromosomal pair consists of one maternally derived (M) and one paternally derived (P) chromosome. When the synapsed chromosomal

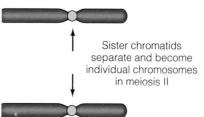

Sister chromatids separate and become individual chromosomes in meiosis II

■ **Assortment** The random distribution of members of homologous chromosomal pairs during meiosis.

■ **Recombination** The exchange of genetic material between homologous chromosomes. Also known as crossing over.

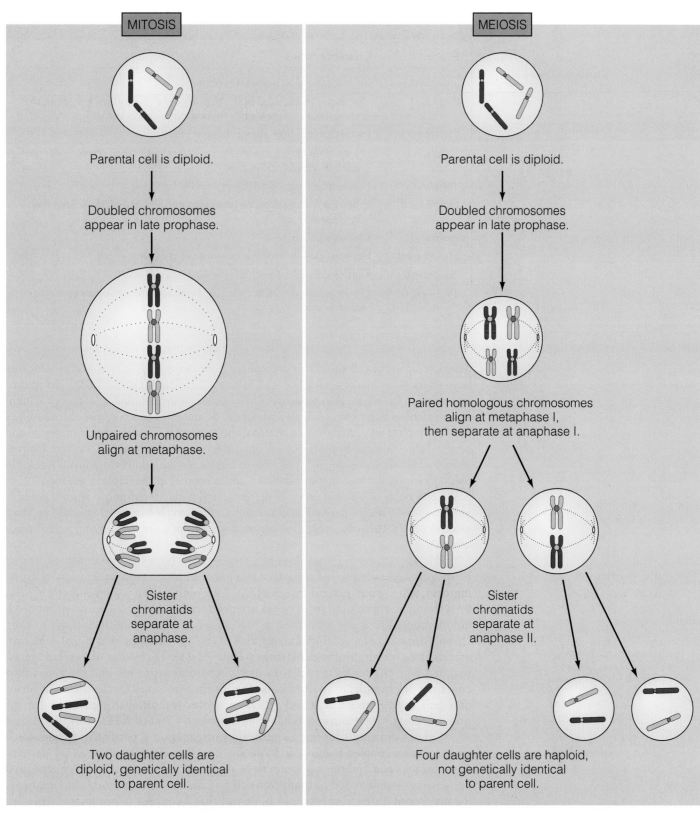

▶ **FIGURE 2.19** A comparison of the events in mitosis and meiosis. In mitosis (left) a diploid parental cell undergoes chromosomal replication and then enters prophase. The chromosomes appear doubled during late prophase, and unpaired chromosomes align at the middle (equator) of the cell during metaphase. In anaphase, the centromeres split, converting the sister chromatids into chromosomes. The result is two daughter cells, each of which is genetically identical to the parental cell. In meiosis I (right), the parental diploid cell undergoes chromosome replication and then enters prophase. Homologous chromosomes pair, and each chromosome appears doubled, except at the centromeres. Paired homologues align at the equator of the cell during metaphase, and members of a chromosome pair separate during anaphase. In meiosis II, the unpaired chromosomes in each cell align at the equator of the cell. During anaphase II, the centromeres split, and one copy of each chromosome is distributed to daughter cells. The result is four haploid daughter cells, which are not genetically equivalent to the parental cell.

Table 2.6   Summary of Meiosis

Stage	Characteristics
Interphase I	Chromosome replication takes place.
Prophase I	Chromosomes become visible, homologous chromosomes pair, and sister chromatids become visible. Recombination takes place.
Metaphase I	Paired chromosomes align at equator of cell.
Anaphase I	Homologous chromosomes separate. Members of each chromosome pair move to opposite poles.
Telophase I	Cytoplasm divides, producing two cells.
Interphase II	Following a brief pause, chromosomes uncoil slightly. This is not a real interphase as such.
Prophase II	Chromosomes re-coil.
Metaphase II	Unpaired chromosomes become aligned at equator of cell.
Anaphase II	Centromeres split. Daughter chromosomes pull apart.
Telophase II	Chromosomes uncoil, nuclear membrane reforms, cytoplasm divides, and meiosis is complete.

pairs line up at the cell equator in metaphase I, the maternal or paternal chromosomes have no fixed pattern of arrangement (▶ Figure 2.20). The alignment of any chromosomal pair can be maternal to paternal or paternal to maternal. As a result, each daughter cell is more likely to receive an assortment of maternal and paternal chromosomes than a complete set of maternal or paternal chromosomes. The number of maternal and paternal chromosome combinations produced by meiosis is equal to 2^n, where 2 represents the chromosomes in each pair, and n represents the number of chromosomes in the haploid set. Because humans have 23 chromosomes in the haploid set, then 2^{23} or 8,388,608 different combinations of maternal and paternal chromosomes are possible in haploid cells. If each parent has this many possible combinations, they could produce more than 7×10^{13} offspring, each carrying a different combination of parental chromosomes.

This astronomic number does not take into account the variability generated by the physical exchange of chromosomal parts, known as recombination or crossing-over, which takes place during meiosis I. Recombination is a second mechanism by which new combinations of genes are produced in meiosis. During prophase I, homologous chromosomes pair with each other and the sister chromatids of each chromosome become visible. During this stage, the arms of two nonsister chromatids can overlap, forming chiasma (▶ Figure 2.21). These sites of overlap are associated with the physical exchange of chromosome segments and the genes they contain. For example, if a pair of homologous chromosomes carries different forms of a gene (e.g., A and a or B and b), crossing-over reshuffles this genetic information and creates new chromosomal combinations (▶ Figure 2.22).

Without crossing-over, the combination of genes on a particular chromosome would remain coupled together indefinitely. Crossing-over allows new and perhaps advantageous combinations of genes to be produced. When the variability generated by crossing-over is added to that produced by random chromosomal combinations, the number of different genetic combinations that a couple can produce in their offspring has been estimated at 80^{23}. Obviously, the offspring of a couple represents only a very small fraction of all these possible gamete combinations. For this reason, it is almost impossible for any two children (aside from identical twins) to be genetically identical.

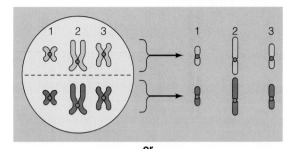

or

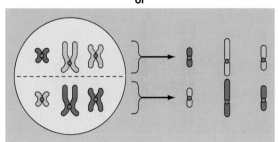

or

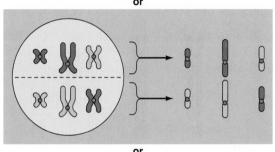

or

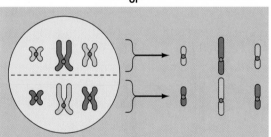

▶ **FIGURE 2.20** The orientation of members of a chromosome pair at meiosis is random. Here three chromosomes (1, 2, and 3) have four possible alignments (maternal members of each chromosome pair are light blue; paternal members are dark blue). There are eight possible combinations of maternal and paternal chromosomes in the resulting haploid cells.

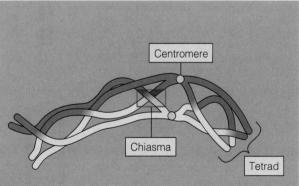

▶ **FIGURE 2.21** Crossing-over is the physical exchange of chromosome parts between homologous chromosomes. The X-shaped regions are the site of crossing-over.

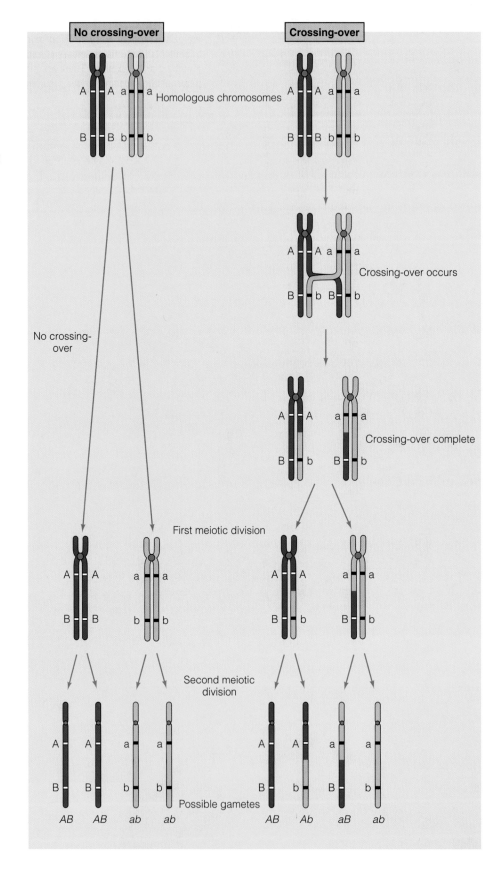

▶ **FIGURE 2.22** Crossing-over increases genetic variation by combining genes from both parents on the same chromatid. At left is shown the combination of maternal and paternal genes when no crossing-over occurs. At right, new combinations (Ab, aB) are produced by crossing-over, increasing genetic variability in the haploid cells that will form gametes.

SPERMATOGENESIS AND OOGENESIS ARE PROCESSES THAT FORM GAMETES

Gametogenesis is the formation of mature ova (eggs) and spermatozoa. In males the production of sperm, known as spermatogenesis, occurs in the testis. Cells called **spermatogonia** line the tubules of the testis and divide mitotically from puberty until death, producing daughter cells called **primary spermatocytes** (▶ Figure 2.23). The spermatocytes undergo meiosis, and the four haploid cells that result are known as **spermatids**. Each spermatid undergoes a period of development into mature sperm. During this period, the nucleus that contains 23 chromosomes (sperm carry 22 autosomes and an X or a Y chromosome) becomes condensed and forms the head of the sperm. In the cytoplasm, a neck and whiplike tail develop, and most of the remaining cytoplasm is lost. In human males, meiosis and sperm production begin at puberty and continue throughout life. The entire process of spermatogenesis takes about 48 days: 16 for meiosis I, 16 for meiosis II, and 16 to convert the spermatid into the mature sperm. The tubules within the testis contain large numbers of spermatocytes, and large numbers of sperm are always in production. A single ejaculate may contain 200 to 300 million sperm, and over a lifetime a male produces billions of sperm.

In females the production of gametes, known as oogenesis, takes place in the ovary. Cells in the ovary, called **oogonia**, divide by mitosis and form **primary oocytes** that undergo meiosis (▶ Figure 2.24). However, the cytoplasmic cleavage

■ **Spermatogonia** Mitotically active cells in the gonads of males that give rise to primary spermatocytes.

■ **Primary spermatocytes** Cells in the testis that undergo meiosis.

■ **Spermatids** The four haploid cells produced by meiotic division of a primary spermatocyte.

■ **Oogonia** Mitotically active cells that produce primary oocytes.

■ **Primary oocytes** Cells in the ovary that undergo meiosis.

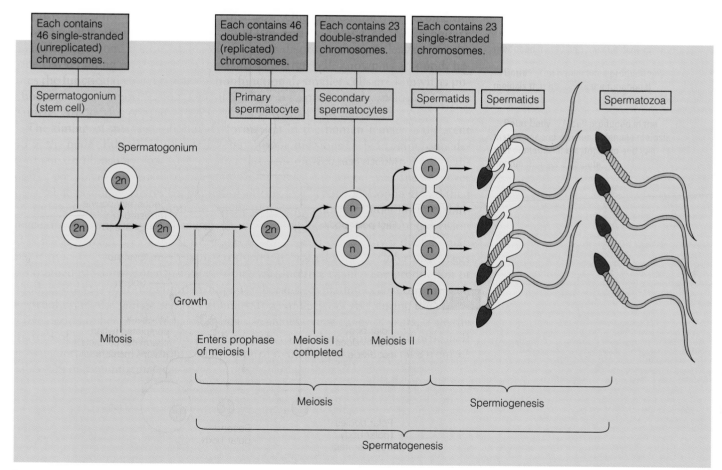

▶ **FIGURE 2.23** The process of spermatogenesis. Germ cells (spermatocytes) divide by mitosis, and beginning at puberty some cells produced in this way enter meiosis as primary spermatocytes. After meiosis I, the secondary spermatocytes contain 23 double-stranded chromosomes. After meiosis II, the haploid spermatids contain 23 single-stranded chromosomes. Spermatids undergo a series of developmental changes (spermiogenesis) and are converted into mature spermatozoa.

Case Studies

CASE 1

It is May 1989 and the scene is a crowded research laboratory with beakers, flasks, and pipettes covering the lab bench. People and equipment take up every possible space. One researcher, Joe, passes a friend staring into a microscope. Another student wears gloves while she puts precisely measured portions of various liquids into tiny test tubes. Joe stops suddenly just before he gets to his desk. Something is wrong with these traces. There it is, a three-base pair deletion—a type of genetic mutation in a DNA sequence. The genetic information required to describe one amino acid in a protein chain is missing, as if one bead had fallen from a precious necklace. Instead of sitting down, Joe rushes to tell his supervisor Dr. Tsui (pronounced "Choy") that he has found a 3-base pair deletion in a person with cystic fibrosis (CF). Cystic fibrosis is a fatal disease that kills about one out of every 2,000 Caucasians (mostly children), but he does not see this same deletion in a normal person's genes.

Dr. Tsui examines the findings and is impressed but wants more evidence to prove that the result is real. He has had false hopes before, so he is not going to celebrate until they check this out carefully. Maybe the difference between the two gene sequences is just a normal variation among individuals.

Five months later, Dr. Tsui and his team identify a "signature" pattern of DNA on either side of the base-pair deletion, and using that as a marker, they compare 100 normal people's genes with the DNA sequence from 100 CF patients. By September 1989, they are sure they have identified the CF gene.

After several more years, Tsui and his team discover that the DNA sequence with the mutation encodes the information for a protein called CFTR (cystic fibrosis transmembrane conductance regulator), a part of the plasma membrane in cells that make mucus. This protein regulates a channel for chloride ions. Proteins are made of long chains of amino acids. The CFTR protein has 1,480 amino acids. Most children with CF are missing one single amino acid in their CFTR. Because of this, their mucus becomes too thick, causing all the other symptoms of CF. Thanks to Tsui's research, scientists now have a much better idea of how the disease works. We can now easily predict when a couple is at risk for having a child with CF. With increasing understanding, scientists may also be able to devise improved treatments for children born with this disease.

Cystic fibrosis (CF) is the most common genetic disease among those of European ancestry. Children who have CF are born with it. Half of them will die before they are 25 and few make it past 30. It affects all the parts of the body that secrete mucus: the lungs, stomach, nose, and mouth. The mucus of children with CF is so thick sometimes they cannot breathe. Why do 1 in 25 Caucasians carry the mutation for CF? Tsui and others think that people who carry it may also have resistance to diarrhea-like diseases.

CASE 2

Jim, a 37-year-old construction worker, and Sally, a 42-year-old business executive, were eagerly preparing for the birth of their first child. They, like more and more couples, chose to wait to have children until they were older and more financially stable. Sally had an uneventful pregnancy with prenatal blood tests and an ultrasound indicating that the baby looked great and everything seemed "normal." Then, a few hours after Ashley was born, they were told she was born with Down syndrome. In shock and disbelief, the couple was devastated. How could something so right go so wrong? What did they do to deserve a disabled child?

It has long been recognized that the risk of having a child with trisomy 21 increases with maternal age. For example, the risk of having a child with Down syndrome when the mother is 30 years old is 1 in 1,000 and at age 40 is 9 in 1,000. Well-defined and distinctive phenotypic features characterize Down syndrome, which is the most frequent form of mental retardation caused by a chromosomal aberration. Most individuals (95%) with Down syndrome or trisomy 21 have 3 copies of chromosome 21; in about 5% of patients, 1 copy is translocated to another chromosome, most often chromosome 14 or 21. Errors in meiosis that lead to trisomy 21 are almost always of maternal origin; only about 5% occur during spermatogenesis. It has been estimated that meiosis I errors account for 76 to 80% of maternal meiotic errors. No one is at fault when a child is born with Down syndrome, but the chances of it occurring increase with advanced maternal age.

Individuals with Down syndrome often have specific major congenital malformations such as those of the heart (30–40% in some studies). Individuals with Down syndrome have an increased incidence (10 to 20 times higher) of leukemia than the normal population. Ninety percent of all Down syndrome patients have significant hearing loss. The frequency of trisomy 21 in the population is 1 in 650 to 1,000 live births.

Summary

1. The cell is the basic unit of structure and function in all organisms, including humans. Because genes control the number, size, shape, and function of cells, the study of cell structure often reveals much about the normal expression of genes.

2. The nucleus contains the genetic information that controls the structure and function of the cell. This genetic information is carried in nuclear structures known as chromosomes. The number, size, and shape of chromosomes are species-specific traits. In humans, 46 chromosomes, the $2n$, or diploid, number, is present in most cells, whereas specialized cells known as gametes contain half of that number, the haploid, or n, number of chromosomes.

3. In humans, chromosomes exist in pairs. Members of a chromosomal pair are known as homologues. One member of each pair is contributed by the mother, and one is contributed by the father. Females carry two homologous X chromosomes, whereas males have an X and a partially nonhomologous Y chromosome. The remaining 22 pairs of chromosomes are known as autosomes.

4. For cells to survive and function, they must contain a complete set of genetic information. This is ensured by replication of each chromosome and by the distribution of a complete chromosomal set in the process of mitosis. Mitosis represents one part of the cell cycle. During the other part, interphase, a duplicate copy of each chromosome is made. The process of mitosis is divided into four stages: prophase, metaphase, anaphase, and telophase. In mitosis, one diploid cell divides to form two diploid cells. Each cell has an exact copy of the genetic information contained in the parental cell.

5. Meiosis is a form of cell division that produces haploid cells containing only the paternal or maternal copy of each chromosome. In meiosis, homologous chromosomes pair with each other. At this time, each chromosome consists of two sister chromatids joined by a common centromere. Chromatids can engage in the physical exchange of chromosome segments, an event known as crossing-over. In metaphase I, paired, homologous chromosomes align at the equator of the cell. In anaphase I, the homologues are separated. In meiosis II, the unpaired chromosomes align at the equator of the cell. In anaphase II, the centromeres divide, and the daughter chromosomes move to opposite poles. The four cells produced in meiosis contain the haploid number (23) of chromosomes.

6. Spermatids, the products of meiosis in males, undergo structural changes to convert them to functional sperm. In female meiosis, division of the cytoplasm is unequal, leading to the formation of one functional gamete and three smaller cells known as polar bodies.

7. Meiosis is a genetically important process. It provides a mechanism for maintaining a constant number of chromosomes from generation to generation. In addition, it generates genetic diversity by reshuffling maternal and paternal chromosomes and by permitting crossing-over.

Questions and Problems

1. Which statement is **not** true about eukaryotic cells?
   a. human cells are eukaryotic
   b. cells have a membrane-bound nucleus
   c. cells have membrane-bound organelles
   d. insect cells are eukaryotic
   e. bacterial cells are eukaryotic

2. Assign a function(s) to the following cellular structures:
   a. plasma membrane
   b. mitochondrion
   c. nucleus
   d. ribosome

3. How many autosomes are present in a human being?

4. Define the following terms:
   a. chromosome
   b. chromatin
   c. chromatid

5. Human haploid gametes (sperm and eggs) contain:
   a. 46 chromosomes, 46 chromatids
   b. 46 chromosomes, 23 chromatids
   c. 23 chromosomes, 46 chromatids
   d. 23 chromosomes, 23 chromatids

6. What is meant by the term homologous chromosomes?

7. What are sister chromatids?

8. List the differences between mitosis and meiosis in the following chart:

	Mitosis	Meiosis
Number of daughter cells produced		
Number of chromosomes per daughter cell		
Number of cell divisions		
Do chromosomes pair? (Y/N)		
Does crossing-over occur? (Y/N)		
Can the daughter cells divide again? (Y/N)		
Do the chromosomes replicate before division? (Y/N)		
Type of cell produced		

9. Identify the stages of mitosis, and describe the important events that occur during each stage.

10. Why is cell furrowing important in cell division? If cytokinesis did not occur, what would be the end result?

11. Which of the following statements is not true when comparing mitosis and meiosis?
    a. twice the number of cells are produced in meiosis versus mitosis
    b. meiosis is involved in the production of gametes, unlike mitosis
    c. crossing-over occurs in meiosis I, not in meiosis II or mitosis
    d. meiosis and mitosis both produce cells that are genetically identical
    e. in both mitosis and meiosis, the parental cell is diploid

12. Match the phase of cell division with the diagrams below. In these cells, 2n = 4.
    a. anaphase of meiosis I
    b. interphase of mitosis
    c. metaphase of mitosis
    d. metaphase of meiosis I
    e. metaphase of meiosis II

13. A cell from a human female has just undergone mitosis. For unknown reasons, the centromere of chromosome 7 failed to divide. Describe the chromosomal contents of the daughter cells.

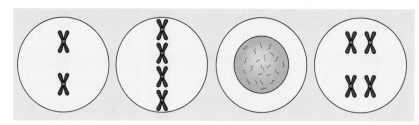

14. During which phases of the mitotic cycle would the terms chromosome and chromatid refer to identical structures?

15. Describe the critical events of mitosis that are responsible for ensuring that each daughter cell receives a full set of chromosomes from the parent cell.

16. Draw the cell cycle. What is meant by the term cycle in the cell cycle? What is happening at the S phase and the M phase?

17. Does the cell cycle refer to mitosis as well as meiosis?

18. Is it is possible that an alternative mechanism for generating germ cells could have evolved. Consider meiosis in a germ cell precursor (a cell that is diploid but will go on to make gametes). If the S phase were skipped, which meiotic division (meiosis I or meiosis II) would no longer be required?

19. A cell has a diploid number of 6 (2n = 6).
    a. Draw the cell in metaphase of meiosis I.
    b. Draw the cell in metaphase of mitosis.
    c. How many chromosomes are present in a daughter cell after meiosis I?
    d. How many chromatids are present in a daughter cell after meiosis II?
    e. How many chromosomes are present in a daughter cell after mitosis?
    f. How many tetrads are visible in the cell in metaphase of meiosis I?

20. A cell (2n = 4) has undergone cell division. Daughter cells have the following chromosome content. Has this cell undergone mitosis or meiosis I or meiosis II?

    a.
    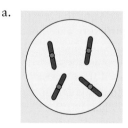

    b.
    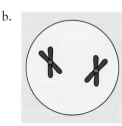

    c.
    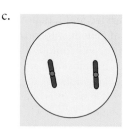

21. Cell division (mitosis) occurs daily in a human being. What type of cells do we need to produce in large quantities on a daily basis?

22. We are following the progress of human chromosome 1 during meiosis. At the end of synapsis, how many chromosomes, chromatids and centromeres are present to ensure that chromosome 1 faithfully traverses meiosis?

23. What is the physical structure that is associated with crossing over?

24. Compare meiotic anaphase I with meiotic anaphase II. Which meiotic anaphase is similar to mitotic anaphase?

25. Speculate on how the Hayflick limit may lead to genetic disorders such as progeria or Werner syndrome.

26. Discuss and compare the products of meiosis in human females and males. How many functional gametes are produced from the daughter cells in each sex?

27. Explain how meiosis leads to genetic variation in diploid organisms.

28. A human female is conceived on April 1, 1949 and is born on January 1, 1950. Onset of puberty occurs on January 1, 1962. She conceives a child on July 1, 1994. How long did it take for the ovum that was fertilized on July 1, 1994 to complete meiosis?

Internet Activities

The following activities use the resources of the World Wide Web to enhance the topics covered in this chapter. To investigate the topics described below, log on to the book's home page at:

http://www.brookscole.com/biology

1. Go to the *Cell Biology Topics Menu* at the University of Texas Internet site. Scroll down and select the topic *Nuclear Organization*. Examine the numerous structures within the nucleus. Take a closer look at the *nuclear pore*. What types of molecules enter and exit through the pores? How do you think the pores might regulate the passage of these molecules into and out of the nucleus?

   Click on the topic *Organization of chromosomes*. Compare heterochromatin and euchromatin, the two major forms of DNA in the nucleus. Heterochromatin is more compact than euchromatin, and is also less transciptionally active. Can you imagine how the compact structure of heterochromatin might contribute to its transcriptional inactivity? Finally, compare the various methods of labeling DNA. Think about what

advantages are provided by each method. For example, which method might be used to quickly and easily determine the total number of chromosomes in a cell? Which methods allow the visualization of specific genetic loci on the chromosomes?

2. For a look at the diversity of various cell types at the microscopic level, access the nanoworld site from the book's homepage. Not only does this site have great photos, but it also illustrates the power of the World Wide Web to allow instantaneous sharing of information across the globe. Look at a variety of images of cells, cell organelles, and cell division to gain an appreciation of this imagery.

3. Use the link on the book's homepage to find the Cells Alive! homepage. Available quick-time movies show a number of cell functions, including bacterial reproduction and phagocytosis. Scroll down to the Inside-Vintage section and click on How Big is a .... Compare the size of a virus, bacterium, red blood cell, lymphocyte, and human sperm by using the line scale provided.

For Further Reading

Bretscher, M. (1985). Molecules of the cell membrane. *Sci. Am. 253:* 100–109.

Cross, P. C. (1993). *Cell and Tissue Ultrastructure: A Functional Perspective.* Upper Saddle River, NJ: Prentice-Hall.

Gosden, R., Krapez, J., and Briggs, D. 1997. Growth and development of the mammalian oocyte. *Bioessays 10:* 875–882.

Kessel, R., & Shih, C. (1974). *Scanning Electron Microscopy in Biology: A Students Atlas of Biological Organization.* New York: Springer-Verlag.

Russell, P. 1998. Checkpoints on the road to mitosis. *Trends Biochem. Sci. 10:* 399–402.

Therman, E., & Susman, M. (1993). *Human Chromosomes: Structure, Behavior, Effects* (3rd ed.). New York: Springer-Verlag.

Travis, J. (1996). What's in the vault? *Sci. News 150:* 56–57.

Wagner, R. P, Maguire, M. P., & Stallings, R. L. (1993). *Chromosomes: A Synthesis.* New York: Wiley-Liss.

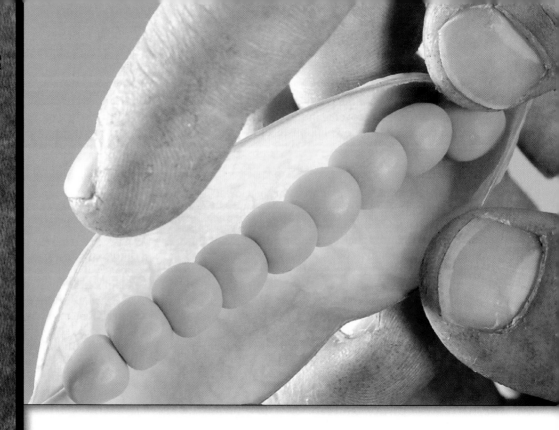

Transmission of Genes from Generation to Generation

*C*harles *Darwin was not the only member of his family to make significant contributions to our understanding of genetics and evolution. Darwin's cousin, Francis Galton, was concerned that Darwin's account of evolution did not explain how small variations in appearance were generated and set out to experimentally test Darwin's idea about the way traits were inherited.*

Darwin originally subscribed to the generally accepted idea of the time that parental traits are blended in the offspring. According to the blending idea, if a plant with red flowers is crossed to a plant with white flowers, the offspring should have pink flowers (a blend of red and white). But Darwin realized that if traits were blended generation after generation, variation would be reduced, not increased. To get around this problem, Darwin decided to support a second idea about the way traits were inherited. This idea was really the ancient theory of pangenesis presented in a slightly different form. According to this concept, instructions to form structures of the body are contained in particles called "gemmules." These particles, formed in various parts of the body, move through the blood to the reproductive organs and are transmitted from there to the offspring. Gemmules from each parent make a contribution to the traits expressed in the offspring.

Galton decided to put the idea of gemmules to a test. He used rabbits that had different coat colors and transfused blood between them. He thought that a transfusion should mix the gemmules from the two rabbits, changing the coat color in the offspring. Blood from black-coated rabbits was transfused into rabbits with white coats, and these white rabbits were bred to each other. If the gemmules from the black rabbits mixed with the gemmules in the white rabbits, then crossing the transfused white rabbits with each other should produce at least some offspring with gray coats (a blend of black and white).

The results of Galton's experiments did not support the idea of pangenesis. No mixing of fur color occurred when transfused rabbits were interbred. He presented his results to the Royal Society, the highest scientific body in England on March 30, 1871, and said: "The conclusion from this large series of experiments is not to be avoided, that the doctrine of pangenesis, pure and simple, as I have interpreted it, is incorrect." The report showed that traits were not transmitted by pangenesis, but it left unanswered the question of how traits are inherited.

Galton went on to make other significant contributions to the study of inheritance. He set up the mathematical basis for studying traits controlled by several genes and pointed out the importance of twin studies in human genetics.

Unfortunately, it appears that neither Galton nor Darwin read the work of Gregor Mendel on the inheritance of traits in the garden pea published in 1866. This work, titled "Experiments in Plant Hybrids," is one of the most important scientific papers ever published. Here Mendel departs from the usual reporting of observations and takes the additional step of fitting his experimental results into a conceptual framework that can be used to explain the mechanism of heredity in any organism, not just the garden pea. In this chapter, we reconstruct the experiments of Mendel and show how he moved from recording his results to drawing conclusions about the principles of heredity.

HEREDITY: HOW DOES IT WORK?

Johann Gregor Mendel was born in 1822 in Hynice, Moravia, a region now part of the Czech Republic. He showed great promise as a student, but poverty prevented him from attending a university. At the age of 21, he entered the Augustinian monastery at Brno as a way of continuing his interests in natural history. After completing his monastic studies, Mendel enrolled at the University of Vienna in the fall of 1851. In his first year he took courses in physics, mathematics, chemistry, and the natural sciences. In his botany courses, Mendel encountered the new concept that all organisms are composed of cells and cells are the fundamental unit of all living things. The cell theory raised several new questions and many controversial issues. One was the question whether both parents contributed equally to the traits of the offspring. Because the female gametes in most plants and animals are so much larger than those of the male, this was a logical and much debated question. Related to this was the question of whether the traits present in the offspring resulted from blending the traits of the parents. If so, did the female contribute 40, 50, 60%, or more of the traits? In 1854 Mendel returned to Brno to teach physics and began a series of experiments that were to resolve these questions.

MENDEL'S EXPERIMENTAL APPROACH RESOLVED MANY UNANSWERED QUESTIONS

Mendel's success in uncovering the mechanisms of inheritance was not blind luck. Instead, it was the result of carefully planned experiments. Having determined what he wanted to investigate, Mendel set about choosing an organism for these experiments. Near the beginning of his landmark paper on inheritance, Mendel wrote:

> The value and validity of any experiment are determined by the suitability of the means as well as by the way they are applied. In the present case as well, it cannot be unimportant which plant species were chosen for the experiments and how these were carried out. Selection of the plant group for experiments of this kind must be made with the greatest possible care if one does not want to jeopardize all possibility of success from the very outset.

Then, he listed the properties that an experimental organism should have. First, it should have a number of differing traits that can be studied. Secondly, the plant should be self-fertilizing and have a flower structure that minimizes accidental contamination with foreign pollen, and third, the offspring of self-fertilized plants should be fully fertile so that further crosses can be made.

He paid particular attention to a plant group known as the legumes because their flower structure allows self-pollination or cross-pollination (with a minimum chance of accidental pollination) by other plants. Among the legumes, he noted that 34 varieties of pea plants with different traits were available from seed dealers. The plant had a relatively short growth period, could be grown in the ground or in pots in the greenhouse, and could be self-fertilized or artificially fertilized by hand (▶ Figure 3.1).

Then, Mendel tested all 34 varieties of pea plants for two years to ensure that the traits they carried were true-breeding, that is, that self-fertilization gave rise to the same traits in all offspring, generation after generation. From these, 22 varieties were planted annually for the next eight years to provide plants for his experiments

▶ **FIGURE 3.1** The study of the way traits such as flower color in pea plants and pod shape are passed from generation to generation provided the material for Mendel's work on heredity.

(▶ Figure 3.2). For his work, he selected seven characters that affected the seeds, pods, flowers, and stems of the plant (Table 3.1). Each character he studied was represented by two distinct forms or traits: plant height by tall and short, seed shape by wrinkled and smooth, and so forth.

To avoid errors caused by small sample sizes, he planned experiments on a large scale. Over the next eight years, Mendel used some 28,000 pea plants in his experiments. In all of his experiments, he kept track of each character separately over several generations. He began by studying one pair of traits at a time and repeated his experiments for each of the traits to confirm his results. Using his training in physics and mathematics, Mendel analyzed his data according to the principles of probability and statistics. His methodical and thorough approach to his work and his lack of preconceived notions were the secrets of his success.

▶ **FIGURE 3.2** The monastery garden where Mendel carried out his experiments in plant genetics.

Table 3.1	Traits Selected for Study by Mendel	
Structure Studied	**Dominant**	**Recessive**
SEEDS		
Shape	Smooth	Wrinkled
Color	Yellow	Green
Seed coat color	Gray	White
PODS		
Shape	Full	Constricted
Color	Green	Yellow
FLOWERS		
Placement	Axial (along stems)	Terminal (top of stems)
STEMS		
Length	Long	Short

CROSSING PEA PLANTS: THE PRINCIPLE OF SEGREGATION

To show how Mendel developed his ideas about inheritance, we will first describe some of his experiments and outline the results he obtained. Then, we will follow the reasoning Mendel used in reaching his conclusions and outline some of the further experiments that confirmed his ideas.

In the first set of experiments, Mendel followed the inheritance of seed shape. Because this involved only one character (seed shape), he called it a **monohybrid cross**. He took plants that had smooth seeds and crossed them with a variety that had wrinkled seeds. In making this cross, flowers from one variety were fertilized by hand using pollen from the other variety. In this first experiment, Mendel performed 60 fertilizations on 15 plants. The seeds that formed as a result of these fertilizations were all smooth. This result was true whether the pollen was contributed by a plant with smooth peas or by a plant with wrinkled peas. The next year, Mendel planted the smooth seeds from this cross. When the plants matured, the flowers were self-fertilized, and 7324 seeds were collected. Of these, 5474 were smooth and 1850 were wrinkled.

> P1 : Smooth × wrinkled
> F1 : All Smooth
> F2 : 5474 Smooth and 1850 wrinkled

He conducted similar experiments on the inheritance of the other six characters. In his experiments, Mendel designated the parental generation as the P1 generation and the offspring as the F1 (first filial) generation. The second generation, produced by self-fertilizing the F1 plants, was called the F2 (or second filial) generation. The experiments with seed shape are summarized in ❯ Figure 3.3.

Results and Conclusions from Mendel's First Series of Crosses

The results from experiments with all seven characters were similar to those obtained with smooth and wrinkled seeds and are summarized in ❯ Figure 3.4. In all crosses, the following results were obtained:

- The F1 offspring showed only one of the two parental traits and always the same trait.
- In all crosses, it did not matter which variety served as the male parent (that is, which plant donated the pollen). The results were always the same.

■ Monohybrid cross A cross between individuals that differ with respect to a single gene pair.

P1 Smooth × wrinkled
F1 Smooth
Self-fertilize F1 plants
F2 5474 Smooth 1850 wrinkled
Total peas in F2: 7324

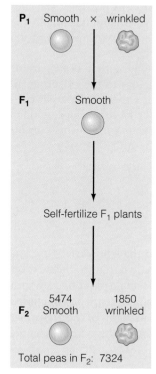

FIGURE 3.3 One of Mendel's crosses. Pure-breeding varieties of peas (smooth and wrinkled) were used as the P1 generation. The offspring in the F1 had all smooth seeds. Self-fertilization of F1 plants gave rise to both smooth and wrinkled progeny in the F2 generation. About three-quarters of the offspring were smooth, and about one-quarter were wrinkled.

Trait Studied
Results in F2
Seed shape
5474 smooth
1850 wrinkled
Seed color
6022 yellow
2001 green
Seed coat color
705 gray
224 white
Pod shape
882 inflated
299 constricted
Pod color
428 green
152 yellow
Flower position
651 along stem
207 at tip
Stem length
787 tall
277 dwarf

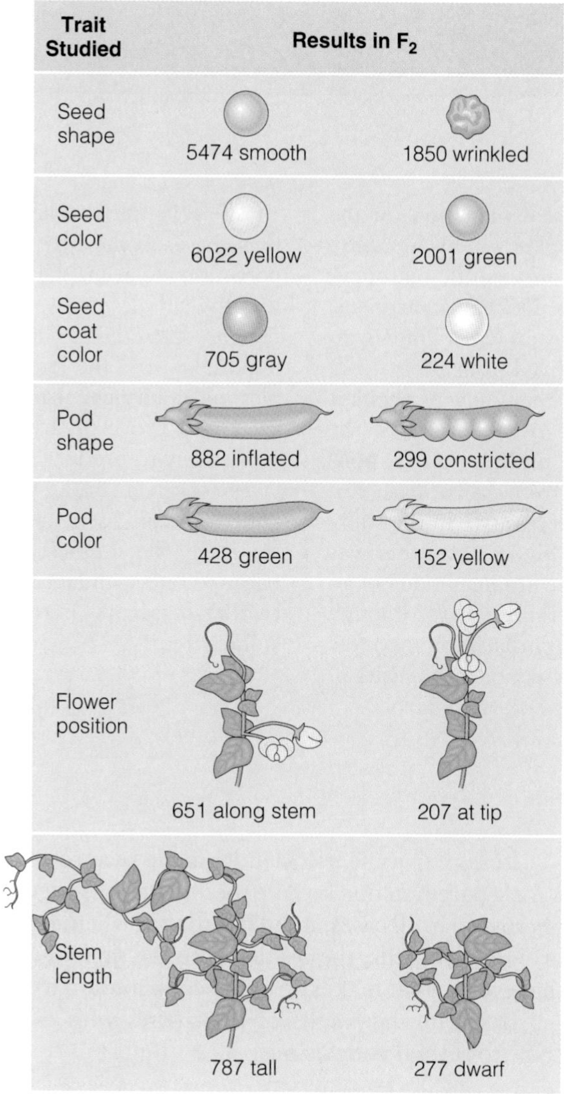

FIGURE 3.4 Results of Mendel's monohybrid crosses in peas. The numbers represent the F2 plants showing a given trait. On average, three-quarters of the offspring showed one trait, and one-quarter showed the other (a 3 : 1 ratio).

SIDEBAR

Why Wrinkled Peas Are Wrinkled

Wrinkled peas were one variety used by Mendel in his experiments. At the time, nothing was known about how peas became wrinkled or smooth. All Mendel needed to know was that a factor controls seed shape and that it has two forms—a dominant one for smooth shape and a recessive one for wrinkled shape.

Recently, scientists have discovered how peas become wrinkled, providing a connection between a gene and its phenotype. While the pea is developing, starch is synthesized and stored as a food source. Starch is a large, branched molecule made up of sugar molecules, and the ability to form branches is controlled by a gene. Normally, starch molecules are highly branched structures. This allows more sugar to be stored in each molecule. In peas that have the wrinkled genotype, the branching gene is inactive. Thus, the developing pea converts sugar into starch very slowly, and excess sugar accumulates. The excess sugar causes the pea to take up large amounts of water, and the seed swells. In a final stage of development, water is lost from the seed. In homozygous wrinkled peas, more water is lost than in the smooth seeds, causing the outer shell of the pea to become wrinkled.

Mendel's contribution was to show that a specific factor controlled a trait and that a given gene could have different forms. Now we know that genes exert their effect on phenotype through the production of a gene product.

- The trait not shown in the F1 offspring reappeared in about 25% of the F2 offspring.

The results of these crosses were the basis for Mendel's first discoveries. His experiments showed that traits remained unchanged as they passed from parent to offspring. Traits did not blend together in any of the offspring. Although traits might be unexpressed, they remained unchanged from generation to generation. This convinced him that inheritance did not work by blending the traits of the parents in the offspring. Rather, traits were inherited as if they were separate units that did not blend together.

Ockham's Razor

*W*hen Mendel proposed the simplest explanation for the number of factors contained in the F1 plants in his monohybrid crosses, he was using a principle of scientific reasoning known as the principle of parsimony, or Ockham's razor.

William of Ockham (also spelled Occam) was a Franciscan monk and scholastic philosopher who lived from about 1300 to 1349. He had a strong interest in the study of thought processes and in logical methods. He is the author of the maxim known as Ockham's razor: "Pluralites non est pondera sine necessitate," which translates from the Latin as "Entities must not be multiplied without necessity." In the study of philosophy and theology of the Middle Ages, this was taken to mean that when constructing an argument, do not go beyond the simplest argument unless it is necessary. Although Ockham was not the first to use this approach, he used this tool of logic so well and so often to dissect the arguments of his opponents that it became known as Ockham's razor.

The principle was adapted to the construction of scientific hypotheses in the 15th century. Galileo used the principle of parsimony to argue that because his model of the solar system was the simplest, it was probably correct. In modern terms, the phrase is taken to mean that in proposing a mechanism or hypothesis, use the least number of steps possible. The simplest mechanism is not necessarily correct, but it is usually the easiest to prove or disprove by doing experiments and the most likely to produce scientific progress.

For a given trait, Mendel concluded that both parents contribute an equal number of factors to the offspring. In this case, the simplest assumption is that each parent contributed one such factor and that the F1 offspring contained two such factors. Further experiments proved this conclusion correct.

In his experiments, Mendel made reciprocal crosses, so that the variety used as a male parent in one set of experiments was used as the female parent in the next set of crosses. In all cases, it did not matter whether the male or female plant had smooth or wrinkled seeds; the results were the same. From these experiments he concluded that each parent makes an equal contribution to the genetic makeup of the offspring.

Based on the results of his crosses with each of the seven characters, Mendel came to several conclusions:

■ **Gene**   The fundamental unit of heredity.

- **Genes** (Mendel called them factors) that determine traits can be hidden or unexpressed. All of the F1 seeds were smooth, but when these seeds were grown and self-fertilized, they produced some plants that had wrinkled seeds. This means that the F1 seeds contained a gene for wrinkled that was present but not expressed. He called the trait that is not expressed in the F1 but is present in the F2 the **recessive trait**. The trait expressed in the F1 is called the **dominant trait**. Mendel called this phenomenon dominance.

■ **Recessive**   The trait unexpressed in the F1 but reexpressed in some members of the F2 generation.

■ **Dominant**   The trait expressed in the F1 (or heterozygous) condition.

- Comparison of the P1 plants with smooth seeds and the F1 plants with smooth seeds showed that despite identical appearances, their genetic makeup must be different. When P1 plants are self-fertilized, they give rise to plants that produce only smooth seeds. But when F1 plants are self-fertilized, they give rise to plants with both smooth and wrinkled seeds. Mendel realized that it was important to make a distinction between the appearance of an organism and its genetic constitution. Now, the term **phenotype** is used to refer to the observed properties or outward appearance of a trait, and the term **genotype** refers to the genetic makeup of an organism. In our example, the P1 and F1 plants that have smooth seeds have identical phenotypes, but must have different genotypes.

■ **Phenotype**   The observable properties of an organism.

■ **Genotype**   The specific genetic constitution of an organism.

- The results of these self-fertilization experiments show that the F1 plants must have contained genes for smooth and wrinkled traits because both types of seeds are present in the F2 generation. The question is, how many genes for seed shape are carried in the F1 plants? Mendel had already reasoned that the male and female parent contributed equally to the traits of the offspring. Therefore, the simplest interpretation is that each F1 plant carried two genes, one for smooth that was expressed, and one for wrinkled that remained unexpressed (see Concepts and Controversies: Ockham's Razor). By extending this reasoning, each P1 and F2 plant must also contain two genes that determine seed shape. Traditionally, uppercase letters are used to represent traits that have a dominant pattern of inheritance, and lowercase letters are used to represent traits that have a recessive pattern of inheritance (S = smooth, and s = wrinkled). Using this shorthand, we can reconstruct the genotypes and phenotypes of the P1 and F1 shown in ▶ Figure 3.5.

Inheritance of a Single Trait: The Principle of Segregation

If genes that determine traits exist in pairs, then some mechanism must exist to prevent these genes from being doubled in each succeeding generation. If each parent has two genes for a given trait, why doesn't the offspring have four? Mendel reasoned that members of a gene pair must separate or segregate from each other during gamete formation. As a result, each gamete receives only one of the two genes that control a given trait. The separation of members of a gene pair during gamete formation is called the **principle of segregation,** or Mendel's First Law.

■ **Segregation** The separation of members of a gene pair from each other during gamete formation.

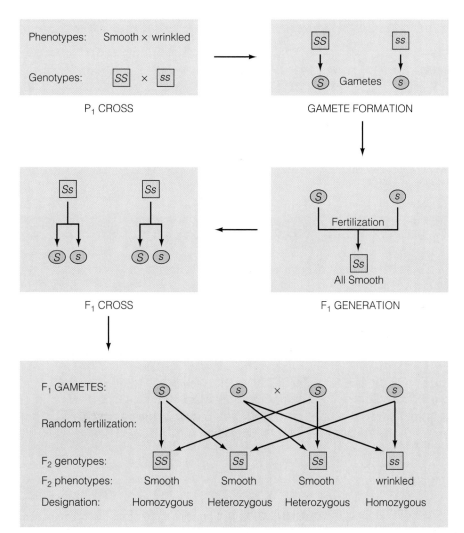

▶ **FIGURE 3.5** The phenotypes and genotypes of the parents and offspring in Mendel's cross involving seed shape.

As shown in ▶ Figure 3.6, members of a gene pair separate (or segregate) from each other so that only one or the other is included in each gamete. In the F1 generation, each of the parents makes two kinds of gametes in equal proportions. At fertilization, the random combination of these gametes produces the genotypic combinations shown in the Punnett square (a method for analyzing genetic crosses devised by R. C. Punnett). The F2 genotypic ratio of 1 *SS* : 2 *Ss* : 1 *ss* is expressed as a phenotypic ratio that is three-quarters dominant and one-quarter recessive. This is usually abbreviated as a 3 : 1 ratio.

Mendel's experiments with the six other sets of traits can also be explained in this way. His reasoning predicts the genotypes of the F2 generation. One-fourth of the F2 plants should carry only smooth factors (*SS*) and give rise only to smooth plants when self-fertilized. One half of the F2 plants should carry factors for both smooth and wrinkled (*Ss*) and give rise to smooth and wrinkled progeny in a three-quarter to one-quarter ratio when self-fertilized (▶ Figure 3.7). Finally, one-fourth of the F2 plants should be homozygous for wrinkled (*ss*) and give rise to all wrinkled progeny if self-fertilized. In fact, Mendel fertilized a number of plants from the F2 generation and five succeeding generations to confirm these predictions.

Mendel carried out his experiments before the discovery of mitosis and meiosis and before the discovery of chromosomes. As we discuss in a later section, his deductions about the way traits are inherited are in fact descriptions of the way chromosomes behave in meiosis. Seen in this light, his discoveries are all the more remarkable.

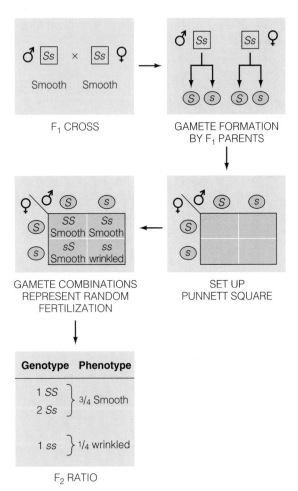

▶ **FIGURE 3.6** How a Punnett square can be used to generate the F2 ratio in a cross from the F1 generation.

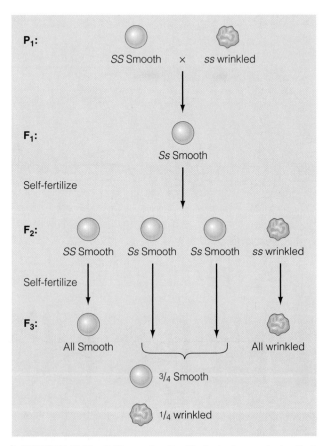

▶ **FIGURE 3.7** Self-crossing the F2 plants demonstrates that there are two different genotypes among the plants with smooth peas in the F2 generation.

Today, we call Mendel's factors genes and refer to the alternate forms of a gene as **alleles**. In the example we have been discussing, the gene for seed shape has two alleles, smooth and wrinkled. Individuals that carry identical alleles of a given gene (*SS* or *ss*) are said to be **homozygous** for the gene in question. Similarly, when two different alleles are present in a gene pair (*Ss*), the individual is said to have a **heterozygous** genotype. The *SS* homozygotes and the *Ss* heterozygotes show dominant phenotypes (because *S* is dominant to *s*), and ss homozygotes show recessive phenotypes.

■ **Allele** One of the possible alternative forms of a gene, usually distinguished from other alleles by its phenotypic effects.

■ **Homozygous** Having identical alleles for one or more genes.

■ **Heterozygous** Carrying two different alleles for one or more genes.

MORE CROSSES WITH PEA PLANTS: THE PRINCIPLE OF INDEPENDENT ASSORTMENT

Mendel realized the need to extend his studies on the inheritance from monohybrid crosses to more complex situations. He wrote:

> In the experiments discussed above, plants were used which differed in only one essential trait. The next task consisted in investigating whether the law of development thus found would also apply to a pair of differing traits.

For this work, he selected seed shape and seed color as traits to be studied, because as he put it: "Experiments with seed traits lead most easily and assuredly to success." A cross that involves two sets of characters is called a dihybrid cross.

Crosses Involving Two Traits

As in the first set of crosses, we will analyze the actual experiments of Mendel, outline the results, and summarize the conclusions he drew from them. From previous crosses it is known that in seeds, smooth is dominant to wrinkled and yellow is dominant to green. In our reconstruction of these experiments, we represent smooth by an uppercase *S*, wrinkled by a lowercase *s*, yellow by an uppercase *Y*, and green by a lowercase *y*.

Methods, Results, and Conclusions

Mendel selected true-breeding plants that had smooth, yellow seeds and crossed them with true-breeding plants that had wrinkled, green seeds (▶ Figure 3.8). The seeds of the F1 plants were all smooth and yellow, confirming that smooth and yellow are dominant traits. Then, he self-fertilized the F1 and produced an F2 generation. These F2 plants produced seeds of four types, often found together in a single pod. From 15 plants, he counted a total of 556 seeds that had the following phenotypic distributions:

315 Smooth and Yellow
108 Smooth and green
101 wrinkled and Yellow
32 wrinkled and green

The F2 phenotypes include the parental phenotypes and also two new combinations (smooth green and wrinkled yellow).

To determine how the two genes in a dihybrid cross were inherited, Mendel first analyzed the results of the F2 for each trait separately, as if the other trait were not present (▶ Figure 3.9). If we consider only seed shape (smooth or wrinkled) and ignore color, we expect three-quarters smooth and one-quarter wrinkled offspring in the F2. Analyzing the actual results, we find that the total number of smooth offspring is 315 + 108 = 423. The total number of wrinkled seeds is 101 + 32 = 133. The

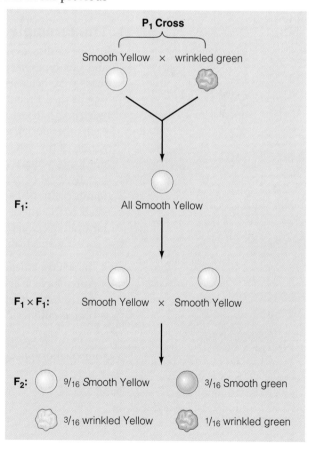

▶ **FIGURE 3.8** The phenotypic distribution in a dihybrid cross. Plants in the F2 generation show the parental phenotypes and two new phenotypic combinations.

FIGURE 3.9 Analysis of a dihybrid cross for the separate inheritance of each trait.

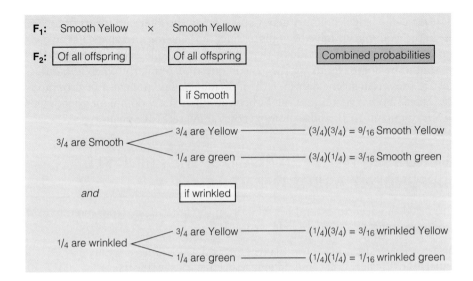

proportion of smooth to wrinkled seeds (423 : 133) is close to a ratio of 3 : 1. Similarly, if we consider only seed color (yellow or green), there are 416 yellow seeds (315 + 101) and 140 green seeds (108 + 32) in the F2 generation. These results are also close to a 3 : 1 distribution.

Once he established a pattern of 3 : 1 inheritance for each trait separately (consistent with the principle of segregation), then Mendel considered the inheritance of both traits simultaneously.

The Principle of Independent Assortment

Before we discuss what is meant by independent assortment, let's see how the phenotypes and genotypes of the F1 and F2 were generated. The F1 plants that had smooth yellow seeds were heterozygous for both seed shape and seed color. This means that the genotype of the F1 plant must have been *SsYy*, and the *S* and *Y* alleles were dominant to *s* and *y*. Mendel had already concluded that members of a gene pair separate or segregate from each other during gamete formation. In this case, when two genes are present, the segregation of the *S* and *s* alleles must have occurred independently from the segregation of the *Y* and *y* alleles (◗ Figure 3.10).

Because each gene pair segregated independently, the gametes formed by the F1 plants contained all combinations of these alleles in equal proportions: *SY*, *Sy*, *sY*, and *sy*. If fertilizations involving the four types of male and female gametes occurred at random (as expected), 16 possible combinations would result (Figure 3.10).

An inspection of the 16 combinations in the Punnett square shows that

- nine have at least one copy of each dominant allele, *S* and *Y*;
- three have at least one copy of the dominant allele *S* and are homozygous for *yy*;
- three have at least one copy of the dominant allele *Y* and are homozygous for *ss*;
- one combination is homozygous for *ss* and *yy*.

In other words, the 16 combinations of fertilization events (genotypes) fall into four phenotypic classes:

smooth and yellow
smooth and green
wrinkled and yellow
wrinkled and green

These phenotypic combinations correspond to the number of phenotypic classes in the F2 generation and to the proportions of progeny in each class (Figure 3.10).

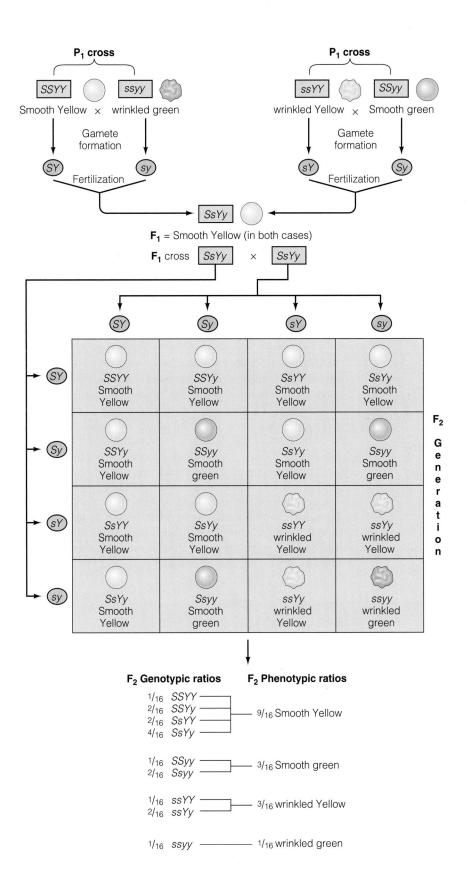

▶ **FIGURE 3.10** Punnett square of the dihybrid cross shown in Figure 3.9. There are two combinations of dominant and recessive traits that result in heterozygous F1 individuals. One is a cross between smooth, yellow and wrinkled, green. The other is a cross between wrinkled, yellow and smooth, green.

Evaluating Results—The Chi-Square Test

One of Mendel's innovations was the application of mathematics and combinatorial theory to biological research. This allowed him to predict the genotypic and phenotypic ratios in his crosses and to follow the inheritance of several traits simultaneously. If the cross involved two alleles of a gene (e.g., A and a), the expected outcome was an F2 phenotypic ratio of $3A:1a$ and a genotypic ratio of $1AA:2Aa:1aa$. What Mendel was unable to analyze mathematically was how well the actual outcome of the cross fulfilled his predictions. He apparently realized this problem and compensated for it by conducting his experiments on a large scale, counting substantial numbers of individuals in each experiment to reduce the chance of error. Shortly after the turn of the century, an English scientist named Karl Pearson developed a statistical test to determine whether the observed distribution of individuals in phenotypic categories is as predicted or occurs by chance. This simple test, regarded as one of the fundamental advances in statistics, is a valuable tool in genetic research. The method is known as the chi-square (χ^2) test (pronounced "kye square"). In use, this test requires several steps:

1. Record the observed numbers of organisms in each phenotypic class.
2. Calculate the expected values for each phenotypic class based on the predicted ratios.
3. If O is the observed number of organisms in a phenotypic class, or category, and E is the expected number, calculate the difference d in each category by subtraction $(O - E) = d$ (Table 3.2).
4. For each phenotypic category, square the difference d, and divide by the number expected (E) in that phenotypic class.
5. Add all the numbers in step 4 to get the χ^2 value.

If there are no differences between the observed and the expected ratios, the value for χ^2 will be zero. The value of χ^2 increases with the size of the difference between the observed and the expected classes. The formula can be expressed in the general form

$$\chi^2 = \Sigma \frac{d^2}{E}$$

Using this formula, we can do what Mendel could not—analyze his data for the dihybrid cross involving wrinkled and smooth seeds and yellow and green cotyledons that produced a 9:3:3:1 ratio. In the F2, Mendel counted a total of 556 peas. The number in each phenotypic class is the observed number (Table 3.2). Using the total of 556 peas, we can calculate that the expected number in each class for a 9:3:3:1 ratio would be 313:104:104:35 (9/16 of 556 is 313, 3/16 of 556 is 104, etc). Substituting these numbers into the formula, we obtain

$$\chi^2 = \frac{2^2}{313} + \frac{4^2}{104} + \frac{3^2}{104} + \frac{3^2}{35} = 0.371$$

The χ^2 value is very low, confirming that there is very little difference between the number of peas observed and the number expected in each class. In other words, the results are close enough to the expectation that we need not reject them.

The question remains, however, how much deviation is permitted from the expected numbers before we decide that the observations do not fit our expectation that a 9:3:3:1 ratio will be fulfilled. To decide this we must have a way of interpreting the χ^2 value. We need to convert this value into a probability and to ask, what is the probability that the calculated χ^2 value is acceptable? In making this calculation, first we must establish something called degrees of freedom, df, which is one less than the number of phenotypic classes, n. In the dihybrid cross we expect four phenotypic classes, so the degrees of freedom are as follows:

$$df = n - 1$$
$$df = 4 - 1$$
$$df = 3$$

Next we can calculate the probability of obtaining the given set of results by consulting a probability chart (Table 3.3). First, find the line corresponding to a df value of 3. Look across on this line for the number corresponding to the χ^2 value. The calculated value is 0.37, which is between the columns headed 0.95 and 0.90. This means that we can expect a deviation of this magnitude at least 90% of the time when we do this experiment. In other words, we can be confident that our expectations of a 9:3:3:1 ratio are correct. In general, a P value of less than .05 means that the observations do not fit the expected distribution into phenotypic classes and that the expectation needs to be reexamined. The acceptable range

Table 3.2   Chi-Square Analysis of Mendel's Data

Seed Shape	Cotyledon Color	Observed Numbers	Expected Numbers (Based on a 9:3:3:1 Ratio)	Difference (*d*) (*O − E*)
Smooth	Yellow	315	313	+2
Smooth	Green	108	104	+4
Wrinkled	Yellow	101	104	−3
Wrinkled	Green	32	35	−3

Table 3.3   Probability Values for Chi-Square Analysis

				Probabilities					
df	0.95	0.90	0.70	0.50	0.30	0.20	0.10	0.05	0.01
1	.004	.016	.15	.46	1.07	1.64	2.71	3.84	6.64
2	.10	.21	.71	1.39	2.41	3.22	4.61	5.99	9.21
3	.35	.58	1.42	2.37	3.67	4.64	6.25	7.82	11.35
4	.71	1.06	2.20	3.36	4.88	5.99	7.78	9.49	13.28
5	1.15	1.61	3.00	4.35	6.06	7.29	9.24	11.07	15.09
6	1.64	2.20	3.83	5.35	7.23	8.56	10.65	12.59	16.81
7	2.17	2.83	4.67	6.35	8.38	9.80	12.02	14.07	18.48
8	2.73	3.49	5.53	7.34	9.52	11.03	13.36	15.51	20.09
9	3.33	4.17	6.39	8.34	10.66	12.24	14.68	16.92	21.67
10	3.94	4.87	7.27	9.34	11.78	13.44	15.99	18.31	23.21

◄——————————————— Acceptable ———————————————► Unacceptable

Note. From *Statistical Tables for Biological, Agricultural and Medical Research* (6th ed.), Table IV, by R. Fisher and F. Yates, Edinburgh: Longman Essex, 1963.

of values is indicated by a line in Table 3.3. The use of *p* = .05 as the border for acceptability has been arbitrarily set.

In the case of Mendel's data, there is very little difference between the observed and expected results (Table 3.2). Several writers have commented that Mendel's results fit the expectations too closely and that perhaps he adjusted his results to fit a preconceived standard.

In human genetics, the χ^2 method is very valuable and has wide applications. It is used in deciding modes of inheritance (autosomal or sex-linked), deciding whether the pattern of inheritance shown by two genes indicates that they are on the same chromosome, and deciding whether marriage patterns have produced genetically divergent groups in a population.

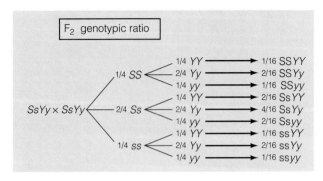

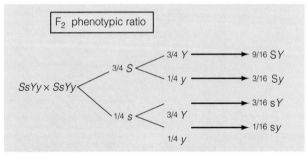

▶ **FIGURE 3.11** The phenotypic and genotypic ratios of a dihybrid cross can be derived using a forked-line method instead of a Punnett square.

■ **Independent assortment** The random distribution of genes into gametes during meiosis.

■ **Genetics** The scientific study of heredity.

For example, 315 of 556 seeds were smooth and yellow, corresponding to about nine-sixteenths of the total number of offspring; 108 of 556 seeds were smooth and green, corresponding to about three-sixteenths of the offspring; and so forth. This distribution of offspring in the F2 corresponds to a phenotypic ratio of 9:3:3:1 (see Concepts and Controversies: Evaluating Results—The Chi Square Test). The results of this cross can be explained by assuming (as Mendel did) that during gamete formation, alleles in one gene pair segregate into gametes independently of the alleles belonging to other gene pairs, resulting in the production of gametes containing all combinations of alleles. This second fundamental principle of genetics outlined by Mendel is called the **principle of independent assortment,** or Mendel's Second Law.

Instead of using a Punnett square to determine the distribution and frequencies of phenotypes and genotypes in the F2, we can also use a branch diagram, or forked-line method, which is based on probability. In the dihybrid F2, the probability that a plant will be smooth is three out of four, and the chance that it will be wrinkled is one in four. Because each trait is inherited independently, each smooth plant has a three out of four chance of being yellow and a one in four chance of being green. The same is true for each wrinkled plant. ▶ Figure 3.11 shows how these probabilities combine to give the phenotypic ratios.

After 10 years of experimentation involving thousands of pea plants, Mendel presented his results in 1865 at the February and March meetings of his local Natural Science Society. The text of these lectures was published in the following year in the Proceedings of the Society. Although his work was cited in several bibliographies and copies of the journal were widely read, the significance of Mendel's findings was unappreciated. Even Charles Darwin, continuing his search for a mechanism to explain heredity and its role in natural selection, failed to realize the significance of Mendel's work. Darwin's own copy of the Proceedings has many notes scribbled in the margin of the paper next to Mendel's, but not one pencil mark anywhere in Mendel's paper.

Finally, in 1900 three scientists, each working independently on the mechanism of heredity, confirmed Mendel's findings and brought his paper to widespread attention. These events stimulated a great interest in the study of what is now called **genetics.** Unfortunately, Mendel died in 1884—unaware that he had founded an entire scientific discipline.

MEIOSIS EXPLAINS MENDEL'S RESULTS: GENES ARE ON CHROMOSOMES

When Mendel performed his experiments, the behavior of chromosomes in mitosis and meiosis was unknown. By 1900, however, cytology was a well-established field, and the principles of mitosis and meiosis had been described. As scientists confirmed that the fundamentals of Mendelian inheritance operated in many organisms, it soon became apparent that genes and chromosomes had much in common (Table 3.4). Both chromosomes and genes occur in pairs. In meiosis, members of a chromosome pair separate from each other. Likewise, members of a gene pair separate from each other during gamete formation. Finally, the fusion of gametes during fertilization restores the diploid number of chromosomes and two copies of each gene to the zygote, producing the genotypes of the next generation.

In 1903 Walter Sutton and Theodore Boveri each noted these similarities and independently proposed the idea that genes are carried on chromosomes. This **chromosome theory of inheritance** has been confirmed by many experiments in the fol-

Table 3.4 Genes, Chromosomes, and Meiosis

Genes	Chromosomes
Occur in pairs (alleles)	Occur in pairs (homologues)
Members of a gene pair separate from each other during meiosis	Members of a pair of homologues separate from each other during meiosis
Members of one gene pair independently assort from other gene pairs during meiosis	Members of one chromosome pair independently assort from other chromosome pairs during meiosis

lowing decades and is one of the foundations of modern genetics. Each gene occupies a specific site (called a **locus**) on a chromosome and each chromosome carries many genes. In humans it is estimated that 50,000 to 100,000 genes are carried on the 23 different chromosomes. Although each gene may have different forms (alleles), normal individuals carry only two such alleles, because he or she has only two copies of each chromosome. Obviously, different individuals can have different combinations of alleles. ▶ Figure 3.12 shows how genes and chromosomes move through meiosis.

▪ **Locus** The position occupied by a gene on a chromosome.

▶ **FIGURE 3.12** Mendel's observations about segregation and independent assortment are explained by the behavior of chromosomes during meiosis. The arrangement of chromosomes at metaphase I is random. As a result, all allelic combinations of the two genes are produced in the gametes.

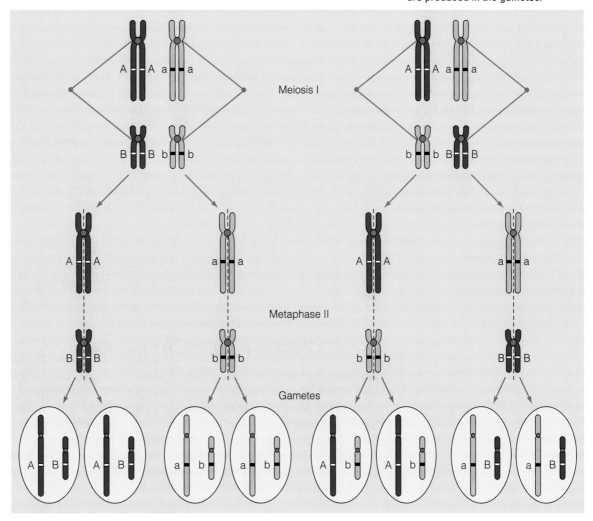

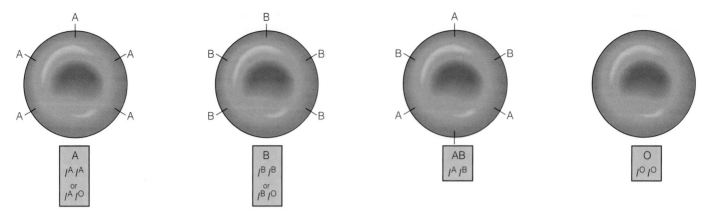

▶ **FIGURE 3.13** Each allele of codominant genes is fully expressed in the heterozygote. Type A blood has A antigens on the cell surface, and type B has B antigens on the surface. In type AB, both the A and the B antigen are present on the cell surface. The *A* and *B* alleles of the *I* gene are codominant. In type O blood, no antigen is present. The *O* allele is recessive to both the *A* and the *B* allele.

MANY GENES HAVE MORE THAN TWO ALLELES

So far, our discussion has been confined to genes that have two alleles. Because alleles represent different forms of a gene, there is no reason why a gene has to have only two alleles. In fact, many genes have more than two alleles. Any *individual* can carry only two alleles of a gene, but in a population, many different alleles of a gene can be present. In humans, the gene that determines ABO blood groups is an example of a gene that has **multiple alleles**. The ABO blood types are determined by molecules (proteins that have polysaccharides attached) on the surface of human red blood cell membranes. These molecules provide the cell with an identity tag recognized by the body's immune system.

ABO blood types are controlled by a single gene, *I*, that has three alleles, I^A, I^B, and I^O. The *A* and *B* alleles control the formation of slightly different forms of a molecule (called an antigen) present on the surface of blood cells and other cells in the body. Individuals homozygous for the *A* allele (*AA*) carry the *A* antigen on cells and have blood type A. Those who are homozygous for the *B* allele (*BB*) carry the *B* antigen and are type B. The third allele, (*O*), does not make any antigen, and individuals homozygous for the third allele for this gene, I^O, carry neither the A nor the B antigen on their cells. The *O* allele is recessive to the *A* and *B* alleles. Because there are three alleles, there are six possible genotypes (▶ Figure 3.13).

Blood type can be determined by a simple test, and it is important to match blood types in transfusions. The ABO blood groups are also used as evidence in paternity cases. Blood typing can provide evidence that rules out a man as the father of a given child.

> ■ **Multiple alleles** Genes that have more than two alleles have multiple alleles.

VARIATIONS ON A THEME BY MENDEL

After Mendel's work became widely known, geneticists turned up cases in the early years of this century in which the phenotypes of the F1 offspring did not resemble those of the parents. In some cases, the offspring had a phenotype intermediate to that of the parents, or a phenotype in which the traits of both parents were expressed. These findings led to a debate as to whether these cases could be explained by Mendelian principles of inheritance, or whether there might be another, separate mechanism of inheritance that did not follow the laws of segregation and independent assortment.

Eventually, experiments with several different organisms showed that these cases were not exceptions to Mendelian inheritance and could be explained by the way in which genes act to produce phenotypes. In this section, we discuss some of these phenotypic variations and show that although the phenotypes may not follow the predicted Mendelian ratios for complete phenotypic dominance, the outcome of crosses involving these traits can be predicted according to the Mendelian distribution of genotypes.

Both Codominant Alleles Are Expressed in the Phenotype

In **codominance,** heterozygotes fully express both alleles. In the ABO blood type, *AB* heterozygotes have both the *A* and *B* antigens on their cell membranes and are blood type AB. In AB heterozygotes, neither allele is dominant over the other, and because each allele is fully expressed, they are said to be codominant (Figure 3.13). As a result, the three alleles of the ABO system have six genotypic combinations that contribute to four phenotypes (Table 3.5).

Incomplete Dominance Has a Distinctive Phenotype

In **incomplete dominance,** the heterozygote has a phenotype intermediate to those of the homozygous parents. An example of this type of inheritance is flower color in snapdragons (▶ Figure 3.14). If a true-breeding variety that bears red flowers is crossed with a variety that produces white flowers, the F1 offspring all have pink flowers. The phenotype of the F1 differs from that of either parent and is intermediate to the phenotypes of the parents. When the F1 plants are self-fertilized, they produce plants with red, pink, and white flowers in a 1 : 2 : 1 ratio. This would not be expected if true blending occurred because crossing plants that have pink flowers should produce offspring that have pink flowers.

Table 3.5

ABO Blood Types

Genotypes	Phenotypes
$I^A I^A$, $I^A I^O$	Type A
$I^B I^B$, $I^B I^O$	Type B
$I^A I^B$	Type AB
$I^O I^O$	Type O

■ **Codominance** Full phenotypic expression of both members of a gene pair in the heterozygous condition.

■ **Incomplete dominance** Expression of a phenotype that is intermediate between those of the parents.

▶ **FIGURE 3.14**
Incomplete dominance in snapdragon flower color. Red-flowered snapdragons crossed with white-flowered snapdragons produce offspring that have pink flowers in the F1. In heterozygotes, the allele for red flowers is incompletely dominant over the allele for white.

The results of this cross can be explained using Mendelian principles. Each genotype in this cross has a distinct phenotype, and the phenotypic ratio of 1 red : 2 pink : 1 white is the same as the genotypic ratio of 1 *RR* : 2 *Rr* : 1 *rr*. It takes two doses of the *R* allele to produce red flowers. One dose results in pink flowers, and as a result, the *R* allele is incompletely dominant over the *r* allele. Because the *r* allele produces no color, the absence of *R* alleles produces white flowers.

The Concepts of Dominance and Recessiveness

Earlier in this chapter, genes were said to have dominant or recessive patterns of inheritance. Dominant alleles were represented by uppercase letters (such as *A*) and recessive alleles were represented by lowercase letters (such as *a*). If a trait is present when inherited from only one parent, the allele is said to be dominant. If the phenotype is shown only when identical alleles have been inherited from each parent, the trait is recessive. Dominant traits are expressed in the heterozygous condition, and recessive traits are not. Referring to alleles as dominant or recessive is a shorthand way of talking about genes. Remember that the terms *dominant* and *recessive* actually refer to phenotypes and not to the gene itself. As we will see in later chapters, the terms *dominant* and *recessive* are somewhat relative. Heterozygotes for recessive traits can often be identified because they produce half as much of a gene product as those that have the homozygous dominant genotype. This does not change the recessive nature of the phenotype because usually only homozygous recessive individuals are phenotypically affected by the mutant condition.

In general, recessive traits have more serious phenotypic effects than those controlled by dominant genes. One reason is that recessive alleles can remain in the population in the heterozygous condition at high frequency, affecting only a small number of homozygotes. On the other hand, if a dominant mutation prevented survival or reproduction, it would quickly be eliminated from the population. As a result, many dominant genetic diseases are more variable and diffuse in their phenotypic effects. If dominant traits have lethal effects, it is usually after reproductive age has been reached. In other words, deleterious dominant traits survive because of low penetrance, variable expressivity, and delayed onset.

▶ **FIGURE 3.15** Albino individuals lack pigment in the skin, hair, and eyes.

MENDELIAN INHERITANCE IN HUMANS

After 1900, the principles of segregation and independent assortment discovered by Mendel were studied in a wide range of organisms. Although some believed that inheritance of traits in humans might be an exception to these principles, the first Mendelian trait (a hand deformity called brachydactyly, MIM/OMIM 112500) was identified in 1905. Since then, over 4500 such traits have been described.

To illustrate how segregation and independent assortment apply to the inheritance of human traits, we will follow the inheritance of a recessive trait called albinism (MIM/OMIM 203100). Individuals who are homozygous (*aa*) for this recessive trait have no pigment in skin, hair and eyes and have very pale, white skin, white hair, and colorless eyes. Actually, the eyes may appear pink because of the blood vessels in the iris (▶ Figure 3.15). The dominant allele (*A*) controls normal pigmentation.

In the example we will consider, both parents have normal pigmentation but are heterozygous for the recessive allele causing albinism (▶ Figure 3.16). In each parent, the dominant and recessive alleles separate or segregate from each other at the time of gamete formation. Because each parent can produce two different types of gametes (one containing the dominant allele *A* and another type carrying the recessive allele *a*), there are four possible combinations of these gametes at fertilization. If enough offspring are produced, these four possible types of fertilization events will show a predicted phenotypic ratio of 3 pigmented : 1 albino offspring and a genotypic ratio of 1 AA : 2Aa : 1aa (Figure 3.16). In other words, segregation of alleles during gamete formation produces the same outcome in both pea plants and humans. This does not mean that there will be one albino child and three normally pigmented children in every such family with four children. It does mean that in a mating between heterozygotes, each child has a 25% chance of being albino and a 75% chance of having normal pigmentation.

The simultaneous inheritance of two traits in humans follows the Mendelian principle of independent assortment (▶ Figure 3.17). To illustrate, let's examine a family in which each parent is heterozygous for albinism (*Aa*) and heterozygous for another recessive trait, hereditary deafness (MIM/OMIM 220290). The normal allele (*D*) is dominant and is expressed in the homozygous dominant (*DD*) or heterozygous condition (*Dd*). During gamete formation, the alleles for skin color and the alleles for hearing assort into gametes independently. As a result, each parent produces equal proportions of four different types of gametes. If each parent produces four types of gametes, there are 16 possible combinations of these gametes at fertilization (four types of gametes in all possible combinations), resulting in four different phenotypic classes (Figure 3.17). An examination of the possible genotypes shows that there is a 1 in 16 chance that a child would be both deaf and an albino.

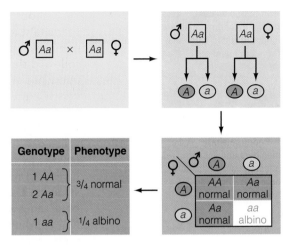

▶ FIGURE 3.16 The segregation of albinism, a recessive trait in humans. As in pea plants, alleles of a gene pair separate from each other during gamete formation.

Pedigree Analysis in Human Genetics

In pea plants and other organisms, such as *Drosophila,* genetic analysis can be performed by experimental crosses. In humans, geneticists must base their work on matings that have already taken place and cannot design crosses to test a hypothesis directly. A basic method of genetic analysis in humans is to follow a trait for several

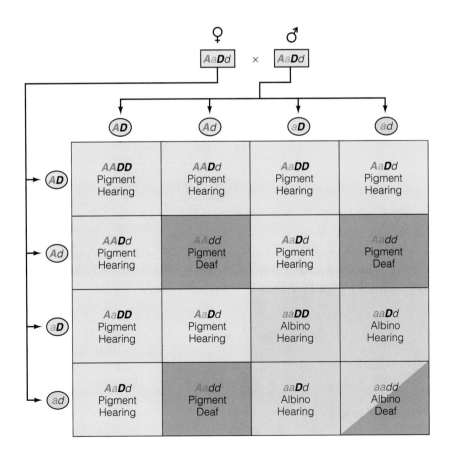

▶ FIGURE 3.17 Independent assortment for two traits in humans follows the same pattern of inheritance as traits in pea plants.

generations in a family to determine how it is inherited. This method is called **pedigree analysis.** A pedigree is the orderly presentation of family information in the form of an easily readable chart. From such a family tree, the inheritance of a trait can be followed through several generations. Using Mendelian principles, the information in the pedigree can be analyzed to determine whether the trait has a dominant or recessive pattern of inheritance and whether the gene in question is located on an autosome or a sex chromosome.

Pedigrees use a set of standardized symbols, many of which are borrowed from genealogy. In constructing a **pedigree chart,** males are represented by squares (□) and females by circles (○). An individual who exhibits a trait in question is represented by a filled symbol (■ or ●). Heterozygotes, when known, are indicated by half-filled symbols (◧ or ◑). The relationships between individuals in a pedigree are indicated by a series of lines. A horizontal line between two symbols represents a mating (□–○). Matings between brother and sister or between close relatives, known as consanguineous matings, are symbolized by a double horizontal line. The offspring, listed from left to right in birth order, are connected to each other by a horizontal line (○ ○ □) and to the parents by a vertical line:

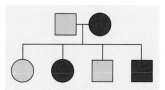

If twins are present as offspring, they are identified either as identical (monozygotic) twins (⤙), arising from a single egg, or nonidentical (dizygotic) twins (⤙), arising from the fertilization of two eggs.

A numbering system is employed to identify individuals in a pedigree. Each generation is indicated by a roman numeral, and within a generation, each individual is numbered by birth order using arabic numbers:

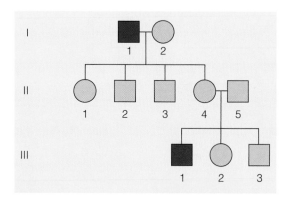

Pedigrees are often constructed after a family member afflicted with a genetic trait has been identified. This individual, known as the **proband,** is indicated on the pedigree by an arrow and the letter P:

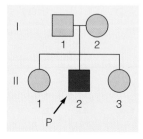

Because pedigree analysis is essentially a reconstruction of a family history, details about earlier generations may be uncertain. If the sex of someone is unknown or unimportant, this is indicated by a diamond shape: (◇). In some cases, spouses in a pedigree are omitted if they are not essential to the inheritance of the trait. If there is doubt that a family member possessed the trait in question, this is indicated by a question mark above the symbol. Many of the symbols and terminology used in constructing pedigrees are presented in ▶ Figure 3.18. A completed pedigree is a form of symbolic communication used by clinicians and researchers in human genetics (▶ Figure 3.19). It contains information that can establish the pattern of inheritance for a trait, identify those at risk of developing or transmitting the trait, and is a resource for establishing biological relationships within a family. Establishing genotypes of parents and predicting the chances of having affected children is part of **genetic counseling,** a topic discussed in Chapter 19.

■ **Genetic counseling** Analysis of genetic risk within families and presentation of available options to avoid or reduce risks.

▶ **FIGURE 3.18** Symbols used in pedigree analysis.

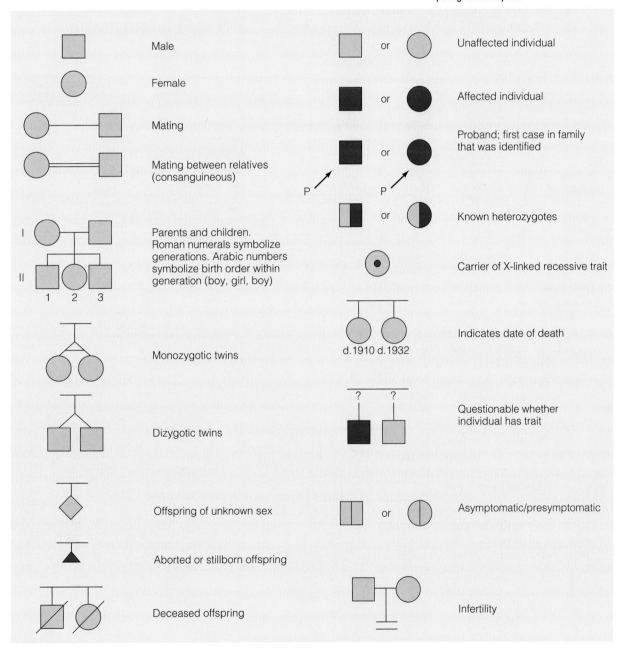

Concepts and Controversies

Solving Genetics Problems

*I*n solving genetics problems, several steps must be followed to ensure success. The process of analyzing and solving these problems depends on several steps: (1) analyze each problem carefully to determine what information is provided and what information is asked for; (2) translate the terms and words of the problems into symbols, and (3) solve the problem using logic.

The most basic problems involving Mendelian inheritance usually provide some information about the parental generation (P1) and ask you to employ your knowledge of Mendelian principles to come to conclusions about the genotypes or phenotypes of the F1 or F2 generation. The solution uses several steps:

1. Carefully read the problem and establish the genotype of each parent, assigning letter symbols if necessary.
2. Based on their genotypes, determine what types of gametes can be formed by each parent.
3. Unite the gametes from the parents in all combinations. Use a Punnett square if necessary. This automatically gives you all possible genotypes and their ratios for the F1 generation.
4. If necessary, use all combinations of F1 individuals as parents for the F2, and repeat steps 2 and 3 to derive the genotypes and phenotypes of the F2 generation.

As an example, consider the following problem. The recessive allele wrinkled (*s*) causes peas to appear wrinkled when homozygous. The dominant allele smooth (*S*) causes peas to appear smooth when homozygous or heterozygous. In the following cross, what phenotypic ratio would you expect in the offspring? One parent is a plant that bears wrinkled seeds. The other is a plant that bears smooth seeds and is the offspring of a cross between true-breeding, smooth and wrinkled parents.

The solution to this problem depends on understanding the principle of segregation and the relationship between dominance and recessiveness. To derive the genotypes of the parental plants, the following are relevant:

1. Because one parental plant bears wrinkled seeds, this plant is homozygous for the recessive allele (*ss*). The other parental plant bears smooth seeds and carries at least one

dominant allele (*S*). Because this parent is the offspring from a cross between true-breeding smooth and true-breeding wrinkled plants, it must have received a wrinkled allele and therefore must be heterozygous (*Ss*).

2. Therefore, the cross is *Ss* x *ss*. The gametes that each parent can make and their combinations in fertilizations are shown below:

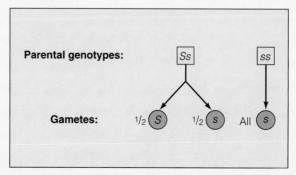

3. In this cross, half of the F1 offspring will be heterozygous smooth individuals (*Ss*), and half will be homozygous wrinkled (*ss*).

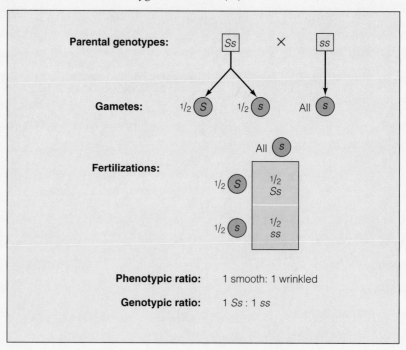

The phenotypic ratio will be 1 smooth : 1 wrinkled.

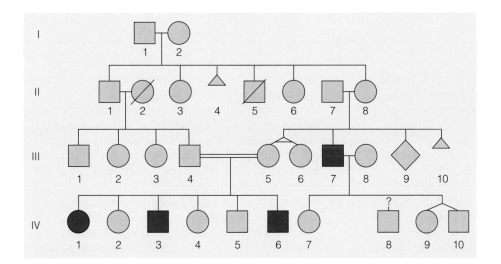

▶ **FIGURE 3.19** A pedigree chart showing the inheritance of a trait in several generations of a family. This pedigree and all those in this book use the standardized set of symbols adopted in 1995 by the American Society of Human Genetics.

Case Studies

CASE 1

Pedigree analysis is a fundamental tool for investigating whether a trait is following a traditional Mendelian pattern of inheritance. It can also be used to help identify individuals within a family that may be at risk for the trait.

Adam and Sarah, a young couple of Eastern European Jewish ancestry, went to a genetic counselor because they were planning a family and wanted to know what their chances were for having a child with a genetic condition. The genetic counselor took a detailed family history from both of them and discovered several traits in their respective families.

Sarah's maternal family history is suggestive of an inherited form of breast and ovarian cancer with an autosomal dominant pattern of cancer from her grandmother to mother and the young ages at which they were diagnosed with their cancers. If an altered gene that predisposed to breast and ovarian cancer were in Sarah's family, she, her sister, and any of her own future children could be at risk for inheriting this gene. The counselor told her that genetic testing is available that may help determine if an altered gene is in her family.

Adam's paternal family history has a very strong pattern of early-onset heart disease. An autosomal dominant condi-tion, known as familial hypercholesterolemia, may be responsible for the number of individuals in the family who have died from heart attacks. Like hereditary breast and ovarian cancer, there is genetic testing available to see if Adam carries this altered gene. Testing may give the couple more information on the chances that their children could inherit the gene. Adam had a first cousin who died from Tay-Sach's disease (TSD), a fatal autosomal recessive condition, that is more commonly found in people who are of Eastern European Jewish descent. In order for his cousin to have TSD, both of his parents must have been carriers for the disease-causing gene. If that is the case, Adam's father could be a carrier as well. If Adam's father and mother were both carriers, each child of Adam and Sarah would have a 25% chance of being afflicted with TSD. If Adam's father has the TSD gene, it is possible that Adam inherited the gene. Since Sarah is also of Eastern European ancestry, she could be a carrier of this gene, although no one in her family has been affected with TSD. A simple blood test performed on both Sarah and Adam could determine if they are carriers of the TSD gene.

Summary

1. In the centuries before Gregor Mendel experimented with the inheritance of traits in the garden pea, several competing theories attempted to explain how traits were passed from generation to generation. In his decade-long series of experiments, Mendel established the foundation for the science of genetics.

2. Mendel studied crosses in the garden pea that involved one pair of alleles and demonstrated that the phenotypes associated with these traits are controlled by pairs of factors, now known as genes. These factors separate or segregate from each other during gamete formation and exhibit dominant/recessive relationships.

3. In later experiments, Mendel discovered that members of one gene pair separate or segregate independently of other gene pairs. This principle of independent assortment leads to the formation of all possible combinations of gametes with equal probability in a cross between two individuals.

4. The principles of segregation and independent assortment proposed by Mendel apply to all sexually reproducing organisms, including humans.

5. Instead of direct experimental crosses, traits in humans are traced by constructing pedigrees that follow a trait through several generations.

6. Because genes for human genetic disorders exhibit segregation and independent assortment, the inheritance of certain human traits is predictable, making it possible to provide genetic counseling to those at risk of having children affected with genetic disorders.

Questions and Problems

1. Explain the difference between the following terms:
   a. gene versus allele
   b. genotype versus phenotype
   c. dominant versus recessive
   d. complete dominance versus incomplete dominance versus codominance

2. Organisms have the genotypes listed below. What types of gametes will these organisms produce and in what proportions?
   a. *Aa*
   b. *AABb*
   c. *AaBb*

3. Suppose that this organism is mated to one with the genotype Aabb. Given the following matings, what are the predicted phenotypic ratios of the offspring?
   a. *Aa* × *aa*
   b. *AABb* × *Aabb*
   c. *AaBb* × *AaBB*

4. A woman is heterozygous for two genes. How many different types of gametes can she produce, and in what proportions?

5. Brown eyes (*B*) are fully dominant over blue eyes (b).
   a. A 3:1 phenotypic ratio of F1 progeny indicates that the parents are of what genotype?
   b. A 1:1 phenotypic ratio of F1 progeny indicates that the parents are of what genotype?

6. An unspecified character controlled by a single gene is examined in pea plants. Only two phenotypic states exist for this trait. One phenotypic state is completely dominant to the other. A heterozygous plant is self-crossed. What proportion of the progeny plants exhibiting the dominant phenotype are homozygous?

7. The following diagram shows a hypothetical diploid cell. The recessive allele for albinism is represented by *a*, and *d* represents the recessive allele for deafness. The

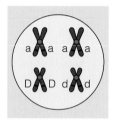

normal alleles for these conditions are represented by *A* and *D*, respectively.
   a. According to the principle of segregation, what is segregating in this cell?
   b. According to Mendel's principle of independent assortment, what is independently assorting in this cell?
   c. How many chromatids are in this cell?
   d. How many tetrads are in this cell?
   e. Write the genotype of the individual from whom this cell was taken.
   f. What is the phenotype of this individual?
   g. What stage of cell division is represented by this cell (prophase, metaphase, anaphase or telophase of meiosis I, meiosis II, or mitosis)?
   h. After meiosis II is complete, how many chromatids and chromosomes will be present in one of the four progeny cells?

8. Use the Punnett square and then the forked-line method to determine the possible genotypes and phenotypes of the F1 offspring from the following cross to show that the two methods achieve the same results.
   P1: *AaBb* × *AaBb*

9. Two traits are simultaneously examined in a cross of two pure-breeding pea-plant varieties. Pod shape can be either swollen or pinched. Seed color can be either green or yellow. A plant with the traits swollen, green

is crossed with a plant with the traits pinched, yellow, and a resulting F1 plant is self-crossed. A total of 640 F2 progeny are phenotypically categorized as follows:

swollen, yellow 360
swollen, green 120
pinched, yellow 120
pinched, green 40

a. What is the phenotypic ratio observed for pod shape? seed color?
b. What is the phenotypic ratio observed for both traits considered together?
c. What is the dominance relationship for pod shape? seed color?
d. Deduce the genotypes of the P1 and F1 generations.

10. Sickle cell anemia (SCA) is a human genetic disorder caused by a recessive allele. A couple plans to marry and wants to know the probability that they will have an affected child. With your knowledge of Mendelian inheritance, what can you tell them if (a) both are normal, but each has one affected parent and the other parent has no family history of SCA; and (b) the man is affected by the disorder, but the woman has no family history of SCA?

11. Consider the following cross in pea plants, where smooth seed shape is dominant to wrinkled and yellow seed color is dominant to green. A plant with smooth, yellow seeds is crossed to a plant with wrinkled, green seeds. The peas produced by the offspring are all smooth and yellow. What are the genotypes of the parents? What are the genotypes of the offspring?

12. Consider another cross involving the genes for seed color and shape. As before, yellow is dominant to green, and smooth is dominant to wrinkled. A plant with smooth, yellow seeds is crossed to a plant with wrinkled, green seeds. The peas produced by the offspring are as follows: one-quarter are smooth, yellow, one-quarter are smooth, green, one-quarter are wrinkled, yellow, and one-quarter are wrinkled, green.
a. What is the genotype of the smooth, yellow parent?
b. What are the genotypes of the four classes of offspring?

13. Stem length in pea plants is controlled by a single gene. Consider the cross of a true-breeding, long-stemmed variety to a true-breeding, short-stemmed variety where long stems are completely dominant.
a. 120 F1 plants are examined. How many plants are expected to be long stemmed? short stemmed?
b. Assign genotypes to both P1 varieties and to all phenotypes listed in (a).
c. A long-stemmed F1 plant is self-crossed. Of 300 F2 plants, how many should be long stemmed? short stemmed?
d. For the F2 plants mentioned in (c), what is the expected genotypic ratio?

14. What are the possible genotypes for the following blood types?

a. type A
b. type B
c. type O
d. type AB

15. A man with blood type A and a woman with blood type B have three children: A daughter with type AB, and two sons, one with type B and one with type O blood. What are the genotypes of the parents?

16. What is the chance that a man with type AB blood and a woman with type A blood whose mother is type O can produce a child that is:
a. type A
b. type AB
c. type O
d. type B

17. Another character of pea plants amenable to genetic analysis is flower color. Imagine that a true-breeding red-flowered variety is crossed to a pure line having white flowers. The progeny are exclusively pink flowered. Diagram this cross, including genotypes for all P1 and F1 phenotypes. What is the mode of inheritance? Let F = red and f = white.

18. In peas, straight stems (S) are dominant to gnarled (s), and round seeds (R) are dominant to wrinkled (r). The following cross (a test cross) is performed: $SsRr \times ssrr$. Determine the expected progeny phenotypes and what fraction of the progeny should exhibit each phenotype.

19. Pea plants usually have white or purple flowers. A strange pea-plant variant is found that has pink flowers. A self-cross of this plant yields the following phenotypes:

red flowers 30
pink flowers 62
white flowers 33

What mode of inheritance can you infer for flower color in this pea plant variant?

20. Determine the possible genotypes of the parents shown below by analyzing the phenotypes of their children. In this case, we will assume that brown eyes (B) is dominant to blue (b) and that right handedness (R) is dominant to lefthandedness (r).
a. Parents: brown eyes, right handed × brown eyes, right handed
Offspring: 3/4 brown eyes, right handed
1/4 blue eyes, right handed
b. Parents: brown eyes, right handed × blue eyes, right handed
Offspring: 6/16 blue eyes, right handed
2/16 blue eyes, left handed
6/16 brown eyes, right handed
2/16 brown eyes, left handed
c. Parents: brown eyes, right handed × blue eyes, left handed
Offspring: 1/4 brown eyes, right handed
1/4 brown eyes, left handed
1/4 blue eyes, right handed
1/4 blue eyes, left handed

21. In the following cross:

    P1: *AABBCCDDEE* × *aabbccddee*
    F1: *AaBbCcDdEe* (self cross to get F2)

    What is the chance of getting an *AaBBccDdee* individual in the F2 generation?

22. In the following trihybrid cross, determine the chance that an individual could be phenotypically A, b, C in the F1 generation.

    P1: *AaBbCc* × *AabbCC*

23. In pea plants, long stems are dominant to short stems, purple flowers are dominant to white, and round seeds are dominant to wrinkled. Each trait is determined by a single, different gene. A plant that is heterozygous at all three loci is self-crossed, and 2048 progeny are examined. How many of these plants would you expect to be long stemmed with purple flowers, producing wrinkled seeds?

24. A plant geneticist is examining the mode of inheritance of flower color in two closely related species of exotic plants. Analysis of one species has resulted in the identification of two pure-breeding lines, one produces a distinct red flower and the other produces no color at all, but he cannot be sure. A cross of these varieties produces all pink-flowered progeny. The second species exhibits similar pure-breeding varieties; that is, one variety produces red flowers, and the other produces an albino flower. A cross of these two varieties, however, produces orange-flowered progeny exclusively. Analyze the mode of inheritance of flower color in these two plant species.

25. Think about this one carefully. Albinism and hair color are governed by different genes. A recessively inherited form of albinism causes affected individuals to lack pigment in their skin, hair and eyes. Hair color itself is governed by a gene where red hair is inherited as a recessive trait, and brown hair is inherited as a dominant trait. An albino woman whose parents both have red hair has two children with a man who is normally pigmented and has brown hair. The brown-haired partner has one parent who has red hair. The first child is normally pigmented and has brown hair. The second child is albino.

    a. What is the hair color (phenotype) of the albino parent?

    b. What is the genotype of the albino parent for hair color?
    c. What is the genotype of the brown-haired parent with respect to hair color? skin pigmentation?
    d. What is the genotype of the first child with respect to hair color and skin pigmentation?
    e. What are the possible genotypes of the second child for hair color? What is the phenotype of the second child for hair color? Can you explain this?

26. Discuss the pertinent features of meiosis that provide a physical correlate to Mendel's abstract genetic laws of random segregation and independent assortment.

27. If you are told that being right- or left-handed is heritable and that a right-handed couple is expecting a child, can you conclude that the child will be right-handed?

28. Define the following pedigree symbols:

    a.
    b.
    c.
    d.
    e.

29. Construct a pedigree given the following information. The proband, Mary, is a 35 year-old woman who is 16 weeks pregnant, and was referred for genetic counseling because of advanced maternal age. Mary has one daughter, Sarah, who is 5 years old. Mary has three older sisters and four younger brothers. The two oldest sisters are married and each have one son. All her brothers are married, but none has any children. Mary's parents are both alive, and she has two maternal uncles and three paternal aunts. Mary's husband John has two brothers, one older and one younger, neither of whom is married. John's mother is alive, but his father is deceased.

30. Draw the following simple pedigree. A man and a woman have three children, a daughter, then two sons. The daughter is married to a man and has monozygotic (identical) twin girls. The youngest son in generation II is affected with albinism.

Internet Activities

The following activities use the resources of the World Wide Web to enhance the topics covered in this chapter. To investigate the topics described below, log on to the book's homepage at:

http://www.brookscole.com/biology

1. *Pea Soup* is an Internet site dedicated to Mendelian genetics. Read about Mendel's life, and about his discoveries. When Mendel began his experiments, how did most scientists think that traits were inherited? What was new about his approach? For a trait to segregate according to Mendel's rules, the trait must be caused by a single gene. Think of an example of a human trait (like a disease) that behaves according to Mendel's rules. What human traits exhibit more complex patterns of inheritance?

   Try the *Interactive Pea Experiment*. Design a cross between any two peas with segregating alleles for pea color and for pea shape. Predict the outcome of the cross, both genotypically and phenotypically. Notice how many distinct genotypes are created by each cross. In which crosses do all four progeny differ from each other genetically? In which crosses are only two different genotypes created? Which crosses give rise to four genetically identical offspring? What about phenotype? Design a cross in which all four offspring differ phenotypically. Can you design a cross in which all four progeny differ genotypically, but have identical phenotypes?

2. Perform fly crosses demonstrating Mendelian principles by linking to the Virtual Fly Lab. By selecting traits and performing crosses, you can demonstrate the concepts of dominant and recessive phenotypes, segregation, and independent assortment.
   a. For an experimental cross, select Designing a Cross. Scroll through the characteristics and choose a mutant strain. Next, click on the Mate Flies button, and then record your results.
   b. Click on Other Places to Go and select Design and Mate Flies. Pick two mutations for two of the nine genes. Click on the Mate Flies button, and then record your results.

3. By connecting to the MendelWeb site, you can read Mendel's original paper, both in English and German. In addition to the original manuscript, the site contains critical interpretations of the work.

For Further Reading

Corcos, A. F., & Monaghan, F. (1985). Role of de Vries in the recovery of Mendel's work. I. Was de Vries really an independent discoverer of Mendel? *J. Heredity* 76: 187–190.

Dahl, H. (1993). Things Mendel never dreamed of. *Med. J. Aust.* 158: 247–252.

Dunn, L. C. (1965). *A Short History of Genetics.* New York: McGraw-Hill.

Edwards, A. W. (1986). Are Mendel's results really too close? *Biol. Rev. Cambridge Philos. Soc.* 61: 295–312.

Finney, D. J. (1980). *Statistics for Biologists.* New York: Chapman & Hall.

Gasking, E. B. (1959). Why was Mendel's work ignored? *J. Hist. Ideas* 20: 62–84.

George, W. (1975). *Gregor Mendel and Heredity.* London: Priory Press.

Hartl, D., & Orel, V. (1992). What did Gregor Mendel think he discovered? *Genetics* 131: 245–253.

Heim, W. G. (1991). What is a recessive allele? *Amer. Biol. Teacher* 53: 94–97.

Mather, K. (1965). *Statistical Analysis in Biology.* London: Methuen.

Monaghan, F. V., & Corcos, A. F. (1985). Mendel, the empiricist. *J. Heredity* 76: 49–54.

Orel, V. (1973). The scientific milieu in Brno during the era of Mendel's research. *J. Heredity* 64: 314–318.

Orel, V. (1984). *Mendel.* New York: Oxford University Press.

Piegorsch, W. W. (1990). Fisher's contributions to genetics and heredity, with special emphasis on the Gregor Mendel controversy. *Biometrics* 46: 915–924.

Pilgrim, I. (1986). A solution to the too-good-to-be-true paradox and Gregor Mendel. *J. Heredity* 77: 218–220.

Sandler, I., & Sandler, L. (1985). A conceptual ambiguity that contributed to the neglect of Mendel's paper. *Publ. Stn. Zool. Napoli* 7: 3–70.

Stern, C., & Sherwood, E. (1966). *The Origins of Genetics: A Mendel Sourcebook.* San Francisco: Freeman.

Voipio, P. (1990). When and how did Mendel become convinced of the idea of general, successive evolution? *Hereditas* 113: 179–181.

Voller, B. R., ed. (1968). *The Chromosome Theory of Inheritance. Classic Papers in Development and Heredity.* New York: Appleton-Century-Crofts.

Weiling, F. (1991). Historical study: Johann Gregor Mendel 1822–1884. *Am. J. Med. Genet.* 40: 1–25.

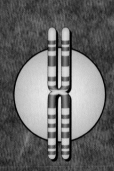

CHAPTER 4

Pedigree Analysis in Human Genetics

Chapter Outline

*W*as Abraham Lincoln, the 16th president of the United States, affected with a genetic disorder? Several writers have speculated that Lincoln had a genetic disease known as Marfan syndrome. The evidence offered to support this idea is based on Lincoln's physical appearance and the report of Marfan syndrome in a distant relative.

Photographs, written descriptions, and medical reports are available to provide ample information about Lincoln's physical appearance. He was 6 ft. 4 in. tall and thin, weighing between 160 and 180 lbs. for most of his adult life. He had long arms and legs and large, narrow hands and feet. Contemporary descriptions of his appearance indicate that he was stoop-shouldered, loose jointed, and walked with a shuffling gait. In addition, he wore eyeglasses to correct a visual problem.

Lincoln's appearance and ocular problems are suggestive of Marfan syndrome, a genetic condition that affects the connective tissue of the body, resulting in a tall, thin individual often with a shifted lens in the eye, blood vessel defects, and loose joints.

In 1960 a man diagnosed with Marfan syndrome was found to have ancestors in common with Lincoln (the common ancestor was Lincoln's great-great-grandfather). Added to the evidence based on physical appearance, the family history suggests to some that Lincoln had Marfan syndrome.

Others strongly disagree with this speculation, arguing that the length of Lincoln's extremities and the proportions of his body were well within the normal limits for tall, thin individuals. In addition, although Lincoln had visual problems, an examination of his eyeglasses indicates that he was farsighted, while those with the classical form of Marfan syndrome are nearsighted. Lastly, Lincoln showed no outward signs of problems with major blood vessels, such as the aorta.

Did Lincoln really have Marfan syndrome, and should we care? Interest in this issue probably grows from our curiousity about the lives of historic figures and our fascination with the intimate details of the lives of the famous and the infamous. At the present time there is no solid evidence to suggest that Lincoln had Marfan syndrome, although molecular testing on bone and hair fragments (discussed below) has been proposed. For now we are left only with speculation and inferential reasoning.

The more important question for this chapter is how we can tell when someone is affected with a genetic disorder? Lincoln's family history shows only one documented case of Marfan syndrome in nine generations. Is this enough to decide that he was also affected with this genetic condition?

In this chapter we show that the principles of inheritance discovered by Mendel in peas also apply to humans. Even though the methods employed in human genetics differ significantly from those used in experimental organisms, the rules for the inheritance of traits in humans are the same as those for pea plants.

HUMANS AS A SUBJECT FOR GENETICS

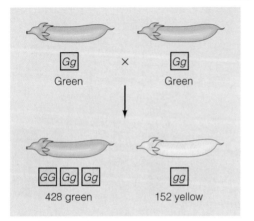

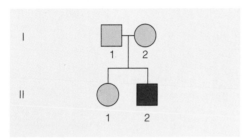

▶ **FIGURE 4.1** Inheritance in pea plants and humans. (a) In pea plants, a cross between two heterozygotes provides enough offspring in each phenotypic class to allow the mode of trait inheritance to be determined. (b) Humans have relatively few offspring, often making it difficult to interpret how a trait is inherited.

Pea plants were selected by Mendel for two primary reasons. First, they can be crossed in any combination. Second, each cross is likely to produce large numbers of offspring, an important factor in understanding how a trait is inherited. If you were picking an organism for genetic studies, humans would not be a good choice. With pea plants, it's easy to carry out crosses between plants with purple flowers and plants with white flowers and to repeat this cross as often as necessary. For obvious reasons, experimental matings in humans are not possible. You can't ask all albino humans to mate with homozygous normally pigmented individuals and have their progeny interbreed to produce an F2. For the most part, human geneticists base their work on matings that have already taken place, whether or not such matings would be the most genetically informative.

Compared with the progeny that can be counted in a single cross with peas, humans produce very few offspring, and these usually represent only a small fraction of the possible genetic combinations. If two heterozygous pea plants are crossed ($Aa \times Aa$), about three-fourths of the offspring will express the dominant phenotype, and the recessive phenotype will be expressed in the remaining one-fourth of the progeny (▶ Figure 4.1). Mendel was able to count hundreds and sometimes thousands of offspring from such a cross, recording progeny in all expected phenotypic classes, clearly establishing a ratio of 3:1 for recessive traits. As a parallel, consider two humans, each of whom is phenotypically normal. Suppose this couple has two children, one of whom is an unaffected daughter, and the other is a son affected with a genetic disorder. The ratio of phenotypes in this case is 1:1. This makes it difficult to decide whether the trait is carried on an autosome or a sex chromosome, whether it is a dominant or recessive trait, and whether it is controlled by a single gene or by two or more genes.

These examples demonstrate that the basic method of genetic analysis in humans is observational rather than experimental and requires reconstructing events that have already taken place rather than designing and executing experiments to test a hypothesis directly. One of the first steps in studying a human trait is to construct a pedigree. Then, the information in the pedigree is used to determine how a trait is inherited.

PEDIGREE CONSTRUCTION IS A BASIC METHOD IN HUMAN GENETICS

As outlined in the previous chapter, a pedigree chart is an orderly presentation of family information, using standardized symbols. Once a pedigree has been constructed, the principles of Mendelian inheritance are used to determine whether the trait is inherited dominantly or recessively and whether the gene in question is located on an autosome or a sex chromosome.

The actual collection of pedigree information is not always straightforward. Knowledge about distant relatives is often incomplete, and recollections about medical conditions can be blurred by the passage of time. Older family members are sometimes reluctant to discuss relatives who had abnormalities or who were placed in institutions. As a result, collecting accurate pedigree information for a large family over several generations can be a challenge for the geneticist. The collection and storage of pedigree information can now be done using software such as Cyrillic (▶ Figure 4.2). These programs give on-screen displays of pedigrees and genetic information that can be used to analyze patterns of inheritance.

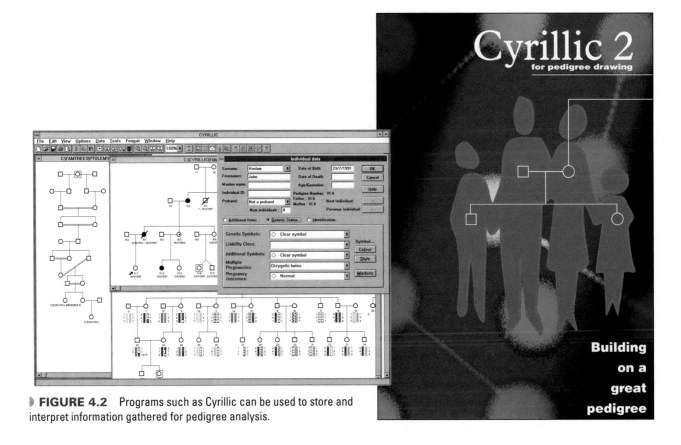

FIGURE 4.2 Programs such as Cyrillic can be used to store and interpret information gathered for pedigree analysis.

Once a pedigree has been constructed, the information in the pedigree is analyzed to determine how the trait is inherited. The modes of inheritance we consider in this chapter include

- Autosomal recessive
- Autosomal dominant
- X-linked dominant
- X-linked recessive
- Y-linked
- Mitochondrial

As outlined above, pedigrees can be difficult to construct. For several reasons, they can also be difficult to analyze. In analyzing a pedigree, a geneticist first forms a hypothesis about how the trait is inherited (for example, is it autosomal dominant?). Then, the pedigree is examined for evidence that supports or rejects this mode of inheritance. Even if the information in the pedigree supports the hypothesis, the analysis of the pedigree is complete only when all possible modes of inheritance have been considered. If only one mode of inheritance is supported by the information in the pedigree, it is accepted as the mode of inheritance for the trait being examined.

It may turn out that the pedigree does not provide enough information to rule out other possible modes of inheritance. For example, analysis of a pedigree may indicate that a trait can be inherited in an autosomal dominant or an X-linked dominant fashion. If this is so, the pedigree is examined to determine whether one mode of transmission is more likely than another. Then, the most likely mode of inheritance is used as the basis for further work.

As a further complication, a genetic disorder can have more than one mode of inheritance. Ehlers-Danlos syndrome (▶ Figure 4.3) (MIM/OMIM 130000 and

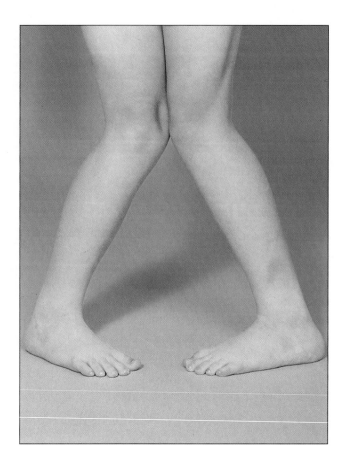

FIGURE 4.3 Ehlers-Danlos syndrome. This disorder can be inherited as an autosomal dominant, autosomal recessive, or X-linked recessive trait. People who have the common autosomal dominant form have loose joints and highly elastic skin, which can be stretched by several inches but returns to its normal position when released.

other numbers), characterized by loose joints and easily stretched skin, can be inherited as an autosomal dominant, autosomal recessive, or an X-linked recessive trait. In other cases, a trait can have a single mode of inheritance but be caused by mutation in any of several genes. Porphyria (MIM/OMIM 176200 and other numbers), a metabolic disorder associated with abnormal behavior, is inherited as an autosomal dominant trait. However, it can be caused by mutation in genes on chromosomes 1, 9, 11, and 14.

It is important for several reasons to establish how a trait is inherited. If the mode of inheritance can be established, it can be used to predict genetic risk in several situations, including

- pregnancy outcome
- adult onset disorders
- recurrence risks in future offspring

THERE IS A CATALOG OF MENDELIAN GENETIC DISORDERS

In this chapter we deal with the six possible modes of inheritance listed previously and use Mendelian principles to analyze pedigrees for these traits. We limit our discussion to traits controlled by a single gene. Near the end of the chapter, we consider factors that can influence gene expression. In the next chapter, we discuss traits that are controlled by two or more genes.

To date, over 5000 genetic traits have been identified in humans. The chromosomal location for a few thousand of these genes has been determined, and the molecular basis of traits associated with deleterious phenotypes is known in a smaller number of cases. Victor McKusick, a geneticist at Johns Hopkins University, and his

Concepts and Controversies

Was Noah an Albino?

The biblical character Noah, along with the ark and its animals, are among the most recognizable figures in the Book of Genesis. His birth is recorded in a single sentence, and although the story of how the ark was built and survived a great flood is told later, there is no mention of Noah's physical appearance. But other sources contain references to Noah that are consistent with the idea that Noah was one of the first albinos mentioned in recorded history.

The birth of Noah is recorded in several sources, including the Book of Enoch the Prophet, written about 200 B.C. This book, quoted several times in the New Testament, was regarded as lost until 1773, when an Ethiopian version of the text was discovered. In describing the birth of Noah, the text relates that his "flesh was white as snow, and red as a rose; the hair of whose head was white like wool, and long, and whose eyes were beautiful."

A reconstructed fragment of one of the Dead Sea Scrolls describes Noah as an abnormal child born to normal parents. This fragment of the scroll also provides some insight into the pedigree of Noah's family, as does the Book of Jubilees. According to these sources, Noah's father (Lamech) and his mother (Betenos) were first cousins. Lamech was the son of Methuselah, and Lamech's wife was a daughter of Methuselah's sister. This is important because marriage between close relatives is sometimes involved in pedigrees of autosomal recessive traits, such as albinism.

If this interpretation of ancient texts is correct, Noah's albinism is the result of a consanguineous marriage, and not only is he one of the earliest albinos on record, but his grandfather Methuselah and Methuselah's sister are the first recorded heterozygous carriers of a recessive genetic trait.

colleagues have compiled a catalog of human genetic traits. The catalog is published in book form as *Mendelian Inheritance in Man: Catalogs of Human Genes and Genetic Disorders*. The catalog is also available at several World Wide Web sites as *Online Mendelian Inheritance in Man*. The online version contains text, pictures, references and links to other databases (▶ Figure 4.4). Each trait is assigned a catalog number (called the MIM or OMIM number). In this chapter and throughout the book, the MIM/OMIM number for each trait discussed is listed. You can use the MIM/OMIM number to obtain more information about these traits. Access to OMIM is available through the book's homepage.

AUTOSOMAL RECESSIVE TRAITS

Although human families are relatively small, analysis of affected and unaffected members over several generations usually provides enough information to determine whether a trait has a recessive pattern of inheritance and is carried on an autosome (as opposed to a sex chromosome). Recessive traits carried on autosomes have several characteristics that can be established by pedigree analysis. Some of these are listed here:

- For recessive traits that are rare or relatively rare, most affected individuals are the children of unaffected parents.
- All of the children of two affected (homozygous) individuals are affected.
- The risk of an affected child from a mating of two heterozygotes is 25%.

▶ **FIGURE 4.4** OMIM is an online database that contains information about human genetic disorders.

[Screenshot: Netscape: Search OMIM — Online Mendelian Inheritance in Man]

National Center for Biotechnology Information

Online Mendelian Inheritance in Man

- Because the trait is autosomal, it is expressed in both males and females, who are affected in roughly equal numbers. The trait can also be transmitted by either male or female parents.
- In pedigrees involving rare traits, the unaffected (heterozygous) parents of an affected (homozygous) individual may be related to each other.

A number of autosomal recessive genetic disorders are listed in Table 4.1. A pedigree illustrating a pattern of inheritance typical of autosomal recessive genes is shown in ▶ Figure 4.5.

Some autosomal recessive traits represent minor variations in phenotype, such as hair color and eye color. Others result in phenotypes that can be life threatening or even fatal. Examples of these more severe phenotypes include cystic fibrosis and sickle cell anemia.

Cystic Fibrosis Is a Recessive Trait

■ **Cystic fibrosis** A fatal recessive genetic disorder associated with abnormal secretions of the exocrine glands.

Cystic fibrosis (CF) (MIM/OMIM 219700) is a disabling and fatal genetic disorder inherited as an autosomal recessive trait. The mutant gene affects the glands that produce mucus, digestive enzymes, and sweat. Because the sweat glands are defective, they release excessive amounts of salt, and the disease is often diagnosed by analyzing the amount of salt in sweat. According to folklore, midwives would lick the forehead of newborns. If the sweat tasted too salty, they predicted that the child would die prematurely of lung congestion.

This disease has far-reaching effects because the affected glands perform a number of vital functions. CF causes the production of thick mucus, which clogs the ducts that carry digestive enzymes from the pancreas to the small intestine, reducing the effectiveness of digestion. As a result, affected children often suffer from malnutrition in spite of an increased appetite and food intake. Eventually, cysts form in the pan-

Table 4.1 Some Autosomal Recessive Traits

Trait	Phenotype	MIM/OMIM Number
Albinism	Absence of pigment in skin, eyes, hair	203100
Ataxia telangiectasia	Progressive degeneration of nervous system	208900
Bloom syndrome	Dwarfism; skin rash; increased cancer rate	210900
Cystic fibrosis	Mucous production that blocks ducts of certain glands, lung passages; often fatal by early adulthood	219700
Fanconi anemia	Slow growth; heart defects; high rate of leukemia	227650
Galactosemia	Accumulation of galactose in liver; mental retardation	230400
Phenylketonuria	Excess accumulation of phenylalanine in blood; mental retardation	261600
Sickle cell anemia	Abnormal hemoglobin, blood vessel blockage; early death	141900
Thalassemia	Improper hemoglobin production; symptoms range from mild to fatal	141900 / 141800
Xeroderma pigmentosum	Lack of DNA repair enzymes, sensitivity to UV light; skin cancer; early death	278700
Tay–Sachs disease	Improper metabolism of gangliosides in nerve cells; early death	272800

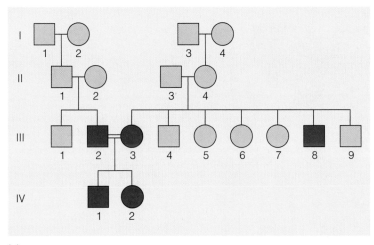

(a)

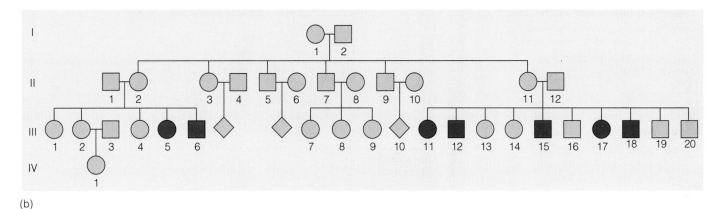

(b)

▶ **FIGURE 4.5** Pedigrees for autosomal recessive traits. (a) This pedigree has many of the characteristics for autosomal recessive inheritance. Most affected individuals have normal parents, about one-fourth of the children in large affected families show the trait, both sexes are affected in roughly equal numbers, and affected parents produce only affected children. (b) A large family pedigree for an autosomal recessive trait.

creas, and the gland degenerates into a fibrous structure, giving rise to the name of the disease. Because CF causes the production of thick mucus that blocks the airways in the lungs, most patients with cystic fibrosis develop obstructive lung diseases and infections that lead to premature death (▶ Figure 4.6).

Many affected individuals are infertile because their reproductive ducts are blocked. As a result, almost all cases of CF are children of phenotypically normal, heterozygous parents. Cystic fibrosis is relatively common in some populations but rare in others (▶ Figure 4.7). Among the U.S. white population, CF has a frequency of 1 in 2000 births, and 1 in 22 members of this group are heterozygous carriers. The disease is less common among the U.S. black population and has a frequency of 1 in 17,000 to one in 19,000. Among U.S. citizens with origins in Asia, CF is a rare disease whose frequency is about 1 in 90,000. Heterozygous carriers are extremely rare in this population.

The underlying defect in CF was identified in 1989 by a team of researchers led by Lap-chee Tsui and Francis Collins. Recombinant DNA techniques were used to localize the gene to region q31 of chromosome 7 (▶ Figure 4.8). This region of the chromosome was explored using several methods of genetic mapping, and the CF gene was identified by comparing the molecular organization of a small segment of chromosome 7 in normal and CF individuals.

▶ **FIGURE 4.6** Organ systems affected by cystic fibrosis. Sweat glands in affected individuals secrete excessive amounts of salt. Thick mucus blocks the transport of digestive enzymes in the pancreas. The trapped digestive enzymes gradually break down the pancreas. The lack of digestive enzymes results in poor nutrition and slow growth. Cystic fibrosis affects both the upper respiratory tract (the nose and sinuses) and the lungs. Thick, sticky mucus clogs the bronchial tubes and the lungs, making breathing difficult. It also slows the removal of viruses and bacteria from the respiratory system, resulting in lung infections. In males, mucus blocks the ducts that carry sperm, and only about 2–3% of affected males are fertile. In women who have cystic fibrosis, thick mucus plugs the entrance to the uterus, lowering fertility.

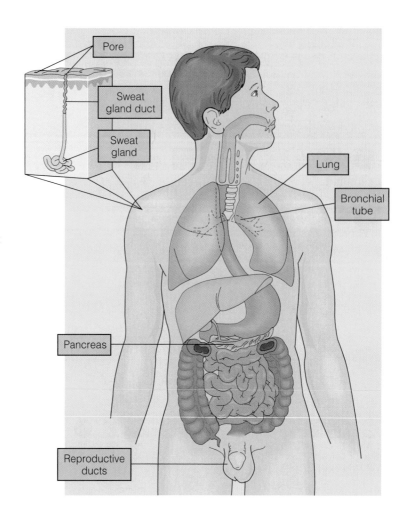

▶ **FIGURE 4.7** About 1 in 25 Americans of European descent, 1 in 46 Hispanics, 1 in 60–65 African-Americans, and 1 in 150 Asian-Americans is a carrier for cystic fibrosis. A crowd such as this may contain a carrier.

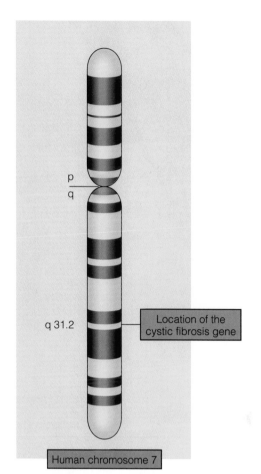

p

q

q 31.2

Location of the cystic fibrosis gene

Human chromosome 7

FIGURE 4.8 Human chromosome 7. The gene for cystic fibrosis (CF) maps to region 7q31.2-31.3, about two-thirds of the way down the long arm of the chromosome.

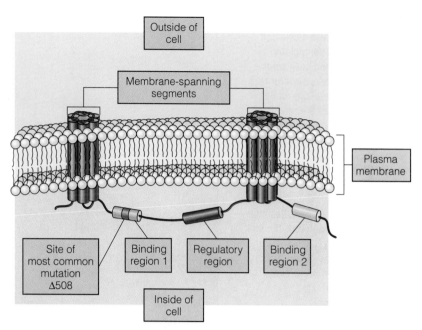

Outside of cell

Membrane-spanning segments

Plasma membrane

Site of most common mutation Δ508

Binding region 1

Regulatory region

Binding region 2

Inside of cell

FIGURE 4.9 The cystic fibrosis gene product. The CFTR protein is located in the plasma membrane of the cell and regulates the movement of chloride ions across the cell membrane. The regulatory region controls the activity of the CFTR molecule in response to signals from inside the cell. In most cases (about 70%), the protein is defective in binding region 1.

The product of the CF gene is a protein that inserts into the plasma membrane of exocrine gland cells. The protein is called the *cystic fibrosis transmembrane conductance regulator,* or CFTR (Figure 4.9). CFTR regulates the flow of chloride ions across the cell's plasma membrane. The CFTR protein is absent or defective in affected individuals. Because fluids move across plasma membranes in response to the movement of ions, a defective CFTR protein causes less fluid to be added to the secretions of exocrine glands. The thickened secretions produce the characteristic symptoms of CF. It is hoped that further studies of the structure of the CF gene and the function of CFTR will lead to the development of new methods for treating this deadly disease.

Sickle Cell Anemia Is a Recessive Trait

Americans whose ancestors lived in parts of West Africa, the lowlands around the Mediterranean Sea, or in parts of the Indian subcontinent have a high frequency of **sickle cell anemia** (MIM/OMIM 141900). Individuals with this recessive genetic disorder produce an abnormal type of hemoglobin, a protein found in red blood cells. Normally, this protein transports oxygen from the lungs to the tissues of the body. Each red cell contains millions of hemoglobin molecules.

In sickle cell anemia, the abnormal hemoglobin molecules pack together to form rods (Figure 4.10). This causes the red blood cells to become crescent- or sickle-shaped

■ **Sickle cell anemia** A recessive genetic disorder associated with an abnormal type of hemoglobin, a blood transport protein.

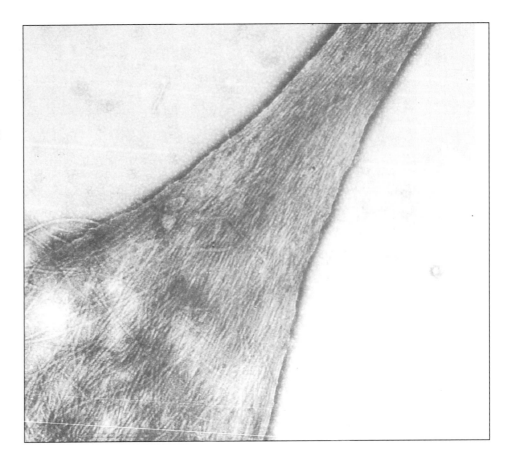

▶ **FIGURE 4.10** Hemoglobin molecules aggregate in sickle cell anemia. The mutant hemoglobin molecules in red blood cells stack together to form rodlike structures. The formation of these aggregates in the cytoplasm causes the red blood cells to deform and become elongated or sickle-shaped.

SIDEBAR

Gas Therapy for Sickle Cell Anemia

In sickle cell anemia (a recessive disorder), the mutant oxygen-carrying hemoglobin molecules polymerize and the red blood cells become sickle-shaped. One approach to treating this disorder is to somehow increase the ability of the mutant hemoglobin molecules (HbS) to take up oxygen. Recently, researchers discovered that if red blood cells from individuals with sickle cell anemia are exposed to low concentrations of nitric oxide, they are able to carry up to 15% more oxygen. In further studies, volunteers with sickle cell anemia breathed air containing 80ppm nitric oxide for 45 min. Red blood cells from these individuals showed an increase in oxygen carrying capacity up to an hour after breathing the nitric oxide. Red blood cells from normal individual's showed no change in oxygen-carrying capacity. These results suggest that low concentrations of nitric oxide may be useful in the treatment of sickle cell anemia.

(▶ Figure 4.11). The deformed red blood cells are fragile and break open as they circulate through the body. New blood cells are not produced fast enough to replace those that are lost, and the oxygen-carrying capacity of the blood is reduced, causing anemia. Individuals with sickle cell anemia tire easily and often develop heart failure because of the increased load on the circulatory system. The deformed blood cells also clog small blood vessels and capillaries, further reducing oxygen transport, sometimes initiating a sickling crisis. As oxygen levels in the circulatory system fall, more and more red blood cells become sickled, bringing on intense pain as blood vessels become blocked. In some affected areas, ulcers and sores appear on the body surface. Blockage of the blood vessels in the brain leads to strokes and can result in partial paralysis.

Because of the number of systems in the body affected and the severity of the effects, untreated sickle cell anemia can be lethal. Some affected individuals die in childhood or adolescence, but aggressive medical treatment allows survival into adulthood. As in cystic fibrosis, most affected individuals are children of phenotypically normal, heterozygous parents.

The high frequency of sickle cell anemia in certain populations is related to the frequency of malaria, an infectious disease. Sickle cell heterozygotes are more resistant to malarial infection than homozygous normal individuals. In the U.S. black population, sickle cell anemia occurs with a frequency of 1 in every 500 births, and approximately 1 in every 12 individuals is heterozygous. The high frequency of this mutation in the black population is a genetic relic of African origins. The same is true for U.S. residents whose ancestral origins are in lowland regions of Italy, Sicily, Cyprus, Greece, and the Middle East. This defect in a hemoglobin molecule has a double effect. It causes sickle cell anemia but also confers resistance to malaria. The molecular basis of this disease is well known and is discussed in later chapters.

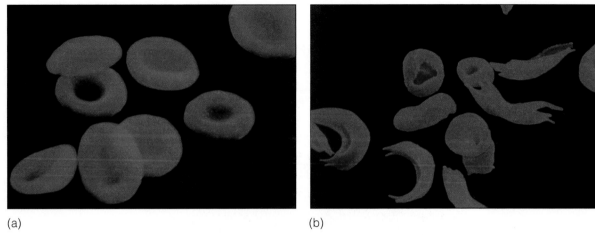

(a)   (b)

▶ **FIGURE 4.11** Red blood cells. (a) Normal red blood cells are flat, disk-shaped cells, indented in the middle on both sides. (b) In sickle cell anemia, the cells become elongated and fragile.

AUTOSOMAL DOMINANT TRAITS

In autosomal dominant disorders, heterozygotes and those with a homozygous dominant genotype express the abnormal phenotype. Unaffected individuals carry two recessive alleles. Careful analysis of pedigrees is necessary to determine whether a trait is caused by a dominant allele.

The Pattern of Inheritance for Autosomal Dominant Traits

Dominant traits that are fully expressed have a distinctive pattern of inheritance:

- Every affected individual should have at least one affected parent. Exceptions can occur in cases where the gene has a high mutation rate. Mutation is the sudden appearance of a heritable trait that was not transmitted by the biological parents.
- Because most affected individuals are heterozygotes who mate with unaffected (homozygous recessive) individuals, there is a 50% chance of transmitting the trait to each child.
- Because the trait is autosomal, the numbers of affected males and females are roughly equal.
- Two affected individuals may have unaffected children, again because most affected individuals are heterozygous. In contrast, two individuals affected with an autosomal recessive trait have only affected children.
- In homozygous dominant individuals, the phenotype is often more severe than the heterozygous phenotype.

A number of human genetic disorders transmitted as autosomal dominant traits are listed in Table 4.2. The pedigree in ▶ Figure 4.12 is typical of the pattern found in autosomal dominant conditions.

Marfan Syndrome Is an Autosomal Dominant Trait

Marfan syndrome (MIM/OMIM 154700) is an autosomal dominant disorder that affects the skeletal system, the eyes, and the cardiovascular system. It was first described in 1896 by a French physician, A. B. Marfan. Affected individuals tend to be tall and thin and have long arms and legs and long, thin fingers. Because of their height and long limbs, those who have Marfan syndrome often excel in sports like

■ **Marfan syndrome** An autosomal dominant genetic disorder that affects the skeletal system, the cardiovascular system, and the eyes.

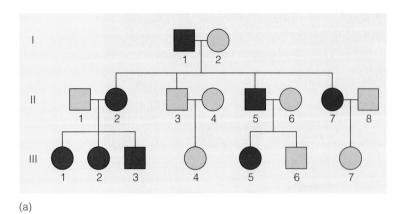

(a)

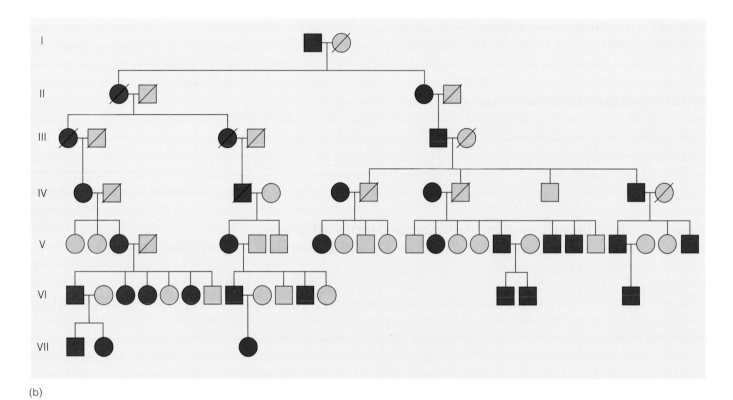

(b)

▶ **FIGURE 4.12** Pedigrees for autosomal dominant traits. (a) This pedigree shows many of the characteristics of autosomal dominant inheritance: Affected individuals have at least one affected parent, about one-half of the children who have one affected parent are affected, both sexes are affected with roughly equal frequency, and affected parents can have unaffected children. (b) Pedigree of Huntington disease, an autosomal dominant disorder, in a large family from Venezuela.

basketball and volleyball, although nearsightedness and defects in the lens of the eye are also common (▶ Figure 4.13).

The most dangerous effects of Marfan syndrome are on the cardiovascular system, especially the aorta. The aorta is the main blood-carrying vessel in the body. As it leaves the heart, the aorta arches back and downward, feeding blood to all major organ systems. Marfan syndrome weakens the connective tissue around the base of the aorta, causing it to enlarge and eventually to split open (▶ Figure 4.14). This is most likely to occur in the first few inches of the aorta, and in cases where the enlargement can be detected, it can be repaired by surgery.

The defective gene that causes Marfan syndrome is located on chromosome 15. The product of the gene is a protein called fibrillin, which is part of the connective

Table 4.2 Some Autosomal Dominant Traits

Trait	Phenotype	MIM/OMIM Number
Achondroplasia	Dwarfism associated with defects in growth regions of long bones	100800
Brachydactyly	Malformed hands with shortened fingers	112500
Camptodactyly	Stiff, permanently bent little fingers	114200
Crouzon syndrome	Defective development of mid-face region, protruding eyes, hook nose	123500
Ehlers–Danlos syndrome	Connective tissue disorder, elastic skin, loose joints	130000
Familial hypercholesterolemia	Elevated levels of cholesterol; predisposes to plaque formation, cardiac disease; may be most prevalent genetic disease	144010
Adult polycystic kidney disease	Formation of cysts in kidneys; leads to hypertension, kidney failure	173900
Huntington disease	Progressive degeneration of nervous system; dementia; early death	143100
Hypercalcemia	Elevated levels of calcium in blood serum	143880
Marfan syndrome	Connective tissue defect; death by aortic rupture	154700
Nail-patella syndrome	Absence of nails, kneecaps	161200
Porphyria	Inability to metabolize porphyrins; episodes of mental derangement	176200

▶ **FIGURE 4.13** Flo Hyman was a star on the U.S. women's volleyball team in the 1984 Olympics. Two years later, she died in a volleyball game from a ruptured aorta caused by Marfan syndrome.

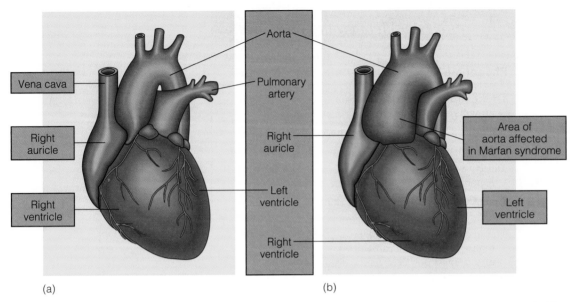

(a)   (b)

▶ **FIGURE 4.14**   The heart and its major blood vessels. Oxygen-rich blood is pumped from the lungs to the left side of the heart. From there, blood is pumped through the aorta to all parts of the body.

tissue in the aorta, the eye, and the sheath covering long bones. The disorder affects males and females with equal frequency and is found in all ethnic groups at a frequency of about 1 in 10,000 individuals. About 25% of affected individuals appear in families which have no previous history of Marfan syndrome, indicating that this gene undergoes mutation at a high rate. As discussed in the chapter introduction, it has been suggested that Abraham Lincoln had Marfan syndrome.

Bone fragments and hair from Lincoln's body are preserved at the National Medical Library in Washington, D.C.. A group of research scientists has requested permission to analyze these samples using recombinant DNA techniques to decide whether Lincoln did, in fact, have Marfan syndrome. A committee appointed to review the request has agreed that this material can be tested, but it has recommended that testing be delayed until more is known about the fibrillin gene.

SEX-LINKED INHERITANCE INVOLVES GENES ON THE X AND Y CHROMOSOMES

As noted in Chapter 2, females have two X chromosomes, and males have an X and a Y chromosome. These two chromosomes are very different in size and appearance. The X chromosome is a medium-sized, metacentric chromosome that has a well-defined banding pattern. The Y chromosome is a much smaller, acrocentric chromosome, only about 25% as large as the X, and has a variable banding pattern (▶ Figure 4.15). At meiosis, the X and Y chromosomes pair only at a small region at the tip of the short arms. The absence of pairing along most of the chromosome suggests that the great majority of genes on the X chromosome are not represented on the Y.

This lack of genetic equivalence between the X and Y chromosomes is responsible for a pattern of transmission known as sex-linked inheritance. For genes on the X chromosome, the pattern is **X-linked**, and for genes on the Y chromosome, the pattern is **Y-linked**. Females carry two copies of all X-linked genes, and can be heterozygous or homozygous for any of them. Males, on the other hand, have only one copy of the X chromosome. Because the Y chromosome does not carry copies of most X-linked genes, a male cannot be heterozygous for genes on the X chromosome.

■ **X-linkage**   The pattern of inheritance that results from genes located on the X chromosome.

■ **Y-linked genes**   Genes located only on the Y chromosome.

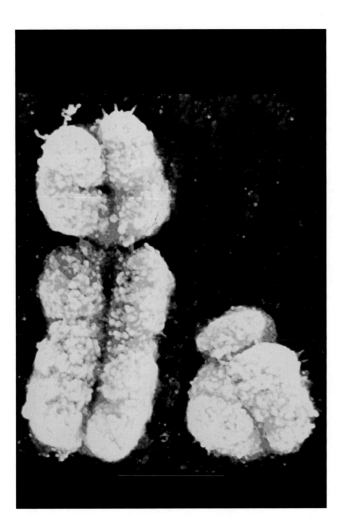

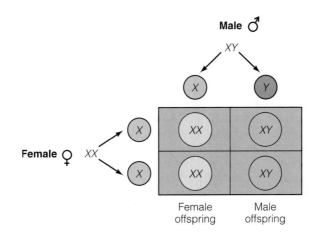

▶ **FIGURE 4.15** The human X chromosome (left) and the Y chromosome (right). This false-color scanning electron micrograph shows the differences between these chromosomes.

This explains why males are affected by X-linked recessive genetic disorders more often than females. The term **hemizygous** is used for genes present in a single dose on the X chromosome in males. Because males cannot be heterozygous for these genes, recessive traits on the X chromosome are expressed in males (traits controlled by genes on the X chromosome are defined as dominant or recessive by their phenotype in females).

Males transmit their X chromosome to all daughters and their Y chromosome to all sons. Females randomly pass on one or the other X chromosome to all daughters and to all sons (▶ Figure 4.16). As a result, the X and Y chromosomes have a distinctive pattern of inheritance. If a trait is X-linked, a male passes it to all of his daughters (who may be heterozygous or homozygous for the condition). If their mother is heterozygous for an X-linked recessive trait, sons have a 50% chance of receiving the recessive allele. In the following sections, we consider examples of sex-linked inheritance and explore the characteristic pedigrees in detail.

■ **Hemizygous gene** A gene present on the X chromosome that is expressed in males in both the recessive and dominant condition.

▶ **FIGURE 4.16** Distribution of sex chromosomes by parents. All children receive an X chromosome from their mothers. Fathers pass their X chromosome to all daughters and a Y chromosome to all their sons. The sex chromosome content of the sperm determines the sex of the child.

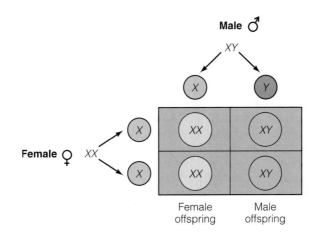

Male ♂
XY

Female ♀ XX

Female offspring   Male offspring

X-Linked Dominant Inheritance

Hypophosphatemia An X-linked dominant disorder. Those affected have low phosphate levels in blood and skeletal deformities.

Only a small number of dominant traits map to the X chromosome. One of these is a phosphate deficiency known as **hypophosphatemia** (MIM/OMIM 307800), which causes a type of rickets, or bowleggedness. Dominant X-linked traits have a distinctive pattern of transmission with three characteristics:

* Affected males produce all affected daughters and no affected sons.
* A heterozygous affected female will transmit the trait to half of her children, and males and females are equally affected.
* On average, twice as many females as males are affected.

As expected, a homozygous female will transmit the trait to all of her offspring. A pedigree for the inheritance of hypophosphatemia is shown in ▶ Figure 4.17b. To determine whether a trait is X-linked dominant or autosomal dominant, the children of affected males should be carefully analyzed. In X-linked dominant traits, affected males transmit the trait only to daughters. In autosomal dominant conditions, affected males pass the trait to both daughters and sons, and about half of the daughters and about half of the sons are affected.

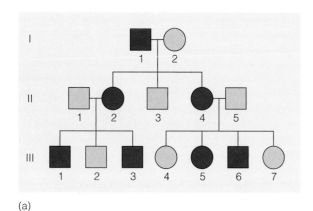

(a)

▶ **FIGURE 4.17** Pedigrees for X-linked dominant traits. (a) This pedigree shows the characteristics of X-linked dominant traits: Affected males produce all affected daughters and no affected sons, affected females transmit the trait to roughly half their children, and males and females are equally affected, and twice as many females as males are affected with the trait. (b) A pedigree for hypophosphatemia, an X-linked dominant trait.

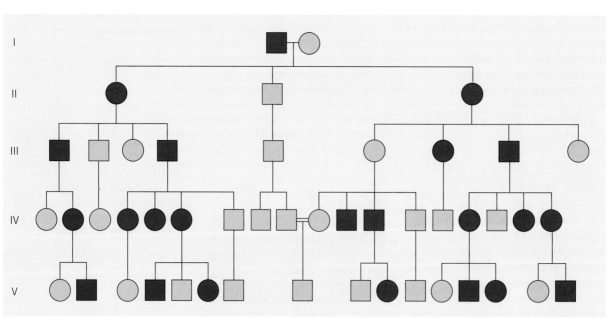

(b)

X-Linked Recessive Inheritance

The X-linked recessive trait for **color blindness** is actually a collection of several abnormalities of color vision. The most common form of color blindness, known as red-green blindness, affects about 8% of the male population in the United States. Red blindness (MIM/OMIM 303900) is characterized by the inability to see red as a distinct color (▶ Figure 4.18). Green blindness (MIM/OMIM 303800) is the inability to see green and other colors in the middle of the visual spectrum (▶ Figure 4.19).

■ **Color blindness** Defective color vision caused by reduction or absence of visual pigments. There are three forms: red, green, and blue blindness.

▶ **FIGURE 4.18** People who are color-blind see colors differently. (a) Those who have normal vision see the red leaves. (b) Someone who is red-green color-blind sees the leaves as gray.

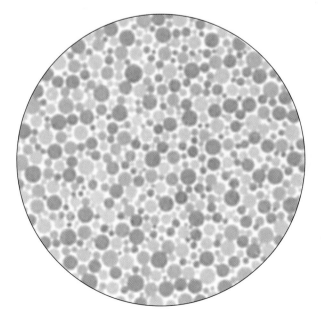

▶ **FIGURE 4.19** People who have normal color vision see the number 29 in the chart, but those who are color-blind cannot see any number.

Both red blindness and green blindness are inherited as X-linked recessive traits. A rare form of blue color blindness (MIM/OMIM 190900) is inherited as an autosomal dominant condition that maps to chromosome 7.

These three genes for color blindness encode different forms of opsins, which are proteins found in the cone cells of the retina (Figure 4.20). Normally, opsins bind to visual pigments in the red-, green-, or blue-cone cells, making the visual pigment/opsin complex sensitive to light of a given wavelength. If the red-opsin gene product is defective or absent, function of the red cones is impaired, and red color blindness results. Similarly, defects in the green or blue opsins produce green and blue blindness.

Pedigrees of color blindness can be used to demonstrate the patterns of transmission for sex-linked recessive traits (Figure 4.21). These patterns can be summarized as follows:

* Both hemizygous males and homozygous females are affected.
* Phenotypic expression is much more common in males than in females, and in the case of rare alleles, males are almost exclusively affected.
* Affected males get the mutant allele from their mothers and transmit it to all of their daughters but not to any sons.
* Daughters of affected males are usually heterozygous and therefore unaffected. Sons of heterozygous females have a 50% chance of receiving the recessive gene.

Some sex-linked recessive conditions are listed in Table 4.3.

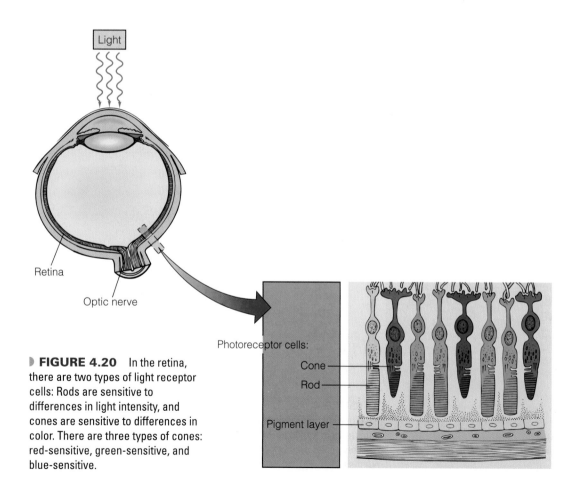

 FIGURE 4.20 In the retina, there are two types of light receptor cells: Rods are sensitive to differences in light intensity, and cones are sensitive to differences in color. There are three types of cones: red-sensitive, green-sensitive, and blue-sensitive.

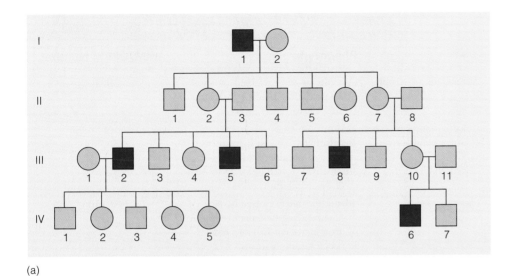

(a)

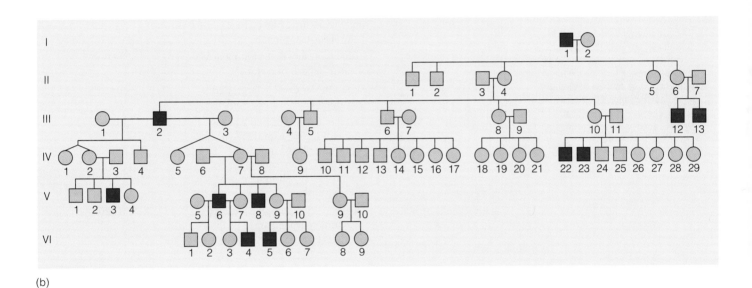

(b)

Muscular Dystrophy Is an X-Linked Recessive Trait

Although often thought of as a single disorder, **muscular dystrophy** is a group of diseases that have common features, including progressive weakness and wasting of muscle tissue. There are autosomal and X-linked forms of muscular dystrophy. Duchenne muscular dystrophy (DMD, MIM/OMIM 310200), an X-linked recessive disorder, is the most common form of muscular dystrophy. In the United States, DMD affects 1 in 3500 males and usually has an onset between 1 and 6 years of age. Progressive muscle weakness is one of the first signs of DMD, and affected individuals use a characteristic set of maneuvers in rising from the prone position (▶ Figure 4.22). The disease progresses rapidly, and by 12 years of age, affected individuals are usually confined to wheelchairs because of muscle degeneration. Death usually occurs by the age of 20 years due to respiratory infection or cardiac failure.

The DMD gene is located in Xp21, a region in the middle of the short arm of the X chromosome, and encodes a protein called *dystrophin*. Normal forms of dystrophin attach to the cytoplasmic side of the plasma membrane in muscle cells and stabilize the membrane during the mechanical strains of muscle contraction. When

■ **Muscular dystrophy** A group of genetic diseases associated with progressive degeneration of muscles. Two of these, Duchenne and Becker muscular dystrophy, are inherited as X-linked, allelic, recessive traits.

Table 4.3   Some X-Linked Recessive Traits

Trait	Phenotype	MIM/OMIM Number
Adrenoleukodystrophy	Atrophy of adrenal glands; mental deterioration; death 1 to 5 years after onset	300100
Color blindness		
Green blindness	Insensitivity to green light; 60 to 75% of color blindness cases	303800
Red blindness	Insensitivity to red light; 25 to 40% of color blindness cases	303900
Fabry disease	Metabolic defect caused by lack of enzyme alpha-galactosidase A; progressive cardiac and renal problems; early death	301500
Glucose-6-phosphate dehydrogenase deficiency	Benign condition that can produce severe, even fatal anemia in presence of certain foods, drugs	305900
Hemophilia A	Inability to form blood clots; caused by lack of clotting factor VIII	306700
Hemophilia B	"Christmas disease"; clotting defect cause by lack of factor IX	306900
Ichthyosis	Skin disorder causing large, dark scales on extremities, trunk	308100
Lesch–Nyhan syndrome	Metabolic defect caused by lack of enzyme hypoxanthine-guanine phosphoribosyl transferase (HGPRT); causes mental retardation, self-mutilation, early death	308000
Muscular dystrophy	Duchenne-type, progressive; fatal condition accompanied by muscle wasting	310200

▶ **FIGURE 4.22** A sign of Duchenne muscular dystrophy. Children who have muscular dystrophy use a characteristic set of movements when rising from the prone position. Once the legs are pulled under the body, children use their arms to push the torso into an upright position.

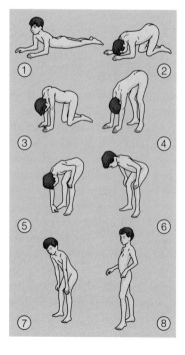

dystrophin is absent or defective, the plasma membranes gradually break down, causing the death of muscle tissue.

Most individuals with DMD have no detectable amounts of dystrophin in their muscle tissue. However, those affected with a second form of X-linked muscular dystrophy, Becker muscular dystrophy (BMD, MIM/OMIM 310200), make a shortened dystrophin that is partially functional. As a result, those with BMD have a later age of onset, milder symptoms, and longer survival compared with DMD. This finding indicates that DMD and BMD represent different allelic forms of the same disease. The gene associated with these two forms of muscular dystrophy has been isolated and cloned using recombinant DNA techniques. The dystrophin gene, one of the largest yet identified, covers some 2 million base pairs of DNA. Future work on the structure and function of dystrophin will hopefully lead to the development of an effective treatment for muscular dystrophy.

Hemophilia and History

*H*emophilia, an X-linked recessive disorder, is characterized by defects in the mechanism of blood clotting. This form of hemophilia, called hemophilia A, occurs with a frequency of 1 in 10,000 males. Because only homozygous recessive females can have hemophilia, the frequency in females is much lower, on the order of 1 in 100 million.

Pedigree analysis indicates that Queen Victoria of England (the granddaughter of King George III) carried this gene. Because she passed the mutant allele on to several of her children (one affected male, two carrier daughters, and one possible carrier daughter), it is likely that the mutation occurred in the X chromosome she received from one of her parents. Although this mutation spread through the royal houses of Europe, the present royal family of England is free of hemophilia because it is descended from Edward VII, an unaffected son of Victoria.

Perhaps the most important case of hemophilia among Victoria's offspring involved the royal family of Russia. Victoria's granddaughter Alix, a carrier, married Czar Nicholas II of Russia. She gave birth to four daughters and then a son, Alexis, who had hemophilia. Frustrated by the failure of the medical community to cure Alexis, the royal couple turned to a series of spiritualists, including the monk Rasputin. While under Rasputin's care, Alexis recovered from several episodes of bleeding, and Rasputin became a powerful adviser to the royal family. Some historians have argued that the czar's preoccupation with Alexis's health and the insidious influence of Rasputin contributed to the revolution that overthrew the throne. Other historians point out that Nicholas II was a weak czar and that revolution was inevitable; but it is interesting to speculate that much of Russian history in the 20th century turns on a mutation carried by an English queen at the beginning of the century.

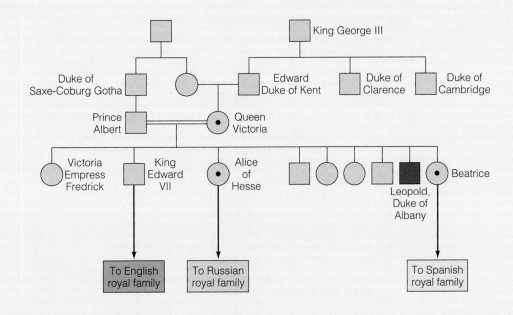

Y-Linked Inheritance Involves Transmission from Male to Male

Genes that occur only on the Y chromosome are said to be Y-linked. Because only males have Y chromosomes, Y-linked traits are present only in males and are passed directly from father to son. Furthermore, every Y-linked trait should be expressed, because males are hemizygous for all genes on the Y chromosome. Although the distinctive pattern of inheritance should make such genes easy to identify, only about three dozen Y-linked traits have been discovered. These include a gene for a protein found in the cell nucleus that may regulate gene expression. Another gene mapped to the Y chromosome, testis-determining factor (TDF/SRY,

Table 4.4 Some of the Genes Mapped to the Y Chromosome

Gene	Product	MIM/OMIM Number
ANT3 ADP/ATP translocase	Enzyme that moves ADP into, ATP out of mitochondria	403000
CSF2RA	Cell surface receptor for growth factor	425000
MIC2	Cell surface receptor	450000
TDF/SRY	Protein involved in early stages of testis differentiation	480000
H-Y antigen	Plasma membrane protein	426000
ZFY	DNA binding protein that may regulate gene expression	490000

MIM/OMIM 480000), is involved in determining maleness in developing embryos. The TDF/SRY gene and early human development are discussed in Chapter 7. Some of the genes mapped to the Y chromosome are listed in Table 4.4.

MITOCHONDRIAL INHERITANCE IS FROM MOTHER TO OFFSPRING

Mitochondria are cytoplasmic organelles that convert energy from food molecules into ATP, a molecule that powers many cellular functions. Billions of years ago, ancestors of mitochondria were probably free-living prokaryotes that formed a symbiotic relationship with primitive eukaryotes. As an evolutionary relic of their free-living ancestry, mitochondria carry DNA molecules that encode information for some thiry-seven mitochondrial genes. Most mitochondria carry five to ten of these DNA molecules, and each cell can contain from several hundred to over a thousand mitochondria (red blood cells are an exception; they have none).

Mitochondria are transmitted through the cytoplasm of the egg (sperm lose all cytoplasm during maturation). As a result, genetic disorders resulting from mutations in mitochondrial genes are maternally inherited. Both males and females can be affected by these disorders, but only females can transmit these mutant genes from generation to generation, producing a distinctive pattern of inheritance (▶ Figure 4.23).

Genetic disorders in mitochondrial DNA are caused by defects in energy conversion and ATP production. Tissues that have the highest energy requirements are most affected. These include the nervous system, skeletal muscle, heart muscle, liver, and kidneys. Some of the disorders associated with mutations in mitochondria genes are listed in Table 4.5.

▶ **FIGURE 4.23** Mitochondrial inheritance. Both males and females can be affected by mitochondrial disorders, but only females can transmit the trait to offspring.

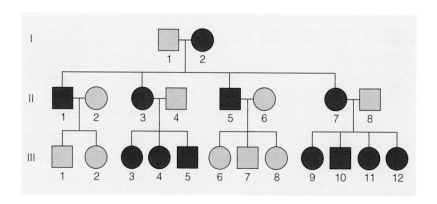

Table 4.5 Some Mitochondrial Traits

Trait	Phenotype	MIM/OMIM Number
Kearns–Sayre syndrome	Short stature; retinal degeneration	530000
Leber optic atrophy (LHON)	Loss of vision in center of visual field; adult onset	535000
MELAS syndrome	Episodes of vomiting, seizures, and stroke-like episodes	540000
MERRF syndrome	Deficiencies in the enzyme complexes associated with energy transfer	545000
Oncocytoma	Benign tumors of the kidney	553000

SEVERAL FACTORS CAUSE VARIATIONS IN GENE EXPRESSION

Many genes have a regular and consistent patterns of expression, but others produce a wide range of phenotypes. In some cases, a mutant genotype may be present but remain unexpressed, resulting in a normal phenotype. Variation in phenotypic expression is caused by a number of factors, including interactions with other genes in the genotype and interactions between genes and the environment.

Temperature and Gene Expression

Siamese cats and Himalayan rabbits have light-colored bodies with dark fur on their paws, nose, ears, and tail (▶ Figure 4.24). In these animals, a gene that controls pigment production is expressed at the lower temperatures found in the extremities, but not at the slightly higher temperatures throughout the rest of the body. All cells of these animals carry the genes for pigment production, but the environment determines the phenotypic pattern of expression.

In humans, a dominantly inherited trait called epidermolysis bullosa of the hands and feet (MIM/OMIM 131800) has a temperature-dependent pattern of expression. Affected individuals develop blisters on the hands and feet mainly in warm weather after working with hand tools or walking long distances. Blistering does not occur in cold weather or when the skin is cooled with ice before working or walking, confirming the role of environmental factors in phenotypic expression.

▶ **FIGURE 4.24** Animals, such as Siamese cats, have dark fur at the tips of the nose, paws, and ears. These colors result from expression of an allele for coat color that is active only at the slightly lower body temperatures found in the extremities.

Nothing in Biology Makes Sense without Evolution

Michael Rose

*T*he best course I took in high school, maybe the best course I've ever taken, was comparative vertebrate anatomy. I loved taking apart the bodies, learning the names of each part, carefully removing material to see little holes and tiny nerves. But before the gore came the scientific theory: fossils, Lamarck, Darwin, Mendel. The theory animated the corpses, made them meaningful, gave them some sense. At the time, I was a big science fiction fan, a genre in which every story had to have a portentous meaning. Evolution is the deep meaning behind life, and without it biology becomes a lot of details.

When I started my doctoral studies, I was given the task of showing experimentally that evolution could make sense of biological aging. At first I was dismayed. Aging, throughout human history, has been a mystery. The people who talk or write most about it are charlatans, quacks, and hustlers. I feared that my career would be aborted by the combination of my failure to accomplish the task and my spattering with the mud of quackery about aging. So far, my fears have proven erroneous.

Consider two genetic diseases, progeria and Huntington disease. It is now thought that each of these is caused by a specific mutation at a single copy of a normal human gene, one gene for each disease. Progeria strikes children between 5 and 10 years of age and is an extremely rare disorder. Huntington disease, on the other hand, strikes almost entirely after the age of 30 and affects thousands of individuals around the world.

Progeria strikes children, prevents their reproduction, and also completely precludes transmission of progeria into the next generation. With 100% success, natural selection screens out new mutations for progeria. With Huntington disease, the patients may already have children before they show any symptoms. Natural selection fails to screen the Huntington mutation out of the population because the gene has effects primarily at later ages, not earlier ages. The key is that the force of natural selection falls with adult age.

Genes that affect the health or development of young animals are sharply scrutinized by natural selection. This fundamental idea was first proposed by evolutionary biologists in the 1930s and 1940s. My role has been to test this idea. When this kind of reproductive pattern is imposed on laboratory populations of the fruit fly, *Drosophila,* they evolve an increased life span over dozens of generations. This happens because we are artificially strengthening natural selection at later ages. Fruit flies that cannot survive to reproduce in their middle age in this experiment are selected against. Thus natural selection alters the genetic basis of aging and increases the life span.

The flies that live longer are interesting beasts. They have a lot of physiological differences that can be related to their ability to live longer. They reproduce less when young, even when they have opportunity to do so. They resist stresses better, including starvation and desiccation. They move around more when they are older, whether walking or flying. They can reproduce more at later ages. Longer lived flies appear, for now, to be superior organisms—'superflies!'

When I present these superflies to audiences, people want to know if I will ever be able to do the same things for them. Literally, the answer is no. But there is the possibility of learning more about the genetics of postponed aging in fruit flies in the hope that we can apply our findings to humans. If this is ever done, it will transform the human life cycle, as aging becomes something that can be controlled instead of merely endured.

Michael Robertson Rose is a professor of evolutionary biology at the University of California, Irvine. He received B.S. and M.S. degrees from Queen's University, Ontario, Canada, and a Ph.D. from the University of Sussex in England. He is a member of the editorial boards of the Journal of Evolutionary Biology and the Journal of Theoretical Biology. In 1992 he won the President's Prize from the American Society of Naturalists.

Age and Gene Expression

Although a large number of genes act before birth or early in development, the phenotypic expression of some genetic disorders is delayed until adulthood. One of the best known examples is **Huntington disease** (HD) (MIM/OMIM 143100), which is inherited as an autosomal dominant trait. The phenotype of this disorder is first expressed between the ages of 30 and 50 years. Affected individuals undergo progressive degeneration of the nervous system, causing mental deterioration and uncontrolled, jerky movements of the head and limbs. The disease progresses slowly, and death occurs some 5 to 15 years after onset. This disorder is particularly insidious

Huntington disease A dominant genetic disorder characterized by involuntary movements of the limbs, mental deterioration, and death within 15 years of onset. Symptoms appear between 30 and 50 years of age.

because onset usually occurs after the affected person has started a family. Because most affected individuals are heterozygotes, each child of an affected parent has a 50% chance of developing the disease. The gene for HD has been identified and cloned using recombinant DNA techniques. This makes it possible to test family members and identify those who will develop the disorder. This disorder is discussed in more detail in Chapter 16, Genetics of Behavior.

Porphyria (MIM/OMIM 176200), an autosomal dominant disorder, is also expressed later in life. This disease is caused by the inability to correctly metabolize porphyrin, a chemical component of hemoglobin. As blood levels of porphyrin increase, some is excreted, producing wine-colored urine. The elevated levels also cause episodes of seizures, intense physical pain, dementia, and psychosis. These symptoms rarely appear before puberty and usually appear in middle age. King George III, the British monarch during the American revolution, may have suffered from porphyria (▶ Figure 4.25). He had a major attack in 1788 at the age of 50 years. He became delirious and suffered convulsions. His physical condition soon improved, but he remained irrational and confused. Early in 1789 his mental functions spontaneously improved, although his physicians took the credit. Later, after two more episodes, the king was replaced on the throne by his son George IV. He died years later, blind and senile. The movie, *The Madness of King George,* is a fictionalized account of how porphyria affected King George, his family, and the politics of Great Britain.

▶ **FIGURE 4.25** King George III of Great Britain (1738–1820) was probably afflicted with porphyria, a genetic disorder that appears in adulthood and affects behavior.

■ **Porphyria** A genetic disorder inherited as a dominant trait that leads to intermittent attacks of pain and dementia. Symptoms first appear in adulthood.

■ **Penetrance** The probability that a disease phenotype will appear when a disease-related genotype is present.

■ **Expressivity** The range of phenotypes resulting from a given genotype.

■ **Camptodactyly** A dominant human genetic trait that is expressed as immobile, bent, little fingers.

Penetrance and Expressivity Are Variations in Gene Expression

The terms *penetrance* and *expressivity* define two different aspects of variation in gene expression. **Penetrance** is the probability that a disease phenotype will appear when a disease-related genotype is present. If all individuals who carry the gene for a dominant disorder have the mutant phenotype, the gene is said to have 100% penetrance. If only 25% of those who carry the mutant gene show the mutant phenotype, the penetrance is 25%. Both genetic and environmental factors can affect penetrance. **Expressivity** refers to the range of phenotypic variation present. The following example shows the relationship between penetrance and expressivity.

The autosomal dominant trait, **camptodactyly** (MIM/OMIM 114200), is caused by the improper attachment of muscles to bones in the little finger. The result is an immobile, bent, little finger. In some people, both little fingers are bent; in others, only one finger is affected; and in a small percentage of cases, neither finger is affected, even though a mutant allele is present (▶ Figure 4.26). Because the trait is dominant, all heterozygotes and homozygotes should be affected on both hands. The

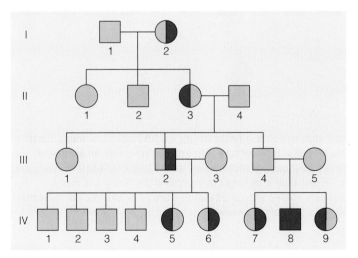

▶ **FIGURE 4.26** Penetrance and expressivity. This pedigree shows the transmission of camptodactyly in a family. Those who have two affected hands are shown as fully shaded symbols. Those affected only in the left hand are indicated by shading the left half of the symbol, and those affected only in the right hand have the right half of the symbol shaded. Symbols with light shading indicate unaffected family members.

pedigree shows that one individual (III-4) is not affected even though he passed the trait to his offspring.

The pedigree in Figure 4.26 shows that nine people must carry the dominant allele for camptodactyly, but phenotypic expression is seen only in eight, giving 8/9, or 88%, penetrance. This is only an estimate because II-1, II-2, and III-1 produced no offspring and could also carry the dominant gene with no penetrance. Many more samples from other pedigrees would be necessary to establish a reliable figure for penetrance of this gene.

Expressivity defines the *degree* of expression for a given trait. In the pedigree for camptodactyly, some individuals are affected on the left hand, and others on the right hand; in one case both hands are affected; in another, neither hand is affected. This variable gene expression results from interactions with other genes and with nongenetic factors in the environment.

The variations in gene expression that we have discussed are all the result of the relationship between a gene and the mechanisms that produce the gene's phenotype. The inheritance of these genes follows the predictable pattern worked out by Mendel for traits in the pea plant, but expression can be complicated by factors that include temperature and age.

GENES ON THE SAME CHROMOSOME ARE LINKED

There are 50,000 to 100,000 genes distributed on the 24 human chromosomes (22 autosomes and the X and the Y chromosome). Obviously, this means that each chromosome carries many genes. Genes on the same chromosome are said to show **linkage**, because they tend to be inherited together. The first case of linkage in humans was discovered by Julia Bell and J.B.S. Haldane in 1936. By pedigree analysis, they demonstrated that the genes for hemophilia (MIM/OMIM 306700) and color blindness (MIM/OMIM 303800) are both on the X chromosome and are therefore linked.

▶ Figure 4.27 shows a pedigree in which linkage is indicated between the gene for the ABO blood group (the *I* locus, (MIM/OMIM 110300) and a condition called nail-patella syndrome (MIM/OMIM 161200). Nail-patella syndrome is an autoso-

Linkage A condition in which two or more genes do not show independent assortment. Rather, they tend to be inherited together. Such genes are located on the same chromosome. By measuring the degree of recombination between linked genes, the distance between them can be determined.

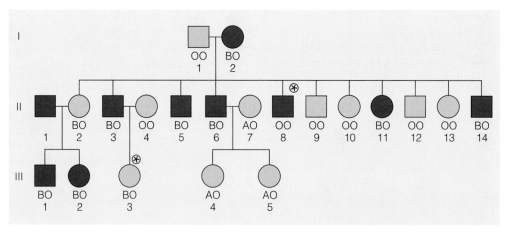

▶ **FIGURE 4.27** Linkage between nail-patella syndrome and the ABO blood type locus. Darkly shaded symbols in this pedigree represent those who have nail-patella syndrome, an autosomal dominant trait. Genotypes for the ABO locus are shown below each symbol. Nail-patella syndrome and the B allele are present in I-2, they tend to be inherited together in this family, and they are identified as linked genes. Individuals marked with an asterisk (II-8 and III-3) inherited the nail-patella allele or the B allele alone. This separation of the two alleles occurred by recombination.

mal dominant disorder associated with deformities in the nails and kneecaps. In this pedigree, the B allele (I^B) and the allele for nail-patella syndrome occur together in individual I-2. In generations II and III, these two alleles tend to be inherited together. Individuals who have type B blood tend to have nail-patella syndrome. In generation II, the B allele and the nail patella allele show linkage in individuals II-1, II-2, II-5, II-6, and II-14. In generation III, the two alleles are linked in individuals III-1 and III-2.

Although linked genes tend to be inherited together, they separate from each other in some offspring because of crossing-over between members of a chromosomal pair. As shown in the pedigree, type B blood and nail-patella syndrome are not always inherited together. Examination of the pedigree shows that 2 of the 16 individuals have inherited *one* of the two alleles but not both (individuals II-8 and III-3). This separation of the two alleles is the result of recombination between the two genes (Figure 4.29).

USING LINKAGE AND RECOMBINATION FREQUENCIES TO MAKE GENETIC MAPS

Early in this century, Alfred Sturtevant, working with the fruit fly, *Drosophila*, realized that the further apart two genes are on a chromosome, the greater the chance that they will be separated by crossing-over. He concluded that the amount of crossing-over between genes can be used to determine the order and distance between genes on a chromosome and produce a genetic map.

In a **genetic map**, genes are arranged in a linear order, and the distance between any two genes is measured by how frequently crossing-over takes place between them (Figure 4.28). In other words, genetic distance is measured by the frequency of recombination between loci on the same chromosome. The units are expressed as a percentage of recombination, where 1 map unit is equal to a frequency of 1% recombination. (This unit is also known as a centimorgan, or cM).

Remember, it is possible to construct genetic maps only because linked genes separate from each other by crossing-over during meiosis. How often this happens depends on the distance between the genes. The farther apart two genes are on a chromosome, the more likely it is that a crossover will separate them. Conversely, the closer together two genes are on a chromosome the less frequently they will be separated by a crossover (Figure 4.29).

Let's use this information to calculate the distance between the gene for nail-patella syndrome and the gene for the ABO blood group by measuring how often recombination occurs between them. In the pedigree in Figure 4.27, the nail-patella gene and the gene for the B blood type have separated in 2 of the 16 individuals (II-8 and III-3). From the frequency of recombination in this pedigree (2/16 = 0.125, or 12.5% recombination), the distance between the gene for the ABO locus and the gene controlling nail-patella syndrome can be calculated as 12.5 map units.

For accuracy, either a much larger pedigree or many more small pedigrees need to be examined to determine the extent of recombination between these two genes. When a large series of families is combined in an analysis, the map distance between the ABO locus and the nail-patella locus is about 10 map units.

▨ **Genetic map** The arrangement and distance between genes on a chromosome deduced from studies of genetic recombination.

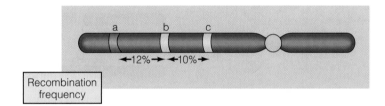

▶ **FIGURE 4.28** Linked genes are carried on the same chromosome. Recombinational frequencies can be used to construct genetic maps, giving the order and distance between genes.

FIGURE 4.29 Crossing-over between homologous chromosomes during meiosis involves the exchange of chromosomal parts. In this case, crossing-over between the genes for blood type (alleles *B, O,* of gene *I*) and nail-patella syndrome (*N*) produces new allelic combinations. The frequency of crossing-over is proportional to the distance between the genes, allowing construction of a genetic map for this chromosomal region.

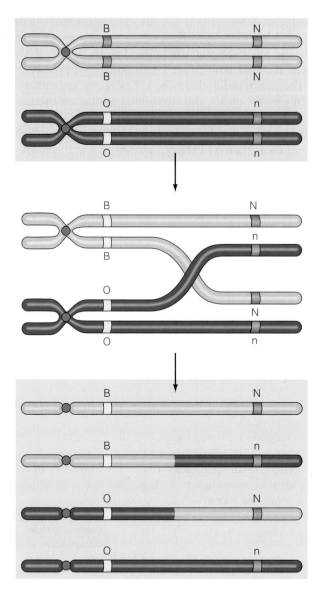

Lod method A probability technique used to determine whether genes are linked.

LINKAGE AND RECOMBINATION CAN BE MEASURED BY A LOD SCORE

In many studies in human genetics, it is difficult to establish linkage and measure the distance between genes because such studies require large pedigrees that have many offspring in three or more generations in which two genetic disorders are present (the two genes being analyzed for linkage). In most cases, pedigrees cover at most three generations—the grandparents, parents, and children, and families that have two genetic disorders that might map to the same chromosome are very rare. To avoid this problem, a statistical technique known as the **lod method** is used to determine whether two genes are linked (carried on the same chromosome) and to measure the distance between them.

Lod scores are calculated using software programs (like LINKMAP) designed to carry out linkage analysis. First, an observed frequency of recombination between two genes is derived from pedigree studies. Then, the program calculates two probabilities, the probability that the observed results would have been obtained if the two genes were linked and the probability that the results would have been obtained even if the two genes were not linked. The results are expressed as the $\log_{10}$ of the

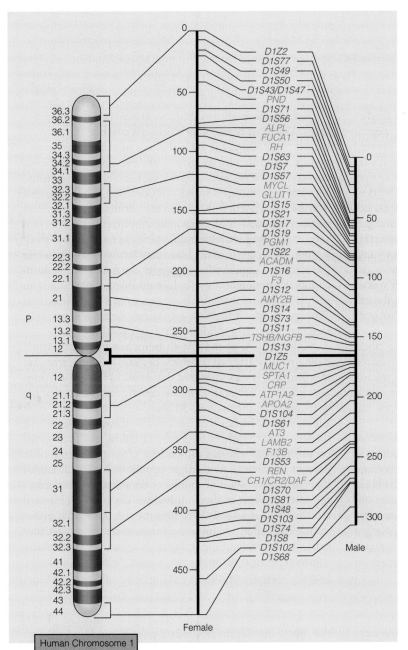

FIGURE 4.30 Genetic map of human chromosome 1. At the left is a drawing of the chromosome. The two vertical lines at the right represent genetic maps derived from studies of recombination in males and females. Between the genetic maps are the order and location of 58 loci, some of which are genes (red) and others (blue) that are genetic markers detected using recombinant DNA techniques. The map in females is about 500 cM long, and in males it is just over 300 cM. This is a result of differences in the frequency of crossing over in males and females. This map provides a framework for locating genes on the chromosome as part of the Human Genome Project.

ratio of the two probabilities, or **lod score** (*lod* stands for the log of the odds). A lod score of 3 means that the odds are 1,000 to 1 in favor of linkage; a score of 4 means that the odds are 10,000 to 1 in favor of linkage. Most geneticists agree that two genes are linked (carried on the same chromosome) when the lod score is 3 or higher.

Using pedigree analysis and lod scores, genetic maps have been constructed for all of the human chromosomes. A genetic map for a human chromosome is shown in ▶ Figure 4.30. The map is drawn next to one of the chromosome itself. The connecting lines show the sites on the actual chromosome where some of the genes are located.

Other methods of mapping use recombinant DNA techniques to map human chromosomes. These methods, along with linkage mapping, are part of the Human Genome Project, an international effort to map all of the genes in the human genome. These methods and the project itself are discussed in Chapter 13.

■ **Lod score** The ratio of the probability that two loci are linked to the probability that they are not linked, expressed as a $\log_{10}$. Scores of 3 or more are taken as establishing linkage.

Case Studies

CASE 1

Florence is an active 44-year-old elementary school teacher who was examined by her doctor because she was experiencing long periods of severe headaches and nausea. She told her physician that her energy level had been dramatically reduced the last few months and her arms and legs felt like they "weighed 100 pounds each," particularly after she worked out in the gym. Her doctor did a complete physical and noticed that she did have reduced strength in her arms and legs and that her left eyelid was droopier than her right eyelid. He referred her to an ophthalmologist who discovered by using an ophthalmoscope that she had an unusual pigment accumulation on her retina, which had not yet affected her vision. She then visited a clinical geneticist, after examining the mitochondria in her muscles, diagnosed her with a mitochondrial myopathy known as Kearns-Sayre syndrome.

The term *mitochondrial myopathy* is used to describe a group of disorders that share the common feature of reduced energy production in the body's tissues, particularly muscles. All tissues are composed of cells. Each cell contains a nucleus, and many mitochondria. Mitochondria are responsible for the conversion of food molecules into energy to meet the cell's energy needs. In mitochondrial myopathies, these biochemical processes are abnormal and energy production is reduced. Muscle tends to be particularly affected because it requires a lot of energy, but other tissues such as the brain may also be involved. Under the microscope, the mitochondria in muscle from people with mitochondrial myopathies look abnormal, and they often accumulate around the edges of muscle fibers. This gives a particular staining pattern, known as a "ragged red" appearance, and this is usually how mitochondrial myopathies are diagnosed.

Mitochondrial myopathies affect people in many ways. The most common problem is a combination of mild muscle weakness in the arms and legs together with droopy eyelids and difficulty in moving the eyes. Some people do not have problems with their eye muscles, but have arm and leg weakness that gets worse after exertion. This weakness may be associated with nausea and headache. Sometimes muscle weakness is obvious in small babies if the illness is severe, and they may have difficulties in feeding and swallowing. Other parts of the body may be involved, including the electrical conduction system of the heart. Most mitochondrial myopathies are mildly disabling, particularly in people who have eye muscle weakness and limb weakness. The age at which the first symptoms develop is variable, ranging from early childhood to late adult life.

About 20% of those with mitochondrial myopathies have similarly affected relatives. Because mothers transmit this disorder, it was suspected that some of these conditions are caused by a mutation in the genetic information of the mitochondria themselves. Mitochondria have their own genes, separate from the genes in the chromosomes of the nucleus. Only mothers pass mitochondria and their genes on to children, whereas the nuclear genes come from both parents. In about one-third of people with mitochondrial myopathies, substantial chunks of the mitochondrial genes are deleted. Most of these individuals do not have affected relatives and it seems likely that the deletions arise either during egg cell production in their mothers, or during very early development of the embryo. Deletions are particularly common in people with eye muscle weakness and the Kearns-Sayre syndrome.

CASE 2

The Smiths had just given birth to their second child and were eagerly waiting to take their newborn home. At that moment, their obstetrician arrived in the hospital room with some news about their daughter's newborn screening tests. The physician told them that the state's mandatory newborn screening test had detected an abnormally high level of phenylalanine in the blood of their daughter. The Smith's asked if this was just a fleeting effect, like newborn jaundice, that would "go away" in a few days. When they were told that this is unlikely, they were even more confused. Their daughter was born looking perfectly "normal" and the pregnancy had progressed without any complications. Mrs. Smith even had a normal amniocentesis early in the pregnancy. The physician asked a genetic counselor to come to their room to explain their daughter's newly diagnosed condition.

The counselor began her discussion with the Smiths by taking a family history from each of them. She explained that phenylketonuria (PKU) is a genetic condition that results when an individual inherits an altered gene from each parent. The counselor wanted to make this point early in the session in case either parent was blaming him or herself or the other spouse for their daughter's condition. She explained that PKU is characterized by an increased concentration of phenylalanine in blood and urine and that mental retardation can be part of this condition if it is not treated at an early age. Therefore, it is important that they treat their daughter as soon as possible. To prevent the development of mental retardation, diagnosis must be done and dietary therapy must be started before the child is 30 days old. Dietary therapy is accomplished by consuming a special diet in which the bulk of protein is substituted for an artificial amino acid mixture low in phenylalanine. The diet must start in the first month of life and should be continued indefinitely to be maximally effective.

PKU is one of several diseases known as the hyperphenyl-alaninemias. Hyperphenylalaninemias occur in 1 in 10,000 births. Classic phenylketonuria accounts for two-thirds of them. It is an autosomal recessive disease, widely distributed among whites and asians. It is rare in blacks. Heterozygous carriers do not show symptoms, but may have slightly increased phenylalanine concentrations. If untreated, children with classic phenylketonuria can experience progressive impairment of cerebral function, seizures, and hyperactivity. EEG abnormalities, mousy odor of the skin, hair, and urine, tendency to hypopigmentation and eczema complete the clinical picture.

Summary

1. The inheritance of single gene traits in humans is often called Mendelian inheritance because of the pattern of segregation within families. These traits produce phenotypic ratios similar to those observed by Mendel in the pea plant. Although the results of studies in peas and humans may be similar, the methods are somewhat different. Instead of direct experimental crosses, human traits are traced by constructing pedigrees that follow a trait through several generations of a family.

2. Information in the pedigree is used to determine the mode of inheritance. These modes include autosomal dominant, autosomal recessive, X-linked dominant, X-linked recessive, Y-linked and mitochondrial.

3. The results of pedigree analysis depend on the distribution of alleles via gametes from parent to child. Because the number of offspring is usually small, large deviations from expected ratios of segregation are often encountered. This effect can be controlled by examining pedigrees from a large number of families and pooling the information from many sources to confirm the mode of inheritance for a given trait.

4. Guidelines for pedigree analysis can be used to determine if a trait is inherited in an autosomal recessive fashion. Other guidelines can be used to establish whether a trait has an autosomal dominant, X-linked recessive, X-linked dominant, y-linked or mitochondrial pattern of inheritance.

5. Several factors can alter an expected pattern of inheritance, including the age of onset, penetrance, expressivity, and environmental interactions.

6. Genes on the same chromosome tend to be inherited together and are called linked genes. Within limits, the amount of recombination between two genes can be used to measure the distance between genes, creating genetic maps of chromosomes. Maps of all the human chromosomes are being prepared as part of the Human Genome Project.

Questions and Problems

1. What is the main reason that pedigree charts are used?
2. Pedigree analysis permits all of the following except
   a. an orderly presentation of family information
   b. the determination of whether a trait is genetic
   c. whether a trait is dominant or recessive
   d. which gene is involved in a heritable disorder
   e. whether a trait is X-linked or autosomal
3. What mode of inheritance suggested by the following pedigree:

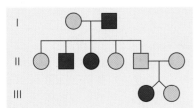

4. Does the indicated individual (III-5) show the trait in question?

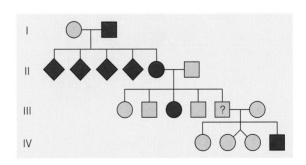

5. a. Construct a pedigree based on the following information:

1. The proband (affected individual that led to the construction of the pedigree) exhibits the trait.
2. Neither her husband nor her only sibling, an older brother, exhibits the trait.
3. The proband has five children by her current husband. The oldest is a boy, followed by a girl, then another boy, and then identical twin girls. Only the second oldest fails to exhibit the trait.
4. The parents of the proband both show the trait.

  b. Determine the mode of inheritance of the trait (go step-by-step to examine each possible mode of inheritance).

  c. Can you deduce the genotype of the proband's husband for this trait?

6. Describe the primary gene/protein defect and the resulting phenotype for the following diseases:
  a. cystic fibrosis
  b. sickle cell anemia
  c. Marfan syndrome
  d. muscular dystrophy DMD and BMD

7. In the following pedigree, assume that the father of the proband is homozygous for a rare trait. Explain a mode of inheritance consistent with this pedigree. In particular, explain the phenotype of the proband.

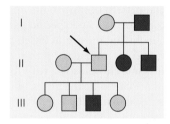

8. Using the following pedigree,
  a. deduce a compatible mode of inheritance.
  b. identify the genotype of the individual in question.

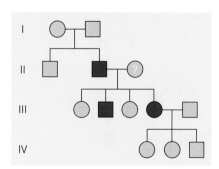

9. List and describe two diseases inherited in the following fashion:
  a. autosomal dominant
  b. autosomal recessive

10. The X and Y chromosomes are structurally and genetically distinct. However, they do pair during meiosis at a small region near the tips of their short arms, indicating that the chromosomes are homologous in this region. If a gene lies in this region, will its mode of transmission be more like a sex-linked gene or an autosomal gene? Why?

11. What is the chance that a colorblind man and a carrier woman will produce:
  a. a colorblind son?
  b. a colorblind daughter?

12. A young boy is color-blind. His one brother and five sisters are not. The boy has three maternal uncles and four maternal aunts. None of his uncles' children or grandchildren are colorblind. One of the maternal aunts married a colorblind man, and half of her children, both male and female, are colorblind. The other aunts married men who have normal color vision. All their daughters have normal vision, but half of their sons are colorblind.
  a. Which of the boy's four grandparents transmitted the gene for colorblindness?
  b. Are any of the boy's aunts or uncles colorblind?
  c. Are either of the boy's parents colorblind?

13. The following is a pedigree for a common genetic trait. Analyze the pedigree to determine whether the trait is inherited as an:
  a. autosomal dominant
  b. autosomal recessive
  c. X-linked dominant
  d. X-linked recessive
  e. Y-linked

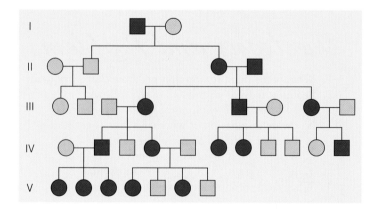

14. As a genetic counselor investigating a genetic disorder in a family, you are able to collect a four-generation pedigree that details the inheritance of the disorder in question. Analyze the information in the pedigree to determine whether the trait is inherited as an:
    a. autosomal dominant
    b. autosomal recessive
    c. X-linked dominant
    d. X-linked recessive
    e. Y-linked

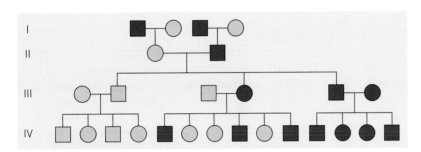

15. What are the unique features of mitochondria that are not present in other cellular organelles in human cells?
16. Define penetrance and expressivity.
17. Suppose that space explorers discover an alien species that has the same genetic principles that apply to humans. Although all 19 aliens analyzed to date carry a gene for a third eye, only 15 display this phenotype. What is the penetrance of the third eye gene in this population?
18. A genetic disorder characterized by falling asleep in genetics lectures is known to be 20% penetrant. All 90 students in a genetics class are homozygous for this gene. How many of the 90 students will theoretically fall asleep during the next lecture?
19. Why are humans difficult subjects for genetic analysis?
20. A proband female suffering from an unidentified disease seeks the advice of a genetic counselor prior to starting a family. Based on the following data, the counselor constructs a pedigree encompassing three generations: (1) The maternal grandfather of the proband suffers from the disease. (2) The mother of the proband is unaffected and is the youngest of five children, the three oldest being male. (3) The proband

has an affected older sister, but the youngest siblings are unaffected twins (boy and girl). (4) All individuals suffering from the disease have been revealed. Duplicate the counselor's feat.

21. The father of 12 children begins to show symptoms of neurofibromatosis.
    a. What is the probability that Sam, the man's second oldest son (II-2), will suffer from the disease if he lives a normal life span? (Sam's mother and her ancestors do not have the disease.)
    b. Can you infer anything about the presence of the disease in Sam's paternal grandparents?

22. Consider the two autosomal recessive conditions: Albinism (A/a) and hereditary deafness (D/d). Given the following data, in which cases are the genes linked? What are the genotypes of the parents? Which alleles are linked to each other (for example, A linked to D or to d)? Assume that no corssing over has occurred.
    a. Case 1
       9 nonalbino, hearing
       3 nonalbino, deaf
       3 albino, hearing
       1 albino, deaf
    b. Case 2
       3 nonalbino, hearing
       albino, deaf

23. In case 2 above, what would be the expected genotypes of the progeny if A was linked to d and a was linked to D instead?
24. A hypothetical human trait is controlled by a single gene. Four alleles of this gene have been identified: a, b, c, and d. Alleles a, b, and c are all codominant; allele d is recessive to all other alleles.
    a. How many phenotypes are possible?
    b. How many genotypes are possible?
25. How can dominant lethal alleles survive in a population?
26. A recombination experiment is carried out in the fruit fly, *Drosophila*, to measure the distance between two genes on chromosome 2. The results indicate that there is 58% recombination between the two loci. Are these genes linked?
27. The frequency of recombination between gene A and gene B is 8%; between gene B and gene C, the frequency is 12%, and between gene A and gene C the frequency is 4%. From this information, deduce the gene order.

28. In the eighteenth century a young boy suffered from a skin condition known as ichthyosis hystrix gravior. The phenotype of this disorder includes thickening of skin and the formation of loose spines that are periodically sloughed off. This 'porcupine man' married and had six sons, all of whom had this same condition. He also had several daughters, all of whom were unaffected. In all succeeding generations, this condition was passed on from father to son. What can you theorize about the location of the gene that causes ichthyosis hystrix gravior.

29. Huntington disease is a rare, fatal disease, usually developing in the fourth or fifth decade of life. It is caused by a single dominant autosomal allele. A phenotypically normal man in his twenties, who has a two-year-old son of his own, learns that his father has developed Huntington disease. What is the probability that he himself will develop the disease? What is the chance that his young son might eventually develop the disease?

Internet Activities

The following activities use the resources of the World Wide Web to enhance the topics covered in this chapter. To investigate the topics described below, log on to the book's homepage at:

http://www.brookscole.com/biology

1. Online Mendelian Inheritance in Man (OMIM) is an on-line catalog of human genetic disorders that is updated daily. For any given genetic disorder, information on clinical features, mode of inheritance, molecular genetics, diagnosis, therapies, and more is presented.
   a. Access OMIM through the homepage, select a genetic disorder mentioned in the chapter, and read the material to supplement the information provided in the chapter.

   b. OMIM also contains information about human traits not associated with disease, such as eye color, handedness, uncontrolled sneezing (the achoo syndrome), alteration of taste sensations, earlobe creases, cleft chins, ear wax, baldness, and many more. Access OMIM and read the material about one or more of these traits.

2. Information about genetic disorders, support groups and organizations is available on the World Wide Web. If you, a member of your family, or someone you know has a genetic disorder, information about the disorder, treatments, and parent groups can be accessed. If you are interested in a genetic disorder, the book's homepage has a link to a listing of genetic support groups. Write to one of these groups to obtain more information about a specific genetic disorder.

For Further Reading

Cawthon, R. M., Weiss, R., Xu, G., Viskochil, D., Culver, M., Stevens, J., Robertson, M., Dunn, D., Gesteland, R., O'Connell, P., & White, R. (1990). A major segment of the neurofibromatosis type 1 gene: cDNA sequence, genomic structure, and point mutations. *Cell* 62: 193–201.

Collins, F. S., Ponder, B. A., Seizinger, B. R., & Epstein, C. J. (1989). The von Recklinghausen neurofibromatosis region on chromosome 17: Genetic and physical maps come into focus. *Am. J. Hum. Genet.* 44: 1–5.

Eaton, W., & Hofrichter, J. (1995). The biophysics of sickle cell hydroxyurea therapy. *Science* 268: 1142–1143.

Embury, S. H. (1986). The clinical pathology of sickle cell disease. *Ann. Rev. Med.* 37: 361–376.

Francomano, C. A., Le, P. L., & Pyeritz, R. E. (1988). Molecular genetic studies in achondroplasia. *Basic Life Sci.* 48: 53–58.

Huntington's Disease Collaborative Research Group. (1993). A novel gene containing a trinucleotide repeat that is expanded and unstable on Huntington's disease chromosomes. *Cell* 72: 971–983.

Kinnear, P. E., Jay, B., & Witkop, C. J., Jr. (1985). Albinism. *Surv. Ophthalmol.* 30: 75–101.

Lucky, P. A., & Nordlund, J. J. (1985). The biology of the pigmentary system and its disorders. *Dermatol. Clin.* 3: 197–216.

Macalpine, I., & Hunter, R. (1969). Porphyria and King George III. *Sci. Am.* 221 (July): 38–46.

Peltonen, L., & Kainulainen, K. (1992). Elucidation of the gene defect in Marfan syndrome. Success by two complementary research strategies. *FEBS Letters* 307: 116–121.

Potts, D.M. and Potts, W.T.W. 1995. *Queen Victoria's Gene: Hemophilia and the Royal Family*. Gloucestershire (England): Alan Sutton Publishing Limited.

Prockop, D. J. (1985). Mutations in collagen genes: Consequences for rare and common diseases. *J. Clin. Invest.* 75: 783–787.

Ramirez, F., Sangiorgi, F. O., & Tsipouras, P. (1986). Human collagens: Biochemical, molecular and genetic features in normal and diseased states. *Horizons Biochem. Biophys.* 8: 341–375.

Rommens, J. M., Iannuzzi, M. C., Bat-Sheva, K., Drumm, M. L., Melmer, G., Dean, M., Rozmahel, R., Cole, J. L., Kennedy, D., Hidaka, N., Zsiga, M., Buchwald, M., Riordan, J. R., Tsui, L. C., & Collins, F. S. (1989). Identification of the cystic fibrosis gene: Chromosome walking and jumping. *Science* 245: 1059–1065.

Rouleau, G. A., Wertelecki, W., Haines, J. L., Hobbs, W., Trofatter, J. A., Seizinger, B., Martuza, R., Superneau, D., Conneally, P. M., & Gusella, J. (1987). Genetic linkage of bilateral acoustic neurofibromatosis to a DNA marker on chromosome 22. *Nature* *329:* 246–248.

Smithies, O. (1993). Animal models of human genetic diseases. *Trends Genet. 9:* 112–116.

Stanbury, J. B., Wyngaarden, J. B., & Fredrickson, D. S. (1983). *The Metabolic Basis of Inherited Disease,* 5th ed. New York: McGraw-Hill.

Stern, C. (1973). *Principles of Human Genetics,* 3rd ed. San Francisco: Freeman.

Stokes, R. W. (1986). Neurofibromatosis: A review of the literature. *J. Am. Osteopathic Assoc. 86:* 49–52.

Tsipouras, P., & Ramirez, F. (1987). Genetic disorders of collagen. *J. Med. Genet. 24:* 2–8.

Wallace, M. R., Marchuk, D. A., Anderson, L. B., Letcher, R., Odeh, H. M., Saulino, A. M., Fountain, J. W., Brereton, A., Nicholson, J., Mitchell, A. L., Brownstein, B. H., & Collins, F. S. (1990). Type 1 neurofibromatosis gene: Identification of a large transcript disrupted in three NF1 patients. *Science 249:* 181–186.

Welsh, M. & Smith, A. (1995). Cystic fibrosis. *Sci. Am. 273:* 52–59.

Worton, R. (1995). Muscular dystrophies: Diseases of the dystrophin-glycoprotein complex. *Science 270:* 755–756.

Yoshida, A. (1982). Biochemical genetics of the human blood group ABO system. *Am. J. Hum. Genet. 34:* 1–14.

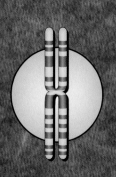

Polygenes and Multifactorial Inheritance

In 1713, a new king was crowned in Prussia. He immediately began one of the largest military buildups of the 18th century. In the space of 20 years, King Frederick William I, ruler of fewer than 2 million citizens, enlarged his army from around 38,000 men to just under 100,000 troops. Compare this to the neighboring kingdom of Austria, which had a population of 20 million and an army of just under 100,000 men, and you will understand why Frederick William was regarded as a military monomaniac. The crowning glory of this military machine was his personal guard, known as the Potsdam Grenadier Guards. This unit was composed of the tallest men obtainable. Frederick William was obsessed with having giants in this guard, and his recruiters used bribery, kidnapping, and smuggling to fill the ranks of this unit. It is said that members of the guard could lock arms while marching on either side of the king's carriage. Many members were close to 7 feet tall. Although someone 7 feet tall is not much of a novelty in the NBA, in 18th-century Prussia anyone taller than about 5 feet 4 inches was above average height.

Frederick William was also rather miserly, and because this recruiting was costing him millions, he decided it would be more economical simply to breed giants to serve in his elite unit. To accomplish this, he ordered that every tall man in the kingdom was to marry a tall, robust woman, expecting that the offspring would all be giants. Unfortunately this idea was a frustrating failure. Not only was it slow, but most of the children were shorter than their parents. While continuing this breeding program, Frederick William reverted to kidnapping and bounties, and he also let it be known that the best way for foreign governments to gain his favor was to send giants to be members of his guard. This human breeding experiment continued until shortly after Frederick William's death in 1740, when his son, Frederick the Great, disbanded the Potsdam Guards.

SOME TRAITS ARE CONTROLLED BY TWO OR MORE GENES

What exactly went wrong with Frederick William's experiment in human genetics? Selecting the tallest men and women as parents should result in tall children. We have the advantage of knowing that when Mendel intercrossed true-breeding tall pea plants, the offspring were all tall. Even when heterozygous tall pea plants are crossed, three-fourths of the offspring are tall.

The problem is that a single gene pair controls height in pea plants, whereas height in humans is determined by several gene pairs and is a **polygenic trait.** The tall and short phenotypes in pea plants are two distinct phenotypes and are examples of **discontinuous variation.** In measuring height in humans, it is difficult to set up only two phenotypes. Instead, height in humans is an example of **continuous variation.** Unlike Mendel's pea plants, people are not either 18 in. or 84 in. tall; they fall into a series of overlapping phenotypic classes. Two or more separate gene pairs often control traits with a gradation of phenotypes.

Understanding the distinction between discontinuous and continuous traits was an important advance in genetics. It is based on accepting the idea that genes interact

■ **Polygenic trait** A phenotype that depends on the action of a number of genes.

■ **Discontinuous variation** Phenotypes that fall into two or more distinct, nonoverlapping classes.

■ **Continuous variation** A distribution of phenotypic characters that is distributed from one extreme to another in an overlapping, or continuous, fashion.

Multifactorial traits Traits that result from the interaction of one or more environmental factors and two or more genes.

with each other and with the environment. All genes can interact with the environment, and they differ only in the degree of this interaction. **Multifactorial traits** are those that involve two or more genes and strong interaction with the environment.

In this chapter, we examine traits controlled by genes at two or more loci and consider how nongenetic factors, such as the environment, affect gene expression. The degree of genetic effects on a trait can be estimated by measuring heritability. We consider this concept and the use of twins as a means of measuring the heritability of a trait. In the last part of the chapter we examine a number of human polygenic traits, some of which have been the subject of political and social controversy.

POLYGENES AND VARIATIONS IN PHENOTYPE

Mendel was not the only scientist in the late 19th century who experimented with the inheritance of traits. In one series of experiments, Josef Kölreuter crossed tall and dwarf tobacco plants. The F1 plants were all intermediate in height to the parents. When self-crossed, the F1 produced an F2 that contained plants of many different heights. Some of the F2 were as tall or short as the parents, but most of the F2 were intermediate in height when compared with the parents (▶ Figure 5.1), and showed continuous variation in the distribution of phenotypes. Mendel's results with pea plants, however, produced evidence of discontinuous variation (Figure 5.1).

Shortly after the turn of the century, it was found that traits in several different plants and animals show continuous variation in phenotype. In each of these cases,

▶ **FIGURE 5.1** A comparison of traits that have continuous and discontinuous phenotypes.
(a) Histograms show the percentage of plants that have different heights in crosses between tall and dwarf strains of tobacco plants carried to the F2 generation. The F1 generation is intermediate to the parents in height, and the F2 shows a range of phenotypes from dwarf to tall. Most plants have a height intermediate to those of the P1 generation.
(b) Histograms show the percentage of plants that have different heights in crosses between tall and dwarf strains of the pea plant. The F1 generation has the tall phenotype, and the F2 has two distinct phenotypic classes; 75% of the offspring are tall, and 25% are dwarf. The differences between tobacco plants and pea plants are explained by the fact that height in tobacco plants is controlled by two or more gene pairs, whereas height in peas is controlled by a single gene.

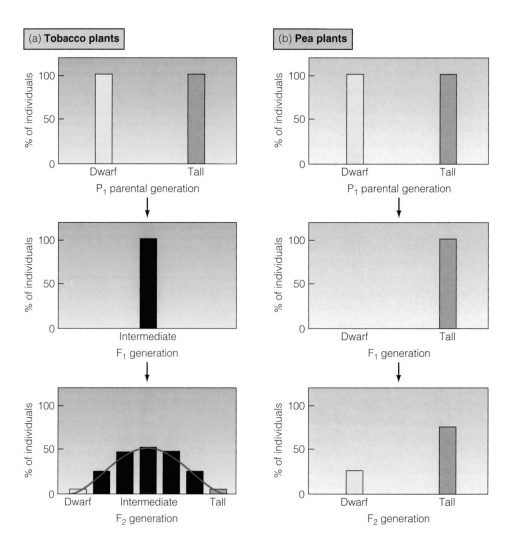

the offspring had a phenotype that seemed to be a blend of the parental traits. Geneticists debated whether continuous variation could be reconciled with the inheritance of Mendelian factors or whether this apparent blending of traits signaled the existence of another mechanism of inheritance. This argument is important to human genetics because many human traits and many genetic disorders show continuous variation.

In the years immediately after the rediscovery of Mendel's work, only a few traits were known in humans, and these were controlled by single genes. At the time, interest in human genetics was largely centered on discovering whether "social" traits, such as alcoholism, feeblemindedness, and criminal behavior, were inherited. Some geneticists simply assumed that these traits were controlled by single genes and constructed pedigrees accordingly. Other geneticists pointed out that these traits were not inherited in the phenotypic ratios observed in experimental organisms and discounted the importance of Mendelian inheritance in humans. In fact, the biomathematician Karl Pearson, who studied polygenic traits in humans, is reported to have said: "there is no truth in Mendelism at all."

Between 1910 and 1930, the controversy over continuous variation was resolved. Experimental work with corn and tobacco plants demonstrated that continuous variation could be explained by Mendelian inheritance. These findings revealed that traits determined by a number of alleles, each of which makes a small contribution to the phenotype, would exhibit a continuous distribution of phenotypes in the F2 generation. This is true even though the inheritance of each gene follows the rules of Mendelian inheritance. This distribution of phenotypes follows a bell-shaped curve and contains a small number of individuals who have extreme phenotypes (very short or very tall, for example). Most individuals, however, have phenotypes between the extremes; their distribution follows what statisticians call a normal curve (▶ Figure 5.2). This pattern of inheritance, known as polygenic or quantitative inheritance, is additive because each allele adds a small, but equal amount to the phenotype.

The continuous variation of phenotypes that results from polygenic inheritance has several distinguishing characteristics:

- Traits are usually quantified by measurement rather than by counting.
- Two or more genes contribute to the phenotype. Each gene contributes additively to the phenotype. The effect of individual additive alleles may be small, and some alleles make no contribution.
- Phenotypic expression of polygenic and multifactorial traits varies within a wide range. This variation is produced by gene interaction and environmental factors and is best analyzed in populations rather than in individuals (▶ Figure 5.3).

▶ **FIGURE 5.2** A bell-shaped or "normal" curve shows the distribution of phenotypes for traits controlled by two or more genes. In a normal curve, few individuals are at the extremes of the phenotype, and most individuals are clustered around the average value. In this case, the phenotype is height measured in a population of human males.

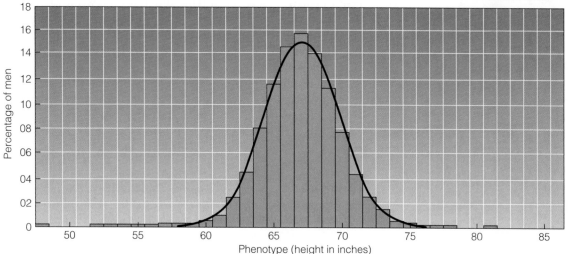

▶ **FIGURE 5.3** Skin color is a polygenic trait controlled by three or four genes, producing a wide range of phenotypes. Environmental factors (exposure to the sun and weather) also contribute to the phenotypic variation, making this a multifactorial trait.

2 loci

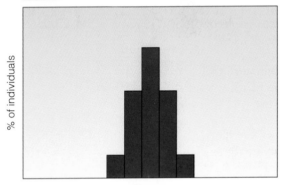

3 loci

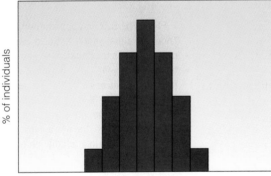

4 loci

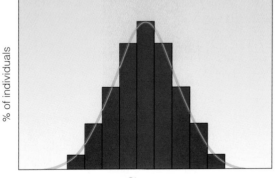

Polygenic inheritance is an important concept in human genetics. Traits, such as height, weight, skin color, and intelligence, are under polygenic control. In addition, congenital malformations, such as neural tube defects, cleft palate, and clubfoot as well as genetic disorders, such as diabetes, hypertension, and behavioral disorders, are polygenic or multifactorial traits.

The distribution of phenotypes and F2 ratios in traits involving two, three, and four genes is shown in ▶ Figure 5.4. If the trait is controlled by two loci, there are five phenotypic classes in the F2, each of which has 4, 3, 2, 1, or 0 dominant alleles. The F2 ratio of 1:4:6:4:1 results from the number of genotypic combinations that produce each phenotype. At each extreme are the homozygous dominant (*AABB*) and homozygous recessive (*aabb*) genotypes that have 4 and 0 dominant alleles, respectively. The largest class (6/16) has 6 genotypic combinations of 2 dominant alleles. As the number of genes that controls the trait increases and the amount of phenotypic variation increases because of environmental factors, the phenotypes blend together, generating a continuous distribution.

As the number of loci that controls a trait increases, the number of phenotypic classes increases. As the number of classes increases, there is less phenotypic difference between each class. This means that there is a greater chance for environmental factors to override the small difference between classes. For example, exposure to sunlight can alter skin color and obscure genotypic differences.

The results of phenotypic measurements in polygenic inheritance are usually expressed in frequency diagrams. ▶ Figure 5.5 shows a frequency distribution for a polygenic trait.

▶ **FIGURE 5.4** The number of phenotypic classes in the F2 generation increases as the number of genes controlling the trait increases. This relationship allows geneticists to estimate the number of genes involved in expressing a polygenic trait. As the number of phenotypic classes increases, the distribution of phenotypes becomes a normal curve.

The Additive Model for Polygenic Inheritance

To explain how polygenes contribute to a trait and how genotypes contribute to variation in phenotypic expression, let's consider a model for polygenic inheritance. To simplify the analysis, the model is examined assuming the following conditions:

- The trait is controlled by several loci. We will assume three loci, each of which has two alleles (*A,a, B,b, C,c*).
- Each of the dominant alleles makes an equal contribution to the phenotype, and the recessive alleles make no contribution.
- The effect of each active (dominant) allele on the phenotype is small and additive.
- The genes controlling the trait are not linked; they assort independently.

This model will be applied to the inheritance of height. In this model, we assume that each dominant allele adds 3 in. to a base height of 5 ft. The recessive alleles *a*, *b*, and *c* add nothing to the base height. An individual who has the genotype *aabbcc* is 5 ft. tall, and someone who has the genotype *AABBCC* is 6 ft. 6 in. tall. If the environment acts equally on all genotypes, those who carry three of the six possible dominant alleles would represent the average height (5 ft. 9 in.).

Figure 5.6a shows the genotypic frequencies and their phenotypes. With three alleles, there are seven phenotypic classes, that have 6, 5, 4, 3, 2, 1, and 0 dominant alleles. In real-life situations, the action of environmental factors blurs the distinction between the phenotypic classes, producing the continuous variation in height that is actually observed (Figure 5.6b). The most frequent phenotype (that has three dominant alleles) produces the average height of 5 ft. 9 in. At either

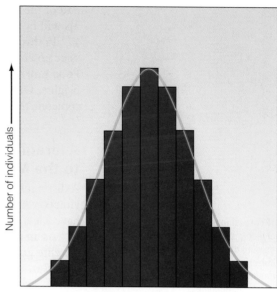

FIGURE 5.5 The results of phenotypic measurements in polygenic inheritance expressed as a frequency diagram, resulting from an interaction of polygenes with environmental factors.

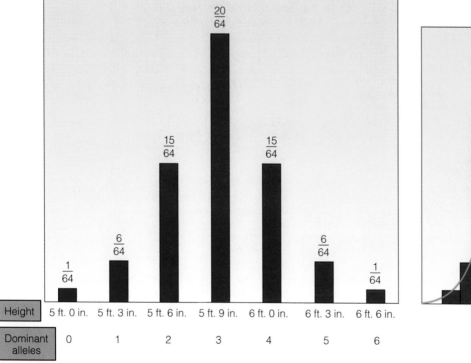

(a)

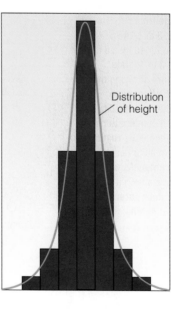

(b)

FIGURE 5.6 (a) The distribution of height in a three-gene model, where each dominant allele adds three inches above a base (homozygous recessive *aabbcc*) level of 5 ft. In this model, the phenotypic extremes are represented by 5 ft. and 6 ft. 6 in. (b) In reality, the interaction of genotypes with the environment produces a continuous distribution of phenotypes (height).

end of the phenotypic range, the frequencies decline, and only 1/64 of the individuals will be as short as 5 ft. or as tall as 6 ft. 6 in.

In this example and in polygenic traits generally, the genotype represents the genetic potential for height. Full expression of the genotype depends on the environment. Poor nutrition during childhood can prevent people from reaching their potential heights. On the other hand, optimal nutrition from birth to adulthood cannot make someone taller than the genotype dictates.

Averaging Out the Phenotype: Regression to the Mean

A distinguishing characteristic of polygenic traits is that most of the offspring of individuals who have a phenotype from one of the extremes (tall × tall, for example) will have a less extreme phenotype. This phenomenon is known as **regression to the mean**.

As an example, let's look at King Frederick William's attempt to breed giants for his elite guard unit. We will assume that this breeding program used individuals at least 5 ft. 9 in. tall. For simplicity, we'll also assume that the dominant alleles *A*, *B*, and *C* each add 3 in. above a base height of 5 ft. 9 in. and the recessive alleles *a*, *b*, and *c* add nothing above the base height. An individual who has the genotype *aabbcc* would be 5 ft. 9 in. tall, and an individual who has the genotype *AABBCC* would be 7 ft. 3 in. tall.

Suppose that a 6 ft. 9 in. member of the guard who carries the genotype *AabCC* mates with a 6 ft. 3 in. woman who has the genotype *AaBbcc*. The possible outcomes are diagrammed in ▶ Figure 5.7. In this case, there are four paternal and four ma-

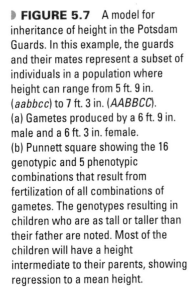

Regression to the mean In a polygenic system, the tendency of offspring of parents who have extreme differences in phenotype to exhibit a phenotype that is the average of the two parental phenotypes.

▶ **FIGURE 5.7** A model for inheritance of height in the Potsdam Guards. In this example, the guards and their mates represent a subset of individuals in a population where height can range from 5 ft. 9 in. (*aabbcc*) to 7 ft. 3 in. (*AABBCC*).
(a) Gametes produced by a 6 ft. 9 in. male and a 6 ft. 3 in. female.
(b) Punnett square showing the 16 genotypic and 5 phenotypic combinations that result from fertilization of all combinations of gametes. The genotypes resulting in children who are as tall or taller than their father are noted. Most of the children will have a height intermediate to their parents, showing regression to a mean height.

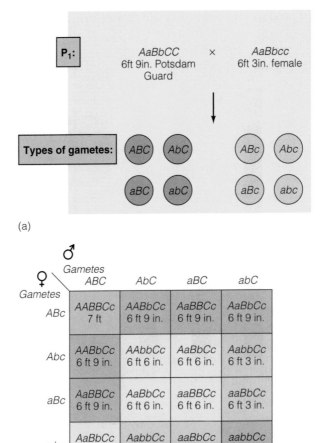

ternal gamete combinations, and 16 possible types of fertilizations with five phenotypic classes.

The phenotypic classes and ratio of possible offspring are diagrammed in ❱ Figure 5.8. As the king discovered (after waiting about 18 years for them to grow up), most of the children tended toward the average height (6 ft. 6 in.) between the two parents. In fact, 11 of the 16 genotypic combinations will result in children shorter than their father. In succeeding generations, further regression to the mean will occur. To make matters worse, many of the Potsdam Grenadier Guards were tall because of endocrine malfunctions and did not have the genotypes to produce tall offspring under any circumstances.

Regression to the mean is brought about not only by dominance and additive effects but also by gene interaction and environmental effects. If these other factors were included in our crosses and if the genotypes included individuals of average height (say, 5 ft. 4 in.), the regression toward the mean would be even more pronounced.

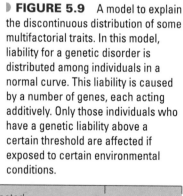

❱ **FIGURE 5.8** Frequency distribution of phenotypes from the possible offspring in Figure 5.7. Height of the offspring shows regression to the mean.

POLYGENES AND THE ENVIRONMENT: MULTIFACTORIAL TRAITS

We have looked at the interaction between genes and the environment in previous chapters but always in the context of single genes. In considering the interaction of polygenes and the environment, we begin by reviewing some basic concepts. Recall that the genotype represents the genetic constitution of an individual. It is fixed at the moment of fertilization and, barring mutation, is unchanging. The phenotype is the sum of the observable characteristics. It is variable and undergoes continuous change throughout the life of the organism. The environment of a gene includes all other genes in the genotype, their effects and interactions, and all nongenetic factors, whether physical or social, that can interact with the genotype (see Concepts and Controversies: Is Autism a Genetic Disorder?).

In assessing the interaction between the genotype and the environment, as in all science, you have to ask the right question. Suppose the question is posed as, "How much of a given phenotype in an individual is caused by heredity and how much by environment?" Because each individual has a unique genotype and has been exposed to a unique set of environmental conditions, it is impossible to evaluate quantitatively the phenotype's genetic and environmental components. Thus, for a given individual, the question as posed cannot be answered. However, in a later section we see that if the question is changed, it is possible to estimate the genotypic contribution to a phenotype.

Threshold Effects and the Expression of Multifactorial Traits

Although the degree of interaction between a genotype and the environment can be difficult to estimate, family studies indicate that such interactions do occur. Some multifactorial traits do not show a continuous distribution of phenotypes; individuals are either affected or not. Congenital birth defects, such as clubfoot or cleft palate, are examples of traits that are distributed discontinuously but are, in fact, multifactorial.

A model has been developed to explain the phenotypic expression of such traits. In this model, genotype frequencies are distributed in a bell-shaped curve, but only a limited number of genotypes express the phenotype (❱ Figure 5.9). This liability is caused by a number of genes, each of which contributes to the liability additively. Those individuals who have a liability above a threshold will develop the genetic disorder

❱ **FIGURE 5.9** A model to explain the discontinuous distribution of some multifactorial traits. In this model, liability for a genetic disorder is distributed among individuals in a normal curve. This liability is caused by a number of genes, each acting additively. Only those individuals who have a genetic liability above a certain threshold are affected if exposed to certain environmental conditions.

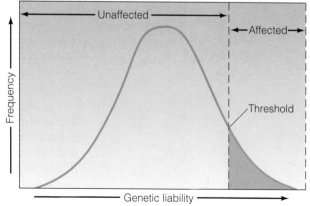

Is Autism a Genetic Disorder?

*A*utism is a developmental disorder characterized by impairment in social interactions, communication, and by narrow and stereotypical patterns of abilities. As depicted in the movie *The Rain Man*, symptoms can include aversion to human contact, language difficulties that show up as bizarre speech patterns, and repetitive body movements. These characteristics seem to be associated with malfunctions of the central nervous system. Symptoms usually appear before the age of 30 months in affected individuals. As outlined in Chapter 4, the information from pedigree construction is used to determine whether a trait is genetically determined, and to determine its mode of inheritance. Although these steps seem simple and clear-cut, in practice, the decisions are often more difficult. To illustrate these difficulties, we will briefly consider two questions of current interest in human genetics: Is autism a genetic disorder, and if so, how is this trait inherited?

Autism affects about 4 in every 10,000 individuals in the general population. There is a much higher frequency of autism in pairs of identical twins than in non-identical twins, and siblings of an autistic child are 75 times more likely to be autistic than are members of the general population. These observations indicate that autism has a strong genetic component. A team of researchers from UCLA and the University of Utah studied the incidence and inheritance of autism, using almost all the families in the state of Utah as a study group. In 187 families, there was a single autistic child, and in 20

families, there were multiple cases. In the multiple case families, simple recessive or dominant Mendelian inheritance does not easily explain the pattern of transmission.

The accuracy of pedigree studies can be affected by several factors. Autism is a behavioral trait, and the phenotype is not always clearly defined. In addition, there may be a number of different diseases that all produce a similar set of symptoms. Some cases may have symptoms that are too mild to be diagnosed, while very severe cases may result in prenatal death, all of which can skew the pedigrees.

To resolve these problems, several laboratories formed the International Molecular Genetic Study of Autism Consortium, and used recombinant DNA technology to search for genes that confer susceptibility to autism. Using pedigree analysis and molecular markers, this team identified loci on several chromosomes that may contain such genes. A locus on the long arm of chromosome 7 gave the most significant results; a region on the short arm of chromosome 16 is also promising, and genes on four other chromosomes may also be involved. These results, as well as those from twin studies are most consistent with a polygenic mode of inheritance for autism or a susceptibility to autism. The role of environmental factors in triggering autism also needs to be evaluated. Further work combining family studies with molecular markers and analysis of environmental factors will be needed to confirm the pattern of inheritance and to identify the genes responsible.

if exposed to certain environmental conditions (Figure 5.9). In other words, environmental conditions are most likely to have the greatest impact on genetically predisposed individuals.

The threshold model is useful in explaining the occurrence of certain disorders and congenital malformations. Evidence for such a threshold in any given disorder is indirect and comes mainly from family studies. The frequency of the trait among relatives of affected individuals is compared with the frequency of the trait in the general population. In a family, first-degree relatives (parent-child) have one-half of their genes in common, second-degree relatives (grandparent-grandchild) have one-fourth of their genes in common, and third-degree relatives (uncle-niece or uncle-nephew, aunt-niece or aunt-nephew) have one-eighth of their genes in common. As the degree of relatedness declines, so does the probability that individuals will share the same combination of alleles at multiple loci.

According to the threshold model, the risk for a disorder should also decrease as the degree of relatedness decreases. The distribution of family patterns for some congenital malformations, shown in Table 5.1, indicates a relationship of declining risk as the degree of relatedness declines. The multifactorial threshold model provides only indirect evidence for the effect of genotype on traits and for the degree of interaction between the genotype and the environment. The model is helpful, however, in genetic counseling for predicting recurrence risks in families that have certain congenital malformations and multifactorial disorders.

Table 5.1 Familial Risks for Multifactorial Threshold Traits

MULTIFACTORIAL TRAIT	Risk Relative to General Population			
	MZ Twins	First-Degree Relatives	Second-Degree Relatives	Third-Degree Relatives
Club foot	300×	25×	5×	2×
Cleft lip	400×	40×	7×	3×
Congenital hip dislocation (females only)	200×	25×	3×	2×
Congenital pyloric stenosis (males only)	80×	10×	5×	1.5×

Estimating the Interaction between Genotype and Environment

To assess accurately the degree of interaction between the genotype and the environment, we must examine the total variation in phenotype that is exhibited by a population of individuals rather than looking at individual members of the population. This phenotypic variation is derived from two sources: (1) the presence of different genotypes in members of the population and (2) the presence of different environments in which all the genotypes have been expressed. Assessing the role of these factors in producing phenotypic variability in a population is embodied in the concept known as **heritability**.

PHENOTYPIC VARIATION IS MEASURED BY HERITABILITY

Variation in phenotypic expression that results from different genotypes is known as **genetic variance**. Any variation in phenotype between individuals of the same genotype is known as **environmental variance**. The heritability of a trait, symbolized by H, is that proportion of the total phenotypic variance caused by *genetic* differences. Heritability is always a variable, and it is not possible to obtain an absolute heritability value for any given trait. The value obtained depends on several factors, including the population being measured and the amount of environmental variability at the time of measurement. Keep in mind that heritability is a population phenomenon and applies to groups, not to individuals. In other words, heritability is a statistical value (expressed as a percent) that defines the genetic contribution to the trait being analyzed in a population of related individuals (see below). In general, if the heritability value is high (it can range up to 100% when $H = 1.0$), the phenotypic variation is largely genetic, and the environmental contribution is low. If the heritability value is low (it can be as low as zero when $H = 0.0$), there is little genetic contribution to the observed phenotypic variation, and the environmental contribution is high.

Heritability Estimates Are Based on Genetic Relatedness

Heritability is calculated from observations made among relatives because we know the fraction of genes shared by related individuals. Parents and children share one-half their genes, grandparents and grandchildren share one-fourth their genes, and so forth. These relationships are expressed as **correlation coefficients**. The half-set of genes received by a child from its parent corresponds to a correlation coefficient of 0.5. The genetic relatedness of identical twins is 100% and is expressed as a correlation coefficient of 1.0. Unless a mother and father are related by descent, they should be genetically unrelated, and the correlation coefficient for this relationship is 0.0.

■ **Heritability** An expression of how much of the observed variation in a phenotype is due to differences in genotype.

■ **Genetic variance** The phenotypic variance of a trait in a population that is attributed to genotypic differences.

■ **Environmental variance** The phenotypic variance of a trait in a population that is attributed to differences in the environment.

■ **Correlation coefficient** A measure of the degree to which variables vary together.

The Biology of Twins

Sir Francis Galton (a cousin of Charles Darwin), the founder of eugenics, a movement to improve humans through selective breeding, pointed out in 1875 the value of studying twins to obtain information about the effects of environment on heredity. Before examining the results of such studies, we need to look briefly at the biology of twinning.

There are two types of twins, **monozygotic (MZ)** (identical) and **dizygotic (DZ)** (fraternal). Monozygotic twins originate from a single fertilization, a single egg fertilized by a single sperm. During mitosis at an early stage of development, two separate embryos are formed. Additional splitting is also possible (see Concepts and Controversies: Twins, Quintuplets, and Armadillos). This separation may take place when the zygote undergoes its first mitotic division or at any time up to the first two weeks of development (▶ Figure 5.13). Conjoined twins (so-called Siamese twins) are thought to arise from an incomplete division of the embryo. Because they arise from a single fertilization, MZ twins share the same genotype, have the same sex, and carry the same genetic markers, such as blood types. Dizygotic twins originate from two separate fertilizations: Two eggs, ovulated in the same menstrual cycle, are fertilized independently. DZ twins are no more related than other pairs of siblings, have

■ **Monozygotic (MZ) twins** Twins derived from a single fertilization involving one egg and one sperm; such twins are genetically identical.

■ **Dizygotic (DZ) twins** Twins derived from two separate and nearly simultaneous fertilizations, each involving one egg and one sperm. Such twins share, on average, 50% of their genes.

▶ **FIGURE 5.13** (a) Monozygotic (MZ) twins result from the fertilization of a single egg by a single sperm. After one or more mitotic divisions, the embryo splits in two and forms two genetically identical individuals. (b) Dizygotic (DZ) twins result from the independent fertilization of two eggs by two sperm during the same ovulatory cycle. Although these two embryos simultaneously share the same uterine environment, they share only about half of their genes.

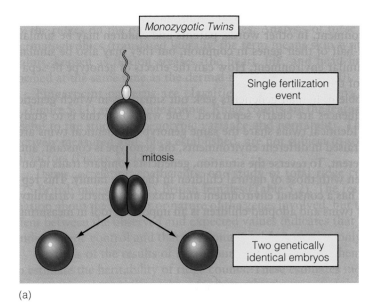

(a)

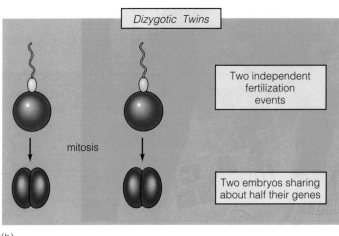

(b)

half of their genes in common, can differ in sex, and may have different genetic markers, such as blood types.

For heritability studies, it is essential to identify a pair of twins as MZ or DZ. An accurate diagnosis of twins as MZ or DZ is made only by extensive tests. Comparison of many traits that correlate absolutely between individuals can be used to identify twins as MZ. Divergence of one or more traits means that the twins are DZ. Among the characters used are blood groups, sex, eye color, hair color, fingerprints, palm and sole prints, DNA fingerprinting, and analysis of other DNA molecular markers.

Concordance and Twins

A simple method for evaluating phenotypic differences between twins is using traits that can be scored as present or absent rather than measured quantitatively. Twins show **concordance** if both have a trait and are discordant if only one twin has the trait. As noted, MZ twins have 100% of their genes in common, whereas DZ twins, on average, have 50% of their genes in common. For any genetically determined trait, the correlation in MZ twins should be higher than that in DZ twins. If the trait is completely heritable, the concordance should be 1.0 in MZ twins and close to 0.5 in DZ twins.

■ **Concordance** Agreement between traits exhibited by both twins.

In evaluating the results of twin studies, it is the degree of difference between concordance in MZ twins versus DZ twins that is important. The greater the difference, the greater the resultant heritability. Concordance values in twins for a variety of traits are listed in Table 5.3. Examination of the table shows that the concordance value for cleft lip in MZ twins is higher than that for DZ twins (42% versus 5%). Although this difference suggests a genetic component to this trait, the value is so far below 100% that environmental factors are obviously important in the majority of cases. In all cases, concordance values must be interpreted cautiously.

Concordance values can be converted to heritability values through a number of statistical methods. Heritability values derived in this manner are given in the last column of Tables 5.2 and 5.5. Remember that heritability is a relative value, valid only for the population measured and only under the environmental conditions in effect at the time of measurement. Heritability determinations made within one group cannot be compared with heritability measurements for the same trait in another group because the two groups differ in genotypes and environmental variables in unknown ways.

Table 5.3 Concordance Values in Monozygotic (MZ) and Dizygotic (DZ) Twins

Trait	Concordance Values (%)	
	MZ	DZ
Blood types	100	66
Eye color	99	28
Mental retardation	97	37
Hair color	89	22
Down syndrome	89	7
Handedness (left or right)	79	77
Epilepsy	72	15
Diabetes	65	18
Tuberculosis	56	22
Cleft lip	42	5

Twins, Quintuplets, and Armadillos

*M*onozygotic twins (MZ) are genetically identical because of the way in which they are formed. The process of embryo splitting that gives rise to MZ twins can be considered a form of human asexual reproduction. In fact, another mammal, the nine-banded armadillo, produces litters of genetically identical, same-sex offspring that arise by embryo splitting. In armadillo reproduction a single fertilized egg splits in two, and daughter embryos can split again, resulting in litters of two to six genetically identical offspring.

Multiple births in humans occur rarely. About 1 in 7500 births are triplets, and 1 in 658,000 births are quadruplets. In many cases, both embryo splitting and multiple fertilizations are responsible for naturally occurring multiple births. Triplets may arise by fertilization of two eggs, where one of them un-

dergoes embryo splitting. The use of hormones to enhance fertility has slightly increased the frequency of multiple births. These drugs work by inducing the production of multiple eggs in a single menstrual cycle. The resulting fertilizations have resulted in multiple births ranging from twins to septuplets.

Embryo splitting in naturally occurring births was documented in the Dionne quintuplets born in May 1934. This was the first case in which all five members of a set of quintuplets survived. Blood tests and physical similarities indicate that these quintuplets arose from a single fertilization followed by several embryo splits.

From this, it seems that MZ twins, armadillos, and the Dionne quintuplets have something in common. They all arise by embryo splitting.

---

available to identify genes that contribute to complex traits in humans (such as obesity). One of these methods is the use of animal model systems (such as the rat or mouse) to identify the contributing genes and to study their modes of action.

Recent breakthroughs in understanding how genes regulate body weight have come from studies in mice. It has been shown that the mutant forms of several genes in the mouse control body weight. The mouse mutants obese (*ob*) and diabetes (*db*) are both obese (▶ Figure 5.15). Using recombinant DNA techniques, these genes and their human equivalents have been isolated, mapped, and studied in detail. The *ob* gene encodes a weight-controlling hormone, **leptin** (from the Greek word for thin)

■ **Leptin** A hormone produced by fat cells that signals the brain and ovary.

▶ **FIGURE 5.15** The obese (*ob*) mouse mutant, shown on the left (a normal mouse is on the right), has provided many clues about the way weight is controlled in humans.

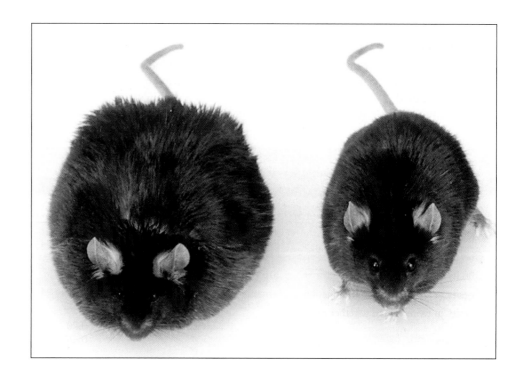

which is produced by fat cells. In mice, the hormone is released from fat cells and travels through the blood to the brain, where cells of the hypothalamus have leptin receptors encoded by the *db* gene. Binding of leptin activates the leptin receptor and initiates a response by the hypothalamus. This response may involve the production of an appetite-suppressing hormone, GLP-1 (glucagon-like protein-1). In normal individuals the control system may regulate the relative amounts of food converted into fat or muscle mass or alter the rate of energy consumption to maintain weight within a relatively narrow range (▶ Figure 5.16). Control of weight in mice may involve other genes, such as tubby (*tb*), which also has an obese phenotype.

The human gene for leptin (MIM/OMIM 164160), which is equivalent to the mouse *ob* gene, maps to chromosome 7q31.1. The gene for the leptin receptor (MIM/OMIM 601007), which is equivalent to the mouse *db* gene, maps to chromosome 1p31. Most obese people do not have defects in the *ob* gene and in fact overproduce leptin. The defect in these individuals seems to be in the receptor or in the control systems activated by the receptor. An understanding of how leptin and its receptor work to regulate weight will hopefully lead to the design of new drugs to treat obesity. Because Americans spend more than $30 billion dollars in weight control products every year, there is a large potential market for leptin and drugs that affect the leptin receptor. A biotech company has paid $20 million dollars for a license to develop drugs based on leptin, and these products may be on the market in a few years.

Although the gene for leptin may regulate some aspects of body weight, in humans, there are few examples of a specific genotype that is directly related to obesity. To search for genes controlling body weight, several groups have initiated genome-wide searches using large, multigenerational families and molecular markers, searching for linkage between the molecular markers and obesity. Using 92 families selected for obesity, one group has found positive evidence for linkage between obesity and one or more genes located on the long arm of chromosome 20. Other groups, using similar methods, have identified loci on chromosome 2 and the long arm of chromosome 11 that may control body weight. Further work on these regions may identify specific genes involved in obesity, and provide insight into the molecular events that regulate body weight.

A SURVEY OF SOME MULTIFACTORIAL TRAITS

Much of the phenotypic diversity in humans results from interactions between genes and environmental factors. Many traits involve a number of genes which, taken one by one, have only a small effect on the phenotype. Each of these genes can interact with the environment, producing a wide range of phenotypes and often obscuring the genetic components associated with a disease.

Some progress has been made in defining the genetic components of the multifactorial traits discussed below. New methods of screening for polygenes that have been developed in the last few years and the identification of genes in the Human Genome Project may help explain how genetic and environmental factors contribute to complex traits.

Cardiovascular Disease Has Genetic and Environmental Components

Two significant factors lead to cardiovascular disease: **hypertension** (MIM/OMIM 145500, high blood pressure) and **atherosclerosis** (MIM/OMIM 143890, deposition

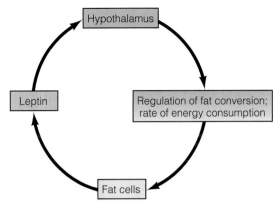

▶ **FIGURE 5.16** The hormone leptin is produced in fat cells, moves through the blood, and binds to receptors in the hypothalamus. Binding presumably activates a control mechanism (still unknown) that controls weight by regulating the conversion of food energy into fat and the rate of energy consumption. As fat levels become depleted, secretion of leptin slows and eventually stops.

SIDEBAR

Leptin and Female Athletes

Leptin, like some other hormones, may have multiple effects. In addition to signaling the hypothalamus about body fat levels, leptin may also signal the ovary. The obese strain of mice in which leptin was discovered has a mutated leptin gene, and these mice do not make any leptin. Females of this strain are infertile and do not ovulate. Injection of leptin into these females leads to ovulation and normal levels of fertility. The discovery that leptin receptors are found in the ovary indicates that leptin may act directly to control ovulation. In mammalian females, including humans, ovulation stops when body fat falls below a certain level. Many female athletes, such as marathon runners, have low levels of body fat and stop menstruating. Since leptin is produced by fat cells, females who have low levels of body fat may have lowered their leptin levels to the point where ovulation and menstruation cease.

■ **Hypertension** Elevated blood pressure, consistently above 140/90 mmHg.

■ **Atherosclerosis** Arterial disease associated with deposition of plaques on inner surface of blood vessels.

Table 5.6
Risk Factors for Cardiovascular Disease

Heredity (history of cardiovascular disease before age 55 in family members)

Being male

Hypertension

High blood cholesterol (high LDL and/or low HDL)

Smoking

Obesity

Lack of exercise

Stress

---

of plaque on artery walls). Both traits have significant environmental contributions (Table 5.6) but also have well-established genetic components.

Hypertension occurs when blood pressure (▶ Figure 5.17) is consistently above 140/90 mmHg (140 is the pressure generated when the heart ventricles contract, and 90 is the pressure when the ventricles are relaxed). At least 10 genes are involved in controlling blood pressure. All of them appear to work by controlling the amount of salt and water that is reabsorbed into the blood by the kidney. One of these is the gene for angiotensinogen (AGT). AGT (MIM/OMIM 106150) is a protein made in the liver and present in blood at a high concentration. In its active form, AGT controls salt and water retention, which in turn controls blood pressure. Some variants of this protein have been linked to a predisposition to hypertension, which is a silent killer since no obvious symptoms appear in early stages of the disease. Some 10-20%

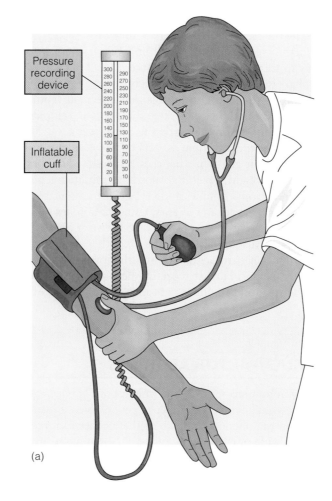

(a)

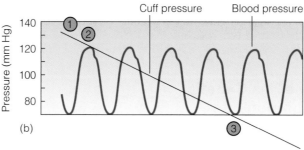

(b)

▶ **FIGURE 5.17** Blood pressure readings. (a) A blood pressure cuff is used to determine blood pressure. As shown in (b), blood pressure rises and falls as the left ventricle of the heart contracts and relaxes. To measure blood pressure, air is pumped into the cuff until it stops blood flow into the lower arm, and no sound can be heard (1) Air is gradually let out of the cuff, and when blood is heard flowing past the cuff, the pressure measured at this point is taken as the systolic pressure—the upper number in a blood pressure reading (2). As pressure in the cuff is gradually released, the artery becomes fully open, and no sound is heard. This is the diastolic pressure—the lower number in a blood pressure reading (3).

of the adult population of the United States suffers from hypertension, making it a serious health problem.

Atherosclerosis is the result of an imbalance between dietary intake and the synthesis, utilization, and breakdown of lipids, especially cholesterol. This imbalance can lead to blockage of blood vessels and the development of cardiovascular disease (▶ Figure 5.18). The most serious consequences come from blockages in the brain and heart, which cause strokes and heart attacks.

Cholesterol is needed to synthesize plasma membranes, steroid hormones (such as estrogen and testosterone), and as a component of bile. The liver synthesizes cholesterol, and it is not a necessary part of the diet. But the accumulation of cholesterol in arterial plaque is an important factor in cardiovascular disease.

Large lipid molecules, such as cholesterol, are not soluble in blood plasma and are wrapped in a coat of proteins and phospholipids for transport. The coat and its

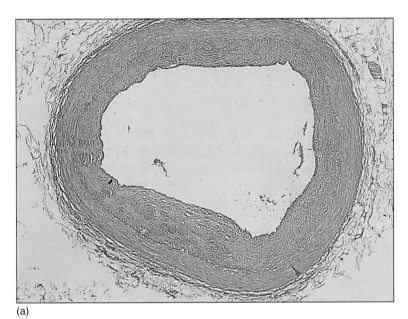

(a)

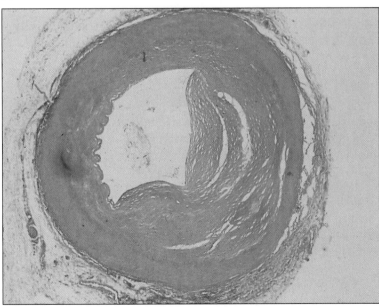

(b)

▶ **FIGURE 5.18** (a) A cross section of a normal artery. (b) A cross section of an artery partially blocked by atherosclerotic plaque. As excess cholesterol accumulates in the body, it accumulates in plaques, leading to cardiovascular disease.

Lipoproteins Particles that have protein and phosopholipid coats that transport cholesterol and other lipids in the bloodstream.

Familial hypercholesteremia (FH) Autosomal dominant disorder with defective or absent LDL receptors. Affected individuals are at increased risk for cardiovascular disease.

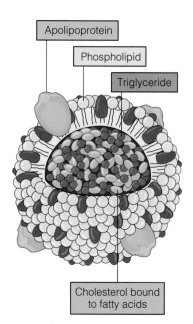

▶ **FIGURE 5.19** Lipoproteins carry cholesterol and other lipids through the blood enclosed in a protein and phosopholipid coating. There are several types of lipoproteins that have different proportions of lipids. LDLs are composed of about 45% cholesterol, and HDLs are made of about 20% cholesterol. High levels of LDLs and a low level of HDLs are risk factors for cardiovascular disease.

contents are known as **lipoproteins** (▶ Figure 5.19). Lipoproteins are classified by their size and density. Cholesterol is carried by both low-density (LDL) and high-density lipoproteins (HDL). LDLs are about 45% cholesterol and HDLs are about 20% cholesterol. LDLs transport cholesterol from the liver to tissues in the body for utilization and breakdown. HDLs transport cholesterol to the liver. The risk of atherosclerosis related to the HDL/total cholesterol ratio. Higher HDL levels mean lower risk.

Genetic studies of cardiovascular disease are directed at finding genetic differences that predispose to disease, identifying environmental factors (such as exercise) that affect progression, and detecting individuals at risk. Many genes, including those that encode apolipoproteins (the part of the lipoprotein that binds to cell receptors) control cholesterol levels in the body. Other genes encode cell receptors that bind to and internalize lipoproteins and enzymes that degrade lipoproteins inside the cell. Mutations in one of the receptor genes are discussed in the next section.

The autosomal dominant disease **familial hypercholesterolemia** (MIM/OMIM 143890) is caused by defective cell surface receptors that regulate the uptake of LDLs. Affected heterozygotes have elevated cholesterol levels in their blood serum and usually develop coronary artery disease between the ages of 40–50. Two general classes of mutants are known: defective receptors and absent receptors. Several types of defective receptors are known, including one that cannot recognize and bind to LDLs, another with reduced ability to bind LDLs, and a third that recognizes and binds LDL but is unable to move the LDL into the cell. Altogether, more than 150 mutations of the gene on chromosome 19 that encodes the LDL receptor have been identified. Each of these mutations results in elevated levels of LDL-derived cholesterol in the blood serum and the deposition of plaques on the inner surface of arteries, leading to premature heart disease.

The frequency of heterozygotes for familial hypercholesterolemia in European, Japanese, and U.S. populations is about 1 in 500, although in regions of Quebec the frequency is 1 in 122. The highest reported frequency is in a community of South Africa, where the frequency is 1 in 71. The average frequency of 1 in 500 makes this disorder one of the most common genetic disorders known, and it is one of the major causes of cardiovascular disease.

Mutations in other genes that control aspects of lipid metabolism are also involved in cardiovascular disease. A recently identified gene, atherosclerosis susceptibility (ATHS), has been mapped to chromosome 19 near the locus for the LDL receptor. ATHS (MIM/OMIM 108725) is associated with an increase in LDL levels in the blood, a decrease in HDL levels, and a threefold increase in heart attack risk.

Skin Color Is a Multifactorial Trait

Charles and Gertrude Davenport, leading figures in the American eugenics movement, first tested the theory of polygenic inheritance in humans. Between 1910 and 1914, they collected information on skin color in black-white marriages in Bermuda and in the Caribbean. Environmental factors also play a role in skin color, making it a multifactorial trait. Exposure to the sun can darken skin color and obscure genotypic differences. The Davenports were aware of this and measured skin color under the upper arms of their subjects.

To measure skin color, they used a top that had a disk composed of colored sectors of various sizes (▶ Figure 5.20). The colors used were black, white, red, and yellow. When the top was spun, the colored sectors blended together and produced a color that could be matched to a given skin color by changing the size of the black-and white-colored sectors. Based on the size of the black sector, individuals were assigned to one of five categories, numbered 0 to 4.

The results of the Davenports' study illustrate several properties of polygenic traits. The offspring (F1) of such marriages have skin color values intermediate be-

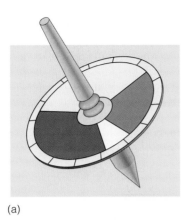

(a)

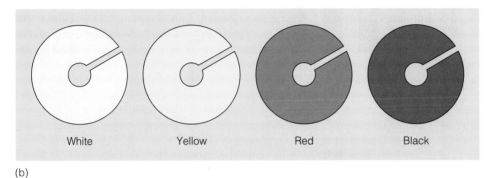

| White | Yellow | Red | Black |

(b)

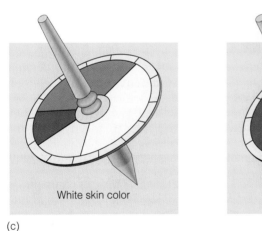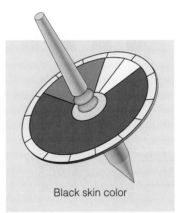

White skin color   Black skin color

(c)

▶ **FIGURE 5.20** (a) The top used by the Davenports to measure skin color with overlaid disks. The color produced when the top is spun can be changed by changing the size of the sectors in the colored disks stacked together on the top. (b) The disks used to measure skin color. Disks of different amounts of black and white can be mounted on the top to produce a color that matches that of the person being studied.
(c) Arrangement of sectors that produce skin colors characteristic of those persons studied by the Davenports.

tween those of their parents. In the F2 generation, a small number of children were as white as one grandparent, a small number were as black as the other grandparent, and most were distributed between these extremes (▶ Figure 5.21). Because individuals in the F2 could be grouped into five classes, the Davenports hypothesized that two gene pairs control skin color. Each class represented a genotype that resulted from the segregation and assortment of two gene pairs. Suppose that these genes are *A* and *B,* respectively. Class 0 that has the lightest skin color represents the genotype *aabb,* Class 1 has the genotype *Aabb* or *aaBb,* and so forth, up to Class 5, which is homozygous dominant (*AABB*) and has the darkest skin.

Later work by other investigators, using a more sophisticated instrument that measures the reflection of light from the skin surface (a reflectometer), has shown that skin color is actually controlled by more than two gene pairs. The current estimate is that between three and six gene pairs are involved. The data are most consistent with a model that involves three or four genes (▶ Figure 5.22).

FIGURE 5.21 Frequency diagrams of skin colors obtained by the Davenports. (a) Skin color distribution in the parents falls into two discontinuous classes. (b) Color values of seven children from the parents in (a) are intermediate to those of their parents. (c) Skin colors of 32 children of the parents in (b) Color values range from one phenotypic extreme to the other, and most are clustered around a mean value. This normal distribution of phenotypes is characteristic of a polygenic trait.

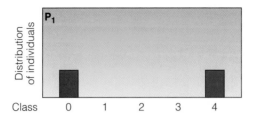

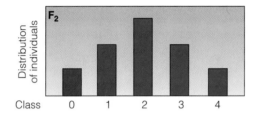

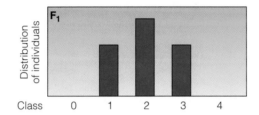

(a)

(b)

(c)

FIGURE 5.22 Distribution of skin color as measured by a reflectometer at a wavelength of 685 nm. The results are shown for an additive model of skin color, with environmental effects, for 1-4 gene pairs. Distributions observed in several populations indicate that 3 or 4 gene pairs control skin color.

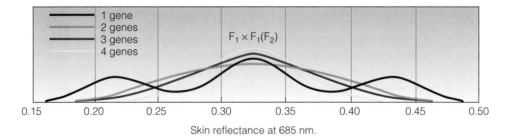

Intelligence and IQ: Are They Related?

The idea that intelligence is a distinct phenotype that can be quantitatively measured arose in the late eighteenth and early nineteenth centuries. Phrenologists believed that the brain is composed of a series of compartments, each of which controls a single function, such as musical ability, courage, or intelligence. Because each area of the brain has an assigned function, they believed that an examination of the shape and size of a particular region of the skull (Figure 5.23) can determine the capacities of an individual. Later, many scientists related intelligence to overall brain size. Then craniometry, the measurement of brain size (by weight or volume), became the dominant means of assessing intelligence. Large brain size was associated with high intelligence, and small brain size with lower intelligence.

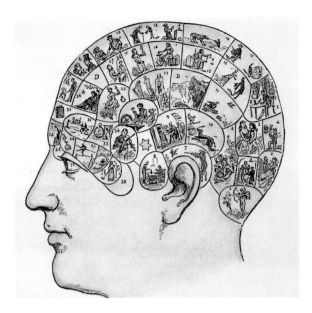

▶ **FIGURE 5.23** Phrenology model showing areas of the head overlying brain regions that control different traits. Intelligence was estimated by measuring the area of the skull overlying the region of the brain thought to control this trait.

At the turn of the 20th century, Alfred Binet, a French psychologist turned to psychological rather than physical methods for measuring intelligence. To identify children whose classroom performance indicated a need for special education, he developed a graded series of tasks related to basic mental processes, such as comprehension, direction (sorting), and correction. Each child began by performing the simplest tasks and progressing in sequence until the tasks became too difficult. The age assigned for the last task performed became the child's mental age, and the intellectual age was calculated by subtracting the mental age from the chronological age. Binet's test became the basis of the Stanford-Binet intelligence tests in use today.

Wilhelm Stern, another psychologist, divided mental age by chronological age, and the number became known as the **intelligence quotient,** or **IQ.** If a child of 7 years (chronological age) were able to successfully perform tasks for a 7-year-old but not tasks for an 8-year-old, a mental age of 7 would be assigned. To determine the IQ for this child, divide mental age by chronological age:

Mental age = 7 divided by chronological age = 7 = 1.0 × 100 = 100 = IQ

▪ **Intelligence quotient (IQ)** A score derived from standardized tests that is calculated by dividing the individual's mental age (determined by the test) by his or her chronological age and multiplying the quotient by 100.

Multiply the quotient by 100 to eliminate the decimal point, and we obtain an IQ of 100 as the average for any given age.

The substitution of psychological for physical methods to measure intelligence does not change the underlying assumption that intelligence has a biological basis and can be quantitatively expressed as a single number. In fact, the use of IQ tests has strengthened the assumption that IQ measures a fundamental genetically determined physiological or biochemical property of the brain related to intelligence. The question is whether psychological methods, such as IQ tests, can measure intelligence any more accurately than the discredited physical methods of phrenology or craniometry.

To determine whether IQ tests or any other method accurately measure intelligence, first we must define intelligence so that it can be objectively measured, in the same way we measure height, weight, or fingerprint ridge counts. Properties such as abstract reasoning, mathematical skills, verbal expression, ability to diagnose and to solve problems, and creativity are often cited as important components of intelligence. Unfortunately, there is no evidence that any of these properties are directly measured by an IQ test, and there is no objective way to quantitate such components of intelligence.

Dozens of distinct neural processes may be involved in mental functions, many of which may contribute to intelligence. In spite of the rapid advances in neurobiology, little or no information is available about the biological basis of any of these processes. Without knowledge of the mechanisms involved, it is difficult to quantitatively measure the outcome of these mechanisms.

The values obtained in IQ measurements, however, do have significant heritable components. The evidence that IQ has genetic components comes from two areas: studies that estimate IQ heritability, and comparison of IQs in groups of individuals raised together (unrelated individuals, parents and children, siblings, and MZ and DZ twins) and individuals raised separately (unrelated individuals, siblings, and MZ twins). Heritability estimates for IQ range from 0.6 to 0.8. The high correlation observed for MZ twins raised together indicates that genetics plays a significant role in determining IQ (▶ Figure 5.24). But rearing MZ twins apart or raising siblings in different environments significantly reduces the correlation and provides evidence that the environment has a substantial role in determining IQ.

The Controversy about IQ and Race

The assumption that intelligence is solely determined by biological factors coupled with the misuse or misunderstanding of the limits of heritability estimates have led to the conclusion that differences in IQ among different racial and ethnic groups are genetically determined. On standardized IQ tests, blacks score an average of 15 points lower than the average white's score of 100. These differences are consistent across different tests, and the scores themselves do not seem to be a serious issue. The controversy is over what causes these differences. Are such differences genetic in origin, do they reflect environmental differences, or are both factors at work? If both, to what degree does inheritance contribute to the differences? The debate about these questions has been renewed by the book *The Bell Curve* by Herrnstein and Murray.

Heritability studies have been used to support the argument that intelligence is mainly innate and inherited, citing heritability values of 0.8 for intelligence. In most cases, however, the reasoning used to support this argument misuses the concept of heritability. Recall that a measured heritability of 0.8 (for example) means that 80% of the phenotypic variation being measured is due to genetic differences within that population. Differences observed between two populations in heritability for a trait simply cannot be compared because heritability measures only variation within a population at the time of measurement and cannot be used to estimate genetic vari-

▶ **FIGURE 5.24** A graphical representation of correlations in IQ measurements in different sets of individuals. The expected correlation coefficients are determined by the degree of genetic relatedness in each set of individuals. The vertical line represents the median correlation coefficient in each case.

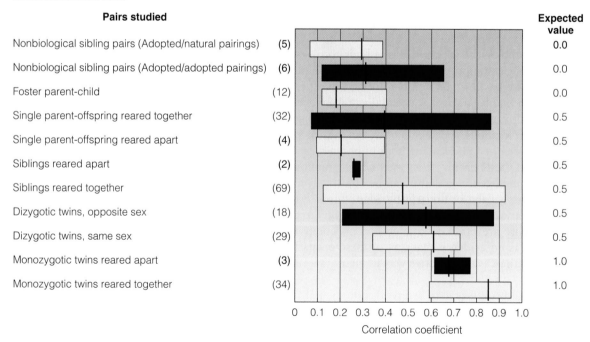

Pairs studied		Expected value
Nonbiological sibling pairs (Adopted/natural pairings)	(5)	0.0
Nonbiological sibling pairs (Adopted/adopted pairings)	(6)	0.0
Foster parent-child	(12)	0.0
Single parent-offspring reared together	(32)	0.5
Single parent-offspring reared apart	(4)	0.5
Siblings reared apart	(2)	0.5
Siblings reared together	(69)	0.5
Dizygotic twins, opposite sex	(18)	0.5
Dizygotic twins, same sex	(29)	0.5
Monozygotic twins reared apart	(3)	1.0
Monozygotic twins reared together	(34)	1.0

Correlation coefficient

ation between populations. In other words, we cannot use heritability differences between groups to conclude that there are genetic differences between those groups.

It is quite evident that both genetic and environmental factors make important contributions to intelligence. Clearly, the relative amount each contributes cannot be measured accurately at this time. Several points about this debate should be kept in mind. First, it is clear that IQ test scores cannot be equated with intelligence. Second, IQ scores are not fixed and can be changed significantly by training in problem solving and, in fact, change somewhat throughout the life of an individual. Variation in IQ scores is quite wide, and measurements in one racial or ethnic group overlap those of other groups.

The problem in discussing the differences in IQ scores arises when the quantitative differences in scores are converted into qualitative judgments used to rank groups as superior or inferior. Genetics, like all sciences, progresses by formulating hypotheses that can be rigorously and objectively tested.

Searching for Genes That Control Intelligence

As discussed in the section on obesity, heritability studies cannot provide information about the number, location, or identity of genes involved in intelligence. To learn about the genes themselves, scientists are using an expanded definition of intelligence that goes beyond IQ and recognizes that many genes are involved in normal cognitive and intellectual function, defined as **general cognitive ability.** Two approaches being used to identify genes associated with intelligence are studies of animal model systems and work on human genes using recombinant DNA techniques. Animal models, including the fruit fly, *Drosophila,* and the mouse have been used to identify single genes that control aspects of learning, memory, and spatial perception. After these genes have been mapped and characterized, the equivalent genes in humans are being identified and mapped.

A second approach uses techniques of molecular genetics and recombinant DNA to identify genes that affect specific polygenic traits, such as reading ability and IQ. These polygenes are also known as **quantitative trait loci** or **QTLs** (and are discussed in more detail in Chapter 16). The search for QTLs has uncovered genes associated with reading disability (developmental dyslexia) on chromosome 6 (MIM/OMIM 600202) and chromosome 15. More recently, a gene associated with the inheritance of IQ has been identified on chromosome 6. Using approaches that combine behavioral genetics, twin studies, and molecular genetics, more genes controlling aspects of cognitive ability are likely to be identified in the near future. As results from the Human Genome Project become available, it will become easier to define the number and actions of genes involved in higher mental processes and provide insight into the genetics of intelligence. Once such genes are identified and isolated, fears may be raised about genetic discrimination against individuals who have certain genotypes and the possibility of using gene therapy to enhance intelligence. These issues and others related to the Human Genome Project are discussed in Chapter 13.

General cognitive ability
Characteristics that include verbal and spatial abilities, memory, speed of perception and reasoning.

Quantitative trait loci (QTLs) Two or more genes that act on a single polygenic trait.

Case Studies

CASE 1

Sue and Tim were referred for genetic counseling after they inquired about the risk of having a child with a cleft lip. Tim was born with a mild cleft lip that was surgically repaired. He expressed concern that his future children could be at risk for a more severe form of clefting. Sue was in her 12th week of pregnancy and both were anxious about this pregnancy because Sue had a difficult time conceiving. The couple stated that they would not consider terminating the pregnancy for any reason, but wanted to be prepared for the possibility of having a child with a birth defect. The genetic counselor took a three-generation family history from both Sue and Tim and found that Tim was the only person to have had a cleft lip. Sue's family history showed no cases of cleft lip. Tim and Sue had several misconceptions about how clefting occurs and the genetic counselor spent time explaining how cleft lips occur and some of the known causes of this birth defect. Outlined below is a summary of the counselor's discussion with this couple.

- Fathers, as well as mothers, can pass on genes that cause clefting.
- Some clefts are caused by environmental factors, which means the condition didn't come from the father or the mother.
- One child in 33 is born with some sort of birth defect. One in 700 is born with a cleft-related birth defect.
- Most clefts occur in boys; however, a girl can be born with a cleft.

- If a person (male or female) is born with a cleft, the chances of that person having a child with a cleft, given no other obvious factor, is seven in one hundred.
- Some clefts are related to identifiable syndromes. Of those, some are autosomal dominant. A person with an autosomal dominant gene has a 50% probability of passing the gene to an offspring.
- Many clefts run in families even when there does not seem to be any identifiable syndrome present.
- Clefting seems to be related to ethnicity, occurring most often among Asians, Latinos, and Native Americans (1:500); next most often among persons of European ethnicity, (1:700) and least often among persons of African origins (1:1,000).
- A cleft condition develops during the 4th to the 8th week of pregnancy. After that critical period, nothing the mother does can cause a cleft. Sometimes a cleft develops even before the mother is aware that she is pregnant.
- Women who smoke are twice as likely to give birth to a child with a cleft. Women who ingest large quantities of Vitamin A or low quantities of folic acid are more likely to have children with cleft.
- In about 70% of cases, the fetal face is clearly visible using ultrasound. Facial disorders have been detected at the 15th gestational week of pregnancy. Ultrasound can be precise and reliable when diagnosing fetal craniofacial conditions.

CASE 2

Louise was an active 27-year-old gymnastic instructor at a local YMCA. She was recently accepted into law school and worked part time in the evenings at the local Gap clothing store. She was very busy and therefore thought nothing of her increasing fatigue until it started to affect her daily activities. She also complained of occasional dizziness, difficulty hearing, constipation, and problems with controlling her bladder. Her general practitioner conducted a series of blood tests to determine what was wrong. All tests showed she was in perfect health but her symptoms were becoming progressively worse. Over a period of months, she was forced to quit teaching gymnastics and was soon confined to a wheelchair. Her doctor finally referred her for a genetic evaluation. The clinical geneticist immediately recognized Louise's symptoms as signs of multiple sclerosis. The geneticist explained that fatigue, almost to the point of being disabling, is the most common symptom of multiple sclerosis (MS). Some medications could help her fatigue but their effect may not last. The best cure for fatigue is to listen to her body and just rest whenever possible.

The geneticist warned her of other possible symptoms of MS, which include: 1) numbness, which is most likely a direct result of the destruction of myelin in the nerves; 2) tingling or pins and needles in her feet (the same feeling you get when your foot "falls asleep"), for which not much can be done other than to massage the affected area and rest; 3) tremors, usually of the hands, also due to myelin destruction; 4) muscle spasms (when a muscle goes into sustained or temporary contractions) in the arms, legs, abdomen, back . . . just about anywhere; 5) depression and mood swings; 6) memory problems; 7) loss of balance with or without dizziness; and 8) bowel and bladder problems. MS is the most common autoimmune disease involving the nervous system, affecting approximately 250,000 individuals in the United States. The cause of MS is unknown but some studies have suggested that the risk to a first-degree relative (sibling, parent, or child) of a patient with MS is at least 15 times that for a member of the general population. Unfortunately, no definite genetic pattern is discernible.

Summary

1. In this chapter we considered genes that affect traits additively or quantitatively. The pattern of inheritance that controls metric characters (those that can be measured quantitatively) is called polygenic inheritance because two or more genes are usually involved.

2. The distribution of polygenic traits through the population follows a bell-shaped, or normal, curve. Parents whose phenotypes are near the extremes of this curve usually have children whose phenotypes are less extreme than the parents' and are closer to the population mean. This phenomenon, known as regression to the mean, is characteristic of systems in which the phenotype is produced by the additive action of many genes.

3. Variations in the expression of polygenic traits are often due to the action of environmental factors. Because of this, organisms that share a common genotype show a range of phenotypic variation.

4. The impact of environment on genotype can cause genetically susceptible individuals to exhibit a trait discontinuously even though there is an underlying continuous distribution of genotypes for the trait.

5. The degree of phenotypic variation produced by a genotype in a given population can be estimated by calculating the heritability of a trait. Heritability is estimated by observing the amount of variation among relatives having a known fraction of genes in common. MZ twins have 100% of their genes in common and when raised in separate environments, provide an estimate of the degree of environmental influence on gene expression.

6. Heritability is a variable, validly calculated only for the population under study and the environmental condition in effect at the time of the study. It provides an estimate of the degree of genetic variance within a population. Heritability values cannot be compared between populations because of differences in genotypes and environmental factors.

7. In twin studies, the degree of concordance for a trait is compared in MZ and DZ twins reared together or apart. MZ twins result from the splitting of an embryo produced by a single fertilization, whereas DZ twins are the products of multiple fertilization. Although twin studies can be useful in determining whether a trait is inherited, they cannot provide any information about the mode of inheritance or the number of genes involved.

8. Many human traits are controlled by polygenes, including skin color, intelligence, and aspects of behavior. The genetics of these traits has often been misused and misrepresented for ideological or political ends.

Questions and Problems

1. Describe why continuous variation is common in humans and provide examples of such traits.

2. The text outlines some of the problems Frederick William I encountered in his attempt to breed tall Potsdam Guards.
   a. Why were the results he obtained so different from those obtained by Mendel with short and tall pea plants?
   b. Why were most of the children shorter than their tall parents?

3. What role might the environment have played in causing Frederick William's problems, especially at a time when nutrition varied greatly from town to town and from family to family?

4. Do you think his experiment would have worked better if he had ordered brother-sister marriages within tall families instead of just choosing the tallest individuals from throughout the country?

5. As it turned out, one of the tallest Potsdam Guards had an unquenchable attraction to short women. During his tenure as guard he had numerous clandestine affairs. In each case children resulted. Subsequently, some of the children, who had no way of knowing that they were related, married and had children of their own. Assume that two pairs of genes determine height. The genotype of the 7-foot-tall Potsdam Guard was $A'A'B'B'$, and the genotype of all of his 5-ft. clandestine lovers was $AABB$, where an A' or B' allele adds 6 inches to the base height of 5 foot conferred by the $AABB$ genotype.
   a. What were the genotypes and phenotypes of all of the F1 children?
   b. Diagram the cross between the F1 offspring, and give all possible genotypes and phenotypes of the F2 progeny.

6. A clubfoot is a common congenital birth defect in the Smith family. This defect is caused by a number of genes but appears to be phenotypically distributed in a non-continuous fashion. Geneticists use the multifactorial threshold model to explain the occurrence of this defect. Explain this model. Explain predisposition to the defect in an individual who has a genotypic liability above the threshold versus an individual who has a liability below the threshold.

7. Describe why there is a fundamental difference between the expression of a trait that is determined by polygenes and the expression of a trait that is determined monogenetically.

8. Define genetic variance.

9. Define environmental variance.

10. How is heritability related to genetic and environmental variance?

11. Why are relatives used in the calculation of heritability?

12. Can conjoined (Siamese) twins be dizygotic twins given the theory that conjoined twins arise from incomplete division of the embryo?

13. Dizygotic twins
    a. are as closely related as monozygotic twins
    b. are as closely related as non-twin siblings
    c. share 100% of their genetic material
    d. share 50% of their genetic material
    e. none of the above

14. Discuss the difficulties in attempting to determine whether intelligence is genetically based.

15. Why are monozygotic twins, reared apart, so useful in the calculation of heritability?

16. Monozygotic (MZ) twins have a concordance value of 44% for a specific trait, whereas dizygotic twins have a concordance value of less than 5% for the same trait. What could explain why the value for MZ twins is significantly less than 100%?

17. What is the importance of the comparison of traits between adopted and natural children in determining of heritability?

18. If monozygotic twins show complete concordance for a trait whether they are reared together or apart, what does this suggest about the heritability of the trait?

19. In this chapter you read about an obesity study conducted in 1974 in which MZ and DZ twins who served in the armed forces were studied at military induction and 25 years later. Results indicated that obesity has a strong genetic component.
    a. What are the problems with this study?
    b. Design a better study to test whether obesity has a genetic component.

20. What does the *ob* gene code for? How does it work? Is this just a gene found in animals, or do humans have it also?

21. If there is no genetic variation within a poulation for a given trait, what is the heritability for the trait in the population?

22. Is cardiovascular disease (hypertension, atherosclerosis and familial hypercholesterolemia) genetic?

23. At the age of nine years, your genetics instructor was able to perform the mental tasks of an 11 year-old. According to Wilhelm Stern's method, calculate his or her IQ.

24. Suppose that a team of researchers analyzes the heritability of SAT scores and assigns a heritability of 0.75 for this skill. This team also determines that a certain ethnic group has a heritability value that is 0.12 lower when compared to other ethnic groups. The group draws the conclusion that there must be a genetic explanation for the differences in scores. Why is this an invalid conclusion?

Internet Activities

The following activities use the resources of the World Wide Web to enhance the topics covered in this chapter. To investigate the topics described below, log on to the book's homepage at:

http://www.brookscole.com/biology

1. Body weight is an example of a multifactorial trait that results from the interaction between multiple genes and the environment. The PBS series Frontline has an Internet site addressing what makes people *Fat* and how fatness is perceived in our society. Go to the site and read the *experts' comments* regarding body weight, especially those comments relating to the various causes—cultural, societal, and genetic—of fatness in people. Does the fact that body weight is largely caused by genetic factors affect your personal feeling about obesity? Certain genes have now been cloned that may play a role in human obesity. Would you support the prospect of correcting mutant genes in humans to prevent, or reverse, obesity? Why or why not? Also, follow one of the links to a chapter in Richard Klein's book *Eat Fat*. Fatness, and even the word fat, used to have very positive connotations, yet this is generally untrue today. Why do you think popular perceptions of fat have changed so much over time? Do you think these perceptions may ever revert to conform with the old value system?

2. Multifactorial traits are those that interact strongly with environmental factors. Two such traits are handedness (MIM/OMIM 139900) and longevity (MIM/OMIM 152430). Access these listings in the OMIM database, and read one or more of the references on the multifactorial nature of these traits. Have twin studies played a role in identifying the genetic components of these traits? If a trait is controlled by more than one gene, and each gene contributes only a small amount to the phenotype, how can such genes be identified and mapped?

For Further Reading

Aldhous, P. (1992). The promise and pitfalls of molecular genetics. *Science* 257: 164–165.

Balmor, M. G. (1970). *The Biology of Twinning in Man.* Oxford, England: Clarendon.

Baringa, M. (1996). Researchers nail down leptin receptor. *Science* 271: 913.

Benirschke, K. (1972). Origin and clinical significance of twinning. *Clin. Obstet. Gynecol.* 15: 220–235.

Bouchard, C., Tremblay, A., Despres, J.-P., Nadeau, A., Lupien, P. J., Theriault, G., Dussault, J., Moorjani, S., Pinault, S., & Fournier, G. (1990). The response to long-term overfeeding in identical twins. *New Engl. J. Med.* 322: 1477–1481.

Bouchard, C., & Aderusse, L. (1993). Genetics of obesity. *Ann. Rev. Nutrition* 13: 337–354.

Bouchard, T. (1994). Genes, environment and personality. *Science* 264: 1700–1701.

Feldman, M. W., & Lewontin, R. (1975). The heritability hangup. *Science* 190: 1163–1166.

Harrison, G. A., & Owens, J. J. T. (1964). Studies on the inheritance of human skin color. *Ann. Hum. Genet.* 28: 27–37.

Horgan, J. (1995). Get smart, take a test. *Sci. Am.* 273: 12–14.

International Molecular Genetic Study of Autism Consortium. (1998). A full genome screen for autism with evidence for linkage to a region on chromosome 7q. *Human Mol. Genet.* 7: 571–578.

Lee, J., Reed, D., Li, W-D., Xu, W., Joo, E-J., Kilker, R., Nanthakumar, E., et al. (1999). Genome scan for human obesity and linkage to markers in 20q13. *Am. J. Hum. Genet.* 64: 196–209.

Mackintosh, N. J. (1986). The biology of intelligence? *Br. J. Psychol.* 77: 1–18.

Moll, P., Burns, T., & Laver, R. (1991). The genetic and environmental sources of body mass index variability: The Muscatine ponderosity family study. *Am. J. Hum. Genet.* 49: 1243–1255.

Sorenson, T. I., Price, R. A., Stunkard, A. J., & Schulsinger, F. (1989). Genetics of obesity in adult adoptees and their biological siblings. *Br. Med. J.* 298: 87–90.

Stern, C. (1970). Model estimates for the number of gene pairs involved in pigmentation variability in Negro-Americans. *Hum. Heredity* 20: 165–168.

Stunkard, A. J., Foch, T., & Hrubec, Z. (1986). A twin study of human obesity. *JAMA* 256: 51–54.

Stunkard, A. J., Harrus, J. R., Pedersen, N. L., & McClearn, G. E. (1990). The body-mass index of twins who have been reared apart. *New Engl. J. Med.* 322: 1483–1487.

CHAPTER

6

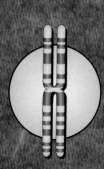

Cytogenetics

*I*n 1953 Jerome Lejeune, a French physician, began working on a condition first described in 1866 by John Langdon Down, an English physician. Affected individuals have a distinctive physical appearance and are mentally retarded. Lejeune suspected that there was a genetic link to this phenotype.

He began by studying the fingerprints and palm prints of children who had this condition, known as Down syndrome, and comparing them with the prints of unaffected children. The prints from Down syndrome children showed a high frequency of abnormalities. Because fingerprints and palm prints are laid down very early in development, they serve as a record of events that take place early in embryogenesis. From his studies on the disturbances in the print patterns of affected children, Lejeune became convinced that such significant changes in print patterns were probably not caused by the action of only one or two genes. Instead he reasoned that many genes must be involved. Perhaps even an entire chromosome might play a role in Down syndrome.

It was a logical step to examine the chromosomes of Down syndrome children, but Lejeune lacked access to the proper equipment and techniques required. In 1957, in cooperation with a colleague, he began culturing cells from Down syndrome children. Lejeune examined the chromosomes in these cells using a microscope that had been discarded by the bacteriology laboratory in the hospital where he worked. He repaired the instrument by inserting a foil wrapper from a candy bar into the gears so that the image could be focused. His chromosome counts indicated that the cells of Down syndrome individuals contained 47 chromosomes, whereas those of unaffected individuals contained 46 chromosomes. He and his colleagues published a short paper in 1959 in which they reported that Down syndrome is caused by the presence of an extra chromosome. This chromosome was later identified as chromosome 21.

This remarkable discovery was the first human chromosome abnormality to be identified and marked an important turning point in human genetics. The discovery made clear that genetic disorders can be associated with changes in chromosome number, not just mutations of single genes inherited in Mendelian fashion.

THE HUMAN CHROMOSOME SET

Human chromosomes are most often studied under a light microscope and are photographed in metaphase of mitosis. For convenience, the chromosomes are cut out from a photograph and are arranged in pairs according to size and centromere location to form a **karyotype** (Figure 6.1). Examination of human

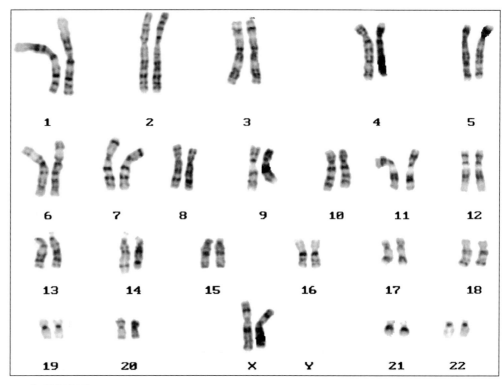

▶ FIGURE 6.1 A human karyotype showing replicated chromosomes from a cell in the metaphase of mitosis. This female has 46 chromosomes, including two X chromosomes.

■ Karyotype The chromosomal complement of a cell line or a person, photographed at metaphase and arranged in a standard sequence.

■ Sex chromosomes Chromosomes involved in sex determination. In humans, the X and Y chromosomes are the sex chromosomes.

■ Autosomes Chromosomes other than the sex chromosomes. In humans, chromosomes 1-22 are autosomes.

karyotypes reveals that one pair of chromosomes does not always involve homologues. Members of this pair are involved in the process of sex determination and are known as **sex chromosomes.** There are two types of sex chromosomes, X and Y. Females have two homologous X chromosomes, and males have a nonhomologous pair consisting of one X and one Y chromosome. Both males and females have two copies of the remaining 22 pairs of chromosomes, known as **autosomes.**

By convention, the chromosomes in the human karyotype are numbered and arranged into seven groups, A through G. Each group is defined by chromosomal size and centromere location. Staining procedures produce patterns of bands that are unique for each chromosome, allowing individual chromosomes to be clearly identified. The standardized G-banding pattern for the human chromosome set is shown in ▶ Figure 6.2. Chromosome banding patterns are used to identify specific regions on each chromosome (▶ Figure 6.3). In this system the short arm of each chromosome is designated the p arm, and the long arm the q arm. Each arm is subdivided into numbered regions beginning at the centromere. Within each region, the bands are identified by number. Thus, any region in the human karyotype can be identified by a descriptive address, such as 1q2.4. The address consists of the chromosome number (1), the arm (q), the region (2), and the band (4) (Figure 6.3).

Karyotypic analysis coupled with banding techniques provides clinicians and researchers with a powerful tool for chromosomal studies (see Concepts and Controversies: Making a Karyotype). Chromosome analysis has many important applications in human genetics, some of which are discussed in later sections. These include prenatal diagnosis, gene mapping, and cancer research.

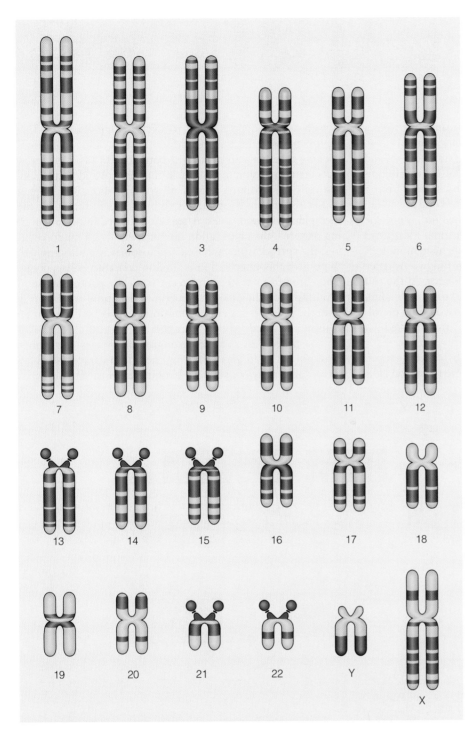

▶ **FIGURE 6.2** A karyogram of the human chromosome set, showing the distinctive banding pattern of each chromosome.

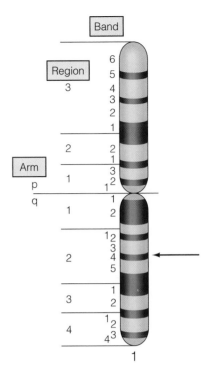

▶ **FIGURE 6.3** The system of naming chromosome bands. Each autosome is numbered from 1-22. The sex chromosomes are X and Y. Within a chromosome, the short arm is the p arm, and the long arm is the q arm. Each arm is divided into numbered regions. Within each region, numbers designate the bands. The area marked by the arrow is designated as 1q2.4 (chromosome 1, long arm, region 2, band 4).

Making a Karyotype

*C*hromosomal analysis can be performed using cells from a number of sources, including white blood cells (lymphocytes), skin cells (fibroblasts), amniotic fluid cells (amniocytes), and chorionic villus cells (placental cells). One of the most common methods begins with collecting a blood sample from a vein in the arm. A few drops of the collected blood are added to a flask containing a nutrient growth medium. Since the lymphocytes in the blood sample do not normally divide, a mitosis-inducing chemical, such as phytohemagglutinin, is added to the flask, and the cells are grown for two or three days at body temperature (37°C) in an incubator. At the end of this time, a drug called colcemid is added to stop dividing cells at metaphase. Over a period of about two hours of colcemid treatment, cells entering mitosis are arrested in metaphase.

Then, the blood cells are concentrated by centrifugation, and a salt solution is added to break open the red blood cells (which are nondividing) and to swell the lymphocytes. This treatment prevents the chromosomes in dividing cells from sticking together and makes the lymphocytes easier to break open. After fixation in a mixture of methanol and acetic acid, the cells are dropped onto a microscope slide. The impact causes the fragile lymphocytes to break open, spreading the metaphase chromosomes. After being stained to reveal banding patterns, the preparation is examined with a microscope, and the spread of metaphase chromosomes is photographed. A print of this photograph is used to construct a karyotype. The chromosomal images are cut from the photograph and arranged according to size, centromere location, and banding pattern. The presence of chromosomal abnormalities can be detected by a trained cytogeneticist.

The process of cutting chromosomal images from photographs by hand and arranging them into karyotypes is tedious and time-consuming. Computer-assisted karyotype

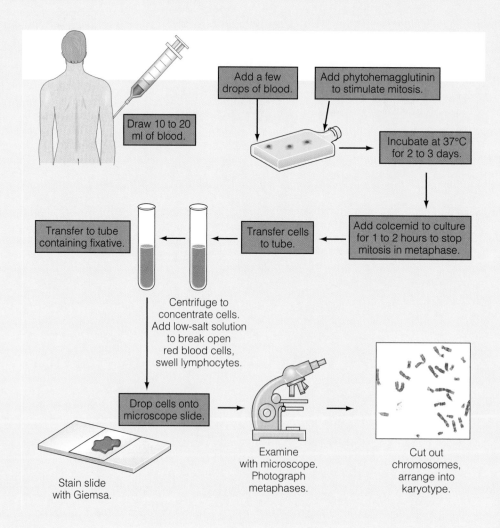

Draw 10 to 20 ml of blood.

Add a few drops of blood.

Add phytohemagglutinin to stimulate mitosis.

Incubate at 37°C for 2 to 3 days.

Add colcemid to culture for 1 to 2 hours to stop mitosis in metaphase.

Transfer cells to tube.

Transfer to tube containing fixative.

Centrifuge to concentrate cells. Add low-salt solution to break open red blood cells, swell lymphocytes.

Drop cells onto microscope slide.

Stain slide with Giemsa.

Examine with microscope. Photograph metaphases.

Cut out chromosomes, arrange into karyotype.

preparation is now used in most cytogenetic laboratories. In this system, a television camera and a computer are linked to a microscope. As metaphase chromosomes are located using the microscope, the image is recorded by the camera, digitized, and transmitted to the computer, where it is processed into a karyotype and printed. In the photographs below, a metaphase array of chromosomes has been printed by such a system. At the right is the computer-derived karyotype. Note that the computer has straightened the chromosomes as they were processed and arranged into a karyotype.

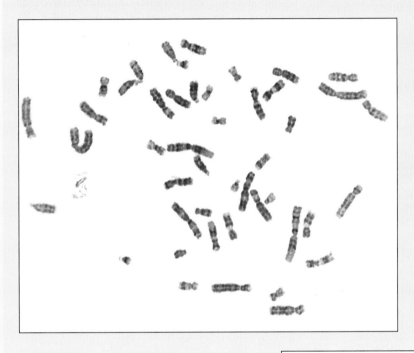

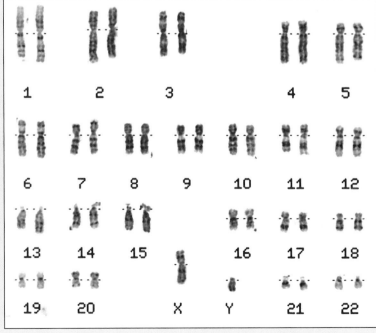

ANALYZING CHROMOSOMES AND KARYOTYPES

Chromosomal banding methods use stains and dyes to produce a pattern of bands that is different for each chromosome (although homologous chromosomes have the same pattern). One of the most common staining methods is G-banding, in which chromosomes are first treated with an enzyme (trypsin) to partially digest some chromosomal proteins, and then are exposed to Giemsa stain (a mixture of dyes). The resulting pattern of bands is visible under the microscope and is used to identify individual chromosomes in cytogenetic analysis. Usually, metaphase chromosomes are stained because this is the stage of maximum chromosomal condensation.

Normally, cytogeneticists use metaphase chromosomes that have about 550 bands. More detailed banding patterns can be obtained by staining chromosomes at early metaphase or late prophase. In these earlier stages, the chromosomes are longer and less condensed, and about 2000 bands can be identified in the normal human karyotype. Some other banding methods are reviewed in ▶ Figure 6.4.

In analyzing karyotypes, a standardized system of terminology is used. In this system, a karyotype is described by (1) the number of chromosomes, (2) the sex chromosome status, (3) the presence or absence of an individual chromosome, and (4) the nature and extent of any structural abnormality. The symbols for structural alterations include **t** for a **translocation, dup** for **duplication,** and **del** for **deletion.** These structural aberrations are discussed later in the chapter. If a male has a deletion in the short arm of chromosome 5, but otherwise is chromosomally normal, this would be represented as 46,XY,del(5p). Table 6.1 lists some chromosomal aberrations using this system of terminology.

Banding technique	Appearance of chromosomes
G-banding — Treat metaphase spreads with enzyme that digests part of chromosomal protein. Stain with Giemsa stain. Observe banding pattern with light microscope.	Darkly stained G bands.
Q-banding — Treat metaphase spreads with the chemical quinacrine mustard. Observe fluorescent banding pattern with a special ultraviolet light microscope.	Bright fluorescent bands upon exposure to ultraviolet light; same as darkly stained G bands.
R-banding — Heat metaphase spreads at high temperatures to achieve partial denaturation of DNA. Stain with Giemsa stain. Observe with light microscope.	Darkly stained R bands correspond to light bands in G-banded chromosomes. Pattern is the reverse of G-banding.
C-banding — Chemically treat metaphase spreads to extract DNA from the arms but not the centromeric regions of chromosomes. Stain with Giemsa stain and observe with light microscope.	Darkly stained C band centromeric region of the chromosome corresponds to region of constitutive heterochromatin.

■ **Translocation** A chromosomal aberration in which a chromosomal segment is transferred to another, nonhomologous chromosome.

■ **Duplication** A chromosomal aberration in which a segment of a chromosome is repeated and therefore is present in more than one copy within the chromosome.

■ **Deletion** A chromosomal aberration in which a segment of a chromosome is missing.

Cytogenetic analysis is a painstaking procedure done by cytogeneticists who study karyotypes stained to reveal black and white banding patterns. To make it easier to spot abnormalities, scientists have introduced a technique called chromosome painting. This method uses fluorescent markers, which attach to chromosome-specific DNA sequences. This labeling can use more than one DNA sequence and more than one fluorescent marker to produce a unique pattern for each of the 24 human chromosomes (22 autosomes and the X and Y chromosomes). Using this method, it is easier to find chromosomal abnormalities (▶ Figure 6.5). Eventually, automated scanning of painted chromosomes may analyze chromosomes.

Getting Cells for Chromosome Studies

Almost any cell that has a nucleus (mature red blood cells have no nuclei) can be used to make a karyotype. In adults, white blood cells (lymphocytes), skin cells (fibroblasts), and cells from biopsies or surgically removed tumor cells are routinely used for chromosomal preparations.

Table 6.1 Chromosomal Aberrations

Chromosomal Abnormality	Syndrome	Phenotype
46,del(4p)	Wolf-Hirschhorn syndrome	Mental retardation; midline facial defects consisting of broad nose, wide-set eyes, small lower jaw, and cleft palate; heart, lung, and skeletal abnormalities common; severely reduced survival
46,del(11)(p13)	WAGR syndrome	Tumors of the kidney (Wilms tumor) and of the gonad (gonadoblastoma); aniridia (absence of the iris); ambiguous genitalia; mental retardation
46,t(9;22)(q34a11)	CML (chronic myelogenous leukemia)	Enlargement of liver and spleen; anemia; excessive, unrestrained growth of white cells (granulocytes) in the bone marrow
46,t(8;14)	Burkitt's lymphoma	Malignancy of B lymphocytes that mature into the antibody-producing plasma cells; solid tumors, typically in the bones of the jaw and organs of the abdomen

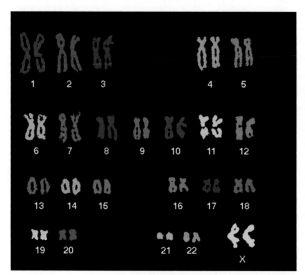

(a)

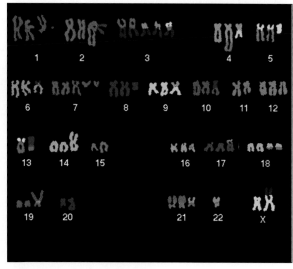

(b)

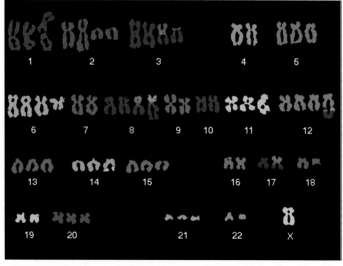

(c)

▶ **FIGURE 6.5** Chromosome painting with five different-colored markers in a normal cell (a) produces a pattern that highlights the translocations and deletions found in cancer cells (b and c).

Using Fetal Cells from the Mother's Blood

*M*ore than a century has passed since placental cells from the fetus were first discovered in the circulatory systems of pregnant women. In 1969, cytogeneticists observed cells that carried a Y chromosome in the blood of women who later gave birth to male infants. Since then, research has been directed at finding ways to recover and use fetal cells from the maternal circulation for prenatal diagnosis. The goal is to carry out chromosomal analysis on fetal cells in a 20-ml sample of maternal blood without using the invasive procedures of amniocentesis and chorionic villus sampling (CVS). Use of fetal blood for prenatal diagnosis would lower the risk of injury to the mother and fetus. Several types of fetal cells enter the maternal circulation, including placental cells (trophoblasts) and white blood cells (lymphocytes). These cells probably enter the bloodstream in detectable amounts between the 6th and 12th week of pregnancy. But since less than 1 in every 100,000 cells in the mother's blood is from the fetus, collecting enough fetal cells from a blood sample is one of the challenges facing those working to develop this technique.

Several methods are being tried to isolate fetal cells from maternal blood, but no single technique has emerged as the best. Even though these methods are still under development, several studies have found it possible to diagnose chromosomal abnormalities in fetal cells collected from maternal blood. These encouraging results have led the National Institute of Child Health and Human Development to begin a clinical trial designed to test the accuracy of chromosomal analysis on fetal blood cells. In this trial, 3000 women who are having amniocentesis or chorionic villus biopsy for a high-risk pregnancy will be asked to donate a blood sample for chromosomal analysis. The results from the fetal blood cells will be compared to those from amniocentesis and CVS. The trial was completed in 1997 and is being evaluated. If successful, fetal blood sampling may gradually replace more invasive procedures for prenatal chromosome analysis.

The development of methods for prenatal diagnosis of chromosomal abnormalities was a major advance in human genetics. Using **amniocentesis** and **chorionic villus sampling**, it is possible to diagnose chromosomal abnormalities before birth. Another method that detects and analyzes fetal cells in the maternal circulation is now undergoing clinical testing and may eventually be the method of choice for prenatal diagnosis (see Concepts and Controversies: Using Fetal Cells from the Mother's Blood).

■ **Amniocentesis** A method of sampling the fluid surrounding the developing fetus by inserting a hollow needle and withdrawing suspended fetal cells and fluid; used in diagnosing fetal genetic and developmental disorders; usually performed in the 16th week of pregnancy.

■ **Chorionic villus sampling** A method of sampling fetal chorionic cells by inserting a catheter through the vagina or abdominal wall into the uterus. Used in diagnosing biochemical and cytogenetic defects in the embryo. Usually performed in the eighth or ninth week of pregnancy.

Amniocentesis Collects Cells from the Fluid Surrounding the Fetus

Until the 1960s, amniocentesis was used to diagnose and to follow the progress of high-risk and potentially fatal disorders, such as fetal hemolytic anemia. In the mid-1960s, amniocentesis was successfully used for cytogenetic analysis. During amniocentesis, the fetus and the placenta are located by ultrasound, and a needle is inserted through the abdominal and uterine walls (avoiding the placenta and fetus) into the amniotic sac (▶ Figure 6.6). Approximately 10 to 30 ml of fluid is withdrawn using a syringe.

The amniotic fluid is composed mostly of fetal urine that contains cells shed from the skin, respiratory tract, and urinary tract of the fetus. The cells can be separated from the fluid by centrifugation (▶ Figure 6.7). The fluid can be tested for biochemical abnormalities that indicate a genetic defect. More than 100 biochemical disorders can be diagnosed by amniocentesis. The amniotic cells can be assayed for biochemical defects or grown in culture for karyotypic analysis following the procedures outlined earlier for metaphase preparations. Once a karyotype has been prepared, it is possible to diagnose the sex of the fetus and chromosomal abnormalities.

Amniocentesis is not usually performed until the 16th week of pregnancy. Before this time, there is very little amniotic fluid, and contamination of the sample with maternal cells is often a problem. Several studies in the United States, Britain, and

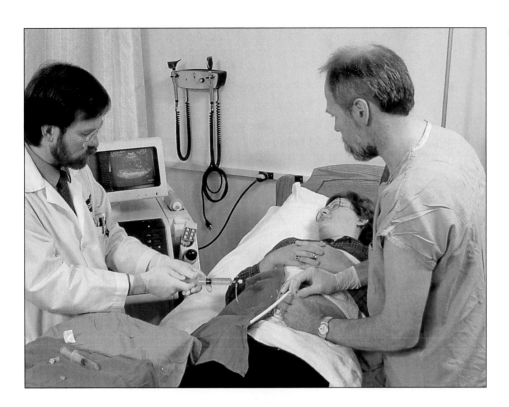

▌**FIGURE 6.6** A patient undergoing amniocentesis.

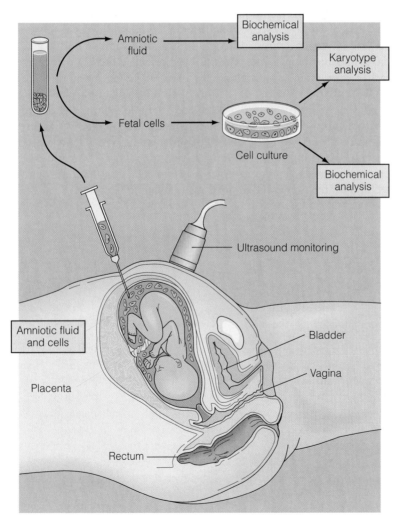

Biochemical analysis

Amniotic fluid

Karyotype analysis

Fetal cells

Cell culture

Biochemical analysis

Ultrasound monitoring

Amniotic fluid and cells

Bladder

Vagina

Placenta

Rectum

▌**FIGURE 6.7** In amniocentesis, a syringe needle is inserted through the abdominal wall and uterine wall to collect a small sample of the amniotic fluid. The fluid contains fetal cells that can be collected and used for prenatal chromosomal or biochemical analysis.

Canada have shown that amniocentesis involves a small but measurable risk to both the fetus and mother. There is a risk of maternal infection and a slight increase (less than 1%) in the probability of a spontaneous abortion. To offset these risks, amniocentesis is normally used only under the following conditions:

- Advanced maternal age. Because the risk for aneuploid offspring increases dramatically after the age of 35, this procedure is recommended when the prospective mother is 35 years or older. This risk factor accounts for the majority of all cases in which amniocentesis is used.
- Previous child with a chromosomal aberration. If a previous child had a chromosome abnormality, the recurrence risk is about 1% to 2%, and amniocentesis is recommended.
- Presence of a balanced chromosomal rearrangement. If either parent carries a chromosomal translocation or other rearrangement that can cause an unbalanced karyotype in the child, amniocentesis should be considered.
- X-linked disorder. If the mother is a carrier of an X-linked biochemical disorder that cannot be diagnosed prenatally and she is willing to abort if the fetus is male, then amniocentesis is recommended.

Chorionic Villus Sampling Retrieves Fetal Tissue from the Placenta

Chorionic villus sampling (CVS) is used for prenatal diagnosis in the first trimester of pregnancy. This technique has several advantages over amniocentesis. CVS can be performed at 8 to 10 weeks of gestation, compared with 15 to 16 weeks for amniocentesis. Cytogenetic results from CVS can be available within a few hours or a few days, and biochemical tests can be performed directly on the sampled fetal tissue rather than after tissue culture, as required following amniocentesis. For CVS, a small flexible catheter is inserted through the vagina or abdomen into the uterus and is guided by ultrasound images. A small sample from the chorionic villi, a fetal tissue that forms part of the placenta is obtained by suction (▶ Figure 6.8). This tissue is mitotically active and can be used immediately to prepare a karyotype. As with amniocentesis, enough material is usually obtained by CVS to allow biochemical testing or preparing DNA samples for molecular analysis. The use of recombinant DNA techniques for prenatal diagnosis is discussed in Chapter 13.

▶ **FIGURE 6.8** The chorionic villus sampling technique. A catheter is inserted into the uterus through the vagina to remove a sample of fetal tissue from the chorion. Cells in the tissue can be used for chromosomal or biochemical analysis.

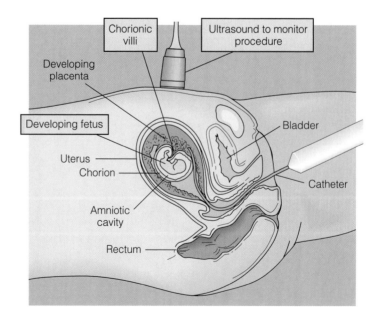

CVS is a technique that is more specialized than amniocentesis and at present is used less often. Although early studies indicated that the procedure posed a higher risk to mother and fetus than amniocentesis, improvements in the instruments and technique have lowered the risk somewhat. CVS offers early diagnosis of genetic diseases, and if termination of pregnancy is elected, maternal risks are lower at 9 to 12 weeks than at 16 weeks.

VARIATIONS IN CHROMOSOMAL NUMBER

At the birth of a child, anxious parents have two questions: Is it a boy or a girl, and is the baby normal? The term normal in this context means free from all birth defects. The causes of such defects, of course, are both environmental and genetic. Among the genetic causes, we have considered disorders, such as sickle cell anemia or Marfan syndrome, caused by the mutation of single genes. Other genetic causes of defects are changes in chromosome number or structural alterations in chromosomes. Such changes may involve entire chromosome sets, individual chromosomes, or alterations within individual chromosomes. Recall that the set of 46 chromosomes in each somatic cell is called the diploid, or $2n$, number of chromosomes. Similarly, the set of 23 chromosomes (constituting the n number) is the haploid set. Together, these normal conditions are called the euploid condition.

A chromosomal complement that contains three or more haploid sets of chromosomes is known as **polyploidy**. A cell that has three sets of chromosomes is triploid, one that has four sets is tetraploid, and so forth. A change in chromosome number that involves less than an entire diploid set of chromosomes is known as **aneuploidy**. The number of chromosomes in aneuploidy is not a simple multiple of the haploid set. The most common forms of aneuploidy in humans involve the gain or loss of a single chromosome. The loss of a chromosome is known as **monosomy** $(2n - 1)$, and the addition of a chromosome to the diploid set is known as **trisomy** $(2n + 1)$. Since the discovery of trisomy 21 in 1959 as the first example of aneuploidy in humans, cytogenetic studies have revealed that alterations in chromosome number are fairly common in humans and are a major cause of reproductive failure. It is now estimated that as many as one in every two conceptions may be aneuploid and that 70% of early embryonic deaths and spontaneous abortions are caused by aneuploidy. About 1 in every 170 live births is at least partially aneuploid, and from 5 to 7% of all deaths in early childhood are related to aneuploidy.

Humans have a rate of aneuploidy that is up to 10 times higher than that of other mammals, including other primates. This difference may represent a considerable reproductive disadvantage for our species. Understanding the causes of aneuploidy in humans remains one of the great challenges in human genetics.

- **Polyploidy** A chromosomal number that is a multiple of the normal haploid chromosomal set.

- **Aneuploidy** A chromosomal number that is not an exact multiple of the haploid set.

- **Monosomy** A condition in which one member of a chromosomal pair is missing; having one less than the diploid number $(2n - 1)$.

- **Trisomy** A condition in which one chromosome is present in three copies whereas all others are diploid; having one more than the diploid number $(2n + 1)$.

Polyploidy Changes the Number of Chromosomal Sets

Abnormalities in the number of haploid chromosomal sets can arise in several ways: (1) errors in meiosis during gamete formation, (2) events at fertilization, or (3) errors in mitosis following fertilization. Polyploidy sometimes occurs when cleavage of the cytoplasm (cytokinesis) fails to occur in the last stage of cell division in either mitosis or meiosis. Recall that mitotic divisions precede meiosis in the ovary and testis. If the chromosomes replicate and sister chromatids divide during one of these mitotic cycles but there is no cytoplasmic division, the result is a tetraploid cell that contains four copies of each chromosome. If this cell undergoes meiosis normally, the result is gametes that contain the diploid instead of the haploid number of chromosomes. Fusion between this genetically unbalanced gamete and a normal haploid gamete produces a triploid zygote (▶ Figure 6.9).

A diploid gamete can also arise through meiotic errors. An error in meiosis I can result in the failure of homologous chromosomes to separate, producing

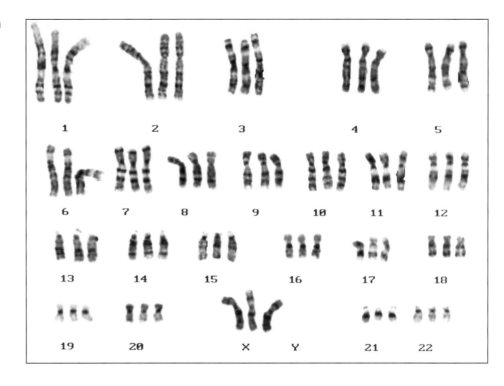

Dispermy Fertilization of haploid egg by two haploid sperm, forming a triploid zygote.

Triploidy A chromosomal number that is three times the haploid number, having three copies of all autosomes and three sex chromosomes.

Tetraploidy A chromosomal number that is four times the haploid number, having four copies of all autosomes and four sex chromosomes.

diploid gametes after meiosis II. Alternatively, if all chromosomes move to the same pole after centromere separation in meiosis II, diploid gametes will also result. Another event that can result in polyploidy is **dispermy**, the simultaneous fertilization of a haploid egg by two haploid sperm. The result is a zygote that contains three haploid chromosome sets, or triploidy.

Triploidy The most common form of polyploidy in humans is **triploidy,** which is found in 15 to 18% of spontaneous abortions. Three types of triploid chromosome sets have been observed: 69,XXY; 69,XXX; and 69,XYY. Approximately 75% of all cases of triploidy have two sets of paternal chromosomes. Accidents in male gamete formation do not occur this often, so most triploid zygotes probably arise as the result of dispermy. Although biochemical changes that accompany fertilization normally prevent such fertilizations, this system is not fail-safe.

Almost 1% of all conceptions are triploid, but over 99% of these die before birth, and only 1 in 10,000 live births is triploid. Survival for triploids is limited, and most die within a month. Triploid newborns have multiple abnormalities including an enlarged head, fusion of fingers and toes (syndactyly), and malformations of the mouth, eyes, and genitals (Figure 6.10). The high rate of embryonic death and failure to survive as a newborn indicates that triploidy is incompatible with life and should be regarded as a lethal condition.

Tetraploidy Tetraploidy is observed in about 5% of all spontaneous abortions and is reported only rarely in live births. The sex chromosome constitution of all tetraploid embryos is either XXXX or XXYY. **Tetraploidy** can arise in the first mitotic division after fertilization if there is no cytoplasmic division (cytokinesis) after chromosomal division. The result is a tetraploid cell. Subsequent rounds of mitosis result in a tetraploid embryo. If tetraploidy arises sometime after the first mitotic division, two different cell lines coexist in the embryo, one a normal diploid line and the other a tetraploid cell line. Such mosaic individuals survive somewhat longer than complete polyploids, but the condition is still life-threatening.

In summary, polyploidy in humans can arise by at least two different mechanisms, but it is inevitably lethal. Polyploidy does not involve the mutation of any

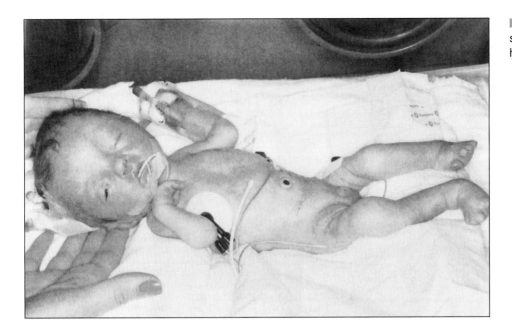

▶ **FIGURE 6.10** A triploid infant, showing the characteristic enlarged head.

genes, but only changes in the number of gene copies. How this quantitative change in gene number is related to lethality in development is unknown.

Aneuploidy Changes the Number of Individual Chromosomes

As defined earlier, aneuploidy is the addition or deletion of individual chromosomes from the normal diploid set of 46. Aneuploidy can be caused in several ways. The most important is **nondisjunction,** a process in which chromosomes fail to separate properly at anaphase. Although this failure can occur in either meiosis or mitosis, nondisjunction in meiosis is the leading cause of aneuploidy in humans. There are two cell divisions in meiosis, and nondisjunction can occur in either the first or second division with different genetic consequences (▶ Figure 6.11).

> ▪ **Nondisjunction** The failure of homologous chromosomes to separate properly during meiosis or mitosis.

If nondisjunction takes place in meiosis I, all of the gametes will be abnormal. These gametes will carry both members of a chromosomal pair or neither member of the pair. Nondisjunction in meiosis II results in two normal gametes and two abnormal gametes (Figure 6.11). Gametes that are missing a copy of a given chromosome will produce a monosomic zygote. Those that contain an extra copy of a chromosome will give rise to a trisomic zygote.

As a group, aneuploid individuals have distinct and characteristic phenotypic features. Those that have a given form of aneuploidy, such as Down syndrome, tend to resemble each other more closely than their own brothers and sisters. The phenotypic effects of aneuploidy range from minor physical symptoms to devastating and lethal deficiencies in major organ systems. Among survivors, phenotypic effects often include behavioral deficits and mental retardation. In the following section, we look at some of the important features of autosomal aneuploid phenotypes. Then we consider the phenotypic effects of sex chromosome aneuploidy.

Monosomy Is Rare

Meiotic nondisjunction during gamete formation should result in equal numbers of monosomic and trisomic embryos. However, autosomal monosomies are only rarely observed among spontaneous abortions and live births. The likely explanation is that the majority of autosomal monosomic embryos are lost very early, even before pregnancy is recognized.

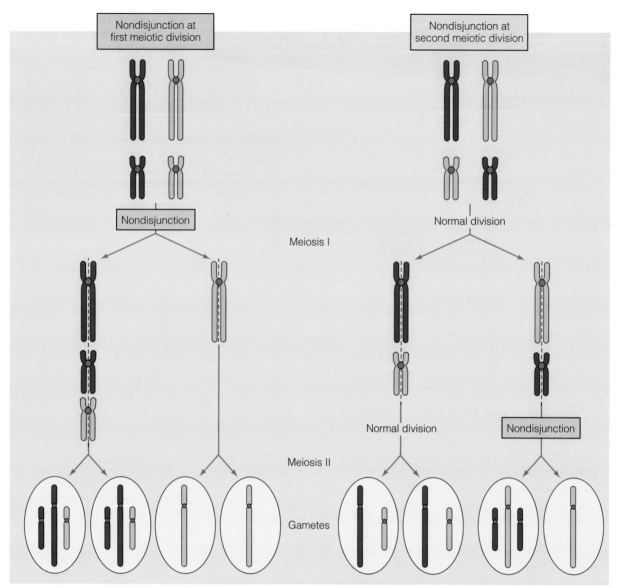

FIGURE 6.11 Nondisjunction in the first meiotic division (left) results in four abnormal gametes. Two gametes carry both members of a chromosome pair, and two are missing one chromosome. Nondisjunction in the second meiotic division (right) produces two normal gametes and two abnormal gametes. One gamete carries both members of a chromosome pair, and one is missing a chromosome.

Trisomy Is Relatively Common

Most autosomal trisomies are lethal conditions; but unlike monosomy, the presence of an extra chromosome allows varying degrees of development. Autosomal trisomy is found in about 50% of all cases of chromosomal abnormalities in fetal death. The findings also indicate that autosomes are differentially involved in trisomy (Figure 6.12). Trisomies for chromosomes 1, 3, 12, and 19 are rarely observed in spontaneous abortions, whereas trisomy for chromosome 16 accounts for almost one-third of all cases. As a group, the acrocentric chromosomes (13-15, 21, and 22) are represented in 40% of all spontaneous abortions. Reasons for this differential involvement include differences in the rate of nondisjunction, in the rate of fetal death before recognition of pregnancy, or a combination of factors. Only a few autosomal trisomies result in live births (trisomy 8, 13, and 18). Trisomy 21 (Down syndrome) is the only autosomal trisomy that allows survival into adulthood.

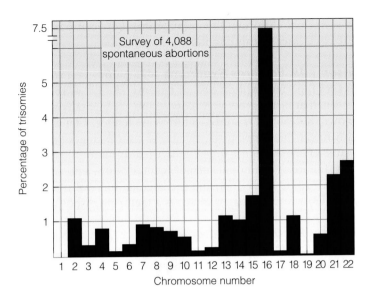

Survey of 4,088 spontaneous abortions

▶ **FIGURE 6.12** The results of a cytogenetic survey of over 4000 spontaneous abortions show a wide variation in the presence of individual chromosomes in trisomic embryos.

Trisomy 13: Patau Syndrome (47, + 13) Trisomy 13 was discovered in 1960 by cytogenetic analysis of a malformed child. The karyotype indicated the presence of 47 chromosomes, and the extra chromosome was identified as chromosome 13 (47, + 13). Only 1 in 15,000 live births involves trisomy 13, and the condition is lethal. Half of all affected individuals die in the first month, and the mean survival time is six months. The phenotype of trisomy 13 involves cleft lip and palate (▶ Figure 6.13), eye defects, polydactyly (extra fingers or toes), and feet that have large protruding heels. Internally, there are usually severe malformations of the brain and nervous system and congenital heart defects. The involvement of so many organ systems indicates that developmental abnormalities begin early in embryogenesis, perhaps as early as the sixth week. Parental age is the only factor known to be related to trisomy 13. The age of parents of children who have trisomy 13 is higher (averaging about 32 years) than the average age of parents who

■ **Trisomy 13** The presence of an extra copy of chromosome 13 that produces a distinct set of congenital abnormalities resulting in Patau syndrome.

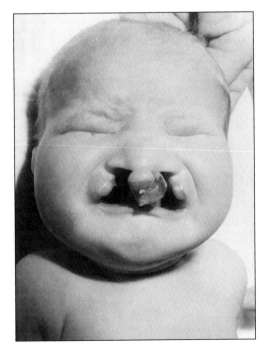

▶ **FIGURE 6.13** An infant who has trisomy 13, showing a cleft lip and palate (the roof of the mouth).

have normal children. The relationship between parental age and aneuploidy is discussed later in this chapter.

Trisomy 18: Edwards Syndrome (47, + 18)

In 1960 John Edwards and his colleagues reported the first case of **trisomy 18** (47, + 18). Infants with this condition are small at birth, grow very slowly, and are mentally retarded. For reasons still unknown, 80% of all trisomy 18 births are female. Clenched fists, with the second and fifth finger overlapping the third and fourth fingers, and malformed feet are also characteristic (Figure 6.14). Heart malformations are almost always present, and heart failure or pneumonia usually causes death. Trisomy 18 occurs with a frequency of 1 in 11,000 live births, and the average survival time is 2 to 4 months. As in trisomy 13, advanced maternal age is a factor predisposing to trisomy 18.

Trisomy 21: Down Syndrome (47, + 21)

The phenotypic features of **trisomy 21**(MIM/OMIM 190685) were first described by John Langdon Down in 1866. He called the condition mongolism because of the distinctive fold of skin, known as an epicanthic fold, in the corner of the eye (Figure 6.15). To remove the racist implications inherent in the term, Lionel Penrose and others changed the designation to Down syndrome. As described in the chapter opening, the presence of an extra copy of chromosome 21 as the underlying cause of Down syndrome, discovered by Jerome Lejeune and his colleagues in 1959, represents the first chromosomal abnormality discovered in humans. Trisomy 21 has also been observed in other primate species, including the chimpanzee.

Down syndrome, one of the most common chromosomal defects in humans, occurs in about 0.5% of all conceptions and in 1 in 900 live births (Figure 6.15). It is a leading cause of childhood mental retardation and heart defects in the United States. Affected individuals have a wide skull that is flatter than normal at the back. The eyelids have an epicanthic fold, and the iris contains spots, known as Brushfield spots. The tongue may be furrowed and protruding, causing the mouth to remain partially open. Physical growth, behavior, and mental development are retarded, and approximately 40% of all affected individuals have congenital heart defects. In addition, children with Down syndrome are prone to respiratory infections and contract leukemia at a rate far above the normal population. In the past

■ **Trisomy 18** The presence of an extra copy of chromosome 18 that results in a clinically distinct set of invariably lethal abnormalities known as Edwards syndrom.

■ **Trisomy 21** Aneuploidy involving the presence of an extra copy of chromosome 21, resulting in Down syndrome.

 FIGURE 6.14 An infant who has trisomy 18 and a karyotype.

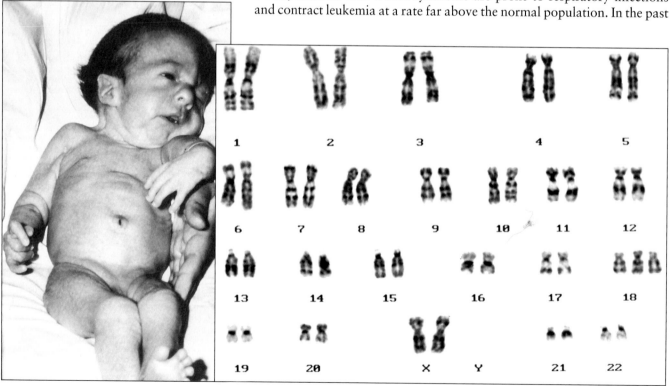

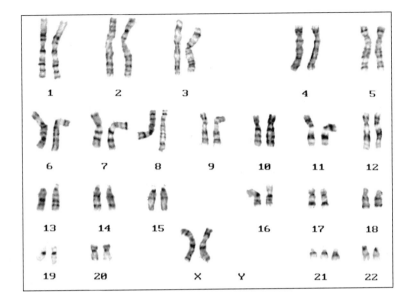

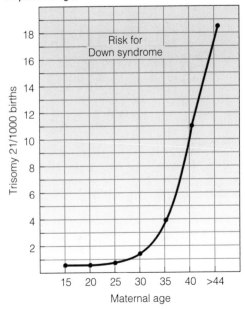

▶ **FIGURE 6.15** A child who has trisomy 21. The karyotype shows this child has three copies of chromosome 21.

decade, improvements in medical care have increased survival rates dramatically, so that many affected individuals survive into adulthood, although few reach the age of 50 years. In spite of these handicaps, many individuals who have Down syndrome lead rich, productive lives and can serve as an inspiration to us all.

WHAT ARE THE RISKS FOR AUTOSOMAL TRISOMY?

The causes of autosomal trisomies such as Down syndrome are unknown, but a variety of genetic and environmental factors have been proposed, including radiation, viral infection, hormone levels, and genetic predisposition. To date the only factor clearly related to autosomal aneuploidy is advanced maternal age. In fact, a relationship between maternal age and Down syndrome was well established 25 years before the chromosomal basis for the condition was discovered.

Maternal Age Is the Leading Risk Factor for Trisomy

The risk of bearing children who have trisomy 21 is low for young mothers but increases rapidly after the age of 35 years. At the age of 20 the incidence of Down syndrome offspring is 0.05%, by age 35 the risk of having a child with Down syndrome has climbed to 0.9%, and at 45 years 3% of all newborns have trisomy 21 (▶ Figure 6.16). The effect of maternal age on other aneuploidies has been documented, and the relationship between advanced maternal age and autosomal trisomy is very striking (▶ Figure 6.17). Paternal age has also been proposed as a factor in trisomy, but the evidence is weak, and no clear-cut link has been demonstrated.

The evidence that advanced maternal age is a risk factor for aneuploid offspring comes from studies on the parental origin of nondisjunction documented by chromosome banding of the affected child and both parents and by studies using recombinant DNA technology. Occasionally, chromosomes have some minor variations in banding patterns. By examining banded chromosomes from the trisomic child and the parents, the parental origin of the nondisjunction can often be determined. For trisomy 21, the nondisjunction is maternal about 94% of the time and paternal about 6% of the time. In other autosomal trisomies studied, the paternal contribution is about 7%. In all trisomies, the great majority of maternal nondisjunctions occur at meiosis I.

▶ **FIGURE 6.16** The relationship between maternal age and the frequency of trisomy 21 (Down syndrome). The risk increases rapidly after 35 years of age.

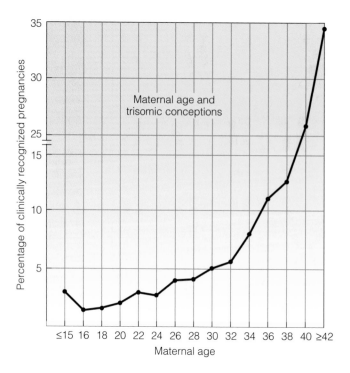

FIGURE 6.17 Maternal age is the major risk factor for autosomal trisomies of all types. By age 42, about 1 in 3 identified pregnancies is trisomic.

Why Is Age a Risk Factor?

One idea about the relationship between maternal age and nondisjunction focuses on the duration of meiosis in females. Recall from Chapter 2 that primary oocytes are formed early in development and enter the first meiotic prophase well before birth. Meiosis I is not completed until ovulation, so that eggs produced at age 40 have been in meiosis I for more than 40 years. During this time, metabolic errors or environmental agents may damage the cell so that aneuploidy results when meiosis resumes. However, it is not yet known whether the age of the ovum is causally related to the increased frequency of nondisjunction.

A second idea proposed to account for the increased risk of aneuploid children in older mothers is related to the interaction between the implanting embryo and the uterine environment. Normally, according to this idea, the embryo-uterine interaction results in the spontaneous abortion of chromosomally abnormal embryos, a process called maternal selection. As women age, this process becomes less effective, allowing more chromosomally abnormal embryos to become implanted and develop.

ANEUPLOIDY OF THE SEX CHROMOSOMES

The incidence of sex chromosome aneuploidy is higher than that for autosomes. The overall incidence of sex chromosome anomalies in live births is 1 in 400 for males and 1 in 650 for females. Unlike the situation with autosomes, where monosomy is always fatal, monosomy for the X chromosome is a viable condition. Monosomy for the Y chromosome (45,Y), however, is always lethal.

Turner Syndrome (45,X)

Monosomy for the X chromosome (45,X) was reported as a chromosomal disorder in 1959, but the cytogenetic findings were only the finishing touch to a larger piece of genetic detective work. In 1954, Paul Polani, who was working on the causes of congenital heart defects, examined three women who suffered from a defect of the aorta usually found only in males. Polani noted that all three women also had phenotypic features originally described by Henry Turner in 1938: short stature, extra folds of skin on the neck, and rudimentary sexual development. Examining cells scraped from inside their cheeks, Polani discovered that, like males, these females lacked a sex chromatin, or Barr body. Because of this, he suspected that females who have **Turner syndrome** might be affected by X-linked traits as frequently as males.

Assembling a large group of females who lacked ovarian development, including those with Turner syndrome, Polani tested them for color blindness, a sex-linked trait. Four of the 25 females tested were color-blind, a result that was much higher than expected in a population of females and similar to that expected in males. In a paper published in 1956, he suggested that Turner syndrome females might have only one X chromosome and in effect be hemizygous for traits on the X chromosome.

After a careful cytogenetic study, Polani and Charles Ford published a paper in 1959 confirming that Turner females are indeed 45,X. The unraveling of the chromosomal basis of Turner syndrome illustrates a basic property of scientific research: Work in one area—in this case, congenital heart disease—often leads to significant findings in another field.

Turner syndrome females are short and wide-chested and have underdeveloped breasts and rudimentary ovaries (▶ Figure 6.18). At birth, puffiness of the hands and feet is prominent, but this disappears in infancy. As reported by Polani, such individuals also have narrowing, or coarctation, of the aorta. There is no mental retardation associated with this syndrome, although evidence suggests that Turner syndrome is associated with reduced skills in interpreting spatial relationships. This chromosomal disorder occurs with a frequency of 1 in 10,000 female births, and although affected newborns suffer no life-threatening problems, 95 to 99% of all 45,X embryos die before birth. It is estimated that 1% of all conceptions are 45,X. Another notable feature of this syndrome is that in 75% of all cases, the nondisjunction apparently originates in the father.

▪ **Turner syndrome** A monosomy of the X chromosome (45,X) that results in female sterility.

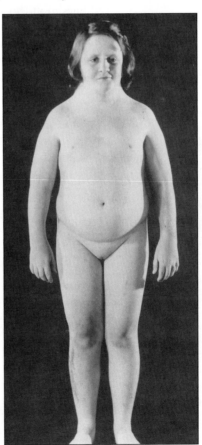

▶ **FIGURE 6.18** A girl who has Turner syndrome and the associated karyotype, containing a single X chromosome. Turner syndrome is characterized by short stature, a broad chest, and lack of sexual development. Stature and sexual development can be treated using hormones.

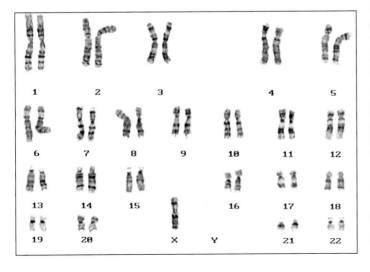

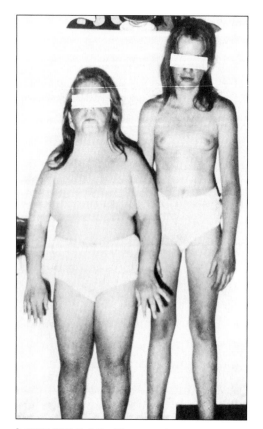

The phenotypic impact of the single X chromosome in Turner syndrome is strikingly illustrated in a case of identical twins, one of them 46,XX and the other 45,X. This situation apparently arose by mitotic nondisjunction after fertilization and twinning. These twins (▶ Figure 6.19) were judged to be identical on the basis of blood types and chromosome banding studies. They show significant differences in height, sexual development, hearing, dental maturity, and performance on tests that measure numerical skills and space perception. Although some environmental factors may contribute to these differences, the major role of the second X chromosome in normal female development is apparent. These and similar results from other studies on individuals with Turner syndrome indicate that a second X chromosome is necessary for normal development of the ovary, normal growth patterns, and development of the nervous system. Complete absence of an X chromosome in the absence or presence of a Y chromosome is always lethal, emphasizing that the X chromosome is an essential component of the karyotype.

Klinefelter Syndrome (47,XXY)

The phenotype of Klinefelter syndrome was first described in 1942, and Patricia Jacobs and John Strong reported the XXY chromosomal constitution in 1959. The frequency of **Klinefelter syndrome** is approximately 1 in 1000 male births. The phenotypic features of this syndrome do not develop until puberty. Affected individuals are male but show poor sexual development and have very low fertility. Some degree of breast development occurs in about 50% of the cases (▶ Figure 6.20). A degree of subnormal intelligence appears in some affected individuals.

▶ **FIGURE 6.19** Monozygotic twins, one of which has Turner syndrome. The twin who has Turner syndrome (left) is 45,X; the normal twin (right) is 46,XX.

■**Klinefelter syndrome** Aneuploidy of the sex chromosomes involving an XXY chromosomal constitution.

A significant fraction of Klinefelter males are mosaics who have XY and XXY cell lines in the body. About 60% of the cases are the result of maternal nondisjunction, and advanced maternal age is known to increase the risk of affected offspring. Other forms of Klinefelter include XXYY, XXXY, and XXXXY. The presence of additional X chromosomes in these karyotypes increases the severity of the phenotypic symptoms and brings on clear-cut mental retardation.

▶ **FIGURE 6.20** A young man who has Klinefelter syndrome and its characteristic karyotype. In some cases, Klinefelter syndrome is associated with breast development.

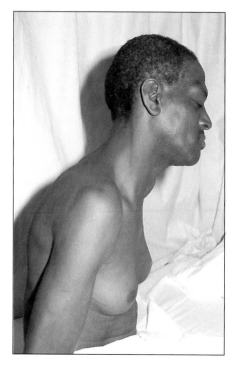

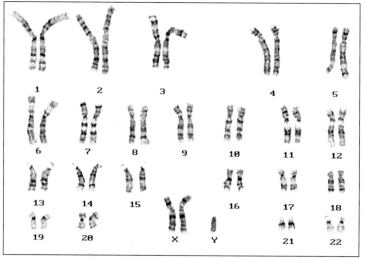

XYY Syndrome (47,XYY)

In 1965 the results of a cytogenetic survey of 197 males institutionalized for violent and dangerous antisocial behavior aroused a great deal of interest in the scientific community and the popular press. The findings indicated that nine of these males (about 4.5% of the institutionalized males in the survey) had an **XYY karyotype** (▶ Figure 6.21). These XYY individuals were all above average in height, all suffered personality disorders, and seven of the nine were of subnormal intelligence. Subsequent studies indicated that the frequency of XYY males in the general population is 1 in 1000 male births (about 0.1% of the males in the general population) and that the frequency of XYY individuals in penal and mental institutions is significantly higher than in the population at large.

Early investigators associated the tendency to violent criminal behavior with the presence of an extra Y chromosome. In effect, this would mean that some forms of violent behavior were brought about by genetic predisposition. In fact, the XYY karyotype has been used on several occasions as a legal defense (unsuccessfully, so far) in criminal trials. The question is this: Is there really a causal relationship between the XYY condition and criminal behavior? There is no strong evidence to support such a link. In fact, the vast majority of XYY males lead socially normal lives. In the United States, long-term studies of the relationship between antisocial behavior and the 47,XYY karyotype have been discontinued. This decision was made because of fear that identifying children who have potential behavioral problems might lead parents to treat them differently, resulting in the development of behavioral problems as a self-fulfilling prophecy.

■ XYY karyotype Aneuploidy of the sex chromosomes involving XYY chromosomal constitution.

Aneuploidy of the Sex Chromosomes: Some Conclusions

Several conclusions can be drawn from the study of sex chromosome disorders. First, at least one copy of an X chromosome is essential for survival. Embryos without any X chromosomes (44,-XX and 45,OY) are inviable and are not observed in studies of spontaneous abortions. They must be eliminated even before pregnancy is recognized, emphasizing the role of the X chromosome in normal development. The second general conclusion is that the addition of extra copies of either sex chromosome

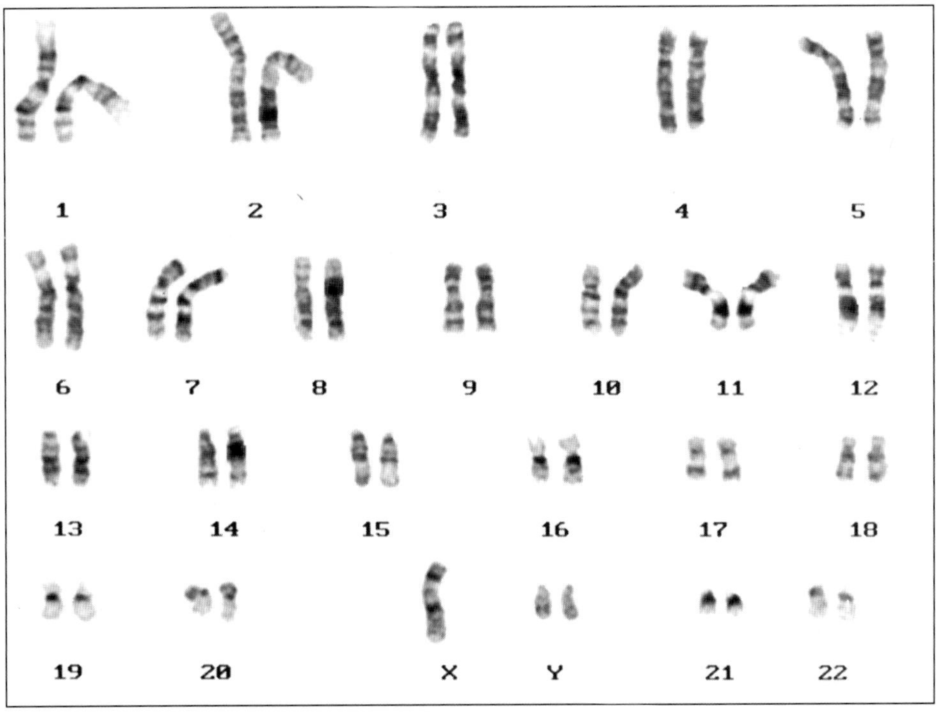

▶ **FIGURE 6.21** The karyotype of an XYY male. Affected individuals are usually taller than normal, and some, but not all, suffer from personality disorders.

interferes with normal development and causes both physical and mental problems. As the number of sex chromosomes in the karyotype increases, the derangements become more severe, indicating that a balance of sex-chromosome gene dosage and gene products is essential to normal development in both males and females.

STRUCTURAL ALTERATIONS WITHIN CHROMOSOMES

Changes in chromosomal structure can involve one, two, or more chromosomes and result from the breakage and reunion of chromosomal parts. In some cases the pieces in chromosome breaks are rejoined to restore the original structure. In others, an array of abnormal chromosomes results. Breaks can occur spontaneously through errors in replication or recombination. Environmental agents, such as ultraviolet light, radiation, viruses, and chemicals, can also produce them. Structural alterations include deletions, the loss of a chromosomal segment; duplications, which are extra copies of a chromosome segment; translocations, which move a segment from one chromosome to another; and **inversions,** which reverse the order of a chromosomal segment (▶ Figure 6.22). Rather than considering how such aberrations are produced, we concentrate on the phenotypic effects of these structural alterations and on how such changes in chromosomal structure can be used to provide information about the location and action of genes.

■ **Inversion**   A chromosomal aberration in which a chromosomal segment has been rotated 180° from its usual orientation.

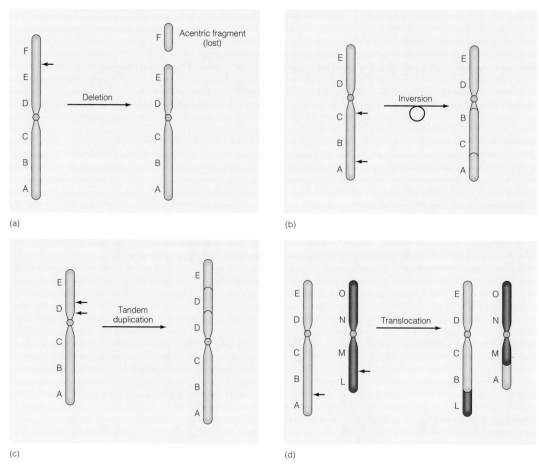

▶ **FIGURE 6.22**   Some of the common structural abnormalities seen in chromosomes. (a) In a deletion, part of the chromosome is lost. This can occur at the tip of the chromosome as shown, or an internal segment can be lost. (b) In an inversion, the order of part of the chromosome is reversed. This does not change the amount of genetic information carried by the chromosome, only its arrangement. (c) A duplication has a chromosomal segment repeated (in this example, gene D and its surrounding region are duplicated). (d) In a translocation, chromosomal parts are exchanged.

Table 6.2 Chromosomal Deletions

Deletion	Syndrome	Phenotype
5p-	Cri du chat syndrome	Infants have catlike cry, some facial anomalies, severe mental retardation
11q-	Wilms tumor	Kidney tumors, genital and urinary tract abnormalities
13q-	Retinoblastoma	Cancer of eye, increased risk of other cancers
15q-	Prader-Willi syndrome	Infants: weak, slow growth; children and adults: obesity, compulsive eating

Deletions Involve Loss of Chromosomal Material

Deletion of more than a small amount of chromosomal material has a detrimental effect on the developing embryo, and deletion of an entire autosome is lethal. Consequently, only a few viable conditions are associated with large-scale deletions. Some of these are listed in Table 6.2.

Cri du chat Syndrome An infant who carried a deletion of the short arm of chromosome 5 was first reported in 1963. This condition occurs in 1 in 100,000 births (▶ Figure 6.23). It is the reduction in gene dosage that is associated with the abnormal phenotype, not the presence of one or more mutant genes. The affected infant is mentally retarded and has defects in facial development, gastrointestinal malformations, and abnormal development of the glottis and larynx. Affected infants have a cry that sounds like a cat meowing, hence the name **cri du chat syndrome** (MIM/OMIM 123450) (▶ Figure 6.24). This deletion of a chromosome segment affects the motor and mental development of affected individuals but does not seem to be life-threatening.

By correlating phenotypes with chromosomal breakpoints in affected individuals, two regions associated with this syndrome have been identified on the short arm of chromosome 5 (Figure 6.23). Losses of chromosomal material in 5p15.3 results in abnormal larynx development, and deletions in 5p15.2 are associated with mental retardation and other phenotypic features of this syndrome. This correlation indicates that genes that control larynx development may be located in 5p15.3, and genes important in the development or function of the nervous system are located in 5p15.2.

■ **Cri du chat syndrome** A deletion of the short arm of chromosome 5 associated with an array of congenital malformations, the most characteristic of which is an infant cry that resembles a meowing cat.

▶ **FIGURE 6.23** A deletion of part of chromosome 5 is associated with cri du chat syndrome. By comparing the region deleted with its associated phenotype, investigators have identified regions of the chromosome that carry genes involved in developing the larynx.

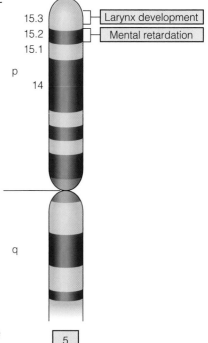

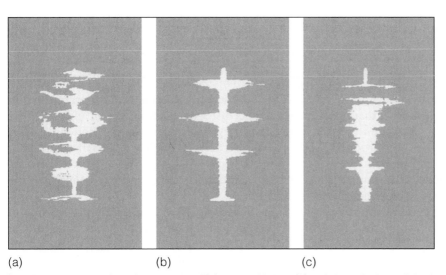

(a)　　　　　　(b)　　　　　　(c)

▶ **FIGURE 6.24** Sound recordings of (a) a normal infant, (b) an infant who has cri du chat syndrome, and (c) a cat. The cry of the affected infant is much closer in sound pattern to that of the cat than a normal infant, giving rise to the name of the syndrome, cry of the cat.

▶ **FIGURE 6.25** Diagram of chromosome 15, showing the region deleted in Prader-Willi syndrome.

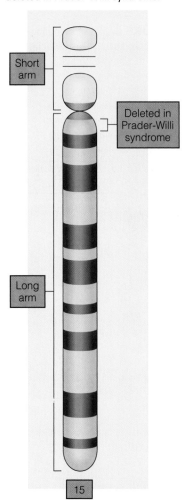

Short arm

Deleted in Prader-Willi syndrome

Long arm

15

Prader-Willi Syndrome A constellation of physical and mental symptoms known as **Prader-Willi syndrome** (MIM/OMIM 176270) has been correlated with deletions in the long arm of chromosome 15. As infants, affected individuals are weak and do not feed well because of a poor sucking reflex. However, by the age of 5 or 6 years, these children develop an uncontrollable compulsion to eat that results in obesity and related health problems, such as diabetes. Left untreated, victims literally eat themselves to death. Other symptoms include poor sexual development in males, behavioral problems, and mental retardation.

Prader-Willi patients have deletions in the long arm of chromosome 15 between bands q11 and q13. The size of the deletion is variable but always includes band 15q11.2 (▶ Figure 6.25). In about 5% of cases, other chromosomal aberrations are found at this site, including duplications and translocations, giving rise to speculation that this region is predisposed to structural instability.

It is estimated that Prader-Willi syndrome affects from 1 in 10,000 to 1 in 25,000 people, and males predominate. The cause of the eating disorder is unknown but may be related to disturbances in endocrine function. Treatment includes behavior modification with constant supervision of access to food.

Translocations Involve Exchange of Chromosomal Parts

Transfer of a chromosomal segment to another nonhomologous chromosome is known as a translocation. There are two major types of translocations: **reciprocal translocations** and **Robertsonian translocations.** If two nonhomologous chromosomes exchange segments, the event is a reciprocal translocation. There is no gain or loss of genetic information in such exchanges, only the rearrangement of gene sequences. In some cases no phenotypic effects are seen, and the translocation is passed through a family for generations. However, during meiosis, cells that contain translocations can produce genetically unbalanced gametes which show duplicated or deleted chromosomal segments. If these gametes participate in fertilization, the result is embryonic death or abnormal offspring.

About 5% of all cases of Down syndrome involve a Robertsonian translocation, most often between chromosomes 21 and 14. In this type of translocation, the centromeres of the two chromosomes fuse, and chromosomal material is lost from the short arms (▶ Figure 6.26). The carrier of such a translocation is phenotypically normal, even though the short arms of both chromosomes may be lost. This carrier is actually aneuploid and has only 45 chromosomes. But because the carrier has two copies of the long arm of chromosome 14 and two copies of the long arm of chromosome 21 (a normal 14, a normal 21, and a translocated 14/21), there is no phenotypic effect. At meiosis the carrier produces six types of gametes in equal proportions (Figure 6.26). Three of these result in zygotes that are inviable. Of the remaining three, one will produce a Down syndrome child.

Although it might seem that the chance of producing future children with Down syndrome is one in three, or 33%, the observed risk is somewhat lower. It is important to remember that this risk does not increase with maternal age. In addition, there is also a 1 in 3 chance of producing a phenotypically normal translocational carrier, who is at risk of producing children who have Down syndrome. For this reason it is important to karyotype a Down syndrome child and his or her parents to determine whether a translocation is involved. This information is essential in counseling parents about future reproductive risks.

Cytogenetic examination of translocations can be used to identify chromosomal regions that are most critical to the expression of aneuploid phenotypes. The study of a large number of translocations involving chromosome 21 has correlated chromosomal regions on the long arm with the symptoms associated with Down syndrome (▶ Figure 6.27). With this information available, efforts are now focused on identifying and characterizing the genes in this chromosomal segment to understand how a change in gene dosage produces such serious phenotypic effects.

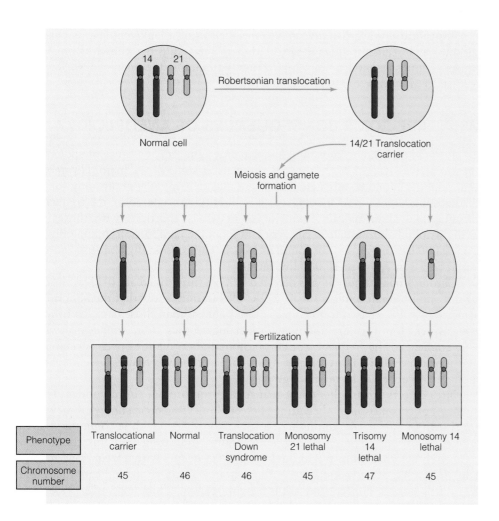

FIGURE 6.26 Segregation of chromosomes at meiosis in a 14–21 translocational carrier. Six types of gametes are produced. When these gametes fuse with those of a normal individual, six types of zygotes are produced. Of these, two (translocational carrier and normal) have a normal phenotype, one is Down syndrome, and three are lethal combinations.

Phenotype	Translocational carrier	Normal	Translocation Down syndrome	Monosomy 21 lethal	Trisomy 14 lethal	Monosomy 14 lethal
Chromosome number	45	46	46	45	47	45

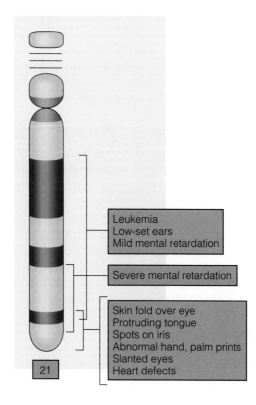

FIGURE 6.27 Chromosome 21, showing regions associated with various phenotypic features of Down syndrome. These assignments have been made by comparing various deletions in the chromosome with the phenotypes they produce.

Leukemia
Low-set ears
Mild mental retardation

Severe mental retardation

Skin fold over eye
Protruding tongue
Spots on iris
Abnormal hand, palm prints
Slanted eyes
Heart defects

21

Translocations are also involved in specific forms of leukemia. The exchange of chromosomal segments alters the regulation of a class of genes known as oncogenes, whose action is essential to developing and maintaining the malignancy. This topic is considered in some detail in Chapter 14.

WHAT ARE SOME CONSEQUENCES OF ANEUPLOIDY?

As indicated earlier, chromosomal abnormalities are a major cause of spontaneous abortions (see Figure 6.12). Table 6.3 summarizes some of the major chromosomal abnormalities found in miscarriages. These include triploidy, monosomy for the X chromosome (45,X), and trisomy 16. It is interesting to compare the frequency of chromosomal abnormalities found in spontaneous abortions with those in live births. Triploidy is found in 17 of every 100 spontaneous abortions but in only about 1 in 10,000 live births; 45,X is found in 18% of chromosomally abnormal abortuses but in only 1 in 7,000 to 10,000 live births.

Chromosomal abnormalities detected by CVS (performed at 10-12 weeks of gestation) and by amniocentesis (at 16 weeks of gestation) show that the abnormalities detected by CVS are two to five times more frequent than those detected by amniocentesis, which in turn are about two times higher than those found in newborns. This gradient in the frequency of chromosomal abnormalities over developmental time is evidence that karyotypically abnormal embryos and fetuses are eliminated by spontaneous abortion throughout gestation.

Birth defects are another consequence of chromosomal abnormalities. The frequencies of chromosomal aberrations detected in cytogenetic surveys of newborns are shown in Table 6.4. Among trisomies, trisomy 16, which is common in spontaneous abortions, is not found among infants, indicating that this condition is completely selected against. Although trisomy 21 occurs with a frequency of about 1 in 900 births, cytogenetic surveys of spontaneous abortions indicate that about two-thirds of such conceptions are lost by miscarriage. Similarly, over 99% of all 45,X conceptions are lost before birth. Overall, while selection against chromosomally abnormal embryos and fetuses is efficient, the high rate of nondisjunction means there is a significant reproductive risk for chromosomal abnormalities. Over 0.5% of all newborns are affected with an abnormal karyotype.

The relationship between the development of cancer in somatic cells of the body and accompanying chromosomal changes is a third consequence of chromosomal abnormalities. A significant number of malignancies, especially leukemias, are associated with specific chromosomal translocations. A wide range of chromosomal abnormalities is present in many solid tumors. New evidence suggests that

Table 6.3 Chromosomal Abnormalities in Spontaneous Abortions

Abnormality	Frequency (%)
Trisomy 16	15
Trisomies, 13, 81, 21	9
XXX, XXY, XYY	1
Other	27
45,X	18
Triploidy	17
Tetraploidy	6
Other	7

Note: Adapted from T. Hassold. (1986). Trends in Genetics 2, 105–110. Adapted with permission of the publisher.

Table 6.4 Chromosomal Abnormalities in Newborns

Abnormality	Approximate Frequency
45,X	1/7500
XXX	1/1200
XXY	1/1000
XYY	1/1100
Trisomy 13	1/15,000
Trisomy 18	1/11,000
Trisomy 21	1/900
Structural abnormalities	1/400

■ **Uniparental disomy** A condition in which both copies of a chromosome are inherited from a single parent.

these abnormalities, which include aneuploidy, translocations, and duplications, may arise during a period of genomic instability that precedes or accompanies the transition of a normal cell into a malignant cell. The mechanisms by which such chromosomal changes can cause or accompany the development of cancer is discussed in Chapter 14.

OTHER FORMS OF CHROMOSOMAL ABNORMALITIES

In some forms of chromosome abnormalities, the karyotype and individual chromosomes appear to be normal. One of these situations is uniparental disomy, where both members of a chromosome pair are derived from one parent, resulting in an abnormal phenotype. Another is the presence of fragile sites, which appear only when cells are grown in the laboratory and certain chemicals are added to the growth medium.

Uniparental Disomy

Normally, meiosis ensures that one member of each chromosomal pair is derived from the mother, and the other member is from the father. On rare occasions, however, a child gets both copies of a chromosome from one parent, a condition known as **uniparental disomy.** This condition can arise in several ways, all of which involve two chromosomal errors in cell division (❱ Figure 6.28). These errors can occur in meiosis or in mitotic divisions following fertilization.

The frequency of uniparental disomy is unknown, but this condition has been identified in some unusual situations including females who are affected with rare X-linked disorders such as hemophilia; in father-to-son transmission of rare, X-linked disorders where the mother is homozygous normal; and where children are affected with rare autosomal recessive disorders, but only one parent is heterozygous. Prader-Willi syndrome and Angelman syndrome (MIM/OMIM 105830) can be caused by deletions in region 15q11.12 or by uniparental disomy. If both copies of chromosome 15 are derived from the mother, Prader-Willi syndrome results. If both copies of chromosome 15 are derived from the father, Angelman syndrome results. The origin of these disorders by uniparental disomy is discussed in detail in Chapter 11.

Fragile Sites

Fragile sites are structural features of chromosomes that become visible when cells are grown under certain conditions. They appear as a gap or a break at a characteristic place on a chromosome and are inherited codominantly. Fragile sites often

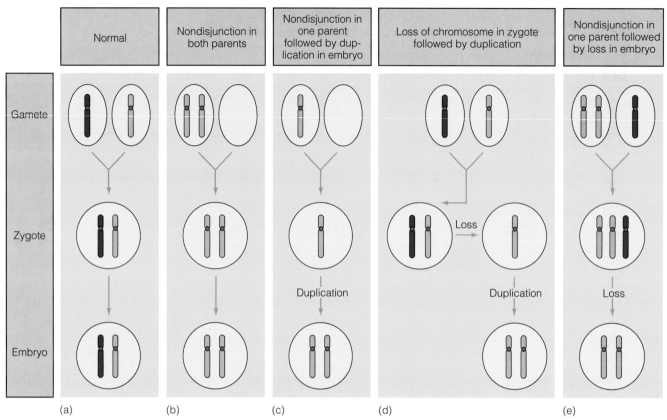

▶ **FIGURE 6.28** Uniparental disomy can be produced by several mechanisms, involving nondisjunction in meiosis or nondisjunction in the zygote or early embryo. (a) Normally, gametes contain one copy of each chromosome, and fertilization produces a zygote carrying two copies of a chromosome, one derived from each parent. (b) Nondisjunction in both parents, where one gamete carries both copies of a chromosome and the other gamete is missing a copy of that chromosome. Fertilization produces a diploid zygote, but both copies of one chromosome are inherited from a single parent. (c) Nondisjunction in one parent, resulting in the loss of a chromosome. This gamete fuses with a normal gamete to produce a zygote monosomic for a chromosome. An error in the first mitotic division results in duplication of the monosomic chromosome, producing uniparental disomy. (d) Normal gametes can fuse to produce a normal zygote, which loses a chromosome at the first mitotic division. A second nondisjunction restores the lost copy, but both copies are derived from one parent. (e) Nondisjunction in one parent produces a gamete that carries both copies of a chromosome. Fusion with a normal gamete produces a trisomic zygote, which loses one copy of the chromosome in nondisjunction, resulting in uniparental disomy.

produce chromosomal fragments, deleted chromosomes, and other alterations in subsequent mitotic divisions. More than seventeen heritable fragile sites have been identified in the human genome (▶ Figure 6.29). The molecular nature of fragile sites is unknown but is of great interest because the sites represent regions that are susceptible to breakage. Almost all studies of fragile sites have been carried out on cells in tissue culture, and it is currently not known whether such sites are expressed in meiotic cells. At least one fragile site is associated with a genetic disorder.

Fragile-X Syndrome Is Associated with Mental Retardation A fragile site near the tip of the long arm of the X chromosome is associated with an X-linked form of mental retardation known as Martin-Bell syndrome, or **fragile-X** (fra-X) (MIM/OMIM 309500) syndrome (▶ Figure 6.30). Males who have this syndrome have long, narrow faces with protruding chins and large ears. They also have enlarged testes and varying degrees of mental retardation. Fragile-X syndrome is the most common form of inherited mental retardation, and about 3 to 5% of the males institutionalized for mental retardation have a fragile-X chromosome. Female carriers show no clear-cut physical symptoms but as a group have a higher rate of mental retardation than normal individuals. The fragile-X syndrome is caused by an alteration in a gene (FMR-1) that is discussed in Chapter 11.

■ **Fragile X** An X chromosome that carries a nonstaining gap, or break, at band q27; associated with mental retardation in males.

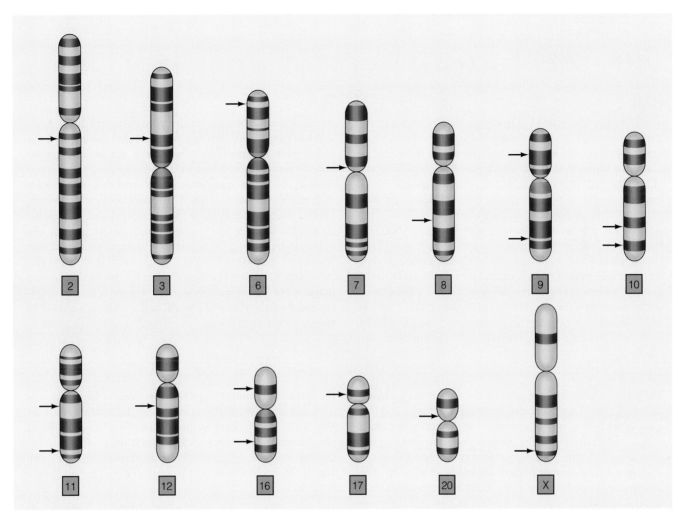

▶ **FIGURE 6.29** The location of the major fragile sites in the human karyotype. These sites appear as gaps or breaks in the chromosome only under certain conditions of cell culture. The site on the X chromosome is associated with mental retardation, and the site on chromosome 3 may be a hot spot for cancer-associated chromosomal rearrangements.

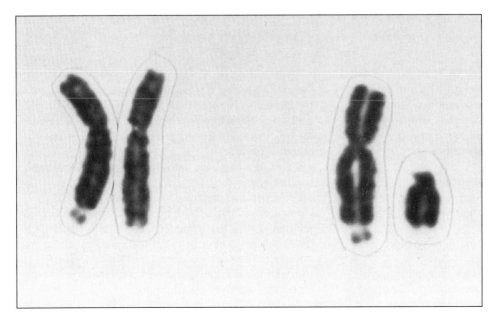

▶ **FIGURE 6.30** The fragile-X syndrome in a carrier female (left) and a male (right). This syndrome causes the lower tip of the X chromosome to appear as a fragile piece hanging by a thread. The nature of the mutation that causes this syndrome is discussed in Chapter 11.

Case Studies

CASE 1

Michelle was a 42-year-old Caucasian woman who had declined counseling and amniocentesis at 16 weeks of pregnancy, but was referred for genetic counseling following an abnormal ultrasound at 20 weeks gestation. Following the ultrasound, a number of findings suggested a possible chromosome abnormality in the fetus. The ultrasound showed nuchal thickening (swelling under the skin at the back of the neck of the fetus), short femur, short humerus, short ear length, and under-development of the middle phalanx of the fifth finger, and echogenic bowel. Michelle's physician then performed an amniocentesis, and a referral was made to the genetics program. The couple felt they did not want genetic counseling before the cytogenetic results were back. This was Michelle's third pregnancy; she and her husband Mike had a 6-year-old daughter and a 3-year-old son. The counselor no-

tified the couple that the results were back and that a problem had been identified (the results revealed trisomy 21). She asked them to come in to discuss the results. The genetic counselor told them the difficult news, explored their understanding of Down syndrome, and elicited their experiences with people with disabilities. The session also considered the clinical concerns revealed by the ultrasound and possible associated anomalies (mild to severe mental retardation, cardiac defects, and kidney problems). The options available to the couple were outlined. They were provided with a booklet written for parents making choices following the prenatal diagnosis of Down syndrome. After a week of careful deliberation with their family, friends, and clergy, they elected to terminate the pregnancy.

CASE 2

The genetic counselor was called to the nursery for a consultation on a newborn who was described as "floppy and had a weak cry." The counselor noted that the newborn's chart indicated that he was having feeding problems and had not gained weight since his delivery 15 days earlier. The counselor noted several other findings during his evaluation. The infant had almond-shaped eyes, a small mouth with a thin upper lip, downturned corners of the mouth and a narrow face. The infant was born with undescended testes and a small penis. The counselor suspected that this child had the genetic disorder known as Prader-Willi syndrome.

The cause of Prader-Willi syndrome is the absence of a small region on the long arm of chromosome 15. It is always the lack of active DNA that should have been contributed by

the father (the mother's copy of this genetic information is normally inactive) that causes Prader-Willi syndrome, and this absence may occur in three different ways: deletion on the paternal chromosome 15, a mutation on the paternal chromosome 15, or maternal uniparental disomy—in other words, both chromosome 15s from the mother and none from the father.

The child and his parents were tested for a deletion in the long arm of chromosome 15 (15q11-q13) by fluorescence in situ hybridization (FISH) and uniparental disomy 15 (polymerase chain reaction (PCR)). In this case, maternal disomy was detected by PCR—which is the cause of Prader-Willi syndrome in about 30% of the cases.

Summary

1. The study of variations in chromosomal structure and number began in 1959 with the discovery that Down syndrome is caused by the presence of an extra copy of chromosome 21. Since then the number of genetic diseases related to chromosomal aberrations has steadily increased.

2. The development of chromosome banding and techniques for identifying small changes in chromosomal structure have contributed greatly to the information that is now available.

3. There are two major types of chromosomal changes: a change in chromosomal number and a change in chromosomal arrangement. Polyploidy, aneuploidy, and changes in chromosomal number are major causes of reproductive failure in humans. Polyploidy is rarely

seen in live births, but the rate of aneuploidy in humans is reported to be more than tenfold higher than in other primates and mammals. The reasons for this difference are unknown, but this represents an area of intense scientific interest.

4. The loss of a single chromosome creates a monosomic condition, and the gain of a single chromosome is called a trisomic condition. Autosomal monosomy is eliminated early in development. Autosomal trisomy is selected against less stringently, and cases of partial development and live births of trisomic individuals are observed. Most cases of autosomal trisomy greatly shorten life expectancy, and only individuals who have trisomy 21 survive into adulthood.

5. Aneuploidy of sex chromosomes shows even more latitude, and monosomy X is a viable condition. Studies of sex chromosome aneuploidies indicate that at least one copy of the X chromosome is required for development. Increasing the number of copies of the X or Y chromosome above the normal range causes progressively greater disturbances in phenotype and behavior, indicating the need for a balance in gene products for normal development.

6. Changes in the arrangement of chromosomes include deletions, duplications, inversions, and translocations. Deletions of chromosomal segments are associated with several genetic disorders, including cri-du-chat and Prader-Willi syndromes. Translocations often produce no overt phenotypic effects but can result in genetically imbalanced and aneuploid gametes. We discussed a translocation resulting in Down syndrome that in effect makes Down syndrome a heritable genetic disease, potentially present in 1 in 3 offspring.

7. Fragile sites appear as gaps, or breaks, in chromosome-specific locations. One of these fragile sites on the X chromosome is associated with a common form of mental retardation that affects a significant number of males.

Questions and Problems

1. Originally, karyotypic analysis relied on size and centromere placement to identify chromosomes. Because many chromosomes are similar in size and centromere placement, the identification of individual chromosomes was difficult, and chromosomes were placed into eight groups, identified by letters A-G. Today, each human chromosome can be readily identified.
   a. What technical advances led to this improvement in chromosome identification?
   b. List two ways this improvement can be implemented.

2. What clinical information does a karyotype give you?

3. A colleague emails you a message that she has identified an interesting chromosome variation at 21q13. In discussing this discovery with a friend who is not a cytogeneticist, explain how you would describe the location, defining each term in the chromosome address 21q13.

4. As a physician, you deliver a baby with protruding heels, clenched fists with the second and fifth fingers overlapping the third and fourth fingers. (a) What genetic disorder do you suspect the baby has? (b) How do you confirm your suspicion?

5. Hypothetical human conditions have been found to have a genetic basis. Suppose a hypothetical genetic disorder is responsible for condition 1 is similar to Marfan syndrome. The defect responsible for condition 2 resembles Edwards syndrome. One of the two conditions results in more severe defects, and death occurs in infancy. The other condition produces a mild phenotypic abnormality and is not lethal. Which condition is most likely lethal and why?

6. Discuss the following sets of terms:

   trisomy and triploidy
   aneuploidy and polyploidy

7. What chromosomal abnormality can result from dispermy?

8. Tetraploidy may result from
   a. lack of cytokinesis in meiosis II
   b. nondisjunction in meiosis I
   c. lack of cytokinesis in mitosis
   d. nondisjunction in mitosis in the early embryo
   e. none of the above

9. A cytology student believes he has identified an individual suffering from monoploidy. The instructor views the cells under the microscope and correctly dismisses the claim. Why was the claim dismissed? What type of cells were being viewed?

10. Describe the process of nondisjunction and when it takes place during cell division.

11. Albinism is caused by an autosomal recessive allele of a single gene. An albino child is born to phenotypically normal parents. However, the paternal grandfather is albino. Exhaustive analysis suggests that neither the mother nor her ancestors carry the allele for albinism. Suggest a mechanism to explain this situation.

12. A geneticist discovers that a girl with Down syndrome has a Robertsonian translocation involving chromosomes 14 and 21. If she has an older brother who is phenotypically normal, what are the chances that he is a translocation carrier?

13. An individual is found to have some tetraploid liver cells but diploid kidney cells. Be specific in explaining how this condition might arise.

14. A spermatogonial cell undergoes mitosis prior to entering the meiotic cell cycle enroute to the production of sperm. However, during mitosis the cytoplasm fails to divide, and only one "daughter" cell is produced. A resultant sperm eventually fertilizes a normal ovum. What is the chromosomal complement of the embryo?

15. A teratogen is an agent that produces non-genetic abnormalities during embryonic or fetal development. Suppose a teratogen is present at conception. As a result, during the first mitotic division the centromeres fail to divide. The teratogen then loses its potency and has no further effect on the embryo. What is the chromosomal complement of this embryo?

16. A woman gives birth to monozygotic twins. One boy has a normal genotype (46,XY), but the other boy has trisomy 13 (47,+13). What events and in what sequence led to this situation?

17. Assume that a meiotic nondisjunction event is responsible for an individual who is trisomic for chromosome 8. If two of the three copies of chromosome 8 are absolutely identical, when during meiosis did the nondisjunction event take place?

18. What is the genetic basis and phenotype for each of the following disorders (use proper genetic notation)?

>   Edwards syndrome
>   Turner syndrome
>   Patau syndrome
>   Kleinfelter syndrome
>   Down syndrome

19. The majority of nondisjunction events leading to Down syndrome are maternal in origin. Based on the duration of meiosis in females, speculate on the possible reasons for females contributing aneuploid gametes more frequently than males.

20. Name and describe the theory, which deals with embryo-uterus interaction, that explains the relationship between advanced maternal age and the increased frequency of aneuploid offspring.

21. Describe the chromosomal alterations and phenotype of cri-du-chat syndrome and Prader-Willi syndrome.

22. If all the nondisjunction events leading to Turner syndrome were paternal in origin, what trisomic condition might be expected to occur at least as frequently?

23. Identify the type of chromosome aberration described in each case:

>   loss of a chromosome segment
>   extra copies of a chromosome segment
>   reversal in the order of a chromosome segment
>   movement of a chromosome segment to another, non-homologous chromosome

24. What are the two prenatal diagnosis techniques used to detect genetic defects in a baby before birth? Which technique can be performed earlier, and why is this an advantage?

25. What are some conditions that warrant prenatal diagnosis?

26. Fragile-X syndrome causes the most common form of inherited mental retardation. What is the chromosomal abnormality associated with this disorder? What is the phenotype of this disorder?

Internet Activities

The following activities use the resources of the World Wide Web to enhance the topics covered in this chapter. To investigate the topics described below, log on to the book's homepage at:

http://www.brookscole.com/biology

1. The University of Washington's Pathology Department maintains a *Cytogallery* with a number of images of human karyotypes and chromosome spreads. Look at the karyotypes from normal males and females, as well as those from individuals with an aneuploidy such as Trysomy 21 (Down Syndrome) or Klinefelter syndrome (47, XXY). Individuals with Klinefelter syndrome have two or more X chromosomes (in addition to a Y chromosome), yet the effects of these excessive X chromosomes in generally much milder than those caused by other polyploid chromosomes. What is unusual about the X chromosome that makes this so? Also, take a look at the karyotypes from individuals with translocated chromosomes, such as the Robertsonian translocation. Translocations, and inversions, do not cause a net gain or loss of genetic material, yet are sometime associated with one or more genetic conditions. How might one of these rearrangements result in a mutant phenotype? (Hint:

think about the chromosomal breakpoints, where the normal chromosomes have been broken and rejoined).

2. Fragile-X syndrome is the leading genetic cause of mental retardation. The Fragile X Research Foundation maintains a Web site that contains information about the disease and also posts information about current research projects. Use the information at this site to supplement what you know about the screening and treatment of fragile-X syndrome.

   a. Describe the features of the fragile-X children shown at this site. What differences can you see between the boys and the girls? Is it worth using physical appearance to identify children who have this syndrome?

   b. What do you think might explain why most affected boys are mentally retarded, but only one-third to one-half of the girls are similarly impaired? Men who inherit the mutant gene (FMR1) that causes fragile-X syndrome but have a normal phenotype are called transmitter males. Try to explain why mothers of transmitter males are normal and have a low risk of bearing fragile-X children, but daughters of transmitter males have a higher risk of bearing affected children (we discuss this topic in Chapter 11).

For Further Reading

Borgaonkar, D. S. (1997). *Chromosome Variation in Man: A Catalogue of Chromosomal Variants and Anomalies*, 8th ed. New York: Wiley-Liss.

Boue, A., Boue, J., & Gropp, A. (1985). Cytogenetics of pregnancy wastage. *Adv. Hum. Genet. 14*: 1–57.

Cassidy, S. B. (1995). Uniparental disomy and genomic imprinting as causes of human genetic disease. *Environ. Mol. Mutagen. 25*: (Suppl. 26) 941–947.

Chu, C. E., & Connor, J. M. (1995). Molecular biology of Turner's syndrome. *Arch. Dis. Child. 72*: 285–286.

Dellarco, V., Voytek, P., & Hollaender, A. (1985). *Aneuploidy: Etiology and Mechanisms*. New York: Plenum.

Faix, R., Barr, M., Jr., & Waterson, J. (1984). Triploidy: Case report of a live-born male and an ethical dilemma. *Pediatrics 74*: 296–299.

Friedmann, T. (1971). Prenatal diagnosis of genetic disease. *Sci. Am. 225* (November): 34–42.

Green, J. E., Dorfmann, A., Jones, S., Bender, S., Patton, L., & Schulman, J. D. (1988). Chorionic villus sampling: Experience with an initial 940 cases. *Obstet. Gynecol. 71*: 208–212.

Grouchy, J. de. (1984). *Clinical Atlas of Human Chromosomes*. New York: Wiley.

Jacobs, P. A., & Hassold, T. J. (1995). The origin of numerical chromosome abnormalities. *Adv. Genet. 33*: 101–133.

Lalande, M. (1996). Parental imprinting and human disease. Annu. Rev. Genet. 30: 173–195.

Linden, M. G., Bender, B. G., & Robinson, A. (1996). Intrauterine diagnosis of sex chromosome aneuploidy. *Obstet. Gynecol. 87*: 468–475.

Lurie, I. W. (1993). Autosomal imbalance syndromes: Genetic interactions and the origin of congenital malformations in aneuploidy syndromes. *Am. J. Med. Genet. 47*: 410–416.

Neely, E. K., & Rosenfeld, R. G. (1994). Use and abuse of human growth hormone. *Ann. Rev. Med. 45*: 407–420.

Oostra, B. A., & Willems, P. J. (1995). A fragile gene. *Bioessays 17*: 941–947.

Philip, J., Bryndorf, T., & Christensen, B. (1994). Prenatal aneuploidy detection in interphase cells by fluorescence in situ hybridization (FISH). *Prenatal Diagn. 14*: 1203–1215.

Rowley, J. D. (1998). The critical role of chromosome translocations in human leukemias. *Annu. Rev. Genet. 32*: 495–519.

Weiss, E., Loevy, H., Saunders, A., Pruzansky, S., & Rosenthal, I. (1982). Monozygotic twins discordant for Ullrich-Turner syndrome. *Am J. Med Genet. 13*: 389–399.

Zinn, A. R., Page, D. C., & Fisher, E. M. (1993). Turner syndrome: The case of the missing chromosome. *Trends Genet. 9*: 90–93.

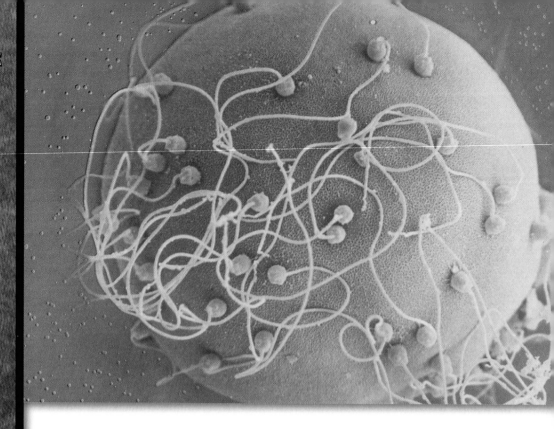

Development and Sex Determination

Louise Brown was born in April 1978 in England. Of the millions of children born that year, Louise represents a landmark in human reproduction and human genetics. She was the first human born after in vitro *fertilization (IVF), a procedure in which gametes are fertilized outside the body and the developing embryo is implanted into the uterus for development. Since then, hundreds of children have been born after IVF, and more than a dozen centers in the United States offer IVF services.*

The development of IVF by Robert Edwards and Patrick Steptoe was a long, slow process conducted over a number of years. After three years of experiments to determine the best conditions for a sperm to fertilize an egg in a petri dish, Edwards began a collaboration with Steptoe, an obstetrician who was an expert in using a laparascope to remove an egg from a woman's ovary. Nine years later, their work progressed to the point where they could remove eggs, fertilize them, and implant the developing embryos into the uterus, but there were no resulting pregnancies. In the fall of 1977, Steptoe recovered an egg from Louise's mother, Lesly, by making a small incision (about one-half inch long) in her abdomen. He inserted the tube-like laparascope to examine the ovary, and another small incision was made through which a second instrument was used to remove an egg. A few minutes after removal, the egg was mixed with semen from Louise's father in a sterile dish, and fertilization took place. The developing embryo was implanted into the uterus via a tube inserted in the vagina, and Louise was born on April 25 of the following year.

This remarkable achievement led to the development of a profusion of reproductive alternatives for infertile couples, some of which are discussed in this chapter. The combination of IVF with recombinant DNA techniques has created a powerful new technology for sex selection, the diagnosis of genetic disorders, and gene transfer. The use of these techniques is covered in Chapter 13.

In this chapter, attention focuses on the biology of human development and the genetics of sex determination and differentiation. The chapter begins with a discussion of the major features of embryonic and fetal development, followed by a section on reproductive technology. The central portion of the chapter covers genetic aspects of sex determination and sex differentiation. The chapter concludes with an examination of how patterns of expression of genes on the X chromosome differ in males and females.

HUMAN DEVELOPMENT FROM FERTILIZATION TO BIRTH

Human reproduction depends on the integrated action of the endocrine system and the reproductive organs. Males and females each possess a pair of gonads and associated accessory glands and ducts. The testes of males produce spermatozoa and sex hormones. The ovaries of females produce eggs or ova and female sex hormones. Within the gonads, cells produced by meiosis mature into gametes.

■ Fertilization The fusion of two gametes to produce a zygote.

Fertilization, the fusion of male and female gametes, usually occurs in the upper third of the oviduct (▶ Figure 7.1). Sperm deposited into the vagina travel through the cervix, up the uterus, and into the oviduct. About 30 minutes after ejaculation, sperm are present in the oviduct (▶ Figure 7.2). Sperm travel this distance by swimming via whiplike contractions of their tails, aided by muscular contractions of the uterus.

Only one sperm fertilizes the egg, but many other sperm assist in this process (▶ Figure 7.3), perhaps by triggering chemical changes in the egg. A single sperm binds

▶ **FIGURE 7.1** Scanning electron micrograph of an oocyte surrounded by sperm. Usually, only one sperm enters the egg.

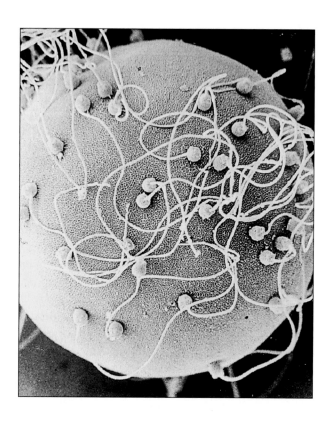

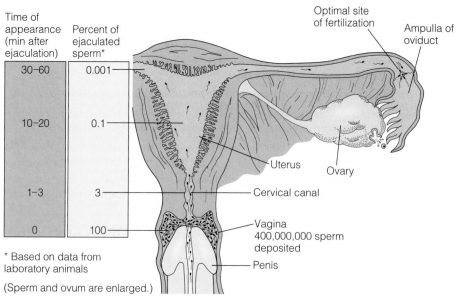

▶ **FIGURE 7.2** The time and relative amounts of sperm transported from the vagina into the oviduct.

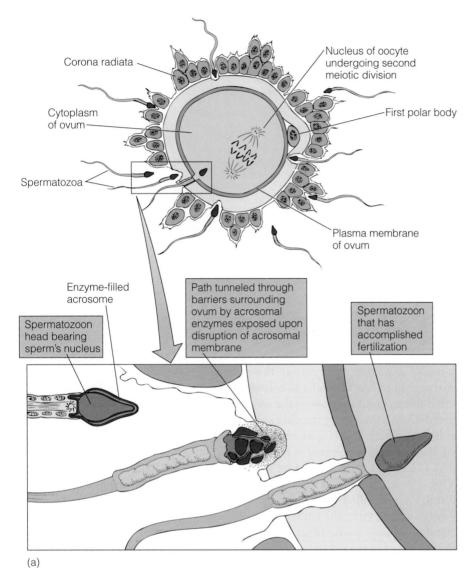

Corona radiata

Cytoplasm of ovum

Spermatozoa

Nucleus of oocyte undergoing second meiotic division

First polar body

Plasma membrane of ovum

Enzyme-filled acrosome

Path tunneled through barriers surrounding ovum by acrosomal enzymes exposed upon disruption of acrosomal membrane

Spermatozoon head bearing sperm's nucleus

Spermatozoon that has accomplished fertilization

(a)

▶ **FIGURE 7.3**   The process of fertilization. The tip of the sperm head, known as the acrosome, contains enzymes that dissolve the outer barriers surrounding the oocyte, allowing the sperm to enter the egg.

to receptors on the surface of the secondary oocyte and fuses with its outer membrane. This attachment triggers a series of chemical changes in the membrane and prevents any other sperm from entering the oocyte. Movement of the sperm into the oocyte cytoplasm initiates the second meiotic division of the oocyte. Fusion of the haploid sperm nucleus with the resulting haploid oocyte nucleus forms a diploid **zygote.**

The zygote travels down the oviduct to the uterus over the next 3 to 4 days. While in the oviduct, the zygote undergoes a series of cell divisions to form a solid ball of cells called a morula. Once the zygote begins to divide, it becomes an embryo. The embryo descends into the uterus and floats unattached in the uterine interior for several days, drawing nutrients from the uterine fluids and continuing to divide to form a **blastocyst** (▶ Figure 7.4).

The blastocyst is made up of the **inner cell mass,** an internal cavity, and an outer layer of cells (the **trophoblast**). During the week or so when the zygote is dividing, the endometrium of the uterus continues to enlarge and differentiate. During implantation, the trophoblast cells adhere to the endometrium and secrete enzymes that allow fingers of trophoblast cells to penetrate the endometrium, locking the embryo into place (Figure 7.4).

■ **Zygote**   The fertilized egg that develops into a new individual.

■ **Blastocyst**   The developmental stage at which the embryo implants into the uterine wall.

■ **Inner cell mass**   A cluster of cells in the blastocyst that gives rise to the embryonic body.

■ **Trophoblast**   The outer layer of cells in the blastocyst that gives rise to the membranes surrounding the embryo.

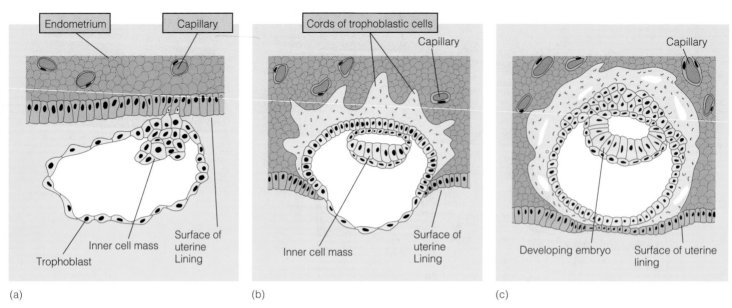

(a)   (b)   (c)

▶ **FIGURE 7.4** The process of implantation. (a) A blastocyst attaches to the endometrial lining of the uterus. (b) As the blastocyst implants, cords of chorionic cells form. (c) When implantation is complete, the blastocyst is buried in the endometrium.

■ **Chorion** A two-layered structure formed from the trophoblast.

At twelve days after fertilization, the embryo is embedded in the endometrium, and the trophoblast has formed a two-layered structure, the **chorion.** One of the first events that follows implantation is the secretion of a peptide hormone, human chorionic gonadotropin (hCG), by the chorion. This hormone prevents breakdown of the uterine lining, which grows and begins secreting hormones to maintain the pregnancy. Excess hCG is eliminated in the urine, and home pregnancy tests work by detecting hCG levels as early as the first day of a missed menstrual period.

As the chorion grows and expands, it forms a series of fingerlike projections that extend into cavities filled with maternal blood. Embryonic capillaries extend into these projections, or villi. The embryonic circulation and the maternal pools of blood are separated by only a thin layer of cells, allowing exchange of nutrients between the embryonic and maternal circulation. These villi are the source of tissue gathered for prenatal diagnosis by chorionic villus sampling (CVS). Further development of this structure forms the placenta. The membranes connecting the embryo to the placenta develop to form the umbilical cord which contains two umbilical arteries and a single umbilical vein (▶ Figure 7.5).

Human Development Is Divided into Three Stages

The period from conception to birth is divided into three trimesters, each about three months (12 weeks) long. During this period of about 38 weeks, the single-celled zygote undergoes about 40–44 rounds of mitosis and produces an infant containing trillions of cells organized into highly specialized tissues and organs.

Organ Formation Occurs in the First Trimester The first 12 weeks of development is a period of radical changes in the size, shape, and complexity of the embryo (▶ Figure 7.6). In the week after implantation, the basic tissue layers are formed, and by the end of the third week, formation of organ systems begins. At the end of the first month the embryo is about 5 mm long, and much of the body is composed of paired segments.

In the second month, the embryo grows dramatically to a length of about 3 cm, and undergoes a 500-fold increase in mass. Most of the major organ systems, including the four chambers of the heart, are formed. The limb buds develop into arms and legs, complete with fingers and toes. The head is very large in relation to the rest of the body because of the rapid development of the nervous system.

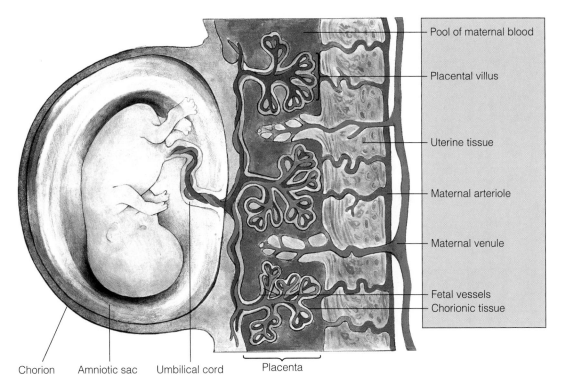

Chorion Amniotic sac Umbilical cord Placenta

▶ **FIGURE 7.5** Maternal and embryonic structures interact to form the placenta.

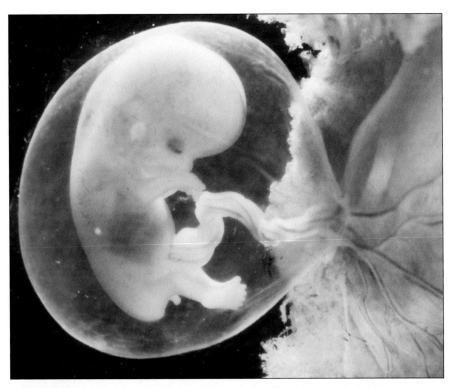

▶ **FIGURE 7.6** A human embryo near the end of the first trimester of development.

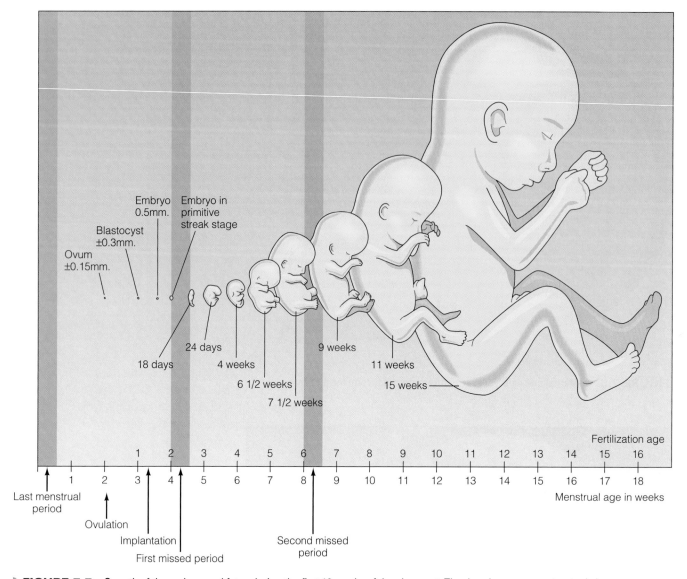

FIGURE 7.7 Growth of the embryo and fetus during the first 16 weeks of development. The drawings represent actual sizes.

By about seven weeks, the embryo is now called a fetus. Although chromosomal sex (XX females and XY males) is determined at the time of fertilization, the fetus is sexually neutral at the beginning of the third month. In the third month, developmental pathways activate different gene sets and initiate sexual development. This process is discussed in detail later in this chapter.

The Second Trimester Is a Period of Organ Maturation In the second trimester, major changes include an increase in size and the further development of organ systems (Figure 7.7). Bony parts of the skeleton begin to form, and the heartbeat can be heard with a stethoscope. Fetal movements begin in the third month, and by the fourth month, the mother can feel movements of the fetus's arms and legs. At the end of the second trimester, the fetus weighs about 700 g (27 ounces) and is 30–40 cm (about 13 inches) long. It has a well-formed face, toes and fingers with nails, and its eyes can open.

Rapid Growth Takes Place in the Third Trimester The fetus grows rapidly in the third trimester, and the circulatory system and the respiratory systems mature to prepare for air breathing. During this period of rapid growth, maternal nutrition is important because a large fraction of the protein the mother eats is used for growth and development of the fetal brain and nervous system. Similarly, much of the calcium in the mother's diet is used to develop the fetal skeletal system.

The fetus doubles in size during the last two months, and chances for survival outside the uterus increase rapidly during this time. In the last month, antibodies from the maternal circulation pass into the fetal circulation, conferring temporary immunity on the fetus. In the first months after birth, the baby's immune system matures and begins to make its own antibodies, and the maternal antibodies disappear. At the end of the third trimester, the fetus is about 50 cm (19 in.) long and weighs from 2.5 to 4.8 kg (5.5 to 10.5 lbs).

Birth Occurs in Stages

Birth is a hormonally induced process. The cervix softens in the last trimester, and the fetus shifts downward. Usually its head is pressed against the cervix. During birth, the cervical opening dilates to allow passage of the fetus, and uterine contractions expel the fetus. The head usually emerges first. If any other body part enters the birth canal first, the result is a breech birth.

A short time after delivery, a second round of uterine contractions begins delivery of the placenta. These contractions separate the placenta from the lining of the uterus, and the placenta is expelled through the vagina.

REPRODUCTION AND TECHNOLOGY

Advances in genetics, physiology, and molecular biology have led to the development of techniques that enhance or reduce the chances of conception. Reproductive technologies can correct defective functions and manipulate the physiology of reproduction.

Contraception Uncouples Sexual Intercourse from Pregnancy

The uncoupling of sexual intercourse from fertilization and pregnancy uses methods that block one of three stages in reproduction: release and transport of gametes, fertilization, and implantation. None of these methods is completely successful in preventing pregnancy or sexually transmitted diseases (STDs) except complete abstinence from sexual intercourse (Figure ▶ 7.8).

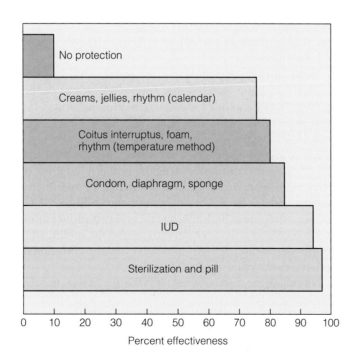

▶ **FIGURE 7.8** Effectiveness of various methods of birth control. The percent effectiveness is a measure of how many women in a group of 100 will not become pregnant in a year when using a given method of birth control.

Tubal ligation A contraceptive procedure in women in which the oviducts are cut, preventing ova from reaching the uterus.

Vasectomy A contraceptive procedure in men in which the vas deferens is cut and sealed to prevent the transport of sperm.

In vitro **fertilization (IVF)** A procedure in which gametes are fertilized in a dish in the laboratory, and the resulting zygote is implanted in the uterus for development.

Aside from abstinence, methods that physically prevent the release and transport of gametes (such as vasectomy and tubal ligation) are the most effective. In **tubal ligation** the oviduct is cut, and the ends are tied off to prevent ovulated eggs from reaching the uterus. In **vasectomy**, the vas deferens is withdrawn through a small incision in the scrotum and cut. The cut ends are sealed to prevent transport of sperm.

Birth control pills are an effective method of birth control. The most common birth control pills contain a combination of hormones that prevent the release of an oocyte from the ovary. Time-release capsules (sold as Norplant) implanted under the skin release hormones slowly and offer long-term suppression of ovulation.

Methods to prevent fertilization include using physical and chemical barriers. Condoms are latex or gut sheaths worn over the penis or pouches inserted into the vagina during intercourse. Only latex condoms prevent sexually transmitted diseases, including AIDS. Diaphragms are caps that fit over the cervix and are inserted before intercourse. Chemical barriers include using spermicidal jelly or foam that kills sperm on contact. They are placed into the vagina just before intercourse. Contraceptive sponges, filled with spermicides, combine physical and chemical barriers. Condoms treated with a chemical spermicide are more successful at preventing pregnancy than either condoms or chemical barriers alone.

RU-486, a drug developed in Europe, is being tested in the United States as a contraceptive. This drug, chemically related to reproductive hormones, interferes with events following fertilization and may inhibit implantation.

Technology Expands Reproductive Choices

About one in six couples are infertile, that is, unable to have children after a year of trying to conceive. Physical and physiological conditions prevent the production of gametes, fertilization, or implantation. Technologies to reduce or overcome these problems have been developed in the last two decades. Blocked oviducts, often the result of untreated STDs, are the leading cause of infertility in females. In males, low sperm count, low motility, and blocked ducts are causes of infertility. Hormone therapy can often be used to increase egg production, and surgical procedures can sometimes be used to open closed ducts in the reproductive tracts. Overall, about 40% of infertility is related to problems in the male reproductive tract, 40% is associated with the female tract, and in 20% of the cases, the infertility is of unknown origin.

One of the first methods of reproductive technology developed was artificial insemination. In the simplest application of this method, the male partner is infertile, and the female receives sperm collected from a donor (Figure 7.9). Sperm can be collected from donors and stored in liquid nitrogen at sperm banks which offer artificial insemination.

Methods to recover and fertilize gametes outside the body, known as *in vitro* fertilization *(IVF)*, are widely used. Several variations of this technology are available (Figure 7.9). IVF can be used when ovulation is normal, but the oviducts are blocked. IVF is mainly offered at reproductive clinics in major medical centers at a cost of about $10,000 per procedure.

Reproductive technology has altered accepted patterns of reproduction. In the United States, variations of surrogate motherhood are a reproductive option. In one version, a woman is artificially inseminated by sperm and carries the child to term. After the child is born, she surrenders the child to the father and his mate. In this case, the surrogate is both the genetic and gestational mother of the child. In another version, a couple provides both the egg and the sperm, and the surrogate is implanted with the developing embryo and serves as the gestational mother but is genetically unrelated to the child she bears.

Following the discovery that it is the age of ova, not the reproductive system, that is responsible for infertility as women age, women are now becoming mothers in their late fifties and early sixties. After hormonal treatment, they receive zygotes for implantation produced by fertilization of donated eggs from younger women. Fertil-

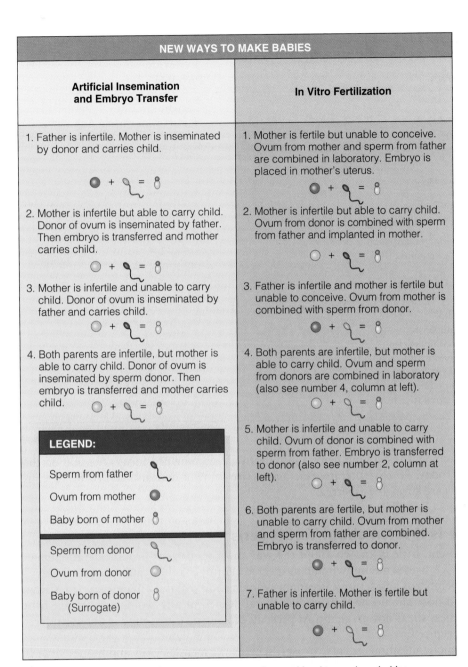

NEW WAYS TO MAKE BABIES

Artificial Insemination and Embryo Transfer	In Vitro Fertilization

Artificial Insemination and Embryo Transfer

1. Father is infertile. Mother is inseminated by donor and carries child.

2. Mother is infertile but able to carry child. Donor of ovum is inseminated by father. Then embryo is transferred and mother carries child.

3. Mother is infertile and unable to carry child. Donor of ovum is inseminated by father and carries child.

4. Both parents are infertile, but mother is able to carry child. Donor of ovum is inseminated by sperm donor. Then embryo is transferred and mother carries child.

LEGEND:

Sperm from father

Ovum from mother

Baby born of mother

Sperm from donor

Ovum from donor

Baby born of donor (Surrogate)

In Vitro Fertilization

1. Mother is fertile but unable to conceive. Ovum from mother and sperm from father are combined in laboratory. Embryo is placed in mother's uterus.

2. Mother is infertile but able to carry child. Ovum from donor is combined with sperm from father and implanted in mother.

3. Father is infertile and mother is fertile but unable to conceive. Ovum from mother is combined with sperm from donor.

4. Both parents are infertile, but mother is able to carry child. Ovum and sperm from donors are combined in laboratory (also see number 4, column at left).

5. Mother is infertile and unable to carry child. Ovum of donor is combined with sperm from father. Embryo is transferred to donor (also see number 2, column at left).

6. Both parents are fertile, but mother is unable to carry child. Ovum from mother and sperm from father are combined. Embryo is transferred to donor.

7. Father is infertile. Mother is fertile but unable to carry child.

▶ **FIGURE 7.9** Some of the ways gametes can be combined to produce babies.

ized eggs can now be collected and frozen for later use. This allows younger women to collect eggs while risks for chromosome abnormalities in the offspring are low and to use them over a period of years, including menopause, to become mothers.

These and other unconventional means of generating a pregnancy have developed more rapidly than the social conventions and laws governing their use. In the process, controversy about the moral, ethical, and legal grounds for using these techniques has arisen but is not yet resolved.

TERATOGENS POSE A RISK TO THE DEVELOPING FETUS

Teratogens are agents that produce abnormalities during embryonic or fetal development. They produce nongenetic birth defects, not heritable changes. In 1960 only four or five agents were known to be teratogens. The discovery that a tranquilizer,

■ **Teratogen** Any physical or chemical agent that brings about an increase in congenital malformations.

Table 7.1   Human Teratogens

Known	Possible
Radiation	Cigarette smoking
Fallout	High levels of vitamin A
X-rays	Lithium
	Zinc deficiency
Infectious Agents	
Cytomegalovirus	
Herpes virus II	
Rubella virus	
Toxoplasma gondii (spread in cat feces)	
Maternal Metabolic Problems	
Phenylketonuria (PKU)	
Diabetes	
Virilizing tumors	
Drugs and Chemicals	
Alcohol	
Aminopterins	
Chlorobiphenyls	
Coumarin anticoagulants	
Diethylstilbestrol	
Tetracyclines	
Thalidomide	

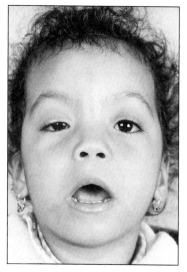

▶ **FIGURE 7.10** A child who has fetal alcohol syndrome has misshapen eyes, flat nose, and face characteristic of this condition.

■ **Fetal alcohol syndrome** A constellation of birth defects caused by maternal drinking during pregnancy.

thalidomide, caused limb defects in unborn children helped focus attention on this field. Today, 30 to 40 teratogenic agents are known, and another 10 to 12 compounds are strongly suspected.

Little Is Known about Teratogens

At present, little is known about the way most teratogens produce fetal damage. Table 7.1 lists some known and suspected teratogens that cause defects in developing embryos and fetuses. Pregnant women should avoid all unnecessary X rays, and no dose should be delivered to the abdomen of any woman of childbearing age unless she is not pregnant. At present, diagnostic ultrasound is not considered teratogenic.

Among infectious agents, viruses, such as the measles virus and herpes virus II, which is associated with genital herpes, can cause severe brain damage and mental retardation in a developing fetus. The damaging effects of the herpes II virus occur only when the mother becomes infected with herpes during pregnancy. There is no damage to the fetus when the mother has been infected before pregnancy and has a recurring attack during pregnancy. Herpes virus I (associated with cold sores) does not appear to be teratogenic. Maternal metabolic abnormalities, such as phenylketonuria (PKU), can also be teratogenic.

Fetal Alcohol Syndrome

Prenatal exposure to alcohol is by far the most serious and the most widespread teratogenic problem. Alcohol consumption during pregnancy results in spontaneous abortion, growth retardation, facial abnormalities (▶ Figure 7.10), and mental retardation. This collection of defects is known as **fetal alcohol syndrome (FAS).** In milder forms, the condition is known as fetal alcohol effects. The incidence of these two conditions is about 1.9 affected infants per thousand births for FAS and 3.5 per thousand for fetal alcohol effects. While this incidence may seem low, the economic and social consequences are significant. In the United States the cost for surgical repair of all facial abnormalities and treatment of all sensory and learning problems and mental retardation is more than $320 million per year, a large part of which is used in treating fetal alcohol syndrome. For example, about 11% of the total budget for treating institutionalized mentally retarded individuals is for those affected by FAS.

The teratogenic effects of alcohol can occur at any time during pregnancy, but weeks 8–12 are particularly sensitive periods. Even in the third trimester, alcohol can seriously impair fetal growth. Consumption of one ounce of absolute alcohol (the amount contained in two mixed drinks) per day in the third trimester of pregnancy reduces the birth weight of the fetus by about 160 gm. Drinking the equivalent of two mixed drinks in one day at any time in the last 3 months of pregnancy will reduce fetus birth weight by 5%. Each day of consumption of a similar amount will reduce the birth weight by another 5%. Low birth weight is associated with high rates of neonatal death. Newborns below 50% of normal birth weight have a 45% mortality rate.

Alcohol produces its effects by constricting the blood vessels of the placenta and umbilical cord, reducing the supply of oxygen to the fetus. Measurements on placentas recovered from normal births show that umbilical blood vessels constrict in the presence of alcohol concentrations as low as 0.05%, the amount contained in 1 to 1.5 drinks. Oxygen deprivation may play a role in the behavioral defects and mental retardation associated with FAS. Low levels of blood alcohol can cause serious problems in fetal development, and pregnant women should avoid all alcohol during pregnancy.

Although the actions of alcohol as a teratogen are now well known, work is needed to resolve the degree of risk involved with chemicals and substances that are suspected teratogens and to identify new teratogens among the thousands of chemicals currently used. More importantly, it is necessary to investigate the genetic basis of susceptibility to teratogenic agents, especially drugs and chemicals.

SEX DETERMINATION IN HUMANS

In humans, as in many other species, there are obvious phenotypic differences between the sexes, a condition known as sexual dimorphism. In some organisms the differences are limited to the gonads; in others, including humans, secondary sex characteristics such as body size, muscle mass, patterns of fat distribution, and amounts and distribution of body hair emphasize the differences between the sexes.

Chromosomes Can Help Determine Sex

What determines maleness and femaleness is a complex interaction between genes and the environment. In some organisms, environmental factors play a major role (◗ Figure 7.11). For example, in some reptiles such as turtles or crocodiles, sex is determined by the temperature at which the eggs develop. Eggs incubated at higher temperatures produce females; those at lower temperatures produce males. In other reptiles, the opposite is true: Higher incubation temperatures produce males, and lower temperatures result in females. In humans, on the other hand, sex determination is primarily associated with the sex chromosomes. As discussed in Chapter 2, females have two X chromosomes (XX) and males have an X and a Y chromosome.

Although the XX-XY mechanism of sex determination seems straightforward, it does not provide all the answers to the question of what determines maleness and femaleness. Is a male a male because he has a Y chromosome or because he does not have two X chromosomes? This question was answered about 25 years ago by the discovery that some humans carry an abnormal number of sex chromosomes. Rarely, individuals who have only 45 chromosomes (45,X) are born, and these individuals

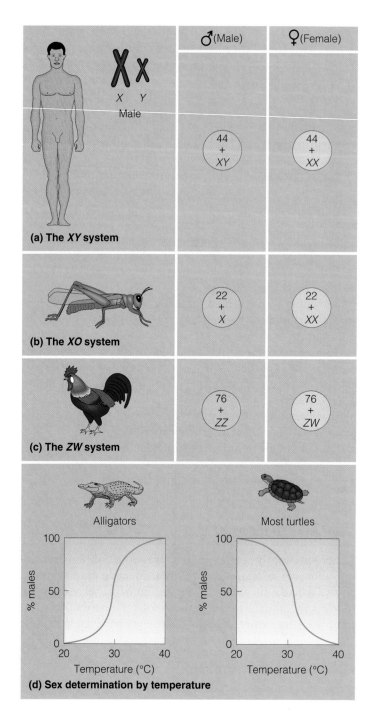

FIGURE 7.11 (a-c) Animals have several mechanisms of sex determination that involve chromosomes. (d) In some reptiles, the temperature at which the egg is incubated determines the sex of the offspring.

♂(Male)   ♀(Female)

(a) The *XY* system

Male

44 + XY   44 + XX

(b) The *XO* system

22 + X   22 + XX

(c) The *ZW* system

76 + ZZ   76 + ZW

Alligators   Most turtles

(d) Sex determination by temperature

are female. At about the same time, males who carry two X chromosomes and a Y chromosome were discovered (47,XXY). From the study of these and other individuals with abnormal numbers of sex chromosomes, it is clear that some females may have only one X chromosome and that some males can have more than one X chromosome. Furthermore, anyone who has a Y chromosome is almost always male, no matter how many X chromosomes he may have.

Chromosome studies have led to the conclusion that under normal circumstances, the male phenotype is associated with the presence of a Y chromosome, and the absence of a Y chromosome results in the female phenotype. However, two X chromosomes are required for normal female development and a single X chromosome is required for normal male development.

The Sex Ratio in Humans

All gametes produced by females contain an X chromosome, whereas males produce roughly equal numbers of gametes that carry an X chromosome and gametes that carry a Y chromosome. Because the male makes two kinds of gametes, he is called the **heterogametic** sex. The female is **homogametic** because she makes only one type of gamete. An egg fertilized by an X-bearing sperm results in an XX zygote that will develop as a female. Fertilization by a Y-bearing sperm will produce an XY, or male, zygote (▶ Figure 7.12). Therefore, the male gamete determines the sex of the offspring.

Because the sex of the offspring is determined by the presence or absence of a Y chromosome and because males produce approximately equal numbers of X- and Y-bearing gametes, males and females should be produced in equal proportions (Figure 7.12). This proportion, known as the **sex ratio,** changes throughout the life cycle. At fertilization, the sex ratio (known as the primary sex ratio) should be 1:1. Although direct determinations are impossible, estimates indicate that more males are conceived than females. The sex ratio at birth, known as the secondary sex ratio, is about 1.05 (105 males for every 100 females). The tertiary sex ratio is the ratio measured in adults. At 20 to 25 years of age, the ratio is close to 1.00; thereafter, females outnumber males in ever-increasing proportions. There are several reasons for the higher death rate among males, including both environmental and genetic factors. Accidents are the leading cause of death among males aged 15 to 35 years, and the expression of deleterious X-linked recessive genes also leads to a higher death rate among males.

■ **Heterogametic** Producing gametes that contain different kinds of sex chromosomes. In humans, males produce gametes that contain X or Y chromosomes.

■ **Homogametic** Producing gametes that contain only one kind of sex chromosome. In humans, all gametes produced by females contain only an X chromosome.

■ **Sex ratio** The relative proportion of males and females belonging to a specific age group in a population.

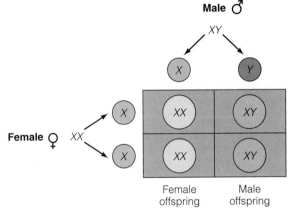

▶ **FIGURE 7.12** The segregation of sex chromosomes into gametes and the random combination of X-bearing or Y-bearing sperm with an X-bearing egg produces, on average, a 1:1 ratio of males to females.

SEX DIFFERENTIATION FOLLOWS SEX DETERMINATION

Sex is chromosomally determined at fertilization, but the expression of this chromosomal state as a phenotype occurs in stages over time. Following is a discussion of factors that influence the course of these events.

Chromosomal Sex and Phenotypic Sex

Sex determination by the XX-XY method provides a genetic framework for the developmental events that guide the zygote toward the acquisition of male or female phenotypes. The process of forming male or female reproductive structures depends on several factors, including gene action, interactions within the embryo, interaction with other embryos that may be in the uterus, and interactions with the maternal environment. As a result of these interactions, the chromosomal sex (XX or XY) of an individual may differ from the phenotypic sex. These differences arise during embryonic and fetal development and can produce a phenotype opposite to the chromosomal sex, intermediate to the phenotypes of the two sexes, or a phenotype that has characteristics and genitalia of both sexes. The sex of an individual can be defined at several levels: chromosomal sex, gonadal sex, and phenotypic sex. In most cases, all of these definitions are consistent, but in others they are not (see Concepts and Controversies: Sex Testing in International Athletics: Is It Necessary?). To understand these variations and the interactions of genes with the environment, we first consider the events in normal sexual development.

Events in Embryogenesis Begin Sex Differentiation

The first step in sex differentiation occurs at fertilization with the formation of a diploid zygote that has an XX or XY chromosomal constitution. Although the chromosomal sex of the zygote is established at fertilization, the embryo that develops is

Concepts and Controversies

Sex Testing in International Athletics—Is It Necessary?

Success in amateur athletics, including the Olympics, is often a prelude to financial rewards and acclaim as a professional athlete. Several methods are used to guard against cheating in competition. Competitors in many international events are required to submit urine samples (collected while someone watches) for drug testing. In other cases, this is done at random in an attempt to eliminate the use of steroids or performance-enhancing drugs. In the 1960s, concern about males attempting to compete as females led the International Olympic Committee (IOC) to institute sex testing for athletes, beginning with the 1968 Olympics. Sex testing may seem reasonable, given that men are usually larger, stronger, and faster than females, and males competing as females would have an unfair advantage.

The IOC test involved analysis of epithelial cells recovered by scraping the inside of the mouth. In genetic females (XX), the inactivated X chromosome forms a Barr body, which can be stained and seen under a microscope. Genetic males (XY) do not have a Barr body. The procedure is noninvasive, and females are not required to submit to a physical examination of their genitals. If sexual identity were called into question as a result of the test, a karyotype was required, and if necessary, a gynecological examination followed.

In practice, the IOC test has been unsuccessful. The test for a Barr body is unreliable and leads to both false positive and false negative results. It fails to take into account the situation where a female can be XY and have testicular feminization and other conditions that result in a discrepancy between chromosomal and phenotypic sex. In addition, the test does not take into account the psychological, social, and cultural factors that enter into one's identity as a male or a female. The test has not identified any men who attempted to compete as females, but has barred several women from competition in every Olympic game since 1968.

An analysis of sex testing of more than 6000 women athletes has found that 1 in 500 had to withdraw from competition as a consequence of failing the sex test. In response to criticism, the IOC and the International Amateur Athletic Federation (IAAF) developed different responses. Since 1991, all IAAF athletes are required to have a physical examination and have their sexual status certified. The IOC has instituted a new test, based on recombinant DNA technology, to detect the presence of the male-determining gene SRY, carried on the Y chromosome. A positive test makes the athlete ineligible to compete as a female. However, this new test remains controversial and fails to recognize several chromosomal combinations that result in a female phenotype, even though an SRY gene is present.

The basic question remains unanswered: Why are such sex tests needed in the first place? The IOF and the IAAF continue to debate the question and review their policies, but at this point are not willing to give up sex testing.

---

sexually ambiguous for the first month or so. The external genitalia of early embryos are neither male nor female, but are indifferent. Internally, both male and female reproductive ducts and associated structures are present. The two internal duct systems are the Müllerian and Wolffian ducts (⬧ Figure 7.13). The Müllerian duct system forms the female reproductive ducts, and the Wolffian system gives rise to the male reproductive duct system.

At 7 weeks of development, diverging developmental pathways activate different sets of genes and establish the gonadal sex of the embryo. This process takes place over the next 4 to 6 weeks. Although it is convenient to think of only two pathways, one leading to males and the other to females, there are many alternate pathways that produce intermediate outcomes in gonadal sex and in sexual phenotypes, some of which we consider in the following paragraphs.

In male development, the presence of a Y chromosome causes the indifferent gonad to begin development as a testis. The selection of this pathway results from the action of genes on the Y chromosome. Products from these activated genes stimulate the growth and differentiation of a small collection of cells within the indifferent gonad, causing the gonad to develop as a testis. A gene called the **sex-determining region of the Y (SRY)** (MIM/OMIM 480000), mapped to a region on the short arm of the Y chromosome, plays a major role in starting the cascade of gene action that

■ **SRY** A gene called the sex-determining region of the Y, located near the end of the short arm of the Y chromosome, that plays a major role in causing the undifferentiated gonad to develop into a testis.

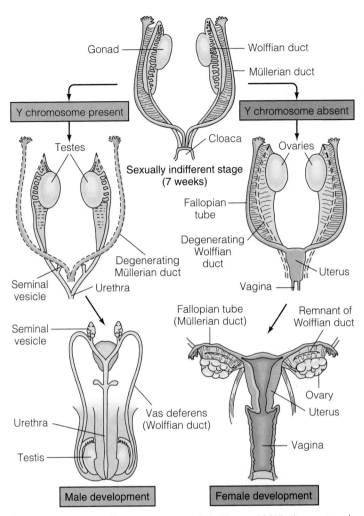

FIGURE 7.13 Two duct systems (Wolffian and Müllerian systems) are present in the early embryo. They enter different developmental pathways in the presence and absence of a Y chromosome.

causes the indifferent gonad to begin testis development. Other genes on the Y chromosome and on autosomes also play important roles at this time.

Once testis development is initiated, cells in the testis secrete two hormones, **testosterone** and the **Müllerian inhibiting hormone (MIH)**. Along with gene expression, these hormones control further sexual development. Testosterone stimulates the development of the male internal duct system (Wolffian ducts), including the epididymis, seminal vesicles, and vas deferens. MIH inhibits further development of female duct structure and causes degeneration of the Müllerian ducts (Figure 7.13).

In female development, the absence of the Y chromosome and the presence of the second X chromosome cause the embryonic gonad to develop as an ovary. Cells along the outer edge of the gonad divide and push into the interior, forming the ovary. In the absence of testosterone, the Wolffian duct system degenerates (Figure 7.13). In the absence of MIH, the Müllerian duct system develops to form the fallopian tubes, uterus, and parts of the vagina. After gonadal sex has been established, the third phase of sexual differentiation, the appearance of sexual phenotype, begins (▶ Figure 7.14). In males, testosterone is metabolized and converted into another hormone, dihydroxytestosterone (DHT), which directs formation of the external genitalia. Under the influence of DHT and testosterone, the genital folds and

■ **Testosterone** A steroid hormone produced by the testis; the male sex hormone.

■ **Müllerian inhibiting hormone (MIH)** A hormone produced by the developing testis that causes the breakdown of the Müllerian ducts in the embryo.

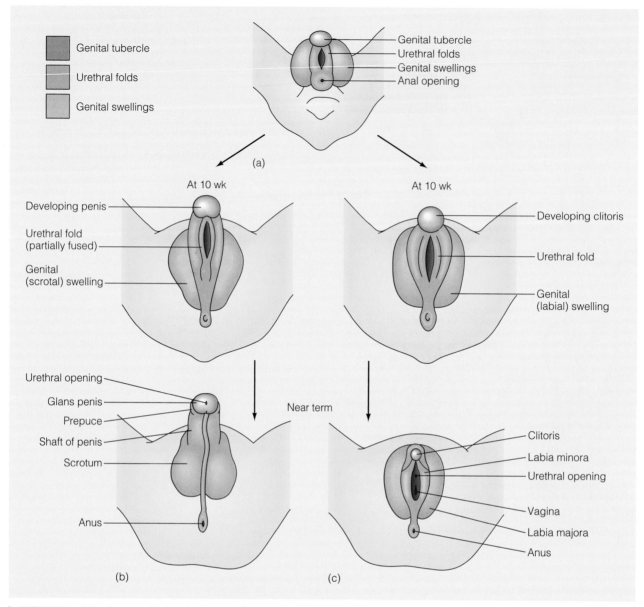

Genital tubercle

Urethral folds

Genital swellings

Genital tubercle
Urethral folds
Genital swellings
Anal opening

(a)

At 10 wk

Developing penis

Urethral fold
(partially fused)

Genital
(scrotal) swelling

At 10 wk

Developing clitoris

Urethral fold

Genital
(labial) swelling

Urethral opening

Glans penis

Prepuce

Shaft of penis

Scrotum

Near term

Clitoris

Labia minora

Urethral opening

Vagina

Labia majora

Anus

Anus

(b)

(c)

▶ **FIGURE 7.14** Steps in the development of phenotypic sex from the undifferentiated stage (a) to the male (b) or female (c) phenotype. The male pathway of development takes place in response to the presence of testosterone and dihydroxytestosterone (DHT). Female development takes place in the absence of these hormones.

genital tubercle develop into the penis, and the surrounding labioscrotal swelling forms the scrotum.

In females, the genital tubercle develops into the clitoris, the genital folds form the labia minora, and the labioscrotal swellings form the labia majora (Figure 7.14).

In terms of gene action, it is important to note that the development of gonadal sex and the sexual phenotype results from different developmental pathways (▶ Figure 7.15). In males, this pathway involves induction by several genes on the Y chromosome, the presence of a single X chromosome, and one or more autosomal genes. In females, this pathway involves the presence of two X chromosomes, the absence of Y chromosome genes, and presumably other autosomal genes. These distinctions indicate that there may be important differences in the way genes in these respective pathways are activated, and they may provide clues in the search for genes that regulate these pathways.

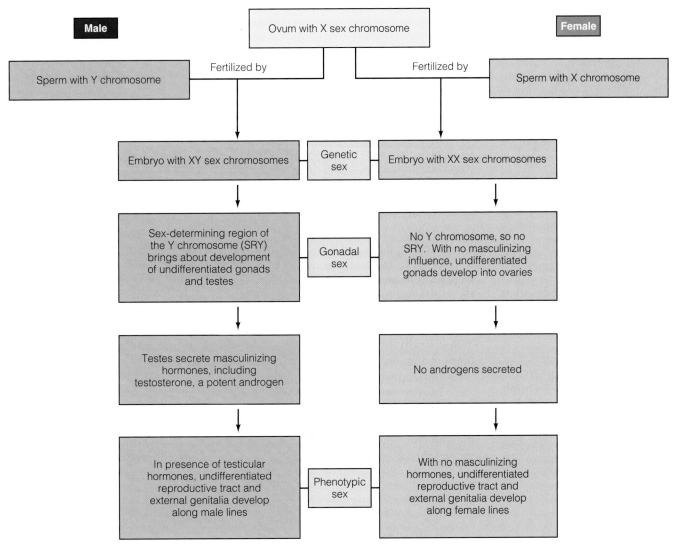

▶ **FIGURE 7.15** The major pathways of sexual differentiation and the stages where genetic sex, gonadal sex, and phenotypic sex are established.

GENETIC CONTROL OF SEXUAL DIFFERENTIATION

Developmental pathways that begin with the indifferent gonad often result in a gonadal and/or sexual phenotype that is at variance with the chromosomal sex of XX for females and XY for males. These different outcomes can result from several causes: chromosomal events that exchange segments of the X and Y chromosomes; mutations that affect the ability of cells to respond to the products of Y chromosome genes; or action of autosomal genes that control events on the X and/or Y chromosome. In addition, interactions between the embryo and maternal hormones in the uterus and the presence of other embryos in the uterus can affect the outcome of both the gonadal sex and the sexual phenotype.

Part of our understanding of sexual development is derived from the study of variations in this process, including single-gene mutations that produce altered sexual phenotypes. We briefly consider three situations in which there is a lack of concordance among chromosomal sex, gonadal sex, and sexual phenotype.

True hermaphrodites are organisms that possess both ovaries and testes and the associated duct systems. In some species, such as the earthworm, this condition is normal. In humans, a true hermaphrodite has ovarian and testicular tissue

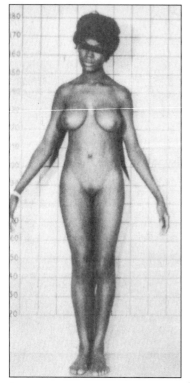

▶ **FIGURE 7.16** A phenotypic female who has an XY chromosomal constitution results from androgen insensitivity.

■ **Androgenic insensitivity** An X-linked genetic trait that causes XY individuals to develop into phenotypic females.

in separate gonads or in a single combined gonad. Cytogenetic examination of several hermaphrodites has shown that they are sex chromosome mosaics: Some cells in the body are XX, and others are XY or XXY. In other cases, only XY cells are found.

Androgen Insensitivity and Phenotypic Sex

Androgen insensitivity (MIM/OMIM 313700) syndrome is an X-linked trait in which chromosomal males develop as females. In this case, chromosomal sex (XY) is opposite from phenotypic sex (▶ Figure 7.16). During sexual development, testis formation is induced normally, and testosterone and MIH production begin as expected. MIH brings about the degeneration of the Müllerian duct system so that no internal female reproductive tract is formed. However, a mutation of an X-chromosome gene that encodes an androgen receptor blocks the ability of cells to respond to testosterone or DHT. As a result, development proceeds as if there were no testosterone or DHT present. The Wolffian duct system degenerates and the indifferent genitalia develop as female structures. Individuals who have this syndrome are phenotypic females who don't menstruate and have developed breasts and very little pubic hair (see Concepts and Controversies: Joan of Arc—Was It Really John of Arc?). The androgen receptor gene (MIM/OMIM 313700) maps to the long arm of the X chromosome in the Xq11-13 region.

Gene Expression and Sexual Phenotype

Pseudohermaphrodites are individuals in whom the phenotypic sex does not match the sex chromosomal constitution. In many cases, these individuals have ambiguous

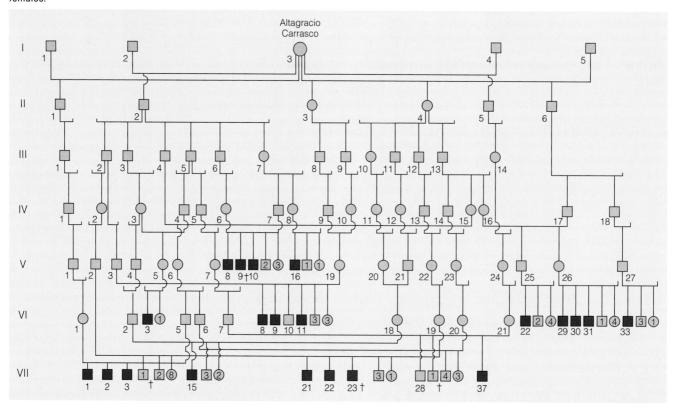

▶ **FIGURE 7.17** Pedigree of a pseudohermaphroditism in several generations of residents of a cluster of villages in the Dominican Republic. Although the condition was first diagnosed in generation V, members of earlier generations were probably affected.

Concepts and Controversies

Joan of Arc—Was It Really John of Arc?

Joan of Arc, the national heroine of France, was born in a village in northeastern France in 1412, during the Hundred Years' War. At the age of 13 or 14 years, she began to have visions that directed her to help fight the English at Orleans. Following victory, she helped orchestrate the crowning of the new king, Charles VII. During a siege of Paris, the English captured Joan, and in 1431 she was tried for heresy. Although her trial was technically a religious one conducted by the English-controlled Church, it was clearly a political trial. Shortly after being sentenced to life imprisonment, she was declared a relapsed heretic, and on May 30, 1431 she was burned at the stake in the marketplace at Rouen.

In 1455 Pope Callistus formed a commission to investigate the circumstances of her trial, and a Trial of Rehabilitation took place over a period of 7 months in 1456. The second trial took testimony from over 100 individuals who knew Joan personally. Extensive documentation from the original

trial and the Trial of Rehabilitation exists. This material has served as the source for the more than 100 plays and countless books written about her life. Although the story of her life is well known, perhaps more remains to be discovered. From an examination of the original evidence, R. B. Greenblatt has proposed that Joan had phenotypic characteristics of testicular feminization. By all accounts, Joan was a healthy female who had well-developed breasts. Those living with her in close quarters testified that she never menstruated, and physical examinations conducted during her imprisonment revealed a lack of pubic hair. While such circumstantial evidence is not enough for a diagnosis, it provides more than enough material for speculation. This speculation also provides a new impetus for those medico-genetic detectives who prowl through history, seeking information about the genetic makeup of the famous, infamous, the notorious, and the obscure.

genitals. One form of **pseudohermaphroditism** (MIM/OMIM 264300) is caused by an autosomal recessive gene. This condition, associated with an XY chromosome constitution, prevents the conversion of testosterone to DHT. In these cases, the Y chromosome initiates the development of testes, and the male duct system and internal organs are properly formed from the Wolffian ducts. MIH secretion prevents the development of female internal structures. However, the failure to produce DHT results in genitalia that are essentially female. The scrotum resembles the labia, a blind vaginal pouch is present, and the penis resembles a clitoris. Although chromosomally male, these individuals are raised as females.

At puberty, however, masculinization takes place. The testes descend into a developing scrotum, and the phallus develops into a functional penis. The voice deepens, a beard grows, and muscle mass increases as in normal males. Biopsy indicates that spermatogenesis is normal. The increased levels of testosterone that accompany puberty mediate these changes. This condition is rare, but in a group of small villages in the Dominican Republic, more than 30 such cases are known. The high incidence of homozygous recessives can be attributed to common ancestry through intermarriage (⯈ Figure 7.17). In 12 of the 13 families, a line of descent can be traced to a single ancestor (I-3). This mutation and the one that produces testicular feminization are evidence for the importance of gene interactions in normal development. Not only is the process of sexual development clearly under genetic control, but the ability to respond to the hormonal environment is critical to normal sexual differentiation.

■ **Pseudohermaphroditism** An autosomal genetic condition that causes XY individuals to develop the phenotypic sex of females.

DOSAGE COMPENSATION AND THE X CHROMOSOME

Because females have two doses of all genes on the X chromosome and males have only one dose of most genes, it would seem that females should have twice as much of these gene products as males.

Expression of Genes on the X Chromosome

In Chapter 4, we discussed hemophilia A, an X-linked genetic disorder in which clotting factor VIII is deficient. Because normal females have two copies of this gene and normal males have only one, should the blood of females contain twice as much clotting factor VIII as that of males? Careful measurements indicate that females have the same amount of this clotting factor as males. In fact, the same is true for all X chromosome genes tested: The level of X-linked gene products is the same in males and females. Somehow, differences in gene dosage are regulated to produce equal amounts of gene products in both sexes. How that is accomplished in humans and how it came to be understood is an interesting story.

Barr Bodies and X Inactivation

An explanation of how **dosage compensation** works in humans is known as the **Lyon hypothesis,** after Mary Lyon, a British geneticist. She discovered that dosage compensation is accomplished by inactivating, or turning off, almost all of the genes on one of the X chromosomes in females. She based this idea on both genetic and cytological evidence.

The genetic evidence came from studies on coat color in mice. In female mice heterozygous for X-linked coat-color genes, Lyon observed that coat color was not like that of either homozygote, nor was it intermediate to the homozygotes. Instead the coat was composed of patches of the two parental colors in a random arrangement. Males, hemizygous for either gene, never showed such patches and had coats of uniform color. This genetic evidence suggested to Lyon that in heterozygous females, both alleles were active but not in the same cells.

The cytological observations were made beginning in 1949 by Murray Barr and his colleagues. Barr, a physiologist, was studying nerve function and used nerve cells obtained from cats. He observed that the nuclei of nerve cells from female cats contained a small, dense mass of chromatin located near the inner surface of the nuclear membrane (◗ Figure 7.18). Nerve-cell nuclei from male cats did not contain this structure, referred to as the Barr body. Based on his work, in the late 1950s Susumo Ohno suggested that the Barr body is actually an inactivated, condensed X chromosome. The Lyon hypothesis summarizes these findings as follows:

* One X chromosome is active in the somatic cells (not the germ cells) of female mammals, and the second X chromosome is randomly inactivated and tightly coiled to form the Barr body.
* The inactive chromosome can originate either paternally or maternally, and one X chromosome is randomly inactivated in each cell of the body.
* Inactivation takes place early in development. After four to five rounds of mitosis following fertilization, each cell of the embryo randomly inactivates one X chromosome.

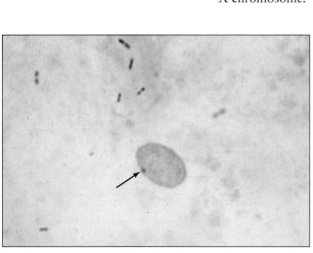

(a)

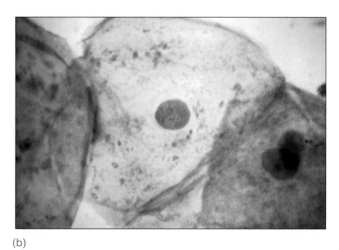

(b)

◗ **FIGURE 7.18** (a) Nucleus from a female cell showing a Barr body (arrow). (b) Nucleus from a male cell shows no Barr body.

- This inactivation is permanent (except in germ cells) and all descendants of a given cell will have the same X chromosome inactivated.
- The random inactivation of one X chromosome in females makes males and females equivalent for the activity of X-linked genes.

Females Are Mosaics for X-Linked Genes

The Lyon hypothesis means that female mammals are actually **mosaics**, constructed of two different cell types: One has the maternal X chromosome active, and one has the paternal X chromosome active. This explains the pattern of coat color that Lyon observed in the heterozygous mice. In females heterozygous for X-linked coat-color genes, patches of one color are interspersed with patches of another color. According to the Lyon hypothesis, each patch represents a group of cells descended from a single cell in which the inactivation event occurred.

Another perhaps more familiar example of this mosaicism is the tortoiseshell cat (▶ Figure 7.19). In cats, a gene for coat color on the X chromosome has two alleles, a dominant mutant allele (O) that produces an orange/yellow coat color and a recessive normal allele (o) that produces a black color. Heterozygous females (O/o) have a tortoiseshell coat with patches of orange/yellow fur mixed with patches of black fur (white fur on the chest and abdomen in such cats is controlled by a different, autosomal gene). Therefore, tortoiseshell cats are invariably female because hemizygous males would either be all orange/yellow or all black.

Mosaicism has also been demonstrated in human females (▶ Figure 7.20). A gene on the X chromosome encodes the information for an enzyme called G6PD (MIM/OMIM 305900) and has two alleles. Each allele produces a distinct and separable form of the enzyme. Heterozygous females were identified by pedigree analysis, and skin cells from these females were isolated and grown individually. Each culture, grown from a single cell, showed only one form or the other of the enzyme.

Random inactivation of the X chromosome in humans occurs very early in embryogenesis, usually at or before the 32-cell stage of development. Given the small number of cells present at the time of inactivation, it is possible that an imbalance in the ratio of inactivated paternal and maternal chromosomes might result by chance, causing females to express X-linked traits for which they are

▶ **FIGURE 7.19** The differently colored patches of fur on this tortoiseshell cat result from X-chromosome inactivation.

■ **Mosaic** An individual composed of two or more cell types of different genetic or chromosomal constitution.

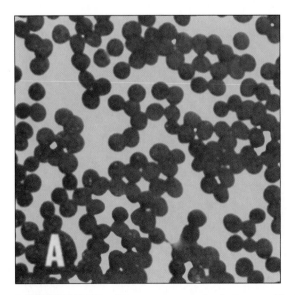 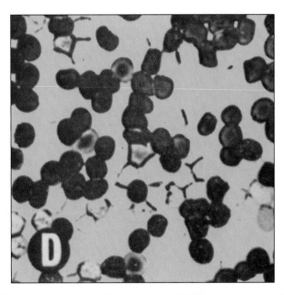

▶ **FIGURE 7.20** Dosage compensation in humans. The X-linked gene that encodes the enzyme G6PD has two alleles. One produces an active form of the enzyme; the other produces an inactive form. (a) Blood cells from an individual who has two active alleles stained to show enzymatic activity. (b) Blood cells from a heterozygous female. About half of the cells are stained, reflecting the presence of an active allele. The unstained cells contain an inactive allele. Heterozygotes are mosaics composed of two cell types. One cell has G6PD enzymatic activity, and one has no activity.

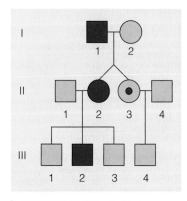

FIGURE 7.21 Pedigree showing monozygotic female twins (II-2 and II-3) discordant for color blindness. The twins inherited the allele for color blindness from their father. Almost all of the active X chromosomes in the color-blind twin carry the mutant allele. Almost all of the active X chromosomes in the twin who has normal vision carry the allele for normal vision.

■ **Sex-influenced genes** Loci that produce a phenotype conditioned by the sex of the individual.

■ **Pattern baldness** A sex-influenced trait that acts like an autosomal dominant trait in males and an autosomal recessive trait in females.

■ **Sex-limited genes** Loci that produce a phenotype in only one sex.

■ **Precocious puberty** An autosomal dominant trait expressed in a sex-limited fashion. Heterozygous males are affected, but heterozygous females are not.

heterozygous. This imbalance has been demonstrated a number of times by observing monozygotic twins, one of whom exhibits an X-linked recessive trait whereas the other does not (❯ Figure 7.21).

Two female identical twins in this pedigree are heterozygotes for red-green color blindness through their color-blind father. One of the twins has normal color vision, and the other has red-green color blindness. The color-blind twin has three sons, two who have normal vision and one who is color-blind (see pedigree).

X inactivation was analyzed in these twins by molecular techniques, allowing the parental origin of active and inactive X chromosomes to be assigned. Testing of skin fibroblasts indicates that in the twin who is color-blind, almost all of the active X chromosomes are paternal X chromosomes that carry the allele for color blindness. In the twin who has normal vision, the opposite situation is observed; almost all of the active X chromosomes are maternal in origin.

The color blindness in one twin and normal color vision in the other twin can be explained by X inactivation associated with twinning. One twin was formed from a small number of cells, most of which had the paternal X inactivated. The other twin originated from a small number of cells, most of which had the maternal X chromosome inactivated.

Sex-Influenced and Sex-Limited Traits

Sex-influenced genes are those that are expressed in both males and females but whose frequencies are much different than would be predicted by Mendelian ratios. These genes are usually autosomal and illustrate the effect of sex on the level of gene expression. **Pattern baldness** (MIM/OMIM 109200) is an example of sex-influenced inheritance. This trait is expressed more often in males than in females. The gene acts as an autosomal dominant in males and as an autosomal recessive in females. In this case, the differential expression of pattern baldness in males and females is related to the different levels of testosterone present. The hormonal environment and the genotype interact in determining expression of this gene.

Sex-limited genes are expressed only in one sex, whether they are inherited in an autosomal or sex-linked pattern. One such gene, an autosomal dominant that controls **precocious puberty** (MIM/OMIM 176410), is expressed in heterozygous males but not in heterozygous females. Affected males undergo puberty at four years of age or earlier. Heterozygous females are unaffected but pass this trait on to half of their sons, making it hard to distinguish this trait from a sex-linked gene. Genes that deal with traits such as breast development in females and facial hair in males are other examples of sex-limited genes, as are virtually all other genes that deal with secondary sexual characteristics.

Case Studies

CASE 1

Jan, a 32-year-old, had been married to her husband, Darryl, for seven years. They had attempted to have a baby on several occasions. Five years ago they had a first trimester miscarriage followed by an ectopic pregnancy later that same year. Jan continued to see her "regular" OB/GYN for infertility problems but was very unsatisfied with the response. After her fourth miscarriage, she went to see a fertility specialist who diagnosed her with severe endometriosis and polycystic ovarian disease (detected by hormone studies). The infertility physician explained that these two conditions were hampering her ability to become pregnant and thus placing her in the category of infertile. She referred Jan to a genetic counselor. At the appointment, the counselor explained to Jan that one form of endometriosis is a genetic disorder, inherited as an autosomal dominant trait, and that polycystic ovarian disease can also be a genetic disorder, and is the most common reproductive disorder among women. The counselor recommended that a detailed family history of both Jan and Darryl would help establish whether Jan's problems have a genetic component, and whether her daughters would be at risk for one or both of these disorders. In the meantime, Jan is in the process of taking hormones and she and Darryl are considering alternative modes of reproduction.

CASE 2

Melissa was referred to genetic counseling because she was 16 weeks pregnant and had a history of epileptic seizures. She takes medication (valproic acid) for her seizures and has not had an attack for the last three years. Her physician became concerned when he learned that she was still taking this medication, against his advice, during her pregnancy. He wanted her to speak to a counselor about the possible effects this medication could have on the developing fetus. The counselor took a detailed family history, which indicated that Melissa was the only individual who had seizures and no other genetic conditions were apparent in the family. The counselor asked Melissa why she continued to take valproic acid during her pregnancy. Melissa stated she was "afraid her child would be like her, if she didn't take her medicine." Melissa went on to say that she was teased as a child when she would have her "fits" and she wanted to prevent that from happening to her children.

With this in mind, the counselor reviewed the process of fetal development and why it is best that a physician should carefully evaluate all medications that a woman takes while she is pregnant. The specific medication that Melissa is taking has been shown to cause spina bifida—affecting almost twice as many children who are exposed to it than children who are not exposed. The counselor explained (with pictures) that spina bifida is a neural tube defect that occurs when the neural tube fails to completely fold during development. The failure to fold exposes part of the spinal area when an infant is born. Valproic acid could also cause problems in the heart and in the genital area.

The counselor explained that prenatal diagnosis using ultrasound, and possibly amniocentesis, could help determine whether the baby's spine has closed properly.

Postscript: Melissa elected to have an ultrasound, which showed that the baby did not have a neural tube defect. However, she was offered an amniocentesis because of possible false negative outcome of the ultrasound. She declined the amniocentesis and Melissa delivered a healthy baby boy.

Summary

1. Human development begins with fertilization and the formation of a zygote. Mitotic divisions of the zygote form an early embryo, called the blastocyst. The embryo implants in the uterine wall, and a placenta develops to nourish the embryo.

2. Human development is divided into three stages, or trimesters, of about 12 weeks each. The first trimester is a period of organ formation and growth to the fetal stage. Growth and maturation of the organ systems takes place in the second trimester. The third trimester is a period of rapid growth.

3. Although mechanisms of sex determination vary from species to species, the presence of a Y chromosome in humans is normally associated with male sexual development, and the absence of a Y chromosome is associated with female development.

4. Early in development, the Y chromosome signals the indifferent gonad to begin development as a testis. Further stages in male sexual differentiation, including the development of phenotypic sex, are controlled by hormones secreted by the testis.

5. In cases of sex-influenced and sex-limited inheritance, the sex of the individual affects whether and the degree to which the trait is expressed. This holds true for both autosomal and sex-linked genes. Sex hormones and the developmental history of the individual are thought to modify expression of these genes, giving rise to altered phenotypic ratios.

Questions and Problems

1. Describe, from fertilization, the major pathways of normal male sexual development, and include the stages where genetic sex, gonadal sex, and phenotypic sex are determined.
2. How does IVF differ from artificial fertilization?
3. Give an example where genetic sex, gonadal sex and phenotypic sex do not coincide. Explain why they do not coincide.
4. Explain the biological basis for a pregnancy test. What body fluid is used for the test?
5. The gestation of a fetus occurs over nine months and is divided into three trimesters. Describe the major events that occur in each trimester. Is there a point where the fetus becomes more "human"?
6. Fetal alcohol syndrome (FAS) is caused by alcohol consumption during pregnancy. It can result in spontaneous abortion, growth retardation, facial abnormalities and mental retardation. How does FAS affect all of us, not just the unlucky children born with this syndrome? What steps need to be taken to prevent this syndrome?
7. Which pathway of sexual differentiation is regarded as the default pathway (male or female)? Why?
8. The absence of a Y chromosome in an early embryo causes:
   a. the embryonic testis to become an ovary
   b. The Wolffian duct system to develop
   c. The Müllerian duct system to degenerate
   d. the indifferent gonad to become an ovary
   e. the indifferent gonad to become testis
9. How can an individual who is XY be phenotypically female?
10. Discuss whether the following individuals are: 1) gonadally male or female, 2) phenotypically male or female (discuss Wolffian/Müllerian ducts and external genitalia), and 3) sterile or fertile:
    a. XY, homozygous for a recessive mutation in the testosterone gene, which renders the gene nonfunctional
    b. XX, heterozygous for a dominant mutation in the testosterone gene, which causes continuous production of testosterone
    c. XY, heterozygous for a recessive mutation in the MIH gene
    d. XY, homozygous for a recessive mutation in the TDF gene
    e. XY, homozygous for a recessive mutation in the MIH gene.
11. Individuals with an XXY genotype are sterile males. If one X is inactivated early in embryogenesis, the genotype of the individual effectively becomes XY. Why should not this individual develop as a normal male?
12. It has been shown that hormones interact with DNA to turn certain genes on and off. Use this fact to explain sex-linked and sex-influenced traits.

13. Assume that human-like creatures exist on Mars. As in the human population on Earth, there are two sexes and even sex-linked genes. The gene for eye color is an example of one such gene. It has two alleles. The purple allele is dominant to the yellow allele. A purple-eyed female alien mates with a purple-eyed male. All of the male offspring are purple-eyed, while half of the female offspring are purple-eyed and half are yellow-eyed. Which is the heterogametic sex?
14. Calico cats are almost invariably female. Why? (Explain genotype and phenotype of calico females and the theory of why calicos are females).
15. How many Barr bodies would the following individuals have:
    a. normal male
    b. normal female
    c. Kleinfelter male
    d. Turner female
16. Males have only one X chromosome and therefore only one copy of all genes on the X chromosome. Each gene is directly expressed, thus providing the basis of hemizygosity in males. Females have two X chromosomes, but one is always inactivated. Therefore females, like males, have only one functional copy of all the genes on the X chromosome. Again, each gene must be directly expressed. Why then are females not considered hemizygous, and why are they not afflicted with sex-linked recessive diseases as often as males?
17. What method of sex testing did the Olympics committee previously use? What method are they currently using? Do either of these methods conclusively test for "femaleness"? Explain.
18. Researchers are currently learning how to transfer sperm-making cells from fertile male mice into infertile male mice in the hopes of learning more about reproductive abnormalities. These donor spermatogonia cells have developed into mature spermatozoa in 70% of the cases and some recipients have gone on to father pups. This new advance opens the way for a host of experimental genetic manipulations. It also offers enormous potential for correcting human genetic disease. One human application that this procedure might be useful in is treating infertile males who wish to be fathers.
    a. Do you foresee any ethical or legal problems with the implementation of this technique? If so, elaborate on them.
    b. Could this procedure have the potential for misuse? If so, explain how.
19. What do you think are the legal and ethical issues surrounding the use of *in vitro* fertilization?
    a. How can these issues be resolved?
    b. What should be done with the extra gametes that are removed from the woman's body but never implanted in her uterus?

Internet Activities

The following activities use the resources of the World Wide Web to enhance the topics covered in this chapter. To investigate the topics described below, log on to the book's homepage at:

http://www.brookscole.com/biology

1. *The Visible Embryo* contains images and descriptions of each stage of human development, from conception to birth. Follow each stage and read about the development of the embryo and the fetus.

   a. Prior to stage 6, the cells of the embryo proliferate and are implanted into the uterine wall, and the embryo proper is essentially an undifferentiated group of cells. At stages 6-7, however, gastrulation begins. What are the three layers produced during gastrulation? Can you name one adult tissue that is derived from each of these layers?

   b. By stage 9, a number of primitive cell types are distinguishable, such as neural cells, muscle cells, and blood cells. Cells undergoing differentiation begin to express a number of cell-type specific genes. What type of genes might be expressed in each of these cell types? Can you imagine what mechanisms might act to limit the expression of these genes to each cell type? What types of genes are likely to be expressed in all cells, regardless of the cell type?

2. Embryonic development of the nervous system, heart, eye, and ear is the focus of the Embryonic Development Web site maintained by the University of Pennsylvania.

   a. Examine the illustrations of the heart as it develops at 30, 40, and 50 days. Describe the changes and new structures formed at each of these stages.

   b. Consider the evolution of these organs for a moment. Can you formulate an evolutionary explanation for the development of a complex sensory organ such as the eye or the ear? What might be the starting point? Do you think the organ evolved in increments or by a revolutionary change? How do you explain these changes at the level of genes? Do you need new genes; if so, where do they come from? Do you adapt genes to new functions: if so, what happens to their old functions?

For Further Reading

Abel, E. L. (1995). An update on the incidence of FAS: FAS is not an equal opportunity birth defect. *Neurotoxicol. Teratol.* 17: 437–443.

Bratton, R. L. (1995). Fetal alcohol syndrome: How you can prevent it. *Postgrad. Med.* 98: 197–200.

Epstein, C. J. (1995). The new dysmorphology: Applications of insights from basic developmental biology to the understanding of human birth defects. *Proc. Nat. Acad. Sci. USA* 92: 8566–8573.

Goodfellow, P. N. (1986). The case of the missing H-Y antigen. *Trends Genet.* 2: 87.

Graves, J. A. (1995). The origin and function of the mammalian Y chromosome and Y-borne genes—an evolving understanding. *Bioessays* 17: 311–320.

Greenblatt, R. B. (1981). Case history: Jeanne d'Arc—Syndrome of feminizing testes. *Br. J. Sex Med.* 8: 54.

Hawkins, J. R. (1994). Sex determination. *Hum. Mol. Genet.* 3 Spec. Nos. 1463–1467.

Hossain, A. M., Barik, S., Rizk, B., and Thorneycroft, I. H. (1998). Preconceptional sex selection: Past, present, and future. *Arch. Androl.* 40: 3–14.

Imperato-McGinley, J., Guerrero, L., Gautier, T., & Peterson, R. G. (1974). Steroid 5-alpha reductase deficiency in man: An inherited form of pseudohermaphroditism. *Science* 186: 1213–1215.

Jiminez, R., and Burgos, M. (1998). Mammalian sex determination: Joining pieces of the genetic puzzle. *Bioessays* 20: 696–699.

Kelley, R. L., & Kuroda, M. I. (1995). Equality for X chromosomes. *Science* 270: 1607–1610.

Lyon, M. F. (1962). Sex chromatin and gene action in the mammalian X-chromosome. *Am. J. Hum. Genet.* 14: 135–148.

Lyon, M. R. (1989). X-chromosome inactivation as a system of gene dosage compensation to regulate gene expression. *Prog. Nucleic Acids Res. Mol. Biol.* 36: 119–130.

Lyon, M. R. (1998). X-chromosome inactivation: A repeat hypothesis. *Cytogenet. Cell Genet.* 80: 133–137.

Nordqvist, K. (1995). Sex differentiation-gonadogenesis and novel genes. *Int. J. Dev. Biol.* 39: 727–736.

Ott, J. (1986). Y-linkage and pseudoautosomal linkage. *Am. J. Hum. Genet.* 38: 891–897.

Sulton, C., Lobaccaro, J., Belon, C., Terraza, A., & Lumbroso, S. (1992). Molecular biology of disorders of sex differentiation. *Horm. Res.* 38: 105–113.

Whitfield, L., Lovell-Badge, R., & Goodfellow, P. (1993). Rapid sequence evolution of the mammalian sex-determining gene SRY. *Nature* 364: 713–715.

Willard, H. F. (1996). X chromosome inactivation, XIST, and the pursuit of the X-inactivation center. *Cell* 86: 5–7.

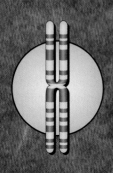

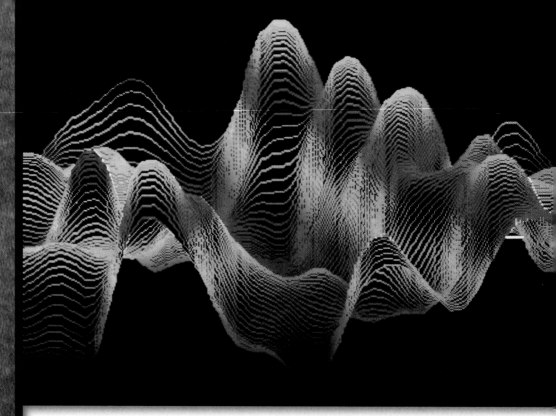

DNA Structure and Chromosomal Organization

Chapter Outline

*E*arly in the 1860s, Frederick Miescher set out to study the chemical composition of human white blood cells to understand something of the nature of cellular mechanisms associated with life. But such cells were difficult to obtain in the quantities needed for chemical studies. Miescher learned that cells present in the pus of infected wounds were derived from the white blood cells. In the days before antibiotics, wound infections were common. Miescher visited local hospitals to collect discarded bandages. He scraped the pus from these bandages and developed a method of separating the pus cells from debris and bandage fragments by washing with a salt solution.

In his early experiments Miescher recovered a chemical substance from the nuclei of the pus cells. To analyze this substance further, he decided to first purify the nuclei. As a first step, he broke open the pus cells by treating them with a protein-digesting substance called pepsin. To obtain pepsin, he prepared extracts of pig stomachs (a good source of pepsin, which functions in digestion). He then treated the pus cells for several hours with the extract of pig stomach. He noted that a gray sediment collected at the bottom of the flask. Under the microscope this sediment turned out to be pure nuclei. Miescher was therefore the first to isolate and purify a cellular organelle.

By chemically extracting the purified nuclei, Miescher obtained a substance he called nuclein. Chemical analysis revealed that it contained hydrogen, carbon, nitrogen, oxygen, and phosphorus. Miescher showed that nuclein was found in other cell types, including kidney, liver, sperm, and yeast. He regarded it as an important component of most cells. Many years later it was shown that his nuclein contained DNA.

At about the same time that Miescher was carrying out his experiments, Mendel outlined the rules for the inheritance of physical traits and developed the notion of what we now call genes. In the 1880s August Weismann and others emphasized the importance of the nucleus in heredity. At the turn of the century Walter Sutton and Theodore Boveri noted that the behavior of genes in inheritance paralleled that of nuclear components (the chromosomes) in meiosis. Later workers confirmed that, in fact, genes are part of chromosomes, and it was generally agreed that the genetic material was to be found in the nucleus. Through all of this work, however, the most basic question remained unanswered: What is the nature of the genetic material?

The answer to this question takes us from the level of the gene as the physical unit of heredity to the level of nucleic acid molecules as the chemical components of cells most closely involved with storing, expressing, and transmitting genetic information. The path to this answer runs from the experiments of Miescher through 80 years to the experiments of Avery and his colleagues in the 1940s, and beyond. Although several lines of evidence provided clues pointing to nucleic acids, especially deoxyribonucleic acid (DNA), as the chemical answer to this problem, most theories were originally based on proteins as the molecular carriers of genetic information. The general requirements for the genetic material, however, were clear and unambiguous. Any chemical structure proposed as the carrier of genetic information must explain the observed properties of genes: replication, information storage, expression of the stored information, and mutation. It was not until the middle of this century that this issue was resolved.

In this chapter we examine the events that led to the confirmation of DNA as the molecule that carries genetic information, and we consider the work of

Watson and Crick on the organization and structure of DNA. We also explore what is known about the way DNA is incorporated into the structure of chromosomes.

DNA CARRIES GENETIC INFORMATION

Research in the first few decades of this century established that genes exist and are carried on chromosomes. But what is a gene? As is often the case in science, the answer to this question came from a completely unexpected direction, the study of an infectious disease.

At the beginning of this century, pneumonia was a serious public health threat and the leading cause of death in the United States. Medical research was directed at understanding the nature of this infectious disease as a step toward developing an effective treatment, perhaps in the form of a vaccine. The unexpected outgrowth of this research was the discovery of the chemical nature of the gene.

Transfer of Genetic Traits in Bacteria

By the 1920s, it was known that pneumonia is caused by a bacterial infection and that one form of pneumonia is caused by *Streptococcus pneumoniae*. Frederick Griffith studied the difference between two strains of this bacterium. In one strain (strain S), the cells were contained in a capsule. This strain was infective and caused pneumonia (that is, was a virulent strain). The other strain (strain R) did not form a capsule and was not infective. The results of Griffith's experiment are straightforward and easily interpreted. Griffith showed that mice injected with living cells of strain R did not develop pneumonia, but mice injected with live cells from strain S developed pneumonia and soon died. Mice injected with heat-killed strain S cells survived and did not develop pneumonia (▶ Figure 8.1).

Mice injected with a mixture of heat-killed strain S cells and live cells from strain R developed pneumonia and died. Griffith recovered live strain S bacteria with capsules from the bodies of the dead mice. When grown in the laboratory, the progeny of these transformed cells were always strain S. After further experiments, Griffith concluded that the living cells from strain R were transformed into strain S cells in the bodies of the mice. Griffith explained his results by proposing that hereditary information had passed from the dead strain S cells into the strain R cells, allowing them to make a capsule and become virulent. He called this process **transformation** and the unknown material the **transforming factor.**

In 1944, after almost a decade of work, a team at the Rockefeller Institute in New York that included Oswald Avery, Colin MacLeod, and Maclyn McCarty discovered that the transforming factor is DNA. The story of this discovery is recounted in a readable memoir by Maclyn McCarty entitled *The Transforming Principle: Discovering That Genes Are Made of DNA* (New York: Norton, 1985).

In a series of experiments that stretched over 10 years. Avery and his colleagues investigated a bacterial strain that causes pneumonia. Griffith had shown that the ability to cause pneumonia is associated with the presence of a thick capsule surrounding the bacterial cell. When an extract of heat-killed smooth bacteria (strain S) is mixed with living

■ **Transformation** The process of transferring genetic information between cells by DNA molecules.

■ **Transforming factor** The molecular agent of transformation; DNA.

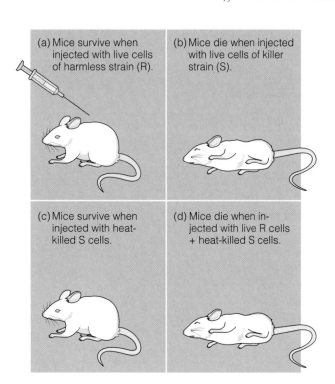

(a) Mice survive when injected with live cells of harmless strain (R).

(b) Mice die when injected with live cells of killer strain (S).

(c) Mice survive when injected with heat-killed S cells.

(d) Mice die when injected with live R cells + heat-killed S cells.

▶ **FIGURE 8.1** Griffith discovered that the ability to cause pneumonia is a genetic trait that can be passed from one strain of bacteria to another. (a) Mice injected with strain R do not develop pneumonia. (b) Mice injected with strain S develop pneumonia and die. (c) When the S strain cells are killed by heat treatment before injection, mice do not develop pneumonia. (d) When mice are injected with a mixture of heat-killed S cells and live R cells, they develop pneumonia and die. Griffith concluded that the live R cells acquired the ability to cause pneumonia from the dead S cells.

DNA as a Commercial Product

The ad for the perfume reads: "Where does love originate? Is it in the mind? Is it in the heart? Or in our genes?" A perfume named DNA has been recently introduced and is marketed in a helix-shaped bottle. There is no actual DNA in the fragrance, but the molecule is invoked to sell the idea that love emanates from the genes. Seem strange? Well, how about jewelry that actually contains DNA from your favorite celebrities? In this line of products, DNA in a single hair or cheek cell is amplified by a process called the polymerase chain reaction (PCR). The resulting solution, containing millions of copies of the DNA, is added to small channels drilled into acrylic earrings, pendants, or bracelets. The liquid can be colored to contrast with the acrylic and be more visible. Just as people wear T-shirts with pictures of Elvis or Einstein, they can now wear jewelry containing DNA from their favorite entertainer, poet, composer, scientist, or athlete. For dead heroes, the DNA can come from a lock of hair; in fact, a single hair will do. How about music composed from the base sequence of DNA? Composers have translated the four bases of

DNA (adenine, guanine, cytosine, and thymine) into musical notes. Long sequences of bases, retrieved from computer databases, are converted into notes, transferred to sheet music, and played by instruments or synthesizers as the music of the genes. Those in the know say that the DNA near chromosomal centromeres sounds much like the music of Bach or other Baroque composers, but that music from other parts of the genome has a contemporary sound.

From a scientific standpoint, this fascination with DNA may be a little difficult to understand, but DNA has clearly captured popular fancy and is being used to sell an ever-increasing array of products. DNA has name recognition. Over the past 40 years, DNA has moved from scientific journals and textbooks to the popular press and even the comic strips. The relationship between genes and DNA is now well known enough to be used in commercials and advertisements. In a few years, this fascination will probably fade and be replaced with another fad, but now, if you want it to sell, relate it to DNA.

rough cells (strain R) a small fraction of the rough cells acquire the ability to form a capsule, grow as smooth colonies, and cause pneumonia. The bacteria that have acquired the ability to form a capsule transmit this trait to all their offspring, indicating that the trait is heritable.

Avery and his colleagues identified DNA as the active component present in the extract from the heat-killed cells. To confirm that DNA was the transforming substance, they treated the preparation with enzymes that destroy protein and RNA before transformation. This treatment removed any residual protein or RNA from the preparation, but did not affect the transforming activity. As a final test, the preparation was treated with deoxyribonuclease, an enzyme that digests DNA, whereupon the transforming activity was abolished.

The work of Avery and his colleagues produced two important conclusions. In the bacterium they studied

- DNA carries genetic information. Only DNA transfers heritable information from one strain to another strain.
- DNA controls the synthesis of specific products. Transfer of DNA also results in the transferring the ability to synthesize a specific gene product (in the form of a capsule).

DNA Is a Part of Chromosomes

Transformation cannot be performed on eukaryotic organisms, so it was not possible to duplicate Avery's results with organisms, such as *Drosophila* or mice. Because it was generally accepted that the chromosomes contain genetic information, indirect evidence was employed to strengthen the link between DNA and chromosomes. A refined method of cytochemical staining indicated that DNA is largely confined to the nucleus and is intimately associated with the chromosomes. Furthermore, DNA is present along the length of chromosomes in a way that corresponds to the distribution of genetic loci. Last, the concentration of DNA within the cells of a given eukaryotic organism is correlated with the number of chromosomes carried by the cell. Most somatic diploid (*2n*) cells contain twice the number of chromosomes as haploid (*n*) gametes. Measurements of DNA concentration indicate that somatic cells have twice as much DNA as gametes. These and other forms of indirect evidence support the idea that DNA is the genetic material of eukaryotic organisms.

Direct evidence for the role of DNA as the genetic material has come from the development of **recombinant DNA technology**. DNA segments from organisms such as humans can be spliced into bacterial DNA, and under the proper conditions this hybrid DNA molecule can direct the synthesis of a human gene product. The synthesis of human proteins in bacteria requires the presence of specific human DNA sequences, providing direct evidence for the role of DNA as the genetic material in higher organisms. In fact, as awareness of DNA and its role in genetics has grown, DNA is even being used to sell products (see Concepts and Controversies: DNA as a Commerical Product).

■ **Recombinant DNA technology** Technique for joining DNA from two or more different organisms to produce hybrid, or recombined, DNA molecules.

WATSON, CRICK, AND THE STRUCTURE OF DNA

Recognition that DNA plays an important role in genetic processes coincided with efforts to understand the chemical structure of nucleic acids. In the years from the mid-1940s through 1953, several laboratories made significant strides in unraveling the structure of DNA, culminating in the Watson-Crick model for the DNA double helix in 1953. The scientific, intellectual, and personal intrigue that characterized the race to discover the structure of DNA has been documented in a number of books, beginning with *The Double Helix* by James Watson. These personal accounts and histories provide a rare glimpse into the ambitions, jealousies, and rivalries that entangled scientists involved in the dash to a Nobel prize.

Reviewing Some Basic Chemistry

The structure of DNA in the Watson-Crick model, and in a later chapter the structure of proteins, is described and drawn using chemical terms and symbols. For this reason, a brief review of the terminology and definition of some terms is in order.

All matter is composed of atoms; the different types of atoms are known as elements (of which there are 114). In nature, atoms are rarely found as separate units. More often, they are combined into molecules, which we can define as units of two or more atoms chemically bonded together. Molecules can be represented by formulas that indicate how many of each type of atom are present. The type of atom is indicated by a symbol for the element it represents: H for hydrogen, N for nitrogen, C for carbon, O for oxygen, and so forth. For example, a water molecule, composed of two hydrogen atoms and one oxygen atom, has its chemical formula represented as H_2O:

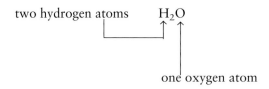

Many molecules in cells are large and have more complex formulas. A molecule of glucose contains 24 atoms and is written as

$$C_6H_{12}O_6$$

The atomic components in molecules are held together by a stable interaction known as a **covalent bond**. In its simplest form, a covalent bond consists of a pair of electrons shared between two atoms. Sharing two or more electrons can form more complex covalent bonds. ▶ Figure 8.2 shows how such bonds are represented in structural formulas of molecules.

■ **Covalent bond** A chemical bond that results from electron sharing between atoms. Covalent bonds are formed and broken during chemical reactions.

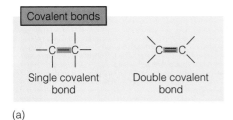

Covalent bonds

Single covalent bond

Double covalent bond

(a)

Hydrogen bonds

(b)

▶ **FIGURE 8.2** Representation of chemical bonds. (a) Covalent bonds are represented as solid lines that connect atoms. Depending on the degree of electron sharing, there can be one (left) or more (right) covalent bonds between atoms. Once formed, covalent bonds are stable and are broken only in chemical reactions. (b) Hydrogen bonds are usually represented as dotted lines that connect two or more atoms. As shown, water molecules form hydrogen bonds with adjacent water molecules. These are weak interactions that are easily broken by heat and molecular tumbling and can be re-formed with other water molecules.

A second type of atomic interaction between molecules involves a weak attraction known as a **hydrogen bond.** In living systems, hydrogen bonds make a substantial contribution to the three-dimensional shape and, therefore, to the functional capacity of biological molecules. Hydrogen bonds are weak interactions between two atoms (one of which is hydrogen), that carry partial but opposite electrical charges. Hydrogen bonds are usually represented in structural formulas as dotted or dashed lines that connect two atoms (Figure 8.2).

Although individual hydrogen bonds are weak and easily broken, they function to hold molecules together by sheer force of numbers. As we see in a following section, hydrogen bonds hold together the two strands in a DNA molecule, and they are also responsible for the three-dimensional structure of proteins (Chapter 9).

Nucleotides: The Building Blocks of Nucleic Acids

There are two types of nucleic acids in biological organisms: **DNA** and **RNA.** Both are made up of subunits known as nucleotides. A **nucleotide** consists of a **nitrogen-containing base** (either a **purine** or a **pyrimidine**), a **pentose sugar** (either ribose or deoxyribose), and a phosphate group. The phosphate groups are strongly acidic and are the source of the designation nucleic acid. In the bases, both purines and pyrimidines have the same six-atom ring, but purines have an additional three-atom ring. The purine bases **adenine** (A) and **guanine** (G) are found in both RNA and DNA (▶ Figure 8.3c). The pyrimidine bases include **thymine** (T), found in DNA; **uracil** (U), found in RNA; and **cytosine** (C), found in both RNA and DNA. RNA has four bases (A, G, U, C), and DNA has four bases (A, G, T, C).

The sugars in nucleic acids have five carbon atoms per molecule. The sugar in RNA is known as **ribose,** and the sugar in DNA is **deoxyribose.** The difference is a single oxygen atom that is present in ribose and absent in deoxyribose (Figure 8.3b).

Nucleotides are molecules composed of a base linked by a covalent bond to a sugar, which in turn is covalently bonded to a phosphate group (Figure 8.3d). Nucleotides are named according to the base and sugar they contain (Table 8.1). Two or more nucleotides can be linked together by a covalent bond between the phosphate group of one nucleotide and the sugar of another nucleotide. Chains of nucleotides called polynucleotides can be formed in this way (▶ Figure 8.4a). Polynucleotides are directional molecules that have slightly different structures at each end of the chain. At one end is a phosphate group; this is the 5′ (pronounced "5 prime") end. At the opposite end is an OH group; this is known as the 3′ ("3 prime") end of the chain. By convention, nucleotide chains are written beginning with the 5′ end, such as 5′-CGATATGCGAT-3′, and are usually labeled to indicate polarity.

Hydrogen bond A weak chemical bonding force between hydrogen and another atom.

Deoxyribonucleic acid (DNA) A molecule consisting of antiparallel strands of polynucleotides that is the primary carrier of genetic information.

Ribonucleic acid (RNA) A nucleic acid molecule that contains the pyrimidine uracil and the sugar ribose. The several forms of RNA function in gene expression.

Nucleotides The basic building blocks of DNA and RNA. Each nucleotide consists of a base, a phosphate, and a sugar.

Nitrogen-containing base A purine or pyrimidine that is a component of nucleotides.

Purine A class of double-ringed organic bases found in nucleic acids.

Pyrimidines A class of single-ringed organic bases found in nucleic acids.

Pentose sugar A five-carbon sugar molecule found in nucleic acids.

Adenine and guanine Purine nitrogenous bases found in nucleic acids.

Cytosine, thymine, and uracil Pyrimidine nitrogenous bases found in nucleic acids.

Deoxyribose and ribose Pentose sugars found in nucleic acids. Deoxyribose is found in DNA, ribose in RNA.

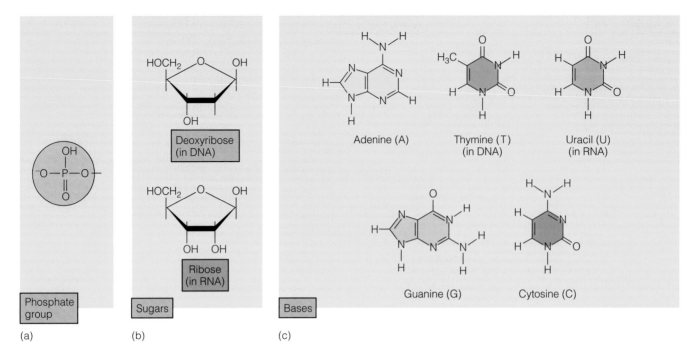

▶ **FIGURE 8.3** DNA is made up of subunits called nucleotides. Each nucleotide is composed of phosphorus (a), a sugar (b), and a base (c). A nucleotide (d).

DNA Is a Double Helix

In the early 1950s James Watson and Francis Crick began to work on a model for the structure of DNA. To build their model, they sifted through and organized the information that was already available about DNA. Their model is based on information about the chemistry of DNA and information on the physical structure of DNA. Information about the physical structure of DNA was based on X-ray crystallography. In this process, molecules are crystallized and placed in a beam of X rays. As the X rays pass through the crystal, some are deflected as they hit the atoms in the crystal. The pattern of X rays emerging from the crystal can be recorded on photographic film and analyzed to produce information about the organization and shape of the crystallized molecule.

Maurice Wilkins and Rosalind Franklin obtained X-ray crystallographic pictures from highly purified DNA samples. These pictures indicated that the DNA molecule has the shape of a helix of constant diameter (▶ Figure 8.5). The X-ray films also permitted measurements of the distances between the stacked bases.

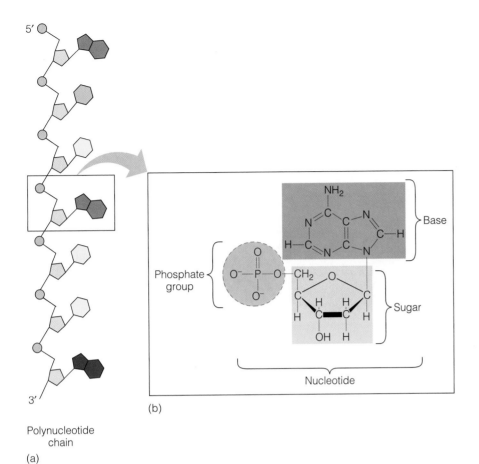

(a)

5′

Polynucleotide
chain

3′

(b)

NH₂

Base

Phosphate
group

Sugar

Nucleotide

▶ **FIGURE 8.4** (a) Nucleotides can be joined together to form chains called polynucleotides. Polynucleotides are polar molecules that have a 5′ end (at the phosphate group) and a 3′ end (at the sugar group). (b) The components of a nucleotide.

Beginning in the late 1940s, Erwin Chargaff and his colleagues extracted DNA from a variety of organisms and separated and analyzed the amounts of the four bases. The results indicated that DNA from all organisms tested had common properties, which became known as Chargaff's rule:

- There is a 1:1 relationship between the amount of adenine and the amount of thymine and a 1:1 relationship between the amount of guanine and the amount of cytosine.

Using the information provided by the physical and chemical studies, Watson and Crick began building wire and metal models of potential DNA structures. They succeeded in producing a model that incorporated all of the information described above (▶ Figure 8.6). Their model has the following features:

- DNA is composed of two polynucleotide chains running in opposite directions.
- The two polynucleotide chains are coiled around a central axis to form a double helix. This part of the model fits the X-ray results of Rosalind Franklin and Maurice Wilkins.
- In each chain, the alternating sugar and phosphate groups are linked to form the backbone of the chain. The bases face inward, where they are paired by hydrogen bonds to bases in the opposite chain.
- In the model, base pairing is highly specific: A pairs only with T in the opposite chain, and C pairs only with G in the opposite chain. Each set of hydrogen-bonded

▶ **FIGURE 8.5** An X-ray diffraction photograph of a DNA crystal. The central x-shaped pattern is typical of helical structures, and the darker areas at the top and bottom indicate a regular arrangement of subunits in the molecule. Watson and Crick used this and other photographs to construct their model of DNA.

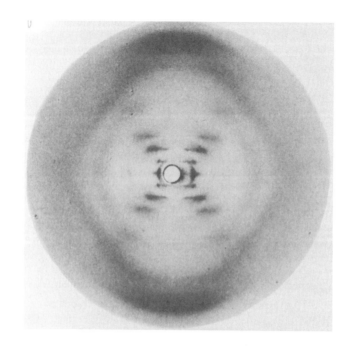

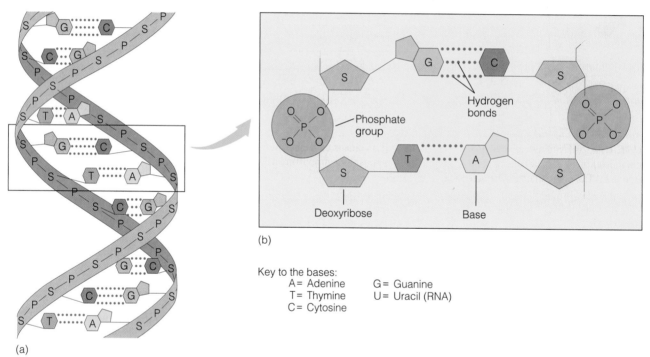

(b)

Key to the bases:
A = Adenine   G = Guanine
T = Thymine   U = Uracil (RNA)
C = Cytosine

(a)

▶ **FIGURE 8.6** (a) The Watson-Crick model of DNA. Two polynucleotide strands are coiled around a central axis, forming a helix. (b) Hydrogen bonds between the bases hold the two strands together. In the molecule, A always pairs with T on the opposite strand, and C always pairs with G.

bases is called a base pair. The pairing of A with T and C with G fits the results obtained by Chargaff.

- The base pairing of the model makes the two polynucleotide chains of DNA complementary in base composition (▶ Figure 8.7). If one strand has the sequence 5'-ACGTC-3', the opposite strand must be 3'-TGCAG-5', and the double-stranded structure would be written as

$$5'\text{-ACGTC-}3'$$
$$3'\text{-TGCAG-}5'$$

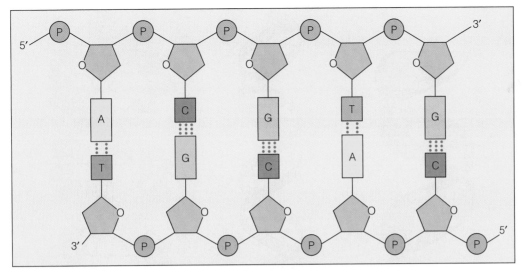

▶ **FIGURE 8.7** The two polynucleotide chains in DNA run in opposite directions. The top strand runs 5′ → 3′ and the bottom strand runs 3′ → 5′. The base sequences in each strand are complementary. An A in one strand pairs with a T in the other strand, and a C in one strand is paired with a G in the opposite strand.

Three important properties of this model should be considered:

- In this model, genetic information is stored in the sequence of bases in the DNA. The linear sequence of bases has a high coding capacity. A molecule n bases long has 4^n combinations. That means that a sequence of 10 nucleotides has 4^{10}, or 1,048,576, possible combinations of nucleotides. The complete set of genetic information carried by an organism (its genome) can be expressed as base pairs of DNA (Table 8.2). Genomic sizes vary from a few thousand nucleotides (in viruses) that encode only a few genes to billions of nucleotides that encode perhaps 50,000 to 100,000 genes (as in humans). The human genome consists of about 3×10^9, or 3 billion, base pairs of DNA, distributed over 24 chromosomes (22 autosomes and two sex chromosomes).
- The model offers a molecular explanation for mutation. Because genetic information can be stored in the linear sequence of bases in DNA, any change in the order or number of bases in a gene can result in a mutation that produces an altered phenotype. This topic is explored in more detail in Chapter 11.
- As Watson and Crick observed, the complementary strands in DNA can be used to explain the molecular basis of the replication of genetic information that takes place before each cell division. In such a model, each strand can be used as

Table 8.2	**Genome Size in Various Organisms**	
Organism	**Species**	**Genome Size in Nucleotides**
Bacterium	*E. coli*	4.2×10^6
Yeast	*S. cerevisiae*	1.3×10^7
Fruit fly	*D. melanogaster*	1.4×10^8
Tobacco plant	*N. tabacum*	4.8×10^9
Mouse	*M. musculus*	2.7×10^9
Human	*H. sapiens*	3.2×10^9

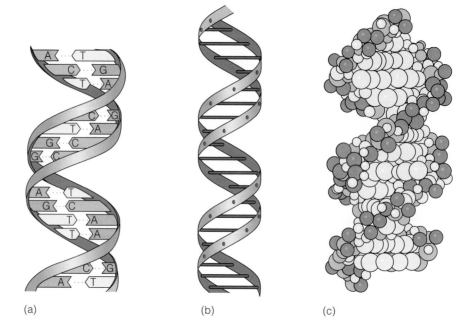

▶ **FIGURE 8.8** Three different ways of representing DNA. (a) A model that shows the complementary base pairing in the interior of the molecule. (b) A model that shows the double helix wound around a central axis. (c) A space-filling model that shows the relative sizes of the atoms in the molecule.

(a)　　　　　　　(b)　　　　　　　(c)

■ **Template** The single-stranded DNA that serves to specify the nucleotide sequence of a newly synthesized polynucleotide strand.

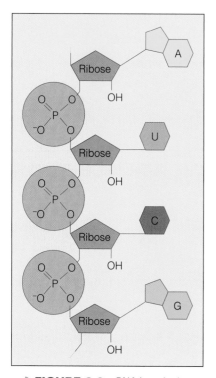

▶ **FIGURE 8.9** RNA is a single-stranded polynucleotide chain. RNA molecules contain a ribose sugar instead of a deoxyribose and have uracil (U) in place of thymine (T).

a **template** to reconstruct the base sequence in the opposite strand. This topic is discussed later in this chapter.

The Watson-Crick model was described in a brief paper in *Nature* in 1953. Although their model was based on the results of other workers, Watson and Crick correctly incorporated the physical and chemical data into a model that also could be used to explain the properties expected of the genetic material (▶ Figure 8.8). Present day applications, including genetic engineering, gene mapping, and gene therapy can be traced directly to this paper.

At the time there was no evidence to support the Watson-Crick model unequivocally, but in subsequent years it has been confirmed by experimental work in laboratories worldwide. The 1962 Nobel prize for Medicine or Physiology was awarded to Watson, Crick, and Wilkins for their work on the structure of DNA. Although much of the X-ray data for the Watson-Crick model was provided by Rosalind Franklin, she did not receive a share of the prize. There has been some controversy over this, but she could not have shared in the prize because it is awarded only to living individuals, and Franklin died of cancer in 1958. Her role in the discovery of the structure of DNA is presented in her biography, *Rosalind Franklin and the Discovery of DNA*.

RNA Is Single-Stranded

RNA (ribonucleic acid) is found in both the nucleus and the cytoplasm. DNA functions as a repository of genetic information, and RNA functions to transfer genetic information from the nucleus to the cytoplasm (in a few viruses, RNA also functions to store genetic information). The nucleotides in RNA differ from those in DNA in two respects: The sugar in RNA nucleotides is ribose (deoxyribose in DNA), and the base uracil is used in place of the base thymine (Table 8.3).

Although two complementary strands of RNA can form a double helix similar to the one found in DNA, the RNA present in most cells is single-stranded, and a complementary strand is not made (▶ Figure 8.9). But RNA molecules can fold back on themselves and form double-stranded regions. The functions of RNA are considered in more detail in Chapter 9.

Table 8.3	**Differences between DNA and RNA**	
	DNA	**RNA**
Sugar	Deoxyribose	Ribose
Bases	Adenine	Adenine
	Cytosine	Cytosine
	Guanine	Guanine
	Thymidine	Uracil

DNA IS COILED WITH PROTEINS TO FORM CHROMOSOMES

Although an understanding of DNA structure represents an important development in genetics, it provides no immediate information about the way a chromosome is organized or about what regulates the cycle of condensation and decondensation of the chromosomes. This problem is significant because the spatial arrangement of DNA may play an important role in regulating the expression of genetic information. In addition, the distribution of DNA into the 46 human chromosomes requires packing a little more than 2 m (2 million μm) of DNA into a nucleus that measures about 5 μm in diameter. Within this cramped environment, the chromosomes unwind and become dispersed during the interphase. In this condition they undergo replication, gene expression, homologous pairing during meiosis, and contraction and coiling to become visible again during the prophase. An understanding of chromosomal organization is necessary to understand these processes. In contrast to what is known about chromosomes in the nucleus, details of the structure and even the nucleotide sequence of the mitochondrial chromosome are well understood, and they are considered first.

The Mitochondrial Chromosome Is a Circular DNA Molecule

As discussed in Chapter 2, mitochondria are organelles involved in energy conversion. Mitochondria contain DNA that encodes genetic information, and this mitochondrial chromosome is transmitted maternally. In recent years, a number of genetic disorders resulting from mutations in mitochondrial DNA have been identified (see Chapter 4), and all offspring of affected mothers in this pattern of inheritance are also affected.

The complete nucleotide sequence of the mitochondrial chromosome is known (▶ Figure 8.10). The chromosome consists of a circular DNA molecule that contains 16,569 base pairs and encodes just about 40 genes. Alterations in the mitochondrial genome can lead to genetic disorders. Most mitochondria contain 5 to 10 copies of the mitochondrial chromosome, and because each cell has hundreds of mitochondria, thousands of mitochondrial genomes are present in each human cell. The DNA of the mitochondrial chromosome is not complexed with proteins and physically resembles the chromosomes of bacteria and other prokaryotes. This similarity reflects the evolutionary history of mitochondria and their transition from free-living prokaryotes to intracellular symbionts.

Nuclear Chromosomes Have a Complex Structure

A combination of biochemical and microscopic techniques developed in recent years has provided a great deal of information about the organization and structure of human chromosomes. In humans and other eukaryotes, each chromosome consists of a

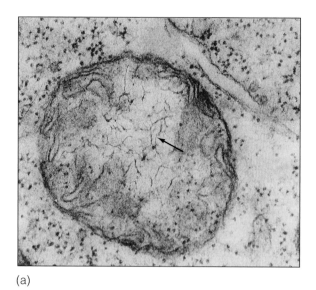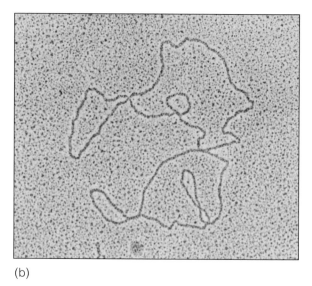

(a)   (b)

▶ **FIGURE 8.10**   Mitochondrial genome. (a) Micrograph that shows filaments of DNA in the mitochondrial matrix (arrow). (b) Mitochondrial DNA occurs in the form of circles 5 to 6 μm in contour length.

■ **Chromatin**   The complex of DNA and proteins that makes up a chromosome.

■ **Histones**   Small DNA-binding proteins that function in the coiling of DNA to produce the structure of chromosomes.

■ **Nonhistone proteins**   The array of proteins other than histones that are complexed with DNA in chromosomes.

■ **Nucleosomes**   A bead-like structure composed of histones wrapped by DNA.

single DNA molecule combined with proteins to form **chromatin**. There are two types of proteins in chromatin: **histones** and **nonhistone proteins.** The histones play a major role in chromosomal structure, and the nonhistones may be involved in gene regulation. Five types of histones are complexed with DNA to form small spherical bodies known as **nucleosomes** that are connected to each other by thin threads of DNA (▶ Figure 8.11). Each nucleosome consists of DNA wound around a core composed of eight histone molecules.

Winding DNA around the histones compresses the length of the DNA molecule by a factor of six or seven. But the level of compaction in the mitotic chromosome is estimated to be 5000- to 10,000-fold. This suggests that there are several levels of organization between the nucleosome and the chromosome, each of which involves folding and/or coiling of the DNA molecule. Electron microscopic observations on mitotic chromosomes have partly clarified the subchromosomal levels of organiza-

▶ **FIGURE 8.11**   Nucleosomes. (a) An electron micrograph that shows the association of DNA with histones to form nucleosomes connected by threads of DNA in chromatin from chicken blood cells. (b) A diagram that shows how DNA coils around the outside of a histone cluster in nucleosomes.

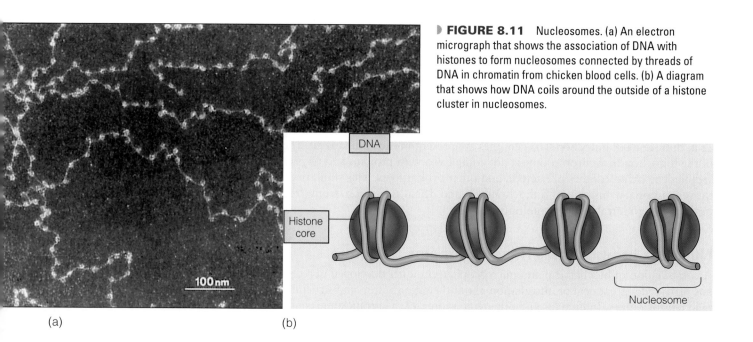

(a)   (b)

tion. ▌Figure 8.12 is a scanning electron micrograph of a mitotic chromosome. At high magnification, individual fibers can be observed. These fibers appear to be the result of supercoiling of the chromatin. A model of chromosomal organization is shown in ▌Figure 8.13.

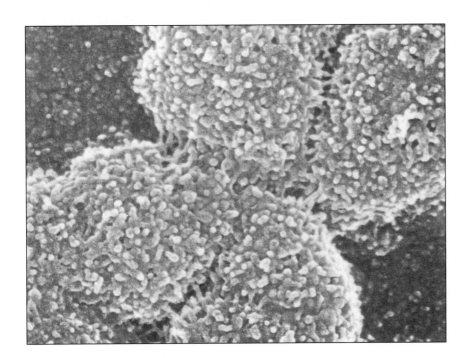

▌**FIGURE 8.12** A scanning electron micrograph of the centromeric region of a human chromosome. Small fibrils whose diameter is about 30 nanometers result from the supercoiling of chromatin.

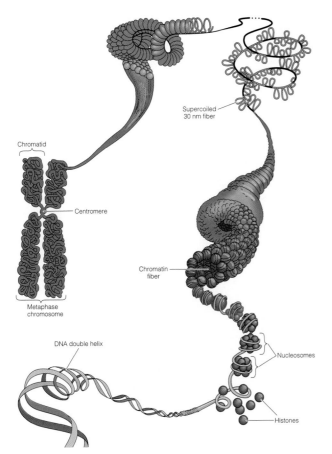

▌**FIGURE 8.13** A model of chromosomal structure beginning with a double-stranded DNA molecule. The DNA is first coiled into nucleosomes. Then, the nucleosomes are coiled again and again into fibers that form the body of the chromosome. Chromosomes undergo cycles of coiling and uncoiling in mitosis and interphase, so their structure is dynamic.

Chromatid

Centromere

Metaphase chromosome

DNA double helix

Supercoiled 30 nm fiber

Chromatin fiber

Nucleosomes

Histones

DNA REPLICATION DEPENDS ON BASE PAIRING

Between divisions, all cells replicate their DNA during the S phase of the cell cycle, so that each daughter cell will receive a complete set of genetic information. In their paper on the structure of DNA, Watson and Crick note that "It has not escaped our notice that the specific pairing we have postulated immediately suggests a possible copying mechanism for the genetic material." In a subsequent paper, they proposed a mechanism for DNA replication that depends on the complementary base pairing in the polynucleotide chains of DNA. If the DNA helix is unwound, each strand can serve as a template or pattern for synthesizing a new, complementary strand (▶ Figure 8.14). This process is known as **semiconservative replication** because one old strand is conserved in each new molecule.

The process of DNA replication in all cells, from bacteria to humans, is a complex process requiring the action of more than a dozen different enzymes. In humans, replication begins at sites called origins of replication that are present all along the length of the chromosome. At these sites, proteins unwind the double helix by breaking the hydrogen bonds between bases in adjacent strands. This process opens the molecule to the action of the enzyme **DNA polymerase**. DNA polymerase links the complementary nucleotides together to form newly synthesized strands.

Each chromosome contains one double-stranded DNA helix running from end to end. When replication is finished, the chromosome consists of two sister chromatids, joined at a common centromere. Each chromatid contains a DNA molecule that consists of one old strand and one new strand. When the centromeres divide at the beginning of the anaphase, each chromatid becomes a separate chromosome that contains an accurate copy of the genetic information in the parental chromosome.

■ **Semiconservative replication** A model of DNA replication that provides each daughter molecule with one old strand and one newly synthesized strand. DNA replicates in this fashion.

■ **DNA polymerase** An enzyme that catalyzes the synthesis of DNA using a template DNA strand and nucleotides.

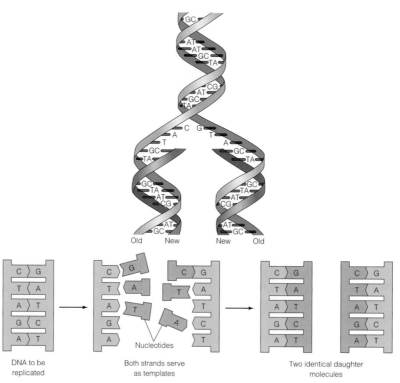

▶ **FIGURE 8.14** In DNA replication, the two polynucleotide strands uncoil, and each is a template for synthesizing a new strand. As the strands uncoil, complementary base pairing occurs with the template strand. These are linked together by DNA polymerase, assisted by other enzymes that help uncoil the DNA and seal up gaps in the new strands. A completed DNA molecule contains one new strand and one old strand.

CASE 1

A 34-year-old woman and her one-month-old newborn were seen by a genetic counselor in the neonatal intensive care unit in a major medical center. The neonatologist was suspicious that the newborn boy had a genetic condition and requested a genetic evaluation. The newborn was very pale, was failing to thrive, had diarrhea, and markedly increased serum cerebrospinal fluid lactate levels. In addition, he had severe muscle weakness with chart notes describing him as "floppy" and he had already had two seizures since birth. The neonatologist reported that the infant was currently suffering from liver failure, which would probably result in his death in the next few days. The panel of tests performed on the infant led the neonatologist and genetic counselor to the diagnosis of Pearson's syndrome. The combination of marked metabolic acidosis, and abnormalities in bone marrow cells is highly suggestive of Pearson's syndrome.

CASE 2

There is considerable debate over the patenting of genes. This controversy is largely fueled by advances made by the Human Genome Project and the biotechnology industry. Despite numerous meetings and publications on the subject, U.S. patent laws have not developed a stance permitting maximum innovation from biotech inventions. Gene patents require a DNA sequence. Humans have 3×10^9 nucleotides that make up our genetic material. A short linear sequence of 25 or more base pairs is usually enough to identify specifically a unique genetic region or gene.

Some years ago, the U.S. National Institutes of Health (NIH) tried to patent genes based on partial DNA sequences, each comprising 150–250 base pairs. This clearly exceeded the 25 that are needed to uniquely identify a gene's transcript. The NIH believed 150–250 base pairs to be sufficient for patentability purposes. However, such partial sequences lack the necessary information to completely describe a gene or its function and relation to disease. The U.S. Patent Office rejected the patent applications, citing lack of utility and novelty, and NIH withdrew its application. As others will undoubtedly file gene patents based on similar data, it is critical to consider gene patent criteria. Given the ability to obtain complete sequences of most human genes over the next three to five years, the stage is set for a gene patent race.

Pearson's syndrome is associated with a large deletion of the mitochondrial (mt) genome. The way the deleted mtDNA molecules are distributed over individual cells during mitosis is not known. However, it is assumed that during cell division daughter cells randomly receive mitochondria carrying wild type (WT) or mutant mtDNA. Mitochondrial DNA is, theoretically, transmitted only to offspring through the mother via the large cytoplasmic component of the oocyte. Nearly all cases of Pearson's syndrome arise from new mutational events. Mitochondria have extremely poor DNA repair mechanisms, and mutations accumulate very rapidly. Most infants die before age 3, often due to infection, or liver failure.

A diagnosis of Pearson's syndrome results in an extremely grave prognosis for the patient. Unfortunately, at this point, treatment can be directed only toward symptomatic relief.

Another genome project, the Human Genome Organization (HUGO) opposes patenting partial and complete DNA sequences that lack specific utility. Furthermore, HUGO recommends that gene sequence information be available for research in an unimpeded manner through international databases. Acceptance of these recommendations would accelerate our knowledge of genes and, in particular, their use in the medical field.

How should utility be defined? The diagnosis of disease certainly meets the definition of utility. Many discoveries have identified disease genes such as those in cystic fibrosis, fragile X syndrome, breast cancer, colon cancer, and obesity. Many of these discoveries have patents based on diagnostic utility. An increasing number of patent applications are being filed for discoveries of hereditary disease genes. These discoveries frequently lack immediate use for practical therapy, however, due to the fact that gene discovery does not always include knowledge of gene function or the development of a disease therapy. As the number of human genes is finite (about 100,000), it may be worth considering gene patents as "core patents" for biology and medicine so that they are broadly available for research. This form of patenting occurs in the electronics industry and benefits both the companies and consumers. This program may be a useful one to consider for gene patents.

Summary

1. At the turn of the century, scientists identified chromosomes as the cellular components that carry genes. This discovery focused efforts to identify the molecular nature of the gene on the chromosomes and the nucleus. Biochemical analysis of the nucleus began around 1870 when Frederick Miescher first separated nuclei from cytoplasm and described nuclein, a protein/nucleic acid complex now known as chromatin.

2. Originally, proteins were regarded as the only molecular component of the cell with the complexity to encode the genetic information. This changed in 1944 when Avery and his colleagues demonstrated that DNA is the genetic material in bacteria.

3. In 1953 Watson and Crick constructed a model of DNA structure that incorporated information from the chemical studies of Chargaff and the X-ray crystallographic work of Wilkins and Franklin. They proposed that DNA is composed of two polynucleotide chains oriented in opposite directions and held together by hydrogen bonding to complementary bases in the opposite strand. The two strands are wound around a central axis in a right-handed helix.

4. The mitochondrial chromosome, carrying genes that can cause maternally transmitted disorders, is a circular DNA molecule.

5. Within chromosomes, DNA is coiled around clusters of histones to form structures known as nucleosomes. Supercoiling of nucleosomes may form fibers that extend at right angles to the axis of the chromosome. The structure of chromosomes must be dynamic to allow the uncoiling and recoiling seen in successive phases of the cell cycle, but the details of this transition are still unknown.

6. In DNA replication, strands are copied to produce semiconservatively replicated daughter strands.

Questions and Problems

1. Until 1944, which cellular component was thought to carry genetic information?
   a. carbohydrate
   b. nucleic acid
   c. protein
   d. chromatin
   e. lipid

2. Why do you think nucleic acids were not originally considered to be carriers of genetic information?

3. The experiments of Avery and his co-workers led to the conclusion that:
   a. bacterial transformation occurs only in the laboratory.
   b. capsule proteins can attach to uncoated cells.
   c. DNA is the transforming agent and is the genetic material.
   d. transformation is an isolated phenomenon in *E. coli*.
   e. DNA must be complexed with protein in bacterial chromosomes.

4. In the experiments of Avery, what was the purpose of treating the transforming extract with enzymes?

5. For this question, read the following experiment and interpret the results to form your conclusion. Experimental data: S bacteria were heat-killed and cell extracts were isolated. The extracts contained cellular components including lipids, proteins, DNA, and RNA. These extracts were mixed with live R bacteria and then injected together into mice along with various enzymes (proteases, RNAses, and DNAses) Proteases degrade proteins, RNAses degrade RNA, and DNAses degrade DNA.

S extract + live R cells	kills mouse
S extract + live R cells + protease	kills mouse
S extract + live R cells + RNAse	kills mouse
S extract + live R cells + DNAse	mouse lives

   Based on these results, what is the transforming principle?

6. Recently, scientists discovered that a rare disorder called polkadotism is caused by a bacterial strain, polkadotiae. Mice injected with this strain (P) develop polkadots on their skin. Heat-killed P bacteria and live D bacteria, a nonvirulent strain, do not produce polkadots when separately injected into mice. However, when a mixture of heat-killed P cells and live D cells are injected together, the mice developed polkadots. What process explains this result? Describe what is happening in the mouse to cause this outcome.

7. List the pyrimidine bases, the purine bases, and the base pairing rules for DNA.

8. In analyzing the base composition of a DNA sample, a student loses the information on pyrimidine content. The purine content is A = 27% and G = 23%. Using Chargaff's rule, reconstruct the missing data, and list the base composition of the DNA sample.

9. The basic building blocks of nucleic acids are:
   a. nucleosides
   b. nucleotides
   c. ribose sugars
   d. amino acids
   e. purine bases

10. Adenine is a:
    a. nucleoside
    b. purine
    c. pyrimidine
    d. nucleotide
    e. base

11. Polynucleotide chains have a 5′ and a 3′ end. Which groups are found at each of these ends?
    a. 5′ sugars, 3′ phosphates
    b. 3′ OH, 5′ phosphates
    c. 3′ base, 5′ phosphate
    d. 5′ base, 3′ OH
    e. 5′ phosphates, 3′ bases
12. DNA contains many hydrogen bonds. Describe a hydrogen bond, and how this type of chemical bond holds DNA together.
13. Watson and Crick received the Nobel prize for:
    a. generating X-ray crystallographic data of DNA structure
    b. establishing that DNA replication is semiconservative
    c. solving the structure of DNA
    d. proving that DNA is the genetic material
    e. showing that the amount of A equals the amount of T
14. State the properties of the Watson-Crick model of DNA in the following categories:
    a. number of polynucleotide chains
    b. polarity (running in same or opposite directions)
    c. bases on interior or exterior of molecule
    d. sugar/phosphate on interior or exterior of molecule
    e. which bases pair with which
    f. right or left handed helix
15. What is the function of DNA polymerase?
    a. degrades DNA in cells
    b. adds RNA nucleotides to new strand
    c. coils DNA around histones to form chromosomes
    d. adds DNA nucleotides to a replicating strand
    e. none of the above
16. Using Figure 8.4 as a guide, draw a dinucleotide composed of C and A. Next to this, draw the complementary dinucleotide in an antiparallel fashion. Connect the dinucleotides with the appropriate hydrogen bonds.
17. Nucleosomes are complexes of:
    a. nonhistone protein and DNA
    b. RNA and histone
    c. histones, non-histone proteins and DNA

    d. DNA, RNA, and protein
    e. amino acids and DNA
18. Which of the following statements is *not* true about DNA replication?
    a. occurs during the M phase of the cell cycle
    b. makes a sister chromatid
    c. denatures DNA strands
    d. occurs semi-conservatively
    e. follows base pairing rules
19. Make the complementary strand to the following DNA template and label both strands as 5′ to 3′, or 3′ to 5′ (P = phosphate in the diagram). Draw an arrow showing the direction of synthesis of the new strand. How many hydrogen bonds are in this double strand of DNA?

    template P - AGGCTCG - OH
    new strand:

20. A beginning genetics student is attempting to complete an assignment to draw a base pair from a DNA molecule, and asks your advice. The drawing is incomplete, and the student does not know how to finish. The assignment sheet shows that the drawing is to contain 3 hydrogen bonds, a purine, and a pyrimidine. From your knowledge of the pairing rules and the number of hydrogen bonds in A/T and G/C base pairs, what base pair do you help the student draw?
21. How does DNA differ from RNA with respect to the following characteristics?
    a. number of chains
    b. bases used
    c. sugar used
    d. function
22. Discuss the levels of chromosomal organization with reference to the following terms:
    a. nucleotide
    b. DNA double helix
    c. histones
    d. nucleosomes
    e. chromatin

Internet Activities

The following activities use the resources of the World Wide Web to enhance the topics covered in this chapter. To investigate the topics described below, log on to the book's homepage at:

   http://www.brookscole.com/biology

1. The Primer on Molecular Genetics produced by the Department of Energy presents basic information about DNA, genes, and chromosomes. Much of this information is applied to mapping the human genome. The introductory part of the Primer can provide you with a good review of DNA, RNA, and proteins (proteins are discussed in Chapter 9).

    a. In what ways are DNA and RNA similar?
    b. In Figure 3 in the DNA section of the Primer, you will find that the Y chromosome is the smallest human chromosome and has about 30 million base pairs of DNA. Does this surprise you? What implications does this suggest about sex determination?
    c. In light of the size and complexity of the human genome (which is present in all cells), does it surprise you to realize that blood cells reproduce at the rate of about 2 million per second?
2. Genetech, one of the world's largest biotechnology companies, has a section on its homepage for teachers

and students called Access Excellence. Scroll through this section and read the descriptions provided for the various sections. Click on the Activities Exchange icon and scroll to the AE Classics Collection. Click on this heading, which links to a number of interesting sites. Scroll to A Visit with Dr. Francis Crick, click, and read the interview with Dr. Crick.

a. It's interesting to note that Dr. Crick studied math and physics until his curiosity motivated him to study the area that separates the living from the nonliving. In your own situation, what genetic topics or diseases did you come to this course curious to learn more about (list at least four)? Why is the discovery of the structure of DNA considered to be one of the most important advances in this century? Crick is now working to understand the biological basis of consciousness. Do you think that he or anyone else will be successful in understanding the structure of consciousness? Are there some similarities that may be present in the search for the structure of DNA versus the structure of consciousness?

b. Are there any biases you can detect in the description of Acess Excellence? Describe some potentially positive and some potentially negative aspects of using educational information provided by a company in an emerging health technology field.

3. The On-line Biology Book provides an interesting overview of *DNA and Molecular Genetics*. Go to the site and read about the discovery of DNA. Who first discovered DNA, and what was it initially called? What enabled the discovery, made 45 years later, that DNA is present in the nucleus of every cell? Our understanding of DNA and genetics contributes fundamentally to our view of the world, yet only 50 years have passed since DNA was shown to be the genetic material. Try to imagine what life was like before that discovery. Would your vision of life be essentially the same? Can you think of any fundamental biological mysteries that have yet to be solved?

For Further Reading

Avery. O. T., Macleod, C. M., & McCarty, M. (1944). *J. Exp. Med.* 70: 137–158.

Felsenfeld, G. (1985). DNA, *Sci. Am.* 253 (October): 58–67.

Howell, N. (1999). Human mitochondrial diseases: Answering questions and questioning answers. Int. Rev. Cytol. 186: 49–116.

Judson, H. F. (1979). *The Eighth Day of Creation: The Makers of the Revolution in Biology.* New York: Simon & Schuster.

Kornberg, A. (1968). The synthesis of DNA. *Sci. Am.* 219 (October): 64–79.

Kornberg, R., & Klug, A. (1981). Nucleosomes. *Sci. Am.* 244 (February): 52–79.

McCarty, M. (1986). *The Transforming Principle: Discovery That Genes Are Made of DNA.* New York: Norton.

Olby, R. (1974). *The Path to the Double Helix.* Seattle: University of Washington Press.

Sayre, A. (1975). *Rosalind Franklin and DNA.* New York: Norton.

Watson, J. D. (1968). *The Double Helix.* New York: Atheneum.

Watson, J. D., & Crick, F. H. C. (1953). Molecular structure of nucleic acids: A structure for deoxyribose nucleic acid. *Nature 171*: 737–738.

Watson, J. D., & Crick, F. H. C. (1953). Genetical implications of the structure of desoxyribonucleic acid. *Nature 171*: 964–967.

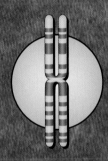

Gene Expression: How Proteins Are Made

*J*ust after World War II, Charles Dent began using the technique of paper chromatography to separate the biochemical components of human urine. To do this, he spotted a sample near the bottom of a piece of filter paper and dried it thoroughly. The paper was placed in a chamber containing a small amount of solvent, and the bottom of the paper touched the solvent. The top of the container was sealed, and gradually, the solvent moved up the filter paper by capillary action. As the solvent moved upward, the chemical components in the urine sample also migrated upward, but at different rates, thus separating from each other. When the solvent reached the top of the paper, the filter paper was removed and sprayed with a reagent to make the amino acids contained in the sample visible as a series of spots.

In conjunction with Harry Harris, Dent used this technique to investigate the amino acids excreted in the urine of patients who had a genetic disorder called cystinuria. In this disorder, a defect in kidney function causes excretion of large quantities of the amino acid cystine in the urine. In some case, the concentration is so high, that cystine crystals or even stones are excreted. Forty years earlier, Archibald Garrod had postulated that this disorder resulted from abnormal metabolism, but until Dent and Harris began their work, the biochemistry was confusing, and there was no good evidence that the disorder was heritable.

Harris and Dent showed that there are two forms of cystinuria, one characterized by the excretion of cystine, lysine, and three other amino acids, and a second form marked by the excretion of just cystine and lysine. Through a combination of genetic and biochemical studies, Harris established that the first type is caused by a homozygous recessive genotype, and the second by a heterozygous condition. This work had three important consequences: (1) it demonstrated that some genetic disorders have biochemical phenotypes; (2) it helped establish human biochemical genetics as part of genetics; and (3) it helped forge the link between genes and phenotypes through biochemical processes.

In this chapter, we consider the relationship between genes and proteins and the role of DNA as a carrier of genetic information. We also discuss the transfer of genetic information from the sequence of nucleotides in DNA into the sequence of amino acids in protein.

THE LINK BETWEEN GENES AND PROTEINS

What are proteins? How and where are they made? How are proteins related to genes and in particular to human genetics? As we see in this chapter, proteins like blood-clotting factors and digestive enzymes are gene products. Proteins are the intermediate between genes and phenotype. The phenotypes of a cell, tissue, and organism are the result of protein function. When these functions are absent or altered, the result is a mutant phenotype, which we describe as a genetic disorder.

At the turn of the twentieth century, Archibald Garrod recognized this relationship based on his genetic and biochemical studies of a condition called **alkaptonuria** (MIM/OMIM 203500). Affected individuals can be identified soon after birth because their urine turns black. Garrod found that alkaptonuria is associated with the excretion of large quantities of a compound he called alkapton (now called ho-

■ **Alkaptonuria** An autosomal recessive condition that alters the metabolism of homogentistic acid. Affected individuals do not produce the enzyme to metabolize this acid.

mogentisic acid). He reasoned that in unaffected individuals, homogentisic acid is metabolized to other products and does not build up in the urine. In alkaptonuria, however, the metabolic pathway is blocked, causing the buildup of homogentisic acid, which is excreted in the urine. He called this condition an "inborn error of metabolism." Garrod also discovered that alkaptonuria is inherited as a recessive Mendelian trait.

Genes and Metabolism

Biochemical reactions in the cell are organized into chains of reactions known as metabolic pathways. The product of one reaction in a pathway serves as the substrate for the following reaction (⏵ Figure 9.1). Enzymes are protein molecules that serve as biological catalysts to carry out specific biochemical reactions. Loss of activity in a single enzyme can disrupt an entire biochemical pathway, causing an alteration in metabolism.

The Relationship between Genes and Enzymes

George Beadle and Edward Tatum established the connection between enzymes and the phenotype in the late 1930s and early 1940s through their work on *Neurospora*, a common bread mold (⏵ Figure 9.2). Using *Neurospora*, they demonstrated that mutation of a single gene caused loss of activity in a single enzyme, resulting in a mutant phenotype. This linkage between mutant genes, mutant enzymes, and a mutant phenotype was an important step in understanding that genes produce phenotypes through the action of proteins, and Beadle and Tatum received the Nobel Prize in 1958 for their work.

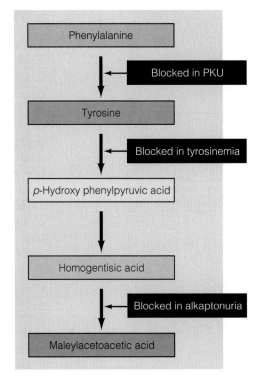

⏵ **FIGURE 9.1** A metabolic pathway beginning with the amino acid phenylalanine. Mutations in humans cause blocks in this pathway, leading to genetic disorders, including phenylketonuria (PKU), tyrosinemia, and alkaptonuria.

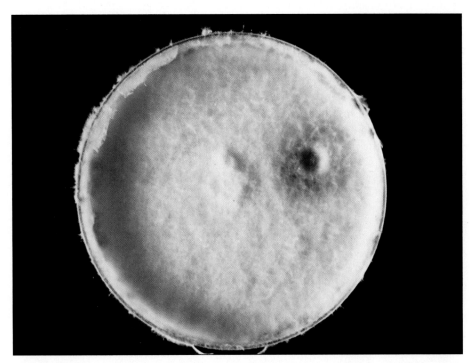

⏵ **FIGURE 9.2** A culture of the mold *Neurospora crassa,* an organism used by George Beadle and Edward Tatum in their work on biochemical genetics.

HOW IS GENETIC INFORMATION STORED IN DNA?

An intriguing feature of the Watson-Crick model of DNA is that it offers an explanation for the way genetic information is encoded in the structure of the molecule. Watson and Crick proposed that genetic information could be represented as the sequence of nucleotide bases in DNA. The amount of information stored in any cell is related to the number of base pairs in the DNA carried within the cell. This number ranges from a few thousand base pairs in some viruses to more than 3 billion base pairs in humans and more than twice that amount in some amphibians and plants. A gene typically consists of hundreds or thousands of nucleotides. Each gene has a beginning and end, marked by specific nucleotide sequences, and a molecule of DNA can contain thousands of genes.

How do genes (in the form of DNA) control the production of proteins? Proteins are linear molecules composed of subunits called amino acids. Twenty different types of amino acids are used to assemble proteins. The diversity of proteins found in nature results from the number of possible combinations of these 20 different amino acids. Because each amino acid position in a protein can be occupied by any of 20 amino acids, the number of different combinations is 20^n, where n is the number of amino acids in the protein. In a protein composed of only 5 amino acids, 20^5, or 3,200,000, combinations are possible, each of which has a different amino acid sequence and a potentially different function. Most proteins are actually composed of several hundred amino acids, so literally billions and billions of combinations are possible. Given that genes are linear sequences of nucleotides and proteins are linear sequences of amino acids, the question is, How is the linear sequence of nucleotides in a gene converted into the linear sequence of amino acids in a protein? In humans and other eukaryotes, the bulk of the cell's DNA is found in the nucleus, and almost all proteins are found in the cytoplasm. This means that the process of information transfer from gene to gene product must be indirect.

THE FLOW OF GENETIC INFORMATION IS A MULTISTEP PROCESS

The transfer of genetic information from DNA nucleotides into the amino acids of a protein has two main steps: **transcription** and **translation** (⏵ Figure 9.3). In **transcription**, a single-stranded molecule of RNA is synthesized at an unwound section of DNA, and one of the DNA strands is a template for the assembly of the RNA. The product is called an RNA transcript or mRNA molecule. In **translation**, mRNA moves to the cytoplasm and interacts with ribosomes to synthesize a protein that contains a linear series of amino acids specified by the sequence of nucleotides in the RNA. The amino acid sequence, in turn, determines the structural and functional characteristics of the protein and its role in phenotypic expression.

■ **Transcription** Transfer of genetic information from the base sequence of DNA to the base sequence of RNA brought about by RNA synthesis.

■ **Translation** The process of converting the information content in an mRNA molecule into the linear sequence of amino acids in a protein.

■ **mRNA** A single-stranded complementary copy of the base sequence in a DNA molecule that constitutes a gene.

⏵ **FIGURE 9.3** The flow of genetic information. One strand of DNA is transcribed into a strand of mRNA. The mRNA moves to the cytoplasm, where it is converted into the amino acid sequence of a protein.

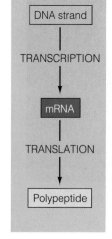

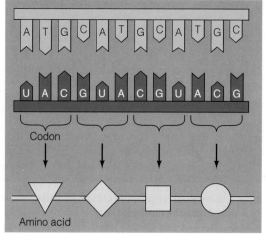

In Figure 9.3, the brackets below the RNA indicate that genetic information is encoded in a sequence of three nucleotides, called **codons**. The four nucleotides of mRNA can be arranged to form 64 codons, each of which contains three letters (4^3 combinations), more than enough to specify the 20 amino acids used in proteins. Codons and the genetic code are examined in a later section.

Transcription Produces Genetic Messages

In the nucleus, transcription begins when a section of a DNA double helix in chromosome unwinds, and one strand acts as a template for the formation of an mRNA molecule. The process of transcription occurs in three stages: **initiation, elongation,** and **termination.** In the first step, an enzyme called **RNA polymerase** binds to a specific nucleotide sequence in the DNA, which marks the beginning of a gene. This sequence is called a **promoter region.** After the polymerase is bound, the adjacent double-stranded DNA unwinds, exposing the DNA strand that will be a template for RNA synthesis. Nucleotides that will make up the mRNA molecule form hydrogen bonds with the complementary nucleotides in the template DNA strand.

In the elongation stage, RNA polymerase links the RNA nucleotides into a polynucleotide chain (▶ Figure 9.4). The rules of base pairing in transcription are the

■ **Codon** Triplets of bases in messenger RNA that encode the information for the insertion of a specific amino acid in a protein.

■ **RNA polymerase** An enzyme that catalyzes the formation of an RNA polynucleotide chain using a template DNA strand and ribonucleotides.

■ **Promoter** A region of a DNA molecule to which RNA polymerase binds and initiates transcription.

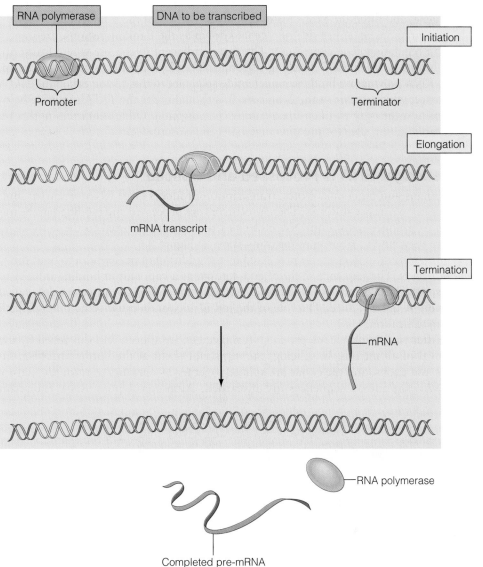

▶ **FIGURE 9.4** Transcription begins when the enzyme RNA polymerase attaches to a promoter sequence that marks the beginning of a gene. One strand of DNA (the template strand) is transcribed into a complementary RNA molecule. Transcription ends when the RNA polymerase reaches a terminator sequence that marks the end of the gene.

same as in DNA replication, with one exception: an A on the DNA template specifies a U in the RNA transcript (recall from Chapter 8 that there is no T in RNA). In humans, 30-50 nucleotides per second are added to an mRNA molecule. In bacteria, up to 500 nucleotides per second are incorporated into mRNA.

The end of the gene is marked by a nucleotide sequence called a **terminator region.** When the RNA polymerase reaches this point, it detaches from the DNA template strand, the mRNA molecule is released, the DNA strands re-form a double helix, and transcription is completed (Figure 9.4). The length of the mRNA transcript depends on the size of the gene. Most transcripts in humans are about 5000 nucleotides long, although lengths up to 200,000 nucleotides have been reported.

Genes Can Have a Complex Internal Organization

Many genes in humans contain nucleotide sequences that are transcribed but not translated into the amino acid sequence of a polypeptide chain. These noncoding sequences, called **introns,** can vary in number from zero to 75 or more. Introns also vary in size, ranging from about 100 nucleotides up to more than 100,000. Most research has shown that introns have no function.

The sequences in a gene that are transcribed and translated into the amino acid sequence of a protein are called **exons.** The internal organization of a typical gene is shown in ▶ Figure 9.5. The combination of exons and introns determines the length of a gene, and often the exons make up only a fraction of the total length.

The regions on either side of a gene, known as the flanking regions, are important in regulating gene expression. The region adjacent to the site where transcription begins is called the **5′ flanking region,** or 5′ end. In the first stage of transcription, RNA polymerase binds to a nucleotide sequence in this region called the **TATA sequence,** which is part of the promoter. It is thought that the TATA sequence helps to position the enzyme to ensure accurate transcription. Other sequences in this region are binding sites for proteins that assist in transcription.

Messenger RNA Is Processed and Spliced

In humans and other eukaryotes, transcription produces large RNA precursor molecules known as **pre-mRNAs.** These precursors are processed and modified in the nucleus to produce mature mRNA molecules that are transported to the cytoplasm, where they bind to ribosomes for translation (▶ Figure 9.6).

Pre-mRNA molecules are processed by the addition of nucleotides to both the 5′ and 3′ ends. The sequence at the 5′ end, known as a **cap,** aids in binding mRNA to ribosomes during translation. At the 3′ end a string of 30-100 A nucleotides, called the **poly-A tail,** is added. The role of the tail is unclear, because some mRNAs lack this modification.

After transcription, the pre-mRNA molecules are spliced and shortened by the removal of all introns. In splicing, the transcript is cut at the borders between introns and exons, and the exons are spliced together by enzymes to form the mature mRNA. The intron sequences are discarded. Although a pre-mRNA transcript

▌**Introns** DNA sequences present in some genes that are transcribed but are removed during processing and therefore are not present in mRNA.

▌**Exons** DNA sequences that are transcribed and joined to other exons during mRNA processing and are translated into the amino acid sequence of a protein.

▌**5′ flanking region** A nucleotide region adjacent to the 5′ end of a gene that contains regulatory sequences.

▌**Cap** A modified base (guanine nucleotide) that is attached to the 5′ end of eukaryotic mRNA molecules.

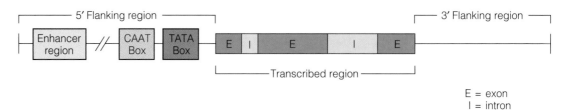

▶ **FIGURE 9.5** The organization of a typical eukaryotic gene. The part of the gene that is transcribed consists of all the introns and exons. Only the sequences in the exons appear in the mature mRNA and are translated into the amino acid sequence of a protein.

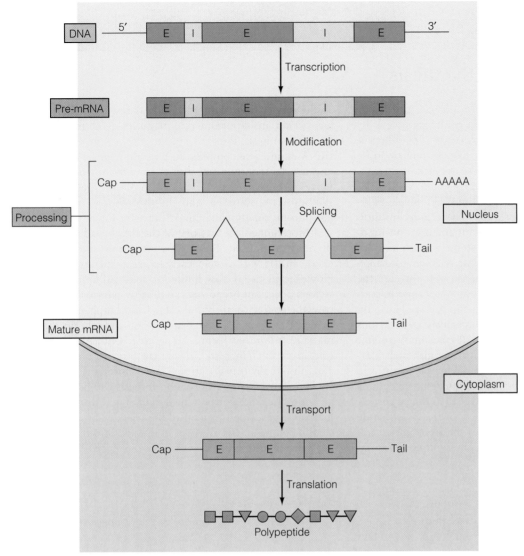

▶ FIGURE 9.6 Steps in the processing and splicing of mRNA. The template strand of DNA is transcribed into a pre-mRNA molecule. The ends of this molecule are modified, and the introns are spliced out to produce a mature mRNA molecule. The mRNA is then moved to the cytoplasm for translation.

might be 5000 nucleotides long, the mature mRNA might be less than 1000 nucleotides long. Proper splicing of pre-mRNA is essential for normal gene function. Several human genetic disorders are the result of abnormal pre-mRNA splicing. In a disorder called β-thalassemia, one or more mutations at the border between introns and exons lowers the efficiency of splicing and results in a deficiency in the amount of the gene product (beta-globin) synthesized.

After processing and splicing, the mRNA moves from the nucleus to the cytoplasm, where the encoded information is translated into the linear series of amino acids in a protein.

Translation Requires the Interaction of Several Components

Proteins are polymers assembled from amino acids. Twenty different amino acids are used in assembling proteins. The structure of each amino acid includes an **amino group** (NH_2), a **carboxyl group** (COOH), and an **R group** (▶ Figure 9.7).

■ **Amino group** A chemical group (NH_2) found in amino acids and at one end of a polypeptide chain.

■ **Carboxyl group** A chemical group (COOH) found in amino acids and at one end of a polypeptide chain.

■ **R group** A term used to indicate the position of an unspecified group in a chemical structure.

Concepts and Controversies

Antibiotics and Protein Synthesis

Antibiotics are chemicals produced by microorganisms as defense mechanisms. The most effective antibiotics work by interfering with essential biochemical or reproductive processes. Many antibiotics block or disrupt one or more stages in protein synthesis. Some of these are listed below.

Tetracyclines are a family of chemically similar antibiotics used to treat a range of bacterial infections. Tetracyclines interfere with the initiation of translation. The tetracycline molecule binds to the small ribosomal subunit and prevents binding of the tRNA anticodon in the first step in initiation. Both eukaryotic and prokaryotic ribosomes are sensitive to the action of tetracycline. This antibiotic cannot pass through the plasma membrane of eukaryotic cells, but it can enter bacterial cells to inhibit protein synthesis and stop bacterial growth.

Streptomycin is an antibiotic used clinically to treat serious bacterial infections. It binds to the small ribosomal subunit but does not prevent initiation or elongation, although it alters the efficiency of protein synthesis. When streptomycin binds to a ribosome, it alters the interaction between codons in the mRNA and anticodons in the tRNA, so that incorrect amino acids are incorporated into the growing polypeptide chain. In addition, streptomycin causes the ribosome to fall off the mRNA at random, preventing the synthesis of complete proteins.

Puromycin is an antibiotic that is not used clinically but has played an important role in studying the mechanism of protein synthesis. Puromycin is about the size and shape of a tRNA-amino acid complex. As a result, it enters the ribosome and can be incorporated into a growing polypeptide chain. Once puromycin has been added to the polypeptide, further synthesis is terminated, and the shortened polypeptide that has puromycin attached falls off the ribosome.

Chloramphenicol was one of the first broad-spectrum antibiotics introduced. Eukaryotic cells are resistant to its actions, and it was widely used to treat bacterial infections. However, its use is now limited to external applications and serious infections because it can destroy cells in the bone marrow, the source of all blood cells. Chloramphenicol binds to the large ribosomal subunit of bacteria and inhibits the enzymatic reaction that forms the peptide bond. Another antibiotic, erythromycin, also binds to the large ribosomal subunit and inhibits the translocational step of protein synthesis.

Almost every step of protein synthesis can be inhibited by an antibiotic, and work on the design of new drugs to fight infections is based on the detailed knowledge of the way the nucleotide sequence in mRNA is converted into the amino acid sequence of a protein.

▶ **FIGURE 9.7** An amino acid, showing the amino group, the carboxyl group, and the chemical side chain known as an R group. The R groups differ in each of the 20 amino acids used in protein synthesis.

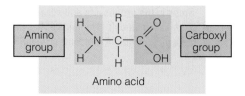

■ **Peptide bond** A covalent chemical link between the carboxyl group of one amino acid and the amino group of another amino acid.

■ **N-terminus** The end of a polypeptide or protein that has a free amino group.

■ **C-terminus** The end of a polypeptide or protein that has a free carboxyl group.

The R groups are side chains that differ for each amino acid. Some R groups are positively charged, some carry a negative charge, and others are electrically neutral. The 20 amino acids typically found in proteins and their abbreviations are listed in Table 9.1.

Amino acids are linked together by forming a covalent **peptide bond** between the amino group of one amino acid and the carboxyl group of another amino acid (▶ Figure 9.8). Two linked amino acids form a dipeptide, three form a tripeptide, and 10 or more make a polypeptide. Each polypeptide (and protein) has a free amino group at one end, known as the **N-terminus,** and a free carboxyl group, called the **C-terminus** at the other.

Converting the information in the mRNA codons into the amino acids of a polypeptide chain is accomplished by the interaction of the mRNA with amino acids

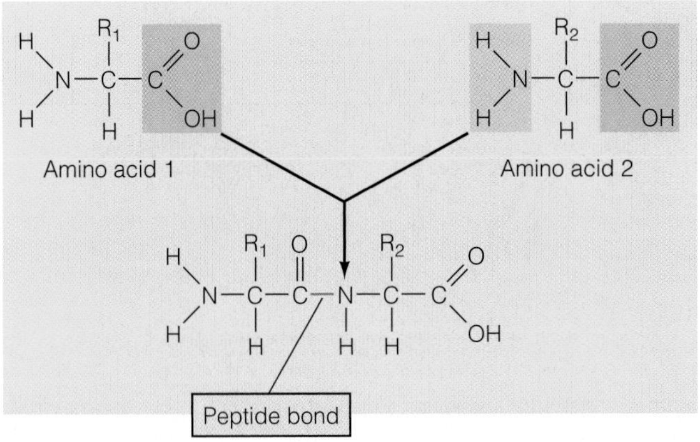

▶ **FIGURE 9.8** The formation of a peptide bond between two amino acids. An equivalent of a water molecule (H_2O) is split off during covalent bond formation. This reaction is catalyzed by an enzyme.

Table 9.1 Amino Acids Commonly Found in Proteins

Amino Acid	Abbreviation
Alanine	ala
Arginine	arg
Asparagine	asn
Aspartic acid	asp
Cysteine	cys
Glutamic acid	glu
Glutamine	gln
Glycine	gly
Histidine	his
Isoleucine	ile
Leucine	leu
Lysine	lys
Methionine	met
Phenylalanine	phe
Proline	pro
Serine	ser
Threonine	thr
Tryptophan	trp
Tyrosine	tyr
Valine	val

and two other components of the cytoplasm, ribosomes and **transfer RNAs (tRNAs)** (Table 9.2).

Ribosomes are cellular organelles composed of two subunits, each of which contains RNA (**rRNA**) combined with proteins. Ribosomes, either free in the cytoplasm or bound to the membranes of the endoplasmic reticulum, are the site of protein synthesis.

Transfer RNA molecules help convert the mRNA codons into the amino acid sequence of a polypeptides. A tRNA molecule is a small (about 80 nucleotides) single-stranded molecule folded back upon itself to form several looped regions (▶ Figure 9.9). Molecules of tRNA act as adapters to match the codons in mRNA with the proper amino acids for incorporation into a polypeptide chain. As adapters, tRNA molecules have two tasks: (1) bind to the appropriate amino acid; and (2) recognize the proper codon in mRNA. The structure of tRNA molecules allows them to perform both tasks. A loop at one end of the molecule contains a triplet nucleotide sequence called an **anticodon loop**. The anticodon recognizes and pairs with a specific codon in an mRNA molecule. The other end of the tRNA contains a site that can bind the appropriate amino acid (Figure 9.9).

There are 20 different amino acids and at least 20 different types of tRNA. Each tRNA carries an anticodon that has a specific nucleotide sequence. For example, if the anticodon in a tRNA molecule is CCC, the amino acid glycine should bind at the other end of the molecule. But by themselves, tRNA molecules cannot recognize amino acids. An enzyme that recognizes a specific amino acid and its proper tRNA carries out this task. Because there are 20 amino acids, there is a family of 20 such

■ **Transfer RNA (tRNA)**   A small RNA molecule that contains a binding site for a specific type of amino acid and a three-base segment known as an anticodon that recognizes a specific base sequence in messenger RNA.

■ **Ribosomal RNA (rRNA)**   The RNA molecules that form part of the ribosome.

Table 9.2 RNA Classes

Class	Function	Number of Different Types	Size (Nucleotides)	% of RNA in Cell
Ribosomal RNA (rRNA)	Structural, functional component of ribosomes	3	120 to 4800	90
Messenger RNA (mRNA)	Carries genetic information from DNA to ribosomes	Many thousands	300 to 10,000	3 to 5
Transfer RNA (tRNA)	Adapter recognizes nucleotide triplets and amino acids. Transports amino acids to ribosomes	50 to 60	75 to 90	5 to 7

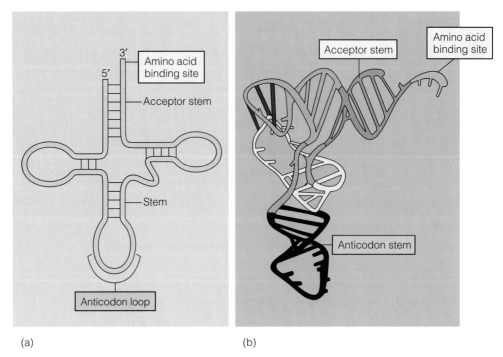

(a)   (b)

▶ **FIGURE 9.9** (a) Transfer RNA acts as a molecular adapter, can recognize mRNA codons (at the tRNA anticodon loop), and can bind the appropriate amino acid (at the amino acid binding site). (b) The three-dimensional structure of a tRNA molecule.

■ Anticodon loop The region of a tRNA molecule that contains the three-base sequence (known as an anticodon) that pairs with a complementary sequence (known as a codon) in an mRNA molecule.

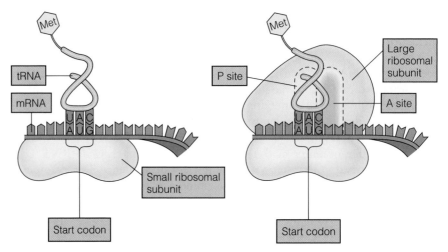

▶ **FIGURE 9.10** Steps in the initiation of translation. An mRNA molecule in the cytoplasm binds to a small ribosomal subunit. A tRNA that carries an amino acid (usually methionine) binds to the first codon (the start codon) in the mRNA. In a second step, a large ribosomal subunit binds to the small subunit, forming the initiation complex.

■ Start codon or initiator codon A codon present in mRNA that signals the location for translation to begin. The codon AUG functions as an initiator codon.

■ Stop codons Codons present in mRNA that signal the end of a growing polypeptide chain. UAA, UGA, and UAG function as stop codons.

enzymes to bind amino acids to the proper tRNAs. In turn, because of base-pairing rules, tRNAs that have the anticodon CCC can bind only to the codon GGG on an mRNA molecule. In this way, the three nucleotides in each mRNA codon are matched with the proper amino acid to form the new polypeptide.

Translation takes place in a series of steps: initiation, elongation, and termination. In the first step, mRNA, the small ribosomal subunit, and a tRNA that carries the first amino acid join together to form an initiation complex (▶ Figure 9.10). Methionine is a unique amino acid in that it is usually inserted first in all polypeptide chains. Each mRNA molecule has a codon that marks the beginning of the message (the **start codon,** also called the initiator codon) and the end of the message (**stop codon**). The small ribosomal subunit binds at the start codon (AUG), and the anticodon (UAC) of a tRNA that carries methionine binds to the start codon. To complete initiation, a large ribosomal subunit binds to the small subunit.

Once initiation is complete, the polypeptide elongates by adding amino acids. Ribosomes have two binding sites for tRNA, called the P site and the A site. The tRNA that carries methionine binds to the P site during initiation. Elongation begins when a tRNA molecule that carries the second amino acid pairs with the mRNA codon in the A site (◗ Figure 9.11). When the second amino acid is in position, an enzyme

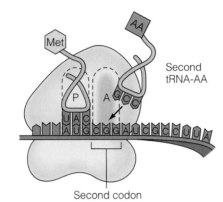

(a) As the first step in elongation, a tRNA-AA complex binds to the codon in the A site.

◗ **FIGURE 9.11** Steps in elongation during translation. (a) After initiation, the anticodon of a second tRNA binds to the second mRNA codon, which occupies the A site of the ribosome. (b) During peptide bond formation, the two amino acids are linked together by a covalent chemical bond. When the peptide bond is formed, the tRNA in the P site is released, leaving this site empty. (c) In translocation, the ribosome shifts down the mRNA by one codon, moving the tRNA that carries the growing polypeptide chain to the P site and moving the next mRNA codon into the A site. (d) A tRNA that carries an amino acid binds to the codon in the A site, and another peptide bond is formed. This process continues until a stop codon in the mRNA (UAA) is reached. When the stop codon occupies the A site, the translation complex comes apart, and the ribosome, mRNA, and completed polypeptide are separated.

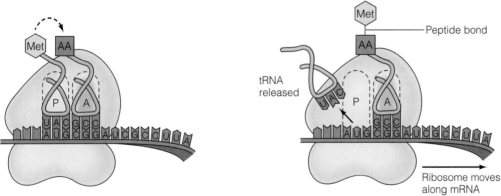

(b) An enzyme catalyzes the formation of a peptide bond between the two amino acids. The dipeptide that forms is attached to the second tRNA. This frees up the first tRNA, which vacates the P site.

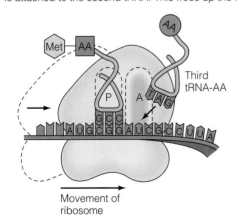

Movement of ribosome

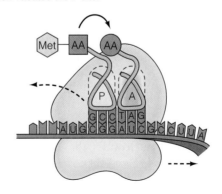

(c) The ribosome moves down the mRNA. The tRNA-dipeptide now occupies the P site and another tRNA-AA complex can occupy the A site.

(d) The dipeptide is linked by a peptide bond to the third amino acid. This frees the second tRNA. The ribosome moves down one more codon, exposing the A site and freeing it up for the addition of another tRNA-AA. This process repeats itself until the terminator codon (UUA) is reached.

The Flow of Genetic Information Is a Multistep Process ● 229

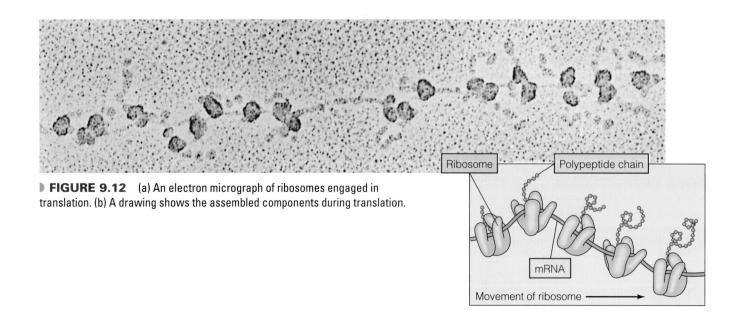

FIGURE 9.12 (a) An electron micrograph of ribosomes engaged in translation. (b) A drawing shows the assembled components during translation.

Ribosome

Polypeptide chain

mRNA

Movement of ribosome →

associated with the ribosome joins the two amino acids together by forming a peptide bond. When this peptide bond is formed, the tRNA in the P site is released and moves away from the ribosome. The tRNA in the A site (with its attached amino acids) is moved to the P site (Figure 9.11). This movement brings the third mRNA codon into the A site, where it is recognized by the anticodon of a tRNA that carries the third amino acid. A peptide bond is formed between the second and third amino acid, and the process repeats itself, adding amino acids to the growing polypeptide chain (▶ Figure 9.12). Elongation continues until the ribosome reaches the stop codon. Stop codons (UAG, UGA, UAA) do not code for amino acids, and there are no tRNA anticodons for stop codons. At this point, polypeptide synthesis is terminated, and the polypeptide, mRNA, and tRNA are released from the ribosome. The role of antibiotics in dissecting the steps in protein synthesis is discussed in Concepts and Controversies: Antibiotics and Protein Synthesis.

THE GENETIC CODE: THE KEY TO LIFE

The information that specifies the order of amino acids in a given polypeptide is encoded in the nucleotide sequence of a gene. Because DNA is composed of only four different nucleotides, at first glance it may seem difficult to envision how the information for literally billions of different combinations of 20 different amino acids can be carried in DNA. How does DNA encode genetic information? If each nucleotide encoded the information for one amino acid, only four different amino acids could be inserted into proteins (four nucleotides, taken one at a time, or 4^1). If a sequence of two nucleotides encoded an amino acid, only 16 combinations would be possible (four nucleotides, taken two at a time, or 4^2). On the other hand, a sequence of three nucleotides allows 64 combinations (four nucleotides, taken three at a time, or 4^3), 44 more than the 22 required.

In a series of experiments using mutants of a bacteriophage called T4, Francis Crick, Sidney Brenner, and colleagues discovered that a sequence of three nucleotides encode the information for one amino acid. They also proposed that some amino acids could be specified by more than one combination of three nucleotides, using most of the remaining 44 combinations. This work established that the genetic code consists of a linear series of nucleotides, read three at a time, and that each triplet specifies an amino acid.

Table 9.3 Codons on Messenger RNA and Their Corresponding Amino Acids

Codon	Amino Acid	Codon	Amino Acid	Codon	Amino Acid	Codon	Amino Acid
AAU, AAC	Asparagine	CAU, CAC	Histidine	GAU, GAC	Aspartic acid	UAU, UAC	Tyrosine
AAA, AAG	Lysine	CAA, CAG	Glutamine	GAA, GAG	Glutamic acid	UAA, UAG	Terminator*
ACU, ACC, ACA, ACG	Threonine	CCU, CCC, CCA, CCG	Proline	GCU, GCC, GCA, GCG	Alanine	UCU, UCC, UCA, UCG	Serine
AGU, AGC	Serine	CGU, CGC	Agrinine	GGU, GGC	Glycine	UCU, UGC	Cystine
AGA, AGG	Arginine	CGA, CGG		GGA, GGG		UGA	Termintor*
						UGG	Tryptophan
AUU, AUC, AUA	Isoleucine	CUU, CUC, CUA, CUG	Leucine	GUU, GUC, GUA, GUG	Valine	UUU, UUC	Phenylalanine
AUG	Methionine**					UUA, UUG	Leucine

*Terminator codons signal the end of the formation of a polypeptide chain.

**Codon has two functions: specifies the amino acid methionine and serves as the start codon, marking the beginning of a polypeptide chain.

The code itself was soon deciphered, and the coding nature of all 64 triplets was established (Table 9.3). By convention, the genetic code is written in mRNA codons. Of these, 61 actually code for amino acids, while three (UAA, UAG, and UGA) serve as stop codons. The AUG codon serves two functions. It specifies the amino acid methionine and serves as the start codon, marking the beginning of a polypeptide chain.

An interesting feature of the genetic code is that the same codons are used for the same amino acids in viruses and life forms from bacteria, algae, fungi, and ciliates to multicellular plants and animals. The universal nature of the genetic code means that the nature of the genetic code was established very early during the evolution of life on this planet. The existence of a universal code is regarded as strong evidence that all living things are closely related and may have evolved from a common ancestor.

There are some rare exceptions to the universal nature of the genetic code that involve stop codons. In some species of the ciliates *Tetrahymena* and *Paramecium*, UAG and UAA encode glutamine. Other more substantial alterations in the genetic code have taken place in mitochondria and point to a divergent evolutionary pathway for this organelle.

THE POLYPEPTIDE PRODUCT OF TRANSLATION

The linear chain of amino acids assembled on the ribosome during translation is known as a **polypeptide**. Once formed, the polypeptide chain can have several fates. If the polypeptide is synthesized on ribosomes associated with endoplasmic reticulum (ER), it is released into the cisterna (the inside of the ER), where it can be chemically modified and transported to the Golgi complex for packaging and secretion from the cell (▶ Figure 9.13). Alternately, if the polypeptide is synthesized on ribosomes free in the cytoplasm, it usually assumes a three-dimensional shape and becomes functional when released from the ribosome. In other cases, the newly synthesized chain interacts with other polypeptides.

▪ **Polypeptide** A polymer made of amino acids joined together by peptide bonds.

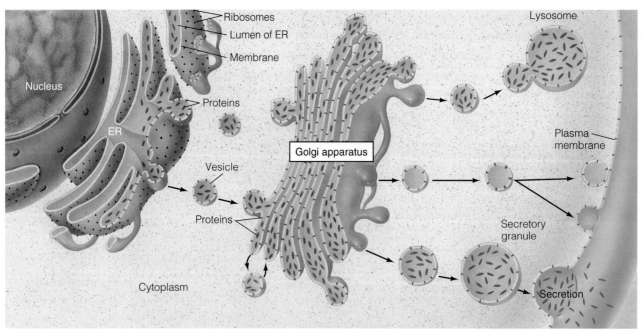

▶ **FIGURE 9.13**   Processing, sorting, and transport of proteins synthesized in a eukaryotic cell. Proteins synthesized on ribosomes attached to ER are transferred into the lumen, where they are chemically modified. Many of these proteins are transferred to the Golgi apparatus in vesicles. In the Golgi the proteins are further modified, sorted, and packaged into vesicles for delivery to other parts of the cell and are incorporated into organelles such as lysosomes, or are transported to the surface for insertion into the plasma membrane. Proteins can also be packaged into vesicles for secretion.

Proteins Have Several Levels of Structure

The amino acid sequence determines the three-dimensional shape and functional capacity of proteins. Four levels of protein structure can be identified. The sequence of amino acids in a protein is known as its **primary structure** (▶ Figure 9.14). The next two levels of structure are determined to a great extent by the R groups of each amino acid. The R groups interact with each other via hydrogen bonds to form a pleated or coiled **secondary structure.** Most proteins have regions that contain each type of secondary structure. Folding the coiled helical or pleated regions back onto themselves forms the **tertiary structure.** Some functional proteins are composed of more than one polypeptide chain, and this level of interaction is known as the **quaternary structure.**

A newly synthesized polypeptide chain can be chemically modified and folded and can interact with other polypeptide chains to achieve a functional state. When it becomes a functional, three-dimensional structure, it is called a protein.

■ **Primary structure**   The amino acid sequence in a polypeptide chain.

■ **Secondary structure**   The pleated or helical structure in a protein molecule that is brought about by the formation of bonds between amino acids.

■ **Tertiary structure**   The three-dimensional structure of a protein molecule brought about by folding on itself.

■ **Quaternary structure**   The structure formed by the interaction of two or more polypeptide chains in a protein.

Proteins Have Many Functions

Proteins are the most abundant class of molecules in the cell and participate in a wide range of functions (summarized in Table 9.4). Myosin and actin are contractile proteins found in muscle cells. Hemoglobin is a transport protein that shuttles oxygen to cells. The immune system depends on protein antibodies to identify and destroy invading organisms such as bacteria. Connective tissue and hair are rich in

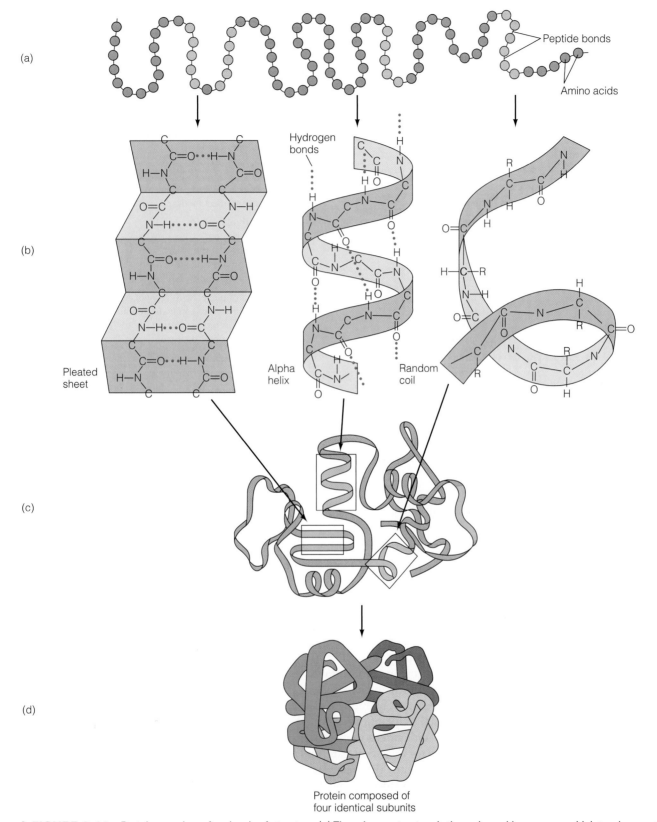

(a)

(b)

Hydrogen bonds

Pleated sheet

Alpha helix

Random coil

Peptide bonds

Amino acids

(c)

(d)

Protein composed of four identical subunits

▶ **FIGURE 9.14** Proteins can have four levels of structure. (a) The primary structure is the amino acid sequence, which to a large extent determines the other levels of structure. (b) The secondary level of structure can be a pleated sheet or a random coil. (c) Folding of the secondary structures into a functional three-dimensional shape is the tertiary level of structure. (d) Some functional proteins are made up of more than one polypeptide chain. These interactions are the quaternary level of protein structure.

Table 9.4 Biological Functions of Proteins

Protein Function	Examples	Occurrence or Role
Catalysis	Lactate dehydrogenase	Oxidizes lactic acid
	Cytochrome C	Transfers electrons
	DNA polymerase	Replicates and repairs DNA
Structural	Viral-coat proteins	Sheath around nucleic acid of viruses
	Glycoproteins	Cell coats and walls
	α-Keratin	Skin, hair, feathers, nails, and hoofs
	β-Keratin	Silk of cocoons and spider webs
	Collagen	Fibrous connective tissue
	Elastin	Elastic connective tissue
Storage	Ovalbumin	Egg-white protein
	Casein	A milk protein
	Ferritin	Stores iron in the spleen
	Gliadin	Stores amino acids in wheat
	Zein	Stores amino acids in corn
Protection	Antibodies	Form complexes with foreign proteins
	Complement	Complexes with some antigen-antibody systems
	Fibrinogen	Involved in blood clotting
	Thrombin	Involved in blood clotting
Regulatory	Insulin	Regulates glucose metabolism
	Growth hormone	Stimulates growth of bone
Nerve impulse transmission	Rhodopsin	Involved in vision
	Acetylcholine receptor protein	Impulse transmission in nerve cells
Motion	Myosin	Thick filaments in muscle fiber
	Actin	Thin filaments in muscle fiber
	Dynein	Movement of cilia and flagella
Transport	Hemoglobin	Transports O_2 in blood
	Myoglobin	Transports O_2 in muscle cells
	Serum albumin	Transports fatty acids in blood
	Transferrin	Transports iron in blood
	Ceruloplasmin	Transports copper in blood

structural proteins such as keratin and collagen. Histones are structural proteins that complex with DNA to help form chromosomes. Many hormones, such as insulin, are proteins.

One of the most important classes of proteins is enzymes, molecules that act as catalysts in biochemical reactions (⊕ Figure 9.15). Enzymes accelerate the rate of a chemical reaction by reducing the energy needed to activate the reaction. The three-dimensional shape of the enzyme generates an active site. Molecules that bind to the active site and undergo a chemical change are known as substrates. Enzymes are usually named for their substrate, and the suffix 'ase' is added. The enzyme that catalyzes the breakdown of the sugar lactose is called lactase, and the enzyme that catalyzes the conversion of the amino acid phenylalanine to tyrosine is called phenylalanine

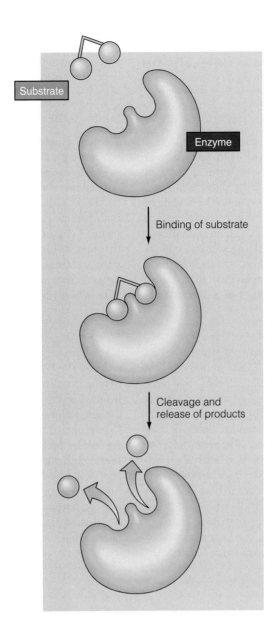

▶ FIGURE 9.15 Substrates bind to enzymes at active sites. The enzyme acts as a catalyst to carry out a chemical reaction, converting the substrate into a product.

Substrate

Enzyme

Binding of substrate

Cleavage and release of products

hydroxylase. The relationship between enzymes and genetic disorders is explored in Chapter 10.

The function of all proteins depends ultimately on the amino acid sequence of the polypeptide chain. The nucleotide sequence of DNA determines the amino acid sequence of proteins. If protein function is to be maintained from cell to cell and from generation to generation, the nucleotide sequence of a gene must remain unchanged. Alterations in the nucleotide sequence of DNA (a mutation) produce mutant gene products that have altered or impaired functions. The phenotypic consequences of changes in DNA are discussed in Chapter 11.

Case Studies

CASE 1

A genetic counselor was called to the newborn nursery to examine an infant who was diagnosed with galactosemia through the state's newborn screening program. The infant was three weeks old and had been admitted to the hospital because of failure to thrive and severe jaundice (yellowing of the skin due to liver problems). Upon examination, the infant had an enlarged liver (hepatomegaly), cataracts, and constant diarrhea and vomiting with milk ingestion. *E. coli* infection is a common cause of death in infants who have galatosemia, and cultures were drawn on this infant. Laboratory results confirmed that the infant had a deficiency of the enzyme galactose-1-phosphate uridyltransferase and was infected with *E. coli*.

The counselor took a detailed family history and explained this condition to the parents. She indicated that this condition is due to the inheritance of an altered gene from both parents (autosomal recessive) and that there is a 25% or 1 in 4 chance that with each pregnancy they have together that they will bear a child who has this condition. The counselor explained that there is wide variability in the condition, ranging from very mild to severe. A blood test could determine which variant of this disease they carry.

CASE 2

Recurring cases of mad cow disease have plagued the United Kingdom during the last three years. What started out as a topic of interest to a limited few cell biologists, has blossomed into a huge public interest story. Mad cow disease is caused by a prion, an infectious participle that consists only of protein. The media reported that cows are dying all over England from some mysterious disease. However, little attention was paid to the effect this condition has had on humans. The British government maintained for ten years that this unusual disease cannot be transmitted to humans. However, in March of 1996, they did an about-face and announced that bovine spongiform encephalopathy, commonly known as "mad cow disease" or BSE, can be transmitted to humans. BSE and a similar condition known as Creutzfeldt-Jakob disease (CJD) eats away at the nervous system, destroying the brain, essentially turning it into a sponge. Victims suffer dementia, confusion, loss of speech, sight, and hearing, convulsions, coma, and death. Spongiform encephalopathies are always fatal, and there is no treatment. Concerns that this condition may become widespread were evident by McDonald's chains in England announcing that they would not sell hamburgers until a non-British source of beef could be found. The precautionary measures that the British took to prevent this disease in humans may have begun too late; many of the victims today may have contracted it a decade ago when the BSE epidemic began, particularly because CJD has an incubation period of 10 to 40 years.

Summary

1. In the first decade of this century, Garrod proposed the idea that genetic disorders result from biochemical alterations.
2. Using *Neurospora,* Beadle and Tatum demonstrated that mutations result in the loss of enzymatic activity, producing a mutant phenotype. Beadle proposed that genes function by controlling the synthesis of proteins and that protein function is responsible for producing the phenotype.
3. Crick later summarized the molecular relationship between DNA and proteins. DNA makes RNA, which in turn makes protein.
4. The processes of transcription and translation require the interaction of many components, including ribosomes, mRNA, tRNA, amino acids, enzymes, and energy sources. Ribosomes are the workbenches upon which protein synthesis occurs. tRNA molecules are adapters that recognize amino acids and the nucleotide sequence in mRNA, the gene transcript.
5. In transcription, one of the DNA strands is a template for synthesizing a complementary strand of RNA. The information transferred to RNA is encoded in triplet sequences of nucleotides. Of the 64 possible triplet codons, 61 code for amino acids, and three are stop codons.

6. Translation requires the interaction of charged tRNA molecules, ribosomes, mRNA, and energy sources. Within the ribosome, anticodons of charged tRNA molecules bind to complementary codons in the mRNA. The ribosome moves along the mRNA, producing a growing polypeptide chain. At termination, this polypeptide is released from the ribosome and undergoes a conformational change to produce a functional protein.

7. Four levels of protein structure are recognized, three of which are results of the primary sequence of amino acids in the backbone of the protein chain. Although proteins perform a wide range of tasks, enzyme activity is one of the primary tasks. Enzymes function by lowering the energy of activation required in biochemical reactions. The products of these biochemical reactions are inevitably involved in producing phenotypes.

Questions and Problems

1. How do mutations in DNA alter proteins?
2. The 5′ promoter and the 3′ flanking regions of genes are important in:
   a. coding for amino acids
   b. gene regulation
   c. structural support for the gene
   d. intron removal
   e. anticodon recognition
3. Define replication, transcription and translation. In what part of the cell does each process occur?
4. Briefly describe the function of the following in protein synthesis:
   a. rRNA
   b. tRNA
   c. mRNA
5. What is the difference between codons and anticodons?
6. If the genetic code used 4 bases at a time, how many amino acids could be encoded?
7. Proteins have many critical functions in the human body. Some of these functions include:
   a. transporting oxygen
   b. hormonal signalling
   c. carrying out enzymatic reactions
   d. destroying invading bacteria
   e. all of the above
8. What are the three modifications made to pre-mRNA molecules before they become mature mRNAs ready to be used in protein synthesis? What is the function of each modification?
9. The pre-mRNA transcript and protein made by several mutant genes were examined. The results are given below. Determine where in the gene a likely mutation lies: the 5′ flanking region, exon, intron, cap on MRNA or ribosome binding site.
   a. normal length transcript, normal length nonfunctional protein
   b. normal length transcript, no protein made
   c. normal length transcript, normal length mRNA, short nonfunctional protein
   d. normal length transcript, longer mRNA, longer nonfunctional protein
   e. transcript never made
10. If the genetic code is a triplet, how many different amino acids can be coded by a repeating RNA polymer composed of UA and UC (UAUCUAUCUAUC. . . . .)
    a. one
    b. two
    c. three
    d. four
    e. five
11. Determine the percent of the gene below that will code for the protein product. Gene length is measured in kilobases (kb) of DNA. Each kilobase is 1000 bases long.

12. How many kilobases of the following DNA strand will code for the protein product?

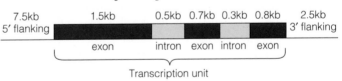

13. The following segment of DNA codes for a protein. The uppercase letters represent exons. The lowercase letters represent introns. The lower strand is the template strand. Draw the primary transcript and the mRNA resulting from this DNA:

GCTAAATGGCAaaattgccggatgacGCACATTGACTCGGaatcgaGGTCAGATGC
CGATTTACCGTtttaacggcctactgCGTGTAACTGAGCCttagctCCAGTCTACG

14. Given the following tRNA anticodon sequence, make the mRNA and the DNA template strand. Also, make the protein that is encoded by this message.
    tRNA: UAC UCU CGA GGC
    mRNA:
    DNA:
    protein:
    How many hydrogen bonds would be present in the DNA segment?

15. Given the following mRNA, write the double stranded DNA segment that served as the template. In addition, indicate both the 5′ and the 3′ ends of both DNA strands. Also make the tRNA anticodons and the protein that is encoded by the mRNA message.
    DNA:
    mRNA: 5′ - CCGCAUGUUCAGUGGGCGUAAACACUGA - 3′
    protein:
    tRNA:

16. The following is a portion of a protein:
    met-trp-tyr-arg-gly-pro-thr-
    Various mutant forms of this protein have been recovered. Using the normal and mutant sequences, determine the DNA and mRNA sequences that code for this portion of the protein and explain each of the mutations.
    a. met-trp-
    b. met-cys-ile-val-val-leu-gln-
    c. met-trp-tyr-arg-ser-pro-thr-
    d. met-trp-tyr-arg-gly-ala-val-ile-ser-pro-thr-

17. The following is the structure of glycine. Draw a tripeptide composed exclusively of glycine. Label the N-terminus and C-terminus. Draw a box around the peptide bonds.

18. Write the anticodon(s) for the following amino acids:
    a. met
    b. trp
    c. ser
    d. leu

19. Each of the following functions either in transcription or in translation. List each item in the proper category, transcription or translation: RNA polymerase, ribosomes, nucleotides, tRNA, pre-mRNA, DNA, A site, TATA box, anticodon, amino acids.

20. Enzyme X normally interacts with substrate A and water to produce compound B.
    a. What would happen to this reaction in the presence of another substance that resembles substrate A and was able to interact with enzyme X?
    b. What is a mutation in enzyme X changed the shape of the active site?

21. Can a mutation change a protein's tertiary structure without changing its primary structure? Can a mutation change a protein's primary structure without affecting its secondary structure?

22. Explain the role of proteins in the relationship between DNA and phenotype.

Internet Activities

The following activities use the resources of the World Wide Web to enhance the topics covered in this chapter. To investigate the topics described below, log on to the book's homepage at:

http://www.brookscole.com/biology

1. *The control of gene expression,* from the On-line Biology page, has much information about how gene expression is controlled. Read about the control of gene expression in bacteria, viruses, and eukaryotes. Often, the expression of multiple genes is controlled by a single protein factor, such as according to the operon model of transcriptional regulation proposed by Jacob and Monod. What benefits might such a mechanism of concerted gene expression provide? Compare the genome sizes for various eukaryotes, including humans, *Drosophila,* toads, and salamanders. What percentage of the average eukaryotic genome encodes proteins? What percentage of the human genome? What function, if any, does the non-coding portion of the genome serve? Gene expression involves multiple steps, from transcription to RNA splicing to translation. Did you know that each of these steps can be regulated according to the particular gene, cell type, and environmental or developmental circumstances?

2. The genome of the bacterium *Haemophilus influenzae* was the first genome to be completely sequenced and mapped. Since then, the genomes of several other prokaryotic organisms have been sequenced. Most of these milestones were accomplished by the staff at TIGR. The TIGR homepage contains information about the DNA and protein sequence and information about the cellular role of various gene products. View the map of *H. influenzae* and compare it with that of *Mycoplasma genitalium,* paying special attention to the relative sizes of the genomes, the number and sizes of the genes in each genome, and the cellular roles of their products.

   a. The genome of *M. genitalum* may be near the minimum size needed for a living organism. How many of the 482 genes from *M. genitalium* are found in *H. influenzae*? How about the reverse?

   b. Can you determine the cellular roles of the three largest genes in each genome?

For Further Reading

Baralle, F. E. (1983). The functional significance of leader and trailer sequences in eucaryotic mRNAs. *Int. Rev. Cytol. 85:* 71–106.

Beadle, G. W., & Tatum, E. L. (1941). Genetic control of biochemical reactions. Neurospora. *Proc. Natl. Acad. Sci. USA 27:* 499–506.

Beato, M., Truss, M., & Chaves, S. (1996). Control of transcription by steroid hormones. *Ann. N.Y. Acad. Sci. 784:* 93–123.

Bielka, H. (Ed.). (1983). *The Eukaryotic Ribosome.* New York: Springer-Verlag.

Chambon, P. (1981). Split genes. *Sci. Am. 244* (May): 60–71.

Crick, F. H. C. (1962). The genetic code. *Sci. Am. 207* (October): 66–77.

Danchin, A., & Slonimski, P. (1985). Split genes. *Endeavour 9:* 18–27.

Darnell, J. (1985). RNA. *Sci. Am. 253* (October): 68–87.

Doolittle, R. F. (1985). Proteins. *Sci. Am. 253* (October): 88–99.

Dreyfuss, G., Hentze, M., & Lammd, A. I. (1996). From transcript to protein. *Cell 85:* 963–972.

Fesenfeld, G. (1996). Chromatin unfolds. *Cell 86:* 3–9.

Garrod, A. (1902). The incidence of alkaptonuria: A study in chemical individuality. *Lancet 2:* 1666–1670.

Hamkalo, B. (1985). Visualizing transcription in chromosomes. *Trends Genet. 1:* 255–260.

Holm, L. & Sander, C. (1996). Mapping the protein universe. *Science 273:* 593–603.

Koshland, D. E. (1973). Protein shape and control. *Sci. Am. 229* (October): 52–64.

Krumm, A., Meulia, T., & Groudine, M. (1993). Common mechanisms for the control of eukaryotic transcriptional elongation. *Bioessays 15:* 659–665.

Noller, H. F. (1985). Structure of ribosomal RNA. *Ann. Rev. Biochem. 53:* 119–162.

Pain, V. M. (1996). Initiation of protein synthesis in eukaryotic cells. *Eur. J. Biochem. 236:* 747–771.

Powers, T. & Walter, P. (1997). A ribosome at the end of the tunnel. *Science 278:* 2072–2073.

Schimmel, P. & Alexander, R. (1998). All you need is RNA. *Science 281:* 658–659.

Sonnenberg, N. (1993). Remarks on the mechanism of ribosome binding to eukaryotic mRNAs. *Gene Expression 3:* 317–323.

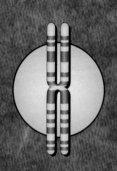

From Proteins to Phenotypes

One of the landmark discoveries in human genetics began with a young Norwegian mother who had two mentally retarded children. The first child, a girl, did not learn to walk until nearly 2 years of age and spoke only a few words. The child also had a musty odor that could not be removed by bathing. Her younger brother was also slow to develop and never learned to walk or talk. He had the same musty odor as his sister. To learn why both her children were retarded and had a musty odor, the mother went from doctor to doctor, but to no avail. Finally, in the spring of 1934, the persistent woman took the two children, then aged 4 and 7 years, to Dr. Asbjorn Fölling, a biochemist and a physician.

Because the urine from these children had a musty odor, Fölling first tested the urine for signs of infection, but there was none. He did find something in the urine that reacted with ferric chloride and produced a green color. Beginning with 20 liters of urine collected from the children, he began a series of biochemical extractions to isolate and identify the unknown substance. After 5 weeks, he was able to purify the compound; he then spent another 6 weeks identifying the substance as phenylpyruvic acid. To confirm that the compound in the urine was indeed phenylpyruvic acid, he synthesized and purified phenylpyruvic acid from organic chemicals and demonstrated that both compounds had the same physical and chemical properties.

With his knowledge of chemistry, Dr. Fölling proposed that the phenylpyruvic acid was produced by a metabolic disorder that affected the breakdown of the amino acid phenylalanine. He further postulated that this biochemical abnormality might cause the phenotype of retardation. To confirm this, he examined the urine of several hundred patients in nursing homes and schools for the retarded and found phenylpyruvic acid in the urine of eight retarded individuals, but never in the urine of normal individuals. Less than six months after he began working on the problem, Dr. Fölling finished a manuscript for publication that described phenylketonuria, a condition now regarded as a prototype for metabolic genetic disorders. His work is the foundation for human biochemical genetics and helped establish the relationship between a gene product and a phenotype.

As we discussed in the last chapter, the discovery of biochemical mutations in fungi by George Beadle and Edward Tatum in 1941 was an important stage in understanding how gene expression results in a phenotype. We also considered how DNA encodes information for the chemical structure of proteins. In this chapter we consider the relationship between proteins and the phenotype, using examples that emphasize the diverse role of proteins in living systems.

THE ROLE OF PROTEINS

As outlined in Chapter 9, proteins are the most numerous and multifunctional macromolecules in the cell. They are essential parts of all structures and biological processes carried out in every cell type. Proteins are components of membrane systems and the internal skeleton of cells. They form the glue that holds cells and tissues together, carry out biochemical reactions, destroy invading microorganisms, and act as hormones (Figure 10.1), receptors, and transport molecules. Even the replication of DNA and the expression of genes depend on the action of proteins.

▶ **FIGURE 10.1** Portrait of a dwarf, by Goya. Genetic forms of dwarfism can be produced by mutations in genes that encode growth factors, hormones, and receptors.

PROTEINS ARE THE LINK BETWEEN GENOTYPE AND PHENOTYPE

The functional diversity of proteins described above is matched only by their structural diversity. Each protein is composed of a specific number of amino acids arranged in a specific order. The amino acid order, or primary structure, is ultimately responsible for the protein's three-dimensional shape and its functional capacity. There is a direct relationship between someone's genotype and the proteins they synthesize. Heritable errors that change the amino acid sequence of a protein can produce phenotypic differences that range from insignificant to lethal. In other words, phenotypes are the result of protein action. If a protein is missing or has an altered function, the phenotype is altered. This chapter examines the direct link between an individual's genotype, the proteins they synthesize, and their phenotype. We will do this by examining how alterations in proteins result in an altered phenotype.

METABOLIC PATHWAYS AND GENETIC DISORDERS

■ **Substrate** The specific chemical compound that is acted upon by an enzyme.

■ **Product** The specific chemical compound that is the result of enzymatic action. In biochemical pathways, a compound can serve as the product of one reaction and the substrate for the next reaction.

■ **Metabolism** The sum of all biochemical reactions by which cells convert and utilize energy.

Because a great deal is known about proteins that function as enzymes, first we consider genetic disorders caused by defects in enzymatic activity. Enzymes are biological catalysts that carry out biochemical reactions. They convert molecular **substrates** into **products** by chemical reactions (▶ Figure 10.2). Enzymatic reactions do not occur at random; they are organized into chains of reactions known as *biochemical pathways* (▶ Figure 10.3). The sum of all biochemical reactions in the cell is called **metabolism,** and sets of biochemical reactions are often called metabolic pathways.

In a metabolic pathway, the product of one reaction serves as the substrate for the next reaction. Failure to carry out one of the reactions in a pathway shuts down all of the reactions beyond that point. It also results in the accumulation of substrates and products in the part of the pathway leading up to the block.

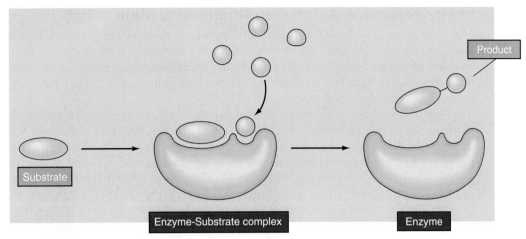

FIGURE 10.2 Each step in a metabolic pathway is a separate chemical reaction catalyzed by an enzyme, in which a substrate is converted to a product.

FIGURE 10.3 The sequence of reactions in a metabolic pathway. In this pathway, compound 1 is converted to compound 2, and 2 is converted to 3. An enzyme that catalyzes the reaction controls each step. Each enzyme is the product of a gene.

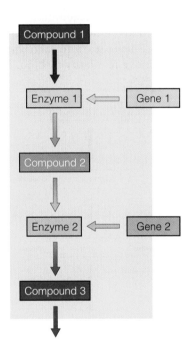

Sir Archibald Garrod first proposed the relationship between human genetic disorders and metabolism in 1901 (see Concepts and Controversies: Garrod and Metabolic Disease). Garrod investigated a disease called **alkaptonuria** (MIM/OMIM 203500), in which large quantities of a compound called homogentisic acid are excreted in urine. Garrod called this heritable condition an **inborn error of metabolism.** His work represented a pioneering study in applying Mendelian genetics to humans and in understanding the relationship between genes and biochemical reactions.

Mutations that cause the loss of activity in a single enzyme can have phenotypic effects in a number of ways. First, the buildup of one or more precursors in the pathway may be detrimental. These accumulated precursors may also overburden alternative pathways, causing a buildup of toxic metabolic products. Second, the pathway may produce an essential component of a cellular process, and lack of this component may be harmful. Mutations that affect the action of enzymes can produce a wide range of phenotypic effects, ranging from inconsequential effects to those that are lethal prenatally or early in infancy.

Alkaptonuria A relatively benign autosomal recessive genetic disorder associated with the excretion of high levels of homogentisic acid.

Inborn error of metabolism The concept advanced by Archibald Garrod that many genetic traits result from alterations in biochemical pathways.

Phenylketonuria: A Defect in Amino Acid Metabolism

Most eukaryotes, including humans, can synthesize most of the 20 amino acids found in proteins. The amino acids that cannot be synthesized in the body are called **essential amino acids** and must be part of the diet. In humans, there are nine essential amino acids: histidine, isoleucine, leucine, lysine, methionine, phenylalanine, threonine, tryptophan, and valine.

Phenylalanine, one of the essential amino acids, is the starting point for a network of metabolic pathways, several of which are associated with human genetic disorders (▶ Figure 10.4). We focus on what happens when there is a block in the first step in the pathway: the conversion of phenylalanine to another amino acid, tyrosine. Failure to convert phenylalanine to tyrosine results in **phenylketonuria** (MIM/OMIM 261600), or PKU, the disorder described at the beginning of the chapter. PKU occurs about once in every 12,000 births and is most often associated with a deficiency of the enzyme, phenylalanine hydroxylase, which converts phenylalanine to tyrosine. Because of the defect in the enzyme, phenylalanine that enters the body as part of the diet accumulates and causes the mental retardation associated with this condition. Alternative biochemical pathways produce compounds including phenylacetic acid (responsible for the musty odor of affected individuals) which also contribute to the phenotype.

The first sign of PKU is the accumulation of excess levels of phenylalanine in blood and other body fluids of newborns. If left untreated, affected individuals become mentally retarded, never learn to talk, and have enhanced reflexes that cause their arms and legs to move in a jerky fashion. Because the skin pigment melanin is

■ **Essential amino acids**   Amino acids that cannot be synthesized in the body and must be supplied in the diet.

■ **Phenylketonuria (PKU)**   An autosomal recessive disorder of amino acid metabolism that results in mental retardation if untreated.

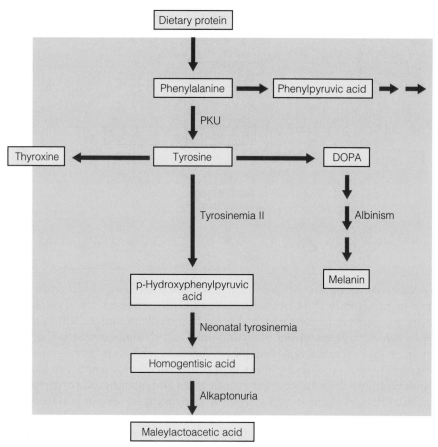

▶ **FIGURE 10.4**   The metabolic pathway that begins with the essential amino acid, phenylalanine. Normally, phenylalanine is converted to tyrosine and from there to many other compounds. A metabolic block, caused by a mutation in the gene encoding the enzyme phenylalanine hydroxylase, prevents the conversion of phenylalanine to tyrosine and in homozygotes produces the phenotype of phenylketonuria (PKU). The diagram also shows other metabolic diseases produced by mutations in genes that encode enzymes in this pathway.

Concepts and Controversies

Garrod and Metabolic Disease

Sir Archibald Garrod was a distinguished physician, Oxford professor, and physician to the royal family. Garrod was widely trained; he was skilled in biology, biochemistry, and clinical medicine. In studying alkaptonuria, he showed that the condition causes the excretion of abnormal amounts of a compound called homogentisic acid, which causes the urine to turn black upon exposure to air.

Garrod proposed that the presence of homogentisic acid in the urine of affected individuals was caused by a metabolic block, which prevented metabolism of the compound. He also speculated that the metabolic block was caused by an enzyme deficiency. Garrod observed that 60% of the affected individuals were the result of first-cousin marriages and that the parents were unaffected. This led him to propose that alkap-

tonuria is inherited as a recessive Mendelian trait. Remember, Garrod did his work just a few years after Mendel's work was rediscovered and widely publicized. Garrod studied other metabolic disorders and published his findings in a book, *Inborn Errors of Metabolism*, that is widely regarded as a milestone in human genetics.

Why was Garrod's work, like that of Mendel, not immediately appreciated? From our perspective, it may seem strange, but Garrod worked in three areas that at the time were almost totally isolated from one another: genetics, medicine, and chemistry. Chemists had no interest in studying heredity, physicians regarded the conditions he studied as too rare to be important, and geneticists cared little about either metabolism or medicine.

also a product of the blocked metabolic pathway (Figure 10.4), most PKU victims usually have lighter hair and skin color than their siblings or other family members.

How does the failure to convert phenylalanine to tyrosine produce mental retardation and the other neurological problems? These phenotypic effects cannot be caused by the failure to convert phenylalanine to tyrosine because adequate amounts of tyrosine are available in food. The problem is caused by the accumulation of high levels of phenylalanine and metabolic by-products (Figure 10.4) when the nervous system is still developing.

The human brain and nervous system continue to develop after birth and require a constant supply of amino acids for protein synthesis. Special proteins in the plasma membrane of cells are responsible for moving amino acids into the cell. Phenylalanine and a group of seven other amino acids are transported by one of these systems. As phenylalanine accumulates in the fluid surrounding cells of the maturing nervous system, it blocks the uptake of the seven other amino acids. It is not clear whether the damage to the nervous system is the result of transporting too much phenylalanine, not enough of other amino acids, or the accumulation of secondary metabolites. The result, however, is brain damage, mental retardation, and the other neurological symptoms that result in the phenotype of PKU.

Dietary Control of the PKU Phenotype   Most people who have PKU are born to heterozygous mothers and are unaffected before birth because maternal enzymes metabolize the excess phenylalanine that accumulates in their bodies during fetal development. PKU homozygotes suffer neurological damage and become retarded only after birth when fed on a diet containing phenylalanine.

PKU can be treated by placing affected individuals on a diet with restricted phenylalanine intake (see Concepts and Controversies: Dietary Management and Metabolic Disorders). This treatment is widely used and has been successful in reducing the effects of this disease. But managing PKU by controlling dietary intake is both difficult and expensive. One major problem is that phenylalanine is an amino acid present in many protein sources, and it is impossible to eliminate all protein from the diet. To overcome this problem, a synthetic mixture of amino acids (with very low levels of phenylalanine) is used as a protein substitute. The challenge is to maintain a level of phenylalanine in the blood that is high enough to permit normal development of the nervous system and yet low enough to prevent mental retardation. Treatment must be started in the first 1 or 2 months after birth. After that time the brain is damaged, and treatment is less effective.

All states now require screening of newborns for PKU, so the number of untreated cases is very low. Screening and treatment with a low phenylalanine diet allows PKU homozygotes to lead essentially normal lives. Some studies suggest that affected individuals can begin a normal diet by about 10 years of age without any effects on intellect or behavior, but some treatment centers recommend that the treatment be continued for life.

PKU Females and Reproduction As PKU children treated with diet therapy have matured and reached reproductive age, an unforeseen problem has developed. If homozygous PKU females are on a regular diet during pregnancy, all of their children are born mentally retarded. Apparently the high levels of phenylalanine in the maternal circulation cross the placenta and damage the nervous system of the developing fetus in a way that is independent of the child's genotype. It is now recommended that PKU females stay on a PKU diet all through life or return to the diet before becoming pregnant and maintain the diet throughout pregnancy. PKU females have other reproductive options, including *in vitro* fertilization and the use of surrogate mothers.

Other Metabolic Disorders in the Phenylalanine Pathway

The mutation in the pathway from phenylalanine to tyrosine is not a unique occurrence. Several other genetic disorders are associated with blocks in the metabolic pathways leading from phenylalanine. For example, the pathway from tyrosine leads to the production of the thyroid hormones thyroxine and triiodothyronine, and a block in this pathway causes the recessive autosomal disorder called genetic goitrous cretinism (Figure 10.4). Newborn homozygotes are unaffected because maternal thyroid hormones cross the placenta and promote normal growth. But in the weeks following birth, physical development is slow, mental retardation occurs, and the thyroid gland greatly enlarges. This condition is caused by the failure to synthesize a metabolic end product (a hormone), not by the accumulation of a metabolic intermediate as in PKU. If diagnosed early, giving the affected infant regular doses of thyroid hormone can treat this disorder.

In this same network of pathways, a lack of enzyme activity that leads to the buildup of homogentisic acid causes alkaptonuria, an autosomal recessive condition. This was the disorder first investigated by Garrod at the turn of the century. He postulated that it was under genetic control. This idea was confirmed in 1958 when the enzyme homogentisic acid oxidase was identified. Because of the metabolic block, excess homogentisic acid accumulates in the body and is excreted in urine. The excess acid is converted to a dark pigment in cartilage areas that are exposed to light. As a result there is often discoloration of the ears, tip of the nose, palate, and whites of the eyes. Deposition in cartilage also produces a form of arthritis in later life.

Defects in Carbohydrate Metabolism

Carbohydrates are organic molecules that include sugars, starches, glycogens, and celluloses. The simplest carbohydrates are sugars known as monosaccharides (▶ Figure 10.5a), and include glucose, galactose, and fructose. These sugars are important as metabolic energy sources. A combination of two monosaccharides produces a disaccharide (Figure 10.5b). Some common disaccharides are maltose (two glucose units, used in brewing beer), sucrose (a glucose and fructose unit, the sugar you buy at the store), and lactose (a glucose and a galactose unit, found in milk). Larger combinations of sugars form long chains known as polysaccharides (glycogen, starch, and cellulose)(Figure 10.5c). In animals, including humans, the principal

■ **Genetic goitrous cretinism** A hereditary disorder in which the failure to synthesize a needed hormone produces physical and mental abnormalities.

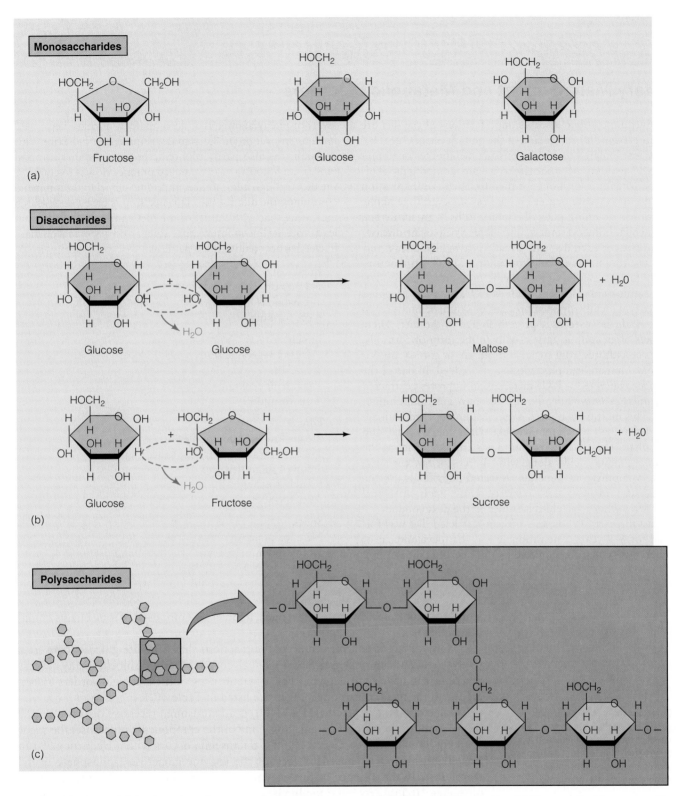

Monosaccharides

Fructose

Glucose

Galactose

(a)

Disaccharides

Glucose + Glucose → Maltose + H₂O

Glucose + Fructose → Sucrose + H₂O

(b)

Polysaccharides

(c)

▶ **FIGURE 10.5**   (a) The formula and structures for four common monosaccharides. (b) Disaccharides. (c) Monosaccharides can combine in pairs to form disaccharides or long chains of polysaccharides.

Concepts and Controversies

Dietary Management and Metabolic Disorders

*I*n several metabolic diseases, modifications of the diet are used to prevent full expression of the mutant phenotype. Diet can be manipulated to replace missing metabolites or to prevent the buildup of toxic intermediate compounds. This method is really a form of gene therapy. Dietary modification is used with varying degrees of success in several metabolic conditions, including phenylketonuria (PKU), galactosemia, tyrosinemia, homocystinuria, and maple syrup urine disease.

The diet for each disorder usually is available in two versions, one for infants and one for children and adults that usually contains higher levels of proteins and other nutrients. For some amino acid disorders, such as PKU, products prepared from enzymatically digested proteins or synthetic mixtures of amino acids are used as a protein source in the diet. These products often contain fats, usually in the form of corn oils, and carbohydrates from sugar, corn starch, or corn syrup. Vitamin and mineral supplements are also added. In one of the products available for PKU diets, casein (a protein extracted from milk) is enzymatically digested into individual amino acids. This mixture is poured over a column packed with activated charcoal, which removes phenylalanine, tyrosine, and tryptophan. These last two amino acids are added back to the remaining amino acids along with sources of fat, carbohydrates, vitamins, and minerals. Affected individuals use the powder at each meal as a source of amino acids. A typical menu for a school-age child is shown in the next column.

Until the early 1980s, this protein-restricted diet was followed for 6–9 years. The rationale was that development of the nervous system is completed by this age and that elevated

levels of phenylalanine that accompany a normal diet would have no impact on intellectual development or behavior. This decision was also partly economic because the diet can cost more than $5,000 a year. Standard practice now is to continue the diet through adolescence, and some clinicians recommend continuing the diet for life. This decision is based on research indicating that withdrawal of the diet can be deleterious and leads to a decline in intellectual ability and abnormal changes in electroencephalographic patterns.

Breakfast
2-3 cup dry rice cereal
1-2 bananas
6 oz. formula

Lunch
1-2 cans vegetable soup
3 crackers
1 cup fruit cocktail
4 oz. formula

Dinner
2 cups low-protein noodles
1-2 cups meatless spaghetti sauce
1 cup of salad (lettuce)
French dressing
4 oz. formula

Snack
1-2 cups popcorn
1 tablespoon margarine

storage form of carbohydrate is glycogen, a molecule composed of long chains of glucose units.

Many different enzymes catalyze the reactions that generate glucose, convert it to glycogen, and later release glucose from glycogen. Metabolic blocks in any of these reactions can have serious phenotypic consequences. Some genetic disorders associated with the metabolism of glycogen are listed in Table 10.1.

Galactosemia (MIM/OMIM 230400) is an autosomal recessive disorder caused by the inability to metabolize galactose, one of the components of lactose, the sugar found in human milk (▶ Figure 10.6). Galactosemia occurs with a frequency of 1 in 57,000 births and is caused by lack of the enzyme galactose-1-phosphate uridyl transferase. In the absence of this enzyme, the metabolic intermediate galactose-1-phosphate accumulates to toxic levels in the body. Homozygous recessive individuals are unaffected at birth but develop symptoms a few days later. They begin with gastrointestinal disturbances, dehydration, and loss of appetite; later symptoms include jaundice, cataract formation, and mental retardation. In severe cases the condition is progressive and fatal, and death occurs within a few months; but mild cases may remain undiagnosed for many years. A galactose-free diet and the use of galactose- and lactose-free milk substitutes and foods lead to a reversal of symptoms. But unless treatment is started within a few days of birth, mental retardation cannot be prevented.

■ **Galactosemia** A heritable trait associated with the inability to metabolize the sugar galactose. If it is left untreated, high levels of galactose-1-phosphate accumulate, causing cataracts and mental retardation.

Table 10.1

Some Inherited Diseases of Glycogen Metabolism

Type	Disease	Metabolic Defect	Inheritance	Phenotype	MIM/OMIM Number
I	Glycogen storage disease—Von-Gierke disease	Glucose-6-phosphatase deficiency	Autosomal recessive	Severe enlargement of liver, often recognized in second or third decade of life; may cause death due to renal disease.	232200
II	Pompe disease	Lysosomal glucosidase deficiency	Autosomal recessive	Accumulation of membrane-bound glycogen deposits. First lyosomal disease known. Childhood form leads to early death.	232300
III	Forbes disease, Cori disease	Amylo 1, 6 glucosidase deficiency	Autosomal recessive	Accumulation of glycogen in muscle, liver. Mild enlargement of liver, some kidney problems.	232400
IV	Amylopectinosis, Andersen disease	Amylo 1, 4 transglucosidase deficiency	Autosomal recessive	Cirrhosis of liver, eventual liver failure, death.	232500

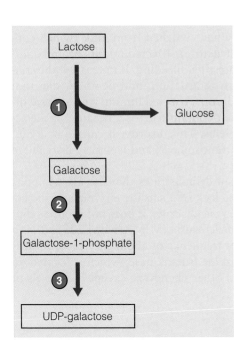

▶ **FIGURE 10.6** Metabolic pathway involving lactose and galactose. Lactose, the main sugar in milk, is enzymatically digested to glucose and galactose in step 1. Step 2 is the conversion of galactose into galactose-1-phosphate. In galactosemia, a mutation in the gene that controls step 3 prevents the conversion of galactose-1-phosphate into UDP-galactose. As a result, the concentration of galactose-1-phosphate rises in the blood, causing mental retardation and blindness.

Dietary treatment does not prevent long-term complications of this disorder, however. Affected individuals have difficulties in balance and impaired motor skills, including problems with handwriting. It is not clear whether this is due to damage to the nervous system that occurred during fetal development or whether secondary metabolites build up to toxic levels. It is also not clear why galactose-1-phosphate is toxic.

Galactosemia is also an example of a multiple-allele system. In addition to the normal allele, *G*, and the recessive allele, *g*, a third allele, known as *GD* (the Duarte allele, named after Duarte, California, the city in which it was discovered), has been found. Homozygous *GD/GD* individuals have only half of the normal enzymatic activity but show none of the symptoms of the disease. The existence of three alleles produces six possible genotypic combinations, and enzymatic activities range from 100% to 0% (Table 10.2). This disease can be detected in newborns, and mandatory screening programs in some states test all newborns for galactosemia.

Table 10.2 Multiple Alleles of Galactosemia

Genotype	Enzyme Activity (%)	Phenotype
G^+/G^+	100	Normal
G^+/G^D	75	Normal
G^D/G^D	50	Normal
G^+/g	50	Normal
G^D/g	25	Borderline
g/g	0	Galactosemia

Frutose Metabolism: A Study of Two Disorders Fructose is a major source of sugar in the human diet. In the U.S., adults eat between 50-100 g of fructose every day. Fructose is found in honey, vegetables, fruits, and table sugar (sucrose).

Two disorders of fructose metabolism provide an insight into the way mutations associated with the metabolism of a single compound can produce very different phenotypes. The first of these, **hereditary fructose intolerance** (MIM/OMIM 229600), is an autosomal recessive trait with a frequency of about 1/20,000 births, which is most often diagnosed in infants. When switched from milk (which contains lactose) to solid foods (which contain fructose), affected infants begin vomiting and develop enlarged livers and gastrointestinal bleeding. If untreated, the condition can progress to malnutrition, growth retardation, and death. A diet that eliminates fructose is used to treat this condition. Affected infants who survive often develop a permanent dislike of foods that contain fructose, because of the associated vomiting and gastrointestinal problems. This disorder is caused by a mutation in the aldolase B gene, located on chromosome 9, and is widely distributed in European populations.

A second disorder of fructose metabolism demonstrates that not all metabolic blocks have devastating effects. Although the lack of a specific enzyme and the accumulation of metabolic intermediates causes this disorder, it does not produce clinically significant symptoms. This condition, **fructosuria** (MIM/OMIM 229800), is a rare autosomal recessive condition that has a frequency of about 1 in 130,000. It is caused by a mutation in the gene for the enzyme fructokinase, which is located on chromosome 3. Affected individuals have no other phenotype except high levels of fructose in blood and urine.

MUTATIONS IN RECEPTOR PROTEINS: FAMILIAL HYPERCHOLESTEROLEMIA

Although many, and perhaps most, proteins function as enzymes, proteins also act in other roles, including signal receptors and transducers. These functions usually take place in the plasma membrane of the cell, and mutations in receptor function can have drastic consequences. For example, in testicular feminization (discussed in Chapter 7), a defect in the ability to detect and bind the hormone testosterone leads to a change in sexual phenotype, causing a genotypic male to develop into a phenotypic female.

Another genetic disorder associated with a defect in a cellular receptor is **familial hypercholesterolemia** (MIM/OMIM 144010), an autosomal dominant condition that has multiple alleles (see Chapter 5). Cholesterol and other fats are not very soluble in the bloodstream. In cells of the intestine, ingested cholesterol is

■ **Hereditary fructose intolerance** An autosomal recessive trait caused by a mutation in an enzyme of fructose metabolism. If untreated by diet changes, it can result in malnutrition, growth retardation, and death.

■ **Fructosuria** An autosomal recessive condition associated with the inability to metabolize the sugar fructose, which accumulates in the blood and urine.

■ **Familial hypercholesterolemia** A dominant autosomal genetic condition associated with a defect in cellular receptors that function in cholesterol metabolism. Affected individuals are susceptible to heart disease and early death.

coated with proteins and packaged into particles called low-density lipoproteins (LDL), which are soluble in the blood. Cells of the body take up LDLs from the circulatory system, using receptor molecules that project from the cell surface. The major site of uptake and metabolism of cholesterol is the liver. Once inside the cell, LDLs are broken down by enzymes, and the cholesterol is used for a variety of purposes, including the synthesis of new plasma membranes. Once uncoupled from the LDL, the receptor is returned to the plasma membrane to bind more LDL (▶ Figure 10.7).

If cell surface receptors are absent or defective, LDL builds up in the blood and is deposited on the artery walls as atherosclerotic plaque which causes heart disease. The gene for familial hypercholesterolemia (*FH*), inherited as an autosomal

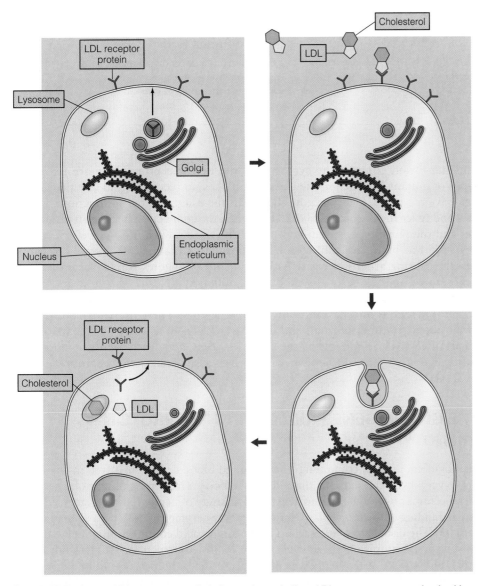

▶ **FIGURE 10.7** LDL receptors and cholesterol metabolism. LDL receptors are synthesized in the endoplasmic reticulum, modified in the Golgi apparatus, and migrate to the outer surface of the plasma membrane. LDL complexed with cholesterol binds to the receptor on the cell surface and is internalized. Once in the cell, the LDL complex associates with a lysosome, which removes the cholesterol for metabolism, and the receptor returns to the cell surface. In familial hypercholesterolemia, the LDL receptor defects prevent this cycle from occurring normally; cholesterol builds up in the bloodstream and is deposited on arterial walls, causing cardiovascular disease.

Table 10.3 Some Heritable Traits Associated with Defective Receptors

Disease	Defective/Absent Receptor	Inheritance	Phenotype	MIM/OMIN Number
Familial hypercholesterolemia	Low-density lipoprotein (LDL)	Autosomal dominant	Elevated levels of cholesterol in blood, atherosclerosis, heart attacks; early death	144010
Pseudohypoparathyroidism	Parathormone (PTH)	X-linked dominant	Short stature, obesity, round face, mental retardation	300800
Diabetes insipidus	Vasopressin receptor defect	X-linked recessive	Failure to concentrate urine; high flow rate of dilute urine, severe thirst, dehydration; can produce mental retardation in infants unless diagnosed early	304800
Testicular feminization	Testosterone/DHT receptor	X-linked recessive	Transformation of genotypic male into phenotypic female; malignancies often develop in intraabdominal testes	313700

dominant trait, controls the synthesis of the LDL receptor. There are two main types of mutant *FH* alleles, one class that makes no receptors and one that makes defective receptors. Heterozygotes have half the usual number of functional receptors and twice the normal level of LDL. In *FH*, the connection between a gene product and the phenotype is relatively straightforward. When blood levels of cholesterol are too high, the cholesterol is deposited as plaque in the walls of blood vessels, causing atherosclerosis. As a result, heterozygotes for *FH* begin to have heart attacks in their early 30s. About 1 in 500 individuals is a heterozygote for an *FH* mutation. Homozygotes (about 1 in a million people) have two faulty genes for receptor synthesis and have no functional receptors. These individuals have LDL levels about six times normal, and heart attacks begin as early as 2 years of age. Heart disease is unavoidable by the age of 20 years, and death occurs in most cases by the age of 30 years. Other genetic disorders associated with receptors are listed in Table 10.3.

THE GLOBIN GENES: UNLOCKING THE PROTEIN-PHENOTYPE LINK

Hemoglobin, an iron-containing protein in red blood cells, is involved in transporting oxygen from the lungs to the cells of the body. The hemoglobin molecule occupies a central position in human genetics. The study of hemoglobin variants led to an understanding of the molecular relationship between genes, proteins, and human disease. (1) The discovery of alterations in the amino acid composition of hemoglobins was the first example of inherited variations in protein structure. (2) The study of hemoglobin in sickle cell anemia was direct proof that mutations result in a change in the amino acid sequence of proteins. (3) The nature of the mutation in sickle cell anemia provided evidence that a change in a single nucleotide is enough to cause a genetic disorder. (4) Understanding the organization of the globin gene clusters has helped scientists understand how genes evolve and how gene expression is regulated. Heritable defects in globin structure or synthesis are well understood at the molecular level and are truly "molecular diseases," as Linus Pauling called them. In this section we consider the structure of the hemoglobin molecule, the organization of the globin genes, and some genetic disorders related to globin structure and synthesis.

Hemoglobin is a protein composed of four globin molecules. Each globin molecule is complexed with a heme group; heme is an organic molecule that contains an iron atom (▶ Figure 10.8). In the lungs, oxygen enters the red blood cell and binds to the iron atom for transport to cells of the body. Although there are several different kinds of globin molecules (and hemoglobins), the heme group is the same in all cases.

Adult hemoglobin (called Hb A) is made up of four globin molecules: two alpha-globins and two beta-globins (▶ Figure 10.9). Genes on different chromosomes encode each of these proteins. Alpha globin is encoded in a gene cluster on chromosome 16, and the beta-globin gene is part of a gene cluster on chromosome 11. Each red blood cell contains about 280 million molecules of hemoglobin, and there are between 4.2 and 5.9 x 10^{12} red blood cells in each liter of blood. Each red blood cell is replaced every 120 days, so hemoglobin production is an important biosynthetic function.

There are two copies of the alpha gene (designated alpha$_1$ and alpha$_2$) in the alpha gene cluster (▶ Figure 10.10), along with three related genes: the zeta gene, the pseudozeta, and pseudoalpha-1 genes. **Pseudogenes** are nonfunctional copies of genes whose nucleotide sequence is similar to a functional gene, but they contain

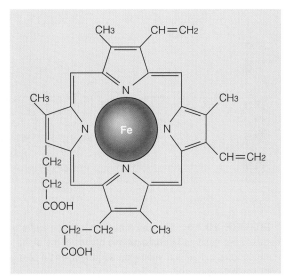

▶ **FIGURE 10.8** A heme group is a flat molecule that is inserted into the folds of a globin polypeptide. Each heme group carries an iron atom, which is important in binding oxygen in the lungs for transport to the cells and tissues of the body.

■ **Pseudogene** A nonfunctional gene that is closely related (by DNA sequence) to a functional gene present elsewhere in the genome.

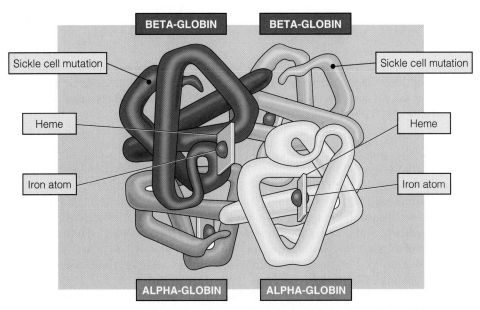

▶ **FIGURE 10.9** The functional hemoglobin molecule is composed of two alpha-globin polypeptides and two beta-globin polypeptides. Each globin molecule has an iron-containing heme group within its folds. The location of the mutation in beta-globin that is responsible for sickle cell anemia is shown near the start of each beta chain.

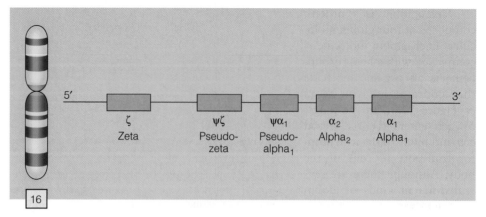

FIGURE 10.10 The chromosomal location and organization of the alpha-globin cluster. Each copy of chromosome 16 contains two copies of the alpha-globin gene (alpha1 and alpha2), two nonfunctional versions (called pseudogenes), and a zeta gene, which is active only during early embryonic development.

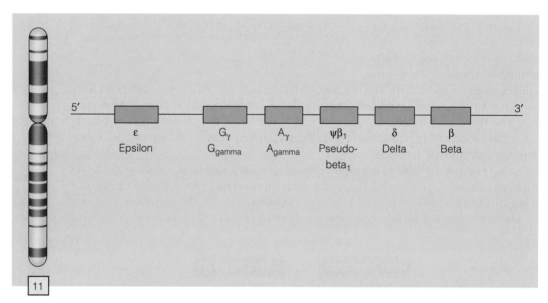

FIGURE 10.11 The chromosomal location and organization of the beta-globin complex on chromosome 11. Each copy of chromosome 11 has an episilon gene, active during embryonic development, two gamma genes (gammaG and gammaA) active in fetal development, and a delta gene and a beta gene, which are transcribed after birth.

Hemoglobin variants Alpha- and beta-globins with variant amino acid sequences.

mutations that prevent their expression. The beta gene cluster located on the short arm of chromosome 11 contains several family members expressed at different times in the life cycle (▶ Figure 10.11).

The alpha- and beta-globin genes have similar organizations, and the alpha- and beta-proteins are similar in size and amino acid composition. Alpha-globin contains 141 amino acids, and the beta-globin molecule has 146 amino acids. The amino acid sequences of the two polypeptides are very similar (▶ Figure 10.12). Because of these similarities, the alpha and beta genes fold into similar configurations, each cradling the heme group inside the folds of the polypeptide chain.

Heritable disorders of hemoglobin fall into two categories: the **hemoglobin variants,** which involve changes in the amino acid sequence of the globin polypeptides, and the **thalassemias,** which are characterized by imbalances in globin synthesis. More than 400 variants of hemoglobin have been described, each originating as the result of genetic mutation. More than 90% of the variants are caused by the substitution of one amino acid for another in the globin chain, and more than 60% of these

α-Globin	V – L S P A D K T N V K A A W G K V G A H A G E Y G A E A L E R M F L S F P T T K T Y F P H F – D L S H
β-Globin	V H L T P E E K S A V T A L W G K V – – N V D E V G G E A L G R L L V V Y P W T Q R F F E S F G D L S T

α-Globin	– – – G S A Q V K G H G K K V A – D A L T N A V A H V D D – M P N A L S A L S D L H A H K L R V D P V N
β-Globin	A V M G N P K V K A H G K K V L – G A F S D G L A H L D N – L K G T F A T L S E L H C D K L H V D P E N

α-Globin	L L S H C L L V T L A A H L P A E F T P A V H A S L D K F L A S V S T V L T S K Y R - 141
β-Globin	L L G N V L V C V L A H H F G K E F T P P V Q A A Y Q K V V A G V A N A L A H K Y H - 146

▶ **FIGURE 10.12** The amino acid sequences of the alpha gene and the beta gene, using single-letter abbreviations for the amino acids. Shaded areas show regions where the amino acid sequences are identical. The two genes are descended from a common ancestor and diverged from each other some 500 million years ago.

Table 10.4 Beta-Globin Chain Variants with Single Amino Acid Substitutions

Hemoglobin	Amino Acid Position	Amino Acid	Phenotype
A₁	6	glu	Normal
S	6	val	Sickle cell anemia
C	6	lys	Hemoglobin C disease
A₁	7	glu	Normal
Siriraj	7	lys	Normal
San Jose	7	gly	Normal
A₁	58	tyr	Normal
HbM Boston	58	his	Reduced O_2 affinity
A₁	145	cys	Normal
Bethesda	145	his	Increased O_2 affinity
Fort Gordon	145	asp	Increased O_2 affinity

are found in beta-globin (Table 10.4). Hemoglobin variants include Hb S and Hb C, which occur with high frequencies in certain populations and are associated with severe (Hb S) or mild (Hb C) disease phenotypes.

Sickle Cell Anemia

Sickle cell anemia (MIM/OMIM 141900) is a hemoglobin variant inherited as an autosomal recessive trait. This disorder is caused by a mutation in the gene for beta-globin. Hemoglobin molecules that contain mutant beta-globin become insoluble after oxygen is released to cells. The insoluble molecules are sticky and polymerize to form long tubular structures that form a rigid gel inside the cell (▶ Figure 10.13). This distorts the membrane of the red blood cell, twisting the cell into a characteristic sickle shape. The deformed blood cells break easily and are removed from circulation. The lowered number of red blood cells results in anemia; the sickled cells also clog blood vessels, producing tissue damage (micrographs of sickled cells are shown on the first page of this chapter.

Linus Pauling and his colleagues were the first to observe that Hb A and Hb S differ in their rate of migration in an electrical field. Because the side chains (the R groups) on many amino acids are electrically charged, they concluded that the amino acid structures of the two molecules must be different. Later, Vernon Ingram showed that the only difference between Hb A and Hb S is a single amino acid change at position 6 in the beta chain. This alteration in a single amino acid is the molecular basis of sickle cell anemia. All of the symptoms of the disease and its inevitably fatal

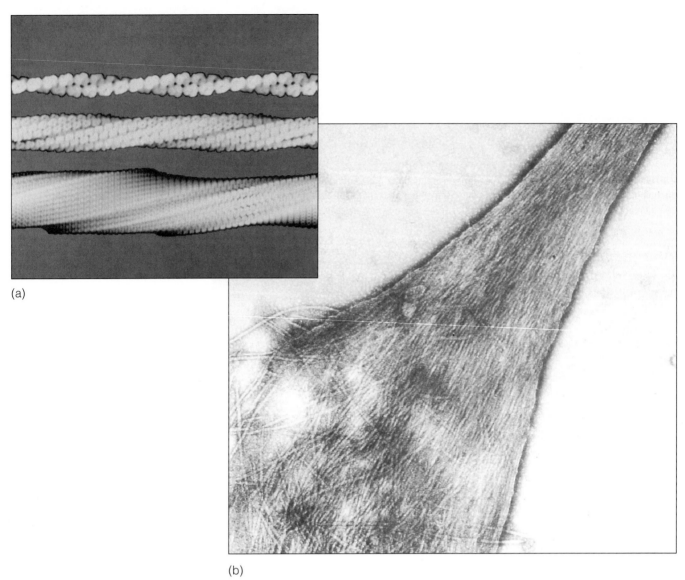

(a)

(b)

▶ **FIGURE 10.13** (a) A computer-generated image of the stages in the polymerization of sickle cell beta-globin to form rods. Top: A pair of intertwined fibers composed of stacked hemoglobin molecules. Middle: Seven pairs of fibers forming the structure that distorts the red blood cell. Bottom: A large fiber composed of many smaller fibers. (b) An electron micrograph of a ruptured sickled red blood cell, showing the internal fibers of polymerized hemoglobin.

outcome, if untreated, derive from this alteration of one amino acid out of the 146 found in beta-globin (▶ Figure 10.14).

Other Hemoglobin Variants

In spite of the devastating effect of changing one amino acid at position 6 in beta-globin, other changes at this position produce milder symptoms. Hb C has a single amino acid change at position 6 in beta-globin, but is only slightly insoluble when it gives up its oxygen load. Hb C forms intracellular crystals that make the erythrocyte membrane more rigid. This produces a mild form of anemia, with accompanying enlargement of the spleen, but there is almost never a need for clinical treatment for this condition. The sequences of the first seven amino acids in Hb A, Hb S, and Hb C are shown below:

> **Hb A:** val-his-leu-thr-pro-glu-glu-
> **Hb S:** val-his-leu-thr-pro-val-glu-
> **Hb C:** val-his-leu-thr-pro-lys-glu-

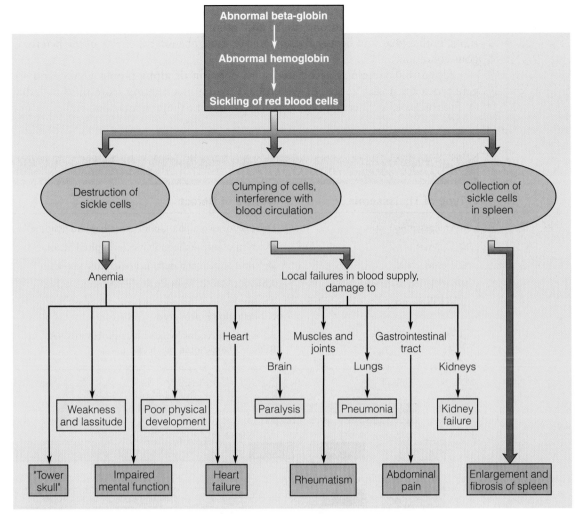

▶ **FIGURE 10.14** The cascade of phenotypic effects resulting from the mutation that causes sickle cell anemia. Affected homozygotes have effects at the molecular, cellular, and organ levels, all resulting from the substitution of a single amino acid in the beta-globin polypeptide chain.

Recall from the previous discussion that not all amino acid substitutions in the globin polypeptides are associated with a genetic disorder. Substitutions at position 7 (Table 10.4) in the beta-globin chain have no detectable effects, although these variations are inherited in the same autosomal recessive fashion as sickle cell anemia.

Thalassemias

The thalassemias are a group of inherited hemoglobin disorders in which a mutant phenotype results from an alteration in the amount of a gene product that is produced. In these conditions, the synthesis of globin chains is reduced or absent, causing the formation of abnormal hemoglobin molecules that do not have two alpha and two beta chains. These altered combinations do not bind oxygen efficiently and can have serious and even fatal effects.

Thalassemias are common in several parts of the world, especially the Mediterranean region and Southeast Asia, where 20 to 30% of the population can be affected. The name "thalassemia" is derived from the Greek word *thalassa*, for "sea," emphasizing that this condition was first described in people living around the Mediterranean Sea.

There are two types of thalassemia: **alpha thalassemia** (MIM/OMIM 141800), that has reduced synthesis or no synthesis of the alpha polypeptide, and **beta thalassemia**

■ **Alpha-thalassemia** Genetic disorder associated with an imbalance in the ratio of alpha- and beta-globin caused by reduced or absent synthesis of alpha-globin.

■ **Beta thalassemia** Genetic disorder associated with an imbalance in the ratio of alpha- and beta-globin caused by reduced or absent synthesis of beta-globin.

(MIM/OMIM 141900), which affects the synthesis of beta chains (Table 10.5). Both conditions have more than one cause, and although inherited as autosomal recessive traits, both alpha- and beta-thalassemia have some phenotypic effects in the heterozygous condition.

Alpha-thalassemia is associated with deletion of alpha-globin genes, and six genotypes are possible, five of which have symptoms ranging from mild to lethal (▶ Figure 10.15). There are several forms of beta-thalassemia, but these do not

Table 10.5 Summary of Thalassemias	
Type of Thalassemia	**Nature of Defect**
α-Thalassemia-1	Deletion of two alpha-globin genes/haploid genome
α-Thalassemia-2	Deletion of one alpha-globin gene/haploid genome
β-Thalassemia	Deletion of beta and delta genes/haploid genome
Nondeletion α-thalassemia	Absent, reduced, or inactive alpha-globin mRNA
β^0-Thalassemia	Absent, reduced, or inactive beta-globin mRNA. No beta-globin produced.
β^+-Thalassemia	Absent, reduced, or inactive beta-globin mRNA. Reduced beta-globin production.

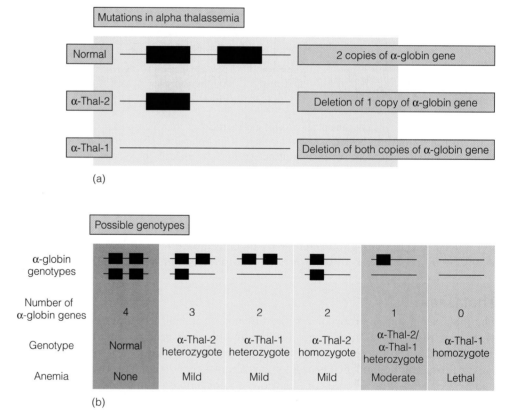

(a)

(b)

▶ **FIGURE 10.15** Deletions and alpha-thalassemia. (a) Normally, each copy of chromosome 16 carries two copies of the alpha-globin gene (normal). One copy is deleted in the alpha-thal-2 allele, and both copies are deleted in the alpha-thal-1 allele. (b) These three alleles can be combined to form six genotypic combinations that have 0–4 copies of the alpha-globin gene. Genotypes that have one copy deleted have moderate anemia and other symptoms, and genotypes that have no copies of the gene are lethal.

usually involve deletion of the gene for the beta polypeptide. In some forms of beta-thalassemia the underlying defect lies in the processing of the beta-globin pre-mRNA into a mature mRNA. In β^0-thalassemia, a mutation at the junction between an intron and an exon interferes with normal mRNA splicing, resulting in very low levels of functional mRNA and, in turn, low levels of beta-globin. Low levels of beta-globin cause an excess of alpha-globin polypeptides, which form abnormal hemoglobin molecules.

Treatment of Hemoglobin Disorders by Gene Switching

If untreated, sickle cell anemia is a fatal disease, and most affected individuals die by the age of 2 years. Even with an understanding of the molecular basis of the disease, treatments are only partially successful in relieving the symptoms. Recently, a discovery made with cancer drugs led to a new and effective treatment. The drug hydroxyurea suppresses cell growth and is used to treat cancer patients. As a side effect, patients have elevated levels of a form of hemoglobin usually seen only in developing fetuses. This hemoglobin is a combination of two alpha polypeptides and two gamma polypeptides. Gamma polypeptides are active in fetal development, and are switched off at birth, when the beta gene is activated (▶ Figure 10.16). Treatment with hydroxyurea reactivates these genes and makes fetal hemoglobin reappear in the red blood cells. When individuals who have sickle cell anemia are treated with hydroxyurea, fetal hemoglobin is produced and replaces the defective hemoglobin in red blood cells. Other drugs, including sodium butyrate, also switch on the synthesis of fetal hemoglobin. In some patients, 25-30% of the hemoglobin in the blood can be in the form of fetal hemoglobin. Because these chemicals are less toxic than hydroxyurea, they are being used in clinical trials to treat both sickle cell anemia and beta-thalassemia by switching on genes that are normally turned off at birth.

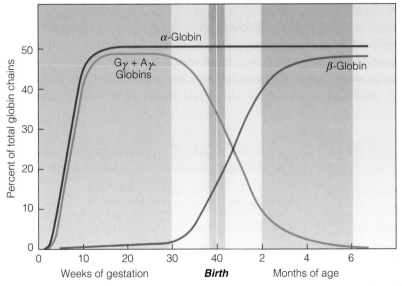

▶ **FIGURE 10.16** Patterns of globin gene expression during development. The alpha genes are switched on early in development and continue throughout life. The gamma G and gamma A members of the beta family are active during fetal development and switch off just before birth. The beta-globin gene is switched on at birth and is active throughout life. Sickle cell anemia and beta-thalassemia are caused by mutations that affect beta-globin. Research aimed at treating these conditions is directed at switching on the gamma genes, producing fetal hemoglobin to correct the conditions.

PHARMACOGENETICS AND ECOGENETICS

■ **Pharmacogenetics**   A branch of genetics concerned with the inheritance of differences in the response to drugs.

■ **Ecogenetics**   A branch of genetics that studies genetic traits related to the response to environmental substances.

Genotypic variations that result in variations in the type and amount of proteins produced result in genetic disorders of metabolism and can also affect the way individuals react to therapeutic drugs. Unlike most metabolic diseases, phenotypic differences in drug reactions appear only when an individual is exposed to the drug. These reactions are often the result of heritable variations in proteins and can be dominant or recessive traits (Table 10.6). A branch of genetics known as **pharmacogenetics** is concerned with genetic variation that underlies drug responses. Differences in drug responses can produce a range of phenotypic responses: drug resistance, toxic sensitivity to low doses, development of cancer after prolonged exposure, or an unexpected reaction to a combination of drugs. Some of these variations are harmless, whereas others can be life-threatening. The scope of pharmacogenetics has expanded to study genetic differences in reactions to chemicals in food, occupational exposure, and industrial pollution and is related to the broader field of **ecogenetics.** In this section, we consider how exposure to chemicals produces a wide range of phenotypes, and where known, describe the role of specific proteins in generating these phenotypes.

Taste and Smell Differences: We Live in Different Sensory Worlds

Shortly after Garrod proposed that we are all biochemicaly unique individuals because of our genotypes, researchers began to demonstrate differences in the way people react to chemicals. The discovery that people have different abilities to taste and smell chemicals and that these differences have a genetic basis was the first indication that there might be important genetic differences in responses to drugs used to treat diseases. Some of these differences are discussed below.

In searching for sugar substitutes, workers at DuPont discovered that some people cannot taste the chemical phenylthiocarbamide (PTC), whereas others find it very bitter or tasting like sand. Shortly thereafter, it was found that the ability to taste PTC depends on a single pair of alleles and that genotypes *TT* and *Tt* represent tasters, whereas those who have genotype *tt* are nontasters. Among U.S. whites, about 70% of the population are tasters, while about 90% of U.S. blacks are tasters. Later work showed that the ability to taste PTC is more complex than originally thought. When PTC solutions at various dilutions are used, a wide, bimodal range of tasters can be detected. It appears that modifying genes affect the threshold of taste sensitivity. The origin and significance of this variation is unknown, but some plant foods do contain PTC and may account for the reason that people don't like certain vegetables.

The ability to smell is mediated by a family of somewhere between 100–1000 different membrane proteins that are distributed over the surface of cells in the nose

Table 10.6 **Some Inherited Pharmacogenetic Conditions**		
Disease or Condition	**Inheritance**	**Drug Causing Reaction**
Aplastic anemia	Autosomal recessive?	Chloramphenicol
Atypical serum cholinesterase	Autosomal recessive	Succinylcholine
Acatalasia	Autosomal recessive	Hydrogen peroxide
Glucose-6-phosphate dehydrogenase	X-linked recessive	Antimalarial drugs Analgesics, sulfonamides, fava beans, many other drugs
Warfarin resistance	Autosomal dominant	Warfarin
Porphyria	Autosomal dominant	Barbiturates

and sinuses. There are many combinations of alleles for these proteins, so that each of us lives in a slightly different sensory world of smell. In fact, some people cannot smell the odor released by skunks (MIM/OMIM 270350).

The flower *Verbena* comes in a variety of colors, including red and pink (▶ Figure 10.17). Blakeslee discovered that people differ in their ability to smell the fragrances produced by these varieties. About two-thirds of the people he tested could smell a fragrance in the pink flowers but not the red ones. The remaining one-third could detect a smell in the red flowers but not the pink ones.

Drug Sensitivities Are Genetic Traits

During the past 50 years, tens of thousands of new drugs have been developed. As these were tested or put into general use, distinctive patterns of response to these chemicals were identified. Many of the differences in drug response are genetically controlled. Over this same period, thousands of new chemicals have been developed for use in manufacturing and agriculture.

Succinylcholine Sensitivity Succinylcholine, a muscle relaxant has been used as a short-acting anesthetic (called suxamethonium) for almost fifty years. Soon after its introduction, it became apparent that some people took hours rather than minutes to recover from a small dose of the drug. Normally the drug is broken down to an inactive form by the enzyme serum cholinesterase. Those who are sensitive to succinylcholine have a variant form of the enzyme, which breaks down the drug very slowly, prolonging the effect of the anesthetic (MIM/OMIM 177400). As a result, use of succinylcholine on sensitive individuals can lead to paralysis of the respiratory muscles and can cause death.

Primaquine Sensitivity and Favism During World War II, several new drugs were developed to protect soldiers from malaria. One of these was primaquine, and another was a related drug, pamaquine. When these drugs were given to members of certain ethnic groups, including U.S. blacks, African blacks, and those from the European shores of the Mediterranean, such as Greeks, the result was a massive destruction of red blood cells (hemolytic anemia). A similar response to eating fava beans (*Vicia fava*) has been known since ancient times in Greece, Italy, and nearby countries.

Both responses are caused by a deficiency in an X-linked enzyme, glucose-6-phosphate dehydrogenase (G6PD). Both antimalarial drugs and fava beans produce peroxides in the blood. The enzyme G6PD, in conjunction with glutathione, another

(a)

(b)

▶ **FIGURE 10.17** Pink and red verbena flowers.

compound in serum, break down and inactivate the peroxides. In those who have reduced or absent G6PD, the high concentration of peroxides causes red blood cells to break down, producing a life-threatening anemia.

There are many other drugs and chemicals that cause lysis of red blood cells in G6PD deficient individuals, including napthalene, the active ingredient in mothballs. More than 400 million people worldwide are affected by G6PD deficiency. The reasons for the high frequency of this genetic variant are discussed in a later chapter.

Pesticide Metabolism

Pests, in the form of insects, weeds, fungi, and other pathogens destroy about 35% of the world's crops. After harvesting, another 10–20% is destroyed in storage. Chemical agents, including herbicides, insecticides, and fungicides, are used to control these pests. In the U.S., around 65% of the insecticides used each year are applied to two crops, cotton and corn.

Agricultural insecticides include a group of chemicals called organophosphates, which includes parathion, an insecticide used for more than fifty years. Exposure to parathion and other organophosphates can occur on the job (agricultural workers and forestry workers) or from eating contaminated food. In the human body, parathion, which is inert, is enzymatically converted to a compound called paraoxon. Paraoxon is a toxic chemical that disrupts the transmission of signals in the nervous system. Paraoxon is broken down by paraoxonase, an enzyme found in blood serum.

The gene for paraoxonase (*PON*1) has two alleles (*A* and *B*). The *A* allele encodes a protein that has high levels of enzymatic activity, and the *B* allele encodes an enzyme that has low levels of activity. The two proteins differ in a single amino acid at position 192. It has been suggested that people homozygous for high activity (*A/A*) are resistant to the effects of parathion. Conversely, those homozygous for low activity (*B/B*) are highly sensitive to parathion poisoning (MIM/OMIM 168820).

Paraoxonase is also involved in metabolizing toxic nerve gas agents, such as somain and sarin (sarin is the nerve gas released in the Tokyo subway in 1995). However, in this case, the allele effects are reversed. The *B/B* homozygote has high activity, and the *A/A* genotype has low activity. The wide range in the frequency of the *A* and *B* alleles in different populations suggests that ethnic groups may differ in their sensitivity to poisoning by insecticides and nerve gases, a rather grim reminder of the role of proteins in producing a phenotype.

The examples outlined in this chapter reinforce the notion of the biological uniqueness of the individual. The constellation of genes present within each person is the result of the random combination of parental genes and the sum of changes brought about by recombination and mutation. This genetic combination confers a distinctive phenotype upon each of us. Garrod referred to this metabolic uniqueness as chemical individuality. Understanding the molecular basis for this individuality remains one of the great challenges of human biochemical genetics.

Case Studies

CASE 1

A couple was referred for genetic counseling because they wanted to know the chances of having a child who was a dwarf. Both had a genetic condition known as achondroplasia, the most common form of short-limbed dwarfism. The couple knew that this condition is autosomal dominant but they were unsure what kind of physical manifestations a child would have if it inherited both genes for the condition. They were each heterozygous for the FGFR3 gene but wanted information on the chances of having a child who was homozygous for the FGFR3 gene. The counselor briefly reviewed the phenotypic features of individuals who have achondroplasia. These include the facial features (large head with prominent forehead, small, flat nasal bridge, and prominent jaw), very short stature, and shortening of the arms and legs. Physical exam and skeletal X rays are used to diagnosis this condition. Final adult height is in the range of 4 feet.

Because achondroplasia is an autosomal dominant condition, a person who has this condition has a 1 in 2 or 50% chance of having children with this condition. However, approximately 75% of individuals who have achrondroplasia are born to parents of average size. In these cases, achondroplasia is due to a new mutation or genetic change. This couple is at risk for having a child with two copies of the changed gene or double homozygosity. Infants who have homozygous achondroplasia are either stillborn or die shortly after birth. The counselor recommended prenatal diagnosis via serial ultrasounds. In addition, a DNA test is available to detect double homozygosity. Achondroplasia occurs in 1 in every 14,000 births.

CASE 2

Tina is 12 years old. Although she was symptomatic since infancy, she was not diagnosed with acid maltase deficiency (AMD) until she was 10 years old. The progression of her disease has been slow and insidious. She has great difficulty walking and breathing due to severe muscle weakness. She relies on a breathing machine for respiratory support. She has developed severe scoliosis, which further compromises her breathing and causes even greater difficulty in walking. She is extremely tired and suffers from constant muscle pain. Although she is very bright and thinks like a normal teenager, her body won't let her function like one. She can no longer attend school. The future is bleak for Tina and other children like her. The cause of death in the childhood form of AMD is frequently due to complications from respiratory infections, which are a constant threat. Life expectancy in the childhood form is only to the second or third decade of life.

Acid maltase deficiency (AMD), glycogen storage disease-type II, or Pompe disease, is an autosomal recessive condition that is genetically transmitted from carrier parents to their child. When both parents are carriers (i.e. have one abnormal and one normal gene), there is a 25% during each pregnancy chance that the child will have two abnormal genes and be affected.

A buildup of glycogen in the muscle cells causes AMD. Glycogen is synthesized from sugars and is stored in the muscle cells for future use. Acid maltase enzyme breaks down the unused glycogen in the muscle cells. The person who has AMD is lacking or deficient in this enzyme. Stored glycogen continues to build up in the muscle tissues and leads to progressive muscle weakness and degeneration. There is no treatment or cure for AMD. Enzyme replacement and gene therapy are tools that may be useful in the future but have been unsuccessful in the past.

Summary

1. In the early part of this century, Sir Archibald Garrod's studies on the human metabolic diseases cystinuria, albinism, and alkaptonuria provided the first hints that gene products control biochemical reactions. These diseases and many others result from mutations that cause metabolic blocks in biochemical pathways. In alkaptonuria, Garrod argued, the normal metabolic reaction is blocked by the lack of a needed enzyme. He speculated that the inability to carry out this reaction results from a recessive Mendelian gene.

2. The work of George Beadle and his colleagues made it clear that single mutations led to the loss of activity in a single enzyme, an idea known as the one gene-one enzyme hypothesis. Subsequent work refined this into today's one gene-one polypeptide hypothesis.

3. Later investigations clearly identified enzyme defects in a large number of human metabolic diseases, and the role of mutations in nonenzymatic proteins became clear. In 1949, James Neel identified sickle cell anemia as a recessive disease, and Linus Pauling began to study the

physical properties of hemoglobin, leading to Vernon Ingram's discovery that the molecular basis of sickle cell anemia is a single change in the amino acid sequence of a polypeptide chain. Logically this might involve changes in the nucleotide sequence of the genetic material, and this idea has been confirmed several times over.

4. Defects in receptor proteins, transport proteins, structural proteins, and other nonenzymatic proteins can cause phenotypic effects in the heterozygous state, and many show an incompletely dominant or dominant pattern of inheritance.

5. Differences in the reactions to therapeutic drugs and chemicals represents a "hidden" set of phenotypes that are not revealed until exposure occurs. Understanding the genetic basis for these differences is the concern of pharmacogenetics and ecogentics.

Questions and Problems

1. Many individuals with metabolic diseases are normal at birth but show symptoms shortly thereafter. Why?

2. Enzymes have all the characteristics below except:
   a. act as biological catalysts
   b. are proteins
   c. carry out random chemical reactions
   d. convert substrates into products
   e. can cause genetic disease

3. Essential amino acids are:
   a. amino acids our body can synthesize
   b. amino acids we need in our diet
   c. amino acids in a box of Frosted Flakes
   d. amino acids that include arginine and glutamic acid
   e. amino acids that cannot harm the body if not metabolized properly

4. Describe the quaternary structure of the blood protein hemoglobin.

5. List the ways in which a metabolic block can have phenotypic effects.

6. A person was found to have very low levels of functional beta globin mRNA, and in turn, very low levels of the beta globin protein. Name the disease and explain what mutation may have occurred in the conversion of pre-mRNA into mRNA.

7. Familial hypercholesterolemia is caused by an autosomal dominant mutation in the gene that produces the LDL receptor. The LDL receptor is present in the plasma membrane of cells and binds cholesterol and helps remove it from the circulatory system for metabolism in the liver. What is the phenotype of the following individuals:

   HH
   Hh
   hh

8. Suppose the gene for the LDL receptor has been isolated by recombinant DNA techniques. Could you treat this disease by producing LDL receptor and injecting it into the bloodstream of affected individuals? Why or why not?

9. Some genes in humans are known to have a multiple phenotypic effects (these are called pleiotropic effects). Does this phenomenon contradict the one gene-one enzyme hypothesis? Explain. Cite examples of human genetic conditions where pleiotropic effects are part of the phenotype.

10. Suppose that in the formation of phenylalanine hydroxylase mRNA, the exons of the pre-mRNA fail to splice together properly, and the resulting enzyme is non-functional. This produces an accumulation of high levels of phenylalanine and other compounds which causes an neurological damage. What phenotype and disease would be produced in the affected individual?

11. The normal enzyme required for converting sugars into glucose is present in cells, but the conversion never takes place, and no glucose is produced. What could have occurred to cause this defect in a metabolic pathway?

12. If an extra nucleotide is present in the first exon of the beta globin gene, what effect would it have on the amino acid sequence of the globin polypeptides? Would the globin most likely be fully functional, partly functional or non-functional? Why?

13. The promoter region in the XP-A gene becomes mutated. How is transcription affected? How would an individual with this mutation be affected?

14. PKU is an autosomal recessive disorder that causes mental retardation. In PKU individuals, high levels of the essential amino acid phenylalanine are present due to a deficiency in the enzyme, phenylalanine hydroxylase. If phenylalanine was not an essential amino acid, would diet therapy (the elimination of phenylalanine from the diet) work?

15. Phenylketonuria and alkaptonuria are both autosomal recessive diseases. If a person with PKU marries a person with AKU, what will the phenotype of their children be?

16. If a chromosomal male has a defect in the cellular receptor that binds the hormone testosterone, what condition results? What is the genotype and phenotype of this individual?

17. Knowing that individuals who are homozygous for the GD allele show no symptoms of galactosemia, is it surprising that galactosemia is a recessive disease? Why?

18. Transcriptional regulators are proteins that bind to promoters (the 5′-flanking regions of genes) to regulate their transcription. Assume a particular transcription regulator normally promotes transcription of gene X. If a mutation makes this regulator gene nonfunctional, would the resulting phenotype be similar to a mutation in gene X itself? Why?

19. Mutations in the alpha thalassemia genes can result in a variety of abnormal phenotypes. If a heterozygous alpha thalassemia-1 man marries an heterozygous alpha thalassemia-2 woman, what are the phenotypes of their offspring? (Refer to Figure 10.15)

Questions 20-22 refer to the following hypothetical pathway in which substance A is converted to substance C by enzymes 1 and 2. Substance B is the intermediate produced in this pathway:

enzyme   enzyme

1   2

A --------→ B --------> C

20. a. If an individual is homozygous for a null mutation in the gene that codes for enzyme 1, what will be the result?
   b. If an individual is homozygous for a null mutation in enzyme 2, what will be the result?
   c. What if an individual is heterozygous for a dominant mutation where enzyme 1 is overactive?
   d. What if an individual is heterozygous for a mutation that abolishes the activity of enzyme 2 (a null mutation)?

21. a. If the first individual in question 18 married the second individual, would their children be able to convert substance A into substance C?
   b. Suppose each of the aforementioned individuals were heterozygous for an autosomal dominant mutation. List the phenotypes of their children with respect to compounds A, B, and C. (Would the compound be in excess, not present, etc.?)

22. An individual is heterozygous for a recessive mutation in enzyme 1 and heterozygous for a recessive mutation in enzyme 2. This individual marries an individual of the same genotype. List the possible genotypes of their children. For every genotype, determine the activity of enzyme 1 and 2, assuming that the mutant alleles have 0% activity and the normal alleles have 50% activity. For every genotype, determine if compound C will be made. If compound C is not made, list the compound that will be in excess.

23. Explain why there are variant responses to drugs and why they act as heritable traits.

Questions 24 to 28 refer to a hypothetical metabolic disease in which protein E is not produced. Lack of protein E causes mental retardation in humans. Protein E's function is not known, but it is found in all cells of the body. Skin cells were taken from eight individuals who cannot produce protein E and were grown in culture. The defect in each of the individuals is due to a single recessive mutation. Each individual is homozygous for her or his mutation. The cells from one individual were grown with the cells from another individual in all possible combinations of two. After a few weeks of growth the mixed cultures were assayed for the presence of protein E. The results are given in the following table. A plus sign means that the two cell types produced protein E when grown together (but not separately), while a minus sign means that the two cell types still could not produce protein E:

	1	2	3	4	5	6	7	8
1	−	+	+	+	+	−	+	+
2		−	+	+	+	+	−	+
3			−	+	+	+	+	−
4				−	+	+	+	+
5					−	+	+	+
6						−	+	+
7							−	+
8								−

24. a. Which individuals seem to have the same defect in protein E production?
   b. If individual 2 married individual 3 would their children be able to make protein E?
   c. If individual 1 married individual 6 would their children be able to make protein E?

25. a. Assuming that these individuals represent all possible mutants in the synthesis of protein E, how many steps are there in the pathway to protein E production?
   b. Compounds A, B, C, and D are known to be intermediates in the pathway for production of protein E. To determine where the block in protein E production occurred in each individual, the various intermediates were given to each individual's cells in culture. After a few weeks of growth with the intermediate, the cells were assayed for the production of protein E. The results for each individual's cells are given in the following table. A plus means that protein E was produced after the cells were given the intermediate listed at the top of the column. A minus means that the cells still could not produce protein E even after being exposed to the intermediate at the top of the column:

	Compounds				
Cells	A	B	C	D	E
1	−	−	+	+	+
2	−	+	+	+	+
3	−	−	−	+	+
4	−	−	−	−	+
5	+	+	+	+	+
6	−	−	+	+	+
7	−	+	+	+	+
8	−	−	−	+	+

26. Draw the pathway leading to the production of protein E.

27. Denote the point in the pathway in which each individual is blocked.

28. a. If an individual who is homozygous for the mutation found in individual 2 and heterozygous for the mutation found in individual 4 marries an individual who is homozygous for the mutation found in individual 4 and heterozygous for the mutation found in individual 2, what will be the phenotype of their children?

b. List the intermediate that would build up each of the types of children who could not produce protein E.

29. Ecogenetics is a branch of genetics that deals with genetic variation that underlies reactive differences to drugs, chemicals in food, occupational exposure, industrial pollution, etc. Cases have arisen where workers claim that exposure to a certain agent has made them feel ill (while other workers are unaffected). While claims like these are not always justified, what are some concrete examples that prove that variation to certain substances exist in the human population?

Internet Activities

The following activities use the resources of the World Wide Web to enhance the topics covered in this chapter. To investigate the topics described below, log on to the book's home page at:

http://www.brookscole.com/biology

1. The Human Genome Project has a site addressing various aspects of *Sickle Cell Disease*. Read about the *causes and effects* of sickle cell disease. What type of mutations can cause red blood cells to sickle? Why does this result in a lower oxygen level in those affected by this disease? Newborns are commonly screened for sickle cell disease. How is the disease diagnosed? How many mutant alleles are known to cause it? Read about the origins and history of sickle cell disease. What advantage is thought to be provided by heterozygosity for the hemoglobin S mutation?

2. The genome of *Haemophilus influenzae* was the first genome to be completely sequenced and mapped. Since then, the genomes of numerous other prokaryotic (and eukaryotic) organisms have been sequenced. *TIGR Microbial Genomes* contains information about the DNA and protein sequence from several of these genomes, as well as information about the cellular role of various gene products. Compare the genome of *H. influenzae* with that of *Mycoplasma genitalium*, paying special attention to the relative sizes of the genomes, the number and sizes of the genes in each genome, and the cellular roles of their products. The genome of *M. genitalium* is significantly smaller than that of *H. influenzae*. Can you identify any genes, or entire categories of genes, that are present in *H. influenzae* but not in *M. genitalium*?

For Further Reading

Antonarakis, S., Kazazian, H., & Orkin, S. (1985). DNA polymorphisms and molecular pathology of the human globin gene clusters. *Hum. Genet.* 69: 1–14.

Beadle, G. W. (1945). Biochemical genetics. *Chem. Rev.* 37: 351.

Bearn, A. G., & Miller, E. D. (1979). Archibald Garrod and the development of the concept of inborn errors of metabolism. *Bull. Hist. Med.* 53: 317–324.

Benson, P. F., & Fenson, A. H. (1985). *Genetic Biochemical Disorders.* New York: Oxford University Press.

Childs, B. (1970). Sir Archibald Garrod's conception of chemical individuality: A modern appreciation. *N. Engl. J. Med.* 282: 4–5.

Cohen, J. (1997). Medicine: Developing prescriptions with a personal touch. *Science* 275: 776–770.

Eisensmith, R., Okano, Y., Dasovitch, M., Wang, T., Güttler, F., Lou, H., Guldberg, P., Lichter-Konecki, U., Konecki, D., et al. (1992). Multiple origins for phenylketonuna in Europe. *Am. J. Hum. Genet.* 51: 1355–1365.

Galjaard, H. (1980). *Genetic Metabolic Diseases: Early Diagnosis and Prenatal Analysis.* New York: Elsevier/North-Holland.

Garrod, A. E. (1902). The incidence of alkaptonuria: A study in chemical individuality. *Lancet 2:* 1616–1620.

Hobbs, H., Brown, M., & Goldstein, J. (1992). Molecular genetics of the LDL receptor gene in familial hypercholesterolemia. *Hum. Mutations 1:* 445–466.

Ingram, V. M. (1957). Gene mutations in human hemoglobin: The chemical differences between normal and sickle cell hemoglobin. *Nature 180:* 326–328.

King, R. A., & Olds, D. P. (1985). Hairbulb tyrosinase activity in oculocutaneous albinism: Suggestions for pathway control and block location. *Am. J. Med. Genet.* 20: 49–55.

Kivirkko, K. (1993). Collagens and their abnormalities in a wide spectrum of diseases. *Ann. Med.* 25: 113–126.

McKusick, V. (1992). *Mendelian Inheritance in Man: Catalogs of Autosomal Dominant, Autosomal Recessive and X-Linked Phenotypes,* 10th ed. Baltimore: Johns Hopkins Press.

Neel, J. V. (1949). The inheritance of sickle cell anemia. *Science 110:* 64–66.

Pauling, L., Itoh, H., Singer, S. J., & Wells, I. C. (1949). Sickle cell anemia: A molecular disease. *Science 110:* 543–548.

Scriver, C. R., Beaudet, A. L., Sly, W. S., & Valle, D. (1989). *The Metabolic Basis of Inherited Disease,* 6th ed. New York: McGraw-Hill.

Scriver, C. R., & Clow, C. L. (1980). Phenylketonuria and other phenylalanine hydroxylation mutants in man. *Ann. Rev. Genet.* 14: 179–202.

Sculley, D., Dawson, P., Emmerson, B., & Gordon, R. (1992). A review of the molecular basis of hypoxanthine-guanine phosphoribosyltransferase (HPRT) deficiency. *Hum. Genet.* 90: 195–207.

Spritz, R. A., Strunk, K. M., Giebel, L. B., & King, R. A. (1990). Detection of mutations in the tyrosinase gene in a patient with Type LA oculocutaneous albinism. *N. Engl. J. Med.* 322: 1724–1728.

Vella, F. (1980). Human hemoglobins and molecular disease. *Biochem. Educ.* 8: 41–53.

Weatherall, D. J. (1993). Molecular medicine: Towards the millenium. *Trends Genet.* 9: 102.

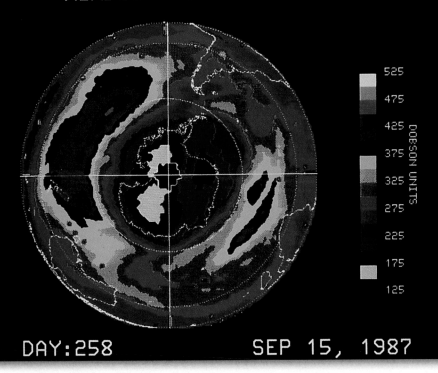

NIMBUS-7 : TOMS OZONE

525
475
425
375
325
275
225
175
125

DOBSON UNITS

DAY:258   SEP 15, 1987

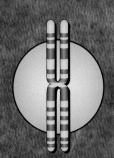

Mutation: The Source of Genetic Variation

Chemist Linus Pauling once recalled that when he first heard a description of how red blood cells change shape in sickle cell anemia, it occurred to him that sickle cell anemia is a molecular disease, involving an abnormality of the hemoglobin molecule determined by a mutated gene.

Early in 1949, Linus Pauling and his student Harvey Itano began a series of experiments to determine whether there is a difference between normal hemoglobin and sickle cell hemoglobin. They obtained blood samples from people who had sickle cell anemia and from unaffected individuals and prepared solutions of hemoglobin from these samples. They placed the hemoglobin in a tube fitted with an electrode at each end and passed an electrical current through the tube. Hemoglobin from individuals who had sickle cell anemia migrated toward the cathode, indicating that it has a positive electrical charge. Samples of normal hemoglobin migrated in the opposite direction (toward the anode), indicating that it has a net negative electrical charge. In the same year, James Neel, working with sickle cell patients in the Detroit area, demonstrated that sickle cell anemia is an autosomal recessive trait by establishing the genetic link to this condition.

Pauling and his colleagues published a paper on their results and incorporated Neel's findings into their discussion. They concluded that a mutant gene involved in the synthesis of hemoglobin caused sickle cell anemia (and the heterozygous condition known as sickle cell trait). The idea that a genetic disorder could be caused by a defect in a single molecule was revolutionary. Pauling's idea about a molecular disease established the foundation for human biochemical genetics and played a key role in understanding the molecular nature of mutations.

Pauling's work on the hemoglobin molecule influenced other researchers. After Watson and Crick worked out the structure of DNA, Crick was anxious to prove that mutant genes produce mutant proteins whose amino acid sequences differ from the normal protein. He persuaded Vernon Ingram to look for such differences. Because of Pauling's work, Ingram settled on hemoglobin as the protein he would analyze. Ingram cut hemoglobin into pieces using the enzyme trypsin and separated the 30 resulting fragments. He noticed that normal hemoglobin and sickle cell hemoglobin differed in only one fragment, a peptide about 10 amino acids long. Ingram then worked out the amino acid sequence in this fragment. In 1956, he reported that there is a difference of only a single amino acid (glutamine in normal hemoglobin and valine in sickle cell hemoglobin) between the two proteins. This finding confirmed the relationship between a mutant gene and a mutant gene product but raised a more basic question: What is the nature of mutation?

THE NATURE OF MUTATION

Mutation can be defined as any heritable change. These changes are the source of all genetic variation in humans and other organisms. The results of mutations can be classified in a number of ways. Mutations that produce dominant alleles are expressed in the heterozygous condition: mutations to recessive alleles are expressed only when homozygous. The effects of mutation can also be ranked in categories, such as the severity of the phenotype or the age of onset. For our purposes, two general categories of mutations can be distinguished: chromosomal aberrations and changes in the nucleotide sequence of a gene. Chromosomal aberrations were discussed in Chapter 6. In this chapter we discuss changes that occur within a single gene, that is, changes in the sequence or number of nucleotides in DNA. First, we consider how mutations are detected and then investigate at what rate these mutations take place. Finally, we examine how mutation works at the molecular level.

DETECTING MUTATIONS

How do we know that a mutation has taken place? In humans, the appearance of a dominant mutation in a family can be detected in a fairly straightforward manner. But mutation of a dominant allele to a recessive allele can be detected only in the homozygous condition, posing a challenge for human geneticists.

If an affected individual appears in an otherwise unaffected family, the first question is whether the trait is caused by genetic or nongenetic factors. For example, if a mother is exposed to the rubella virus (which causes a form of measles) early in pregnancy, the fetus may develop a phenotype that resembles those of inherited metabolic disorders. These symptoms are not the result of mutation but the effect of the virus on the developing fetus. In general, the detection of mutations depends on pedigree analysis and the study of births in a family line.

If a mutant allele is dominant, fully penetrant, and appears in a family that has no history of this condition over several generations, we can presume that a mutation has taken place. In the pedigree shown in ▶ Figure 11.1, severe blistering of the feet appeared in one out of six children, although the parents were unaffected. The trait was transmitted by the affected female to six of her eight children and was passed to the succeeding generation as expected for an autosomal dominant condition. A reasonable explanation for this pedigree is that a mutation to a dominant allele occurred and that individual II-5 was heterozygous for this dominant allele. On the other hand, a number of uncertainties can affect this conclusion. For example, if the child's father is not the husband in the pedigree but is an affected male, then it would only seem that a mutational event had taken place.

If mutation results in an allele that is recessive and sex-linked, it often can be detected by examining males in the family line. But it can be difficult to determine whether a heterozygous female who transmits a trait to her son is the source of the mutation or is only passing on a mutation that arose in an ancestor. The X-linked

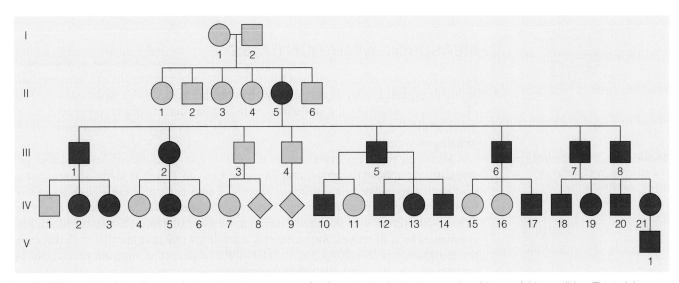

▶ **FIGURE 11.1** A dominant trait, foot blistering, appeared (II-5) in a family that had no previous history of this condition. The trait is transmitted through subsequent generations in an autosomal dominant fashion.

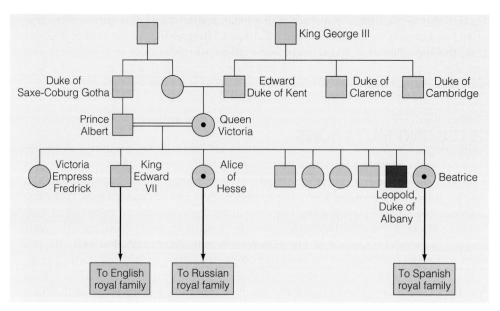

▶ **FIGURE 11.2**   Pedigree of Queen Victoria of Britain, showing her immediate ancestors and children. Because she passed the mutant allele for hemophilia on to three of her children, she was probably a heterozygote rather than the source of the mutation.

form of hemophilia that spread through the royal families of Western Europe and Russia probably originated with Queen Victoria (▶ Figure 11.2; see Concepts and Controversies: Hemophilia and History in Chapter 4). None of the males in generations prior to Victoria's had hemophilia, but one of her sons was affected, and at least two of her daughters were carriers. Because Victoria transmitted the trait to a number of her children, it is reasonable to assume that she was a heterozygous carrier. Her father was not affected, and there is nothing in her mother's pedigree to indicate that she was a carrier. It is, therefore likely that Victoria received a newly mutated allele from one of her parents. We can only speculate as to which parent.

If an autosomal recessive trait appears suddenly in a family, it is usually difficult or impossible to trace the trait through a pedigree to identify the person or even the generation in which the mutation first occurred because only homozygotes are affected. Such a new mutation can remain undetected for generations, as it passes from heterozygote to heterozygote.

MEASURING MUTATION RATES

Pedigree analysis reveals that mutation does take place in the human genome. The available evidence suggests that it is a rare event, but is it possible to measure the rate of mutation? Knowing the underlying rate of mutation would allow geneticists to monitor the rate over time to determine whether it is increasing, decreasing, or remaining the same.

Mutation rates are expressed as the number of mutated alleles per locus per generation. Suppose that for a certain gene, 4 out of 100,000 births show a mutation from a recessive to a dominant allele. Because each of these 100,000 individuals carries two copies of the gene, we have sampled 200,000 copies of the gene. The 4 births represent 4 mutated genes (we are assuming that the newborns are heterozygotes for a dominant mutation who carry only one mutant allele). In this case, the mutation rate is 4/200,000, or 2/100,000. In scientific notation this would be written as 2×10^{-5}/per allele/per generation.

If the locus were X-linked and if 100,000 male births were examined and 4 mutants were discovered, this would represent a sampling of 100,000 copies of the gene

■ **Mutation rate**   The number of events that produce mutated alleles per locus/per generation.

(since the males have only one copy of the X chromosome). Excluding contributions from female carriers, the mutation rate in this case would be 4/100,000, or 4×10^{-5}/per allele/per generation.

Gene-Specific Mutation Rates

Is there a way to measure directly the rate of mutation for a gene? For certain dominant alleles, the answer is yes, under certain conditions. To ensure accuracy, the trait selected must

- never be produced by recessive alleles
- always be fully expressed and completely penetrant so that mutant individuals can be identified
- have clearly established paternity
- never be produced by nongenetic agents such as drugs or infection
- be produced by dominant mutation of only one locus

One dominant mutation, achondroplasia, fulfills most if not all of these requirements. **Achondroplasia** (MIM/OMIM 100800) is a dominant form of dwarfism that produces short arms and legs and an enlarged skull. A diagnosis by X-ray examination can be done shortly after birth. Several population surveys have used this gene to estimate the mutation rate in humans. A recent survey recorded 7 achondroplastic children of unaffected parents in a total of 242,257 births. From these data, the mutation rate for achondroplasia has been calculated at 1.4×10^{-5}.

Although the mutation rate for achondroplasia can be measured directly, it is not clear whether this is a typical rate of mutation for human genes. Maybe this gene has an inherently high rate of mutation that is atypical. For this and other reasons, it is important to measure mutation rates in a number of different genes before making any general statements. As it turns out, two other dominant mutations have widely different rates of mutations. **Neurofibromatosis** (MIM/OMIM 162200), an autosomal dominant condition, is characterized by pigmentation spots and tumors of the skin and nervous system (described in Chapter 4). About 1 in 3000 births are affected. Many of these births (about 50%) occur in families that have no previous history of neurofibromatosis, indicating that this locus has a high mutation rate. In fact, the calculated mutation rate in this disease is as high as 1 in 10,000 (1×10^{-4}), one of the highest rates so far discovered in humans. For Huntington disease (MIM/OMIM 143100), the mutation rate has been calculated as 1×10^{-6}, a rate 100-fold lower than neurofibromatosis and 10-fold lower than achondroplasia.

Some estimates of mutation rates in human genes are given in Table 11.1. These rates average to about 1×10^{-5}. Note that the genes listed in the table are all inherited as autosomal dominant or X-linked traits. It is almost impossible to measure directly the mutation rates in autosomal recessive alleles by inspection of phenotypes, but population surveys using recombinant DNA methods are now providing estimates of the rate and type of mutations found in many human genes. Still, many geneticists feel that to reduce any potential bias, a more conservative estimate of the mutation rate in humans should be used, and 1×10^{-6} is used as the average mutation rate.

Factors That Influence the Mutation Rate

Several factors influence the mutation rate and contribute to its observed variation. Some of these factors include

- *Size of the gene.* Larger genes are bigger targets for mutation. Neurofibromatosis (NF-1) has a high mutation rate and is an extremely large gene. The NF-1 protein contains more than 2000 amino acids, but including the noncoding regions, the gene extends over 300,000 base pairs of DNA. The gene

Achondroplasia A dominantly inherited form of dwarfism.

Neurofibromatosis An autosomal dominant condition characterized by pigmented spots and tumors of the skin and nervous system.

Table 11.1 Mutation Rates for Selected Genes

Trait	Mutants/Million Gametes	Mutation Rate	MIM/OMIM Number
Achondroplasia	10	1×10^{-5}	100800
Aniridia	2.6	2.6×10^{-6}	106200
Retinoblastoma	6	6×10^{-6}	180200
Osteogenesis imperfecta	10	1×10^{-5}	166200
Neurofibromatosis	50–100	$0.5–1 \times 10^{-4}$	162200
Polycystic kidney disease	60–120	$6–12 \times 10^{-4}$	173900
Marfan syndrome	4–6	$4–6 \times 10^{-6}$	154700
Von Hippel-Landau syndrome	<1	1.8×10^{-7}	193300
Duchenne muscular dystrophy	50–100	$0.5–1 \times 10^{-4}$	310200

for Duchenne and Becker muscular dystrophy, the largest gene identified to date in humans, contains more than 2 million base pairs. Both of these genes, have high mutation rates.

- *Nucleotide sequence.* In some genes, short nucleotide repeats are present in the DNA. In the gene for fragile-X syndrome, a CGG repeat is present in 6 to 50 copies in unaffected individuals. Symptoms begin to appear in those who have more than 52 copies and become more severe as the number of CGG repeats increases. The presence of these repeats may predispose a gene to mutate at a higher rate.

- *Spontaneous chemical changes.* Among the bases in DNA, cytosine is especially susceptible to chemicals that can change the nucleotide sequence in DNA. These and other chemical changes are discussed below. Genes rich in G-C base pairs are more likely to undergo spontaneous chemical changes than those rich in A-T pairs.

ENVIRONMENTAL FACTORS THAT INFLUENCE MUTATION RATES

Mutations can occur as the result of internal factors such as errors in DNA replication, by the action of environmental agents such as chemicals, or by the effects of radiation.

Radiation Is One Source of Mutations

Radiation The process by which electromagnetic energy travels through space or a medium such as air.

Radiation is a process by which energy travels through space. There are two main forms of radiation; electromagnetic and corpuscular. Electromagnetic radiation is waves of electrical or magnetic energy, whereas corpuscular radiation is composed of atomic and subatomic particles that move at high speeds. These particles cause damage when they collide with other particles, including biological molecules. Both forms of radiation are known as **ionizing radiation** because they can form reactive ions when they collide with molecules in the body. However, not all mutagenic forms of radiation produce ions. Ultraviolet (UV) light is a powerful mutagen whose energy is lower than that of ionizing radiation. The energy in UV light is absorbed by DNA and produces mutations.

Ionizing radiation Radiation that produces ions during interaction with other matter, including molecules in cells.

Exposure to radiation is unavoidable. Everything in the physical world contains sources of radiation. This includes our bodies, the air we breathe, the food we eat, and the bricks in our houses. Some of this radiation is left over from the birth of the universe, and some has been created by interaction of atoms on earth with cosmic radiation. These natural sources of radiation are called **background radiation.** We are also exposed to manufactured sources of radiation from medical testing, nuclear testing, nuclear power, and consumer goods.

Radiation can cause damage at several levels in biological systems. As radiation strikes the molecules in cells, electrons are removed, creating ions and charged atoms. Such ionized molecules are highly reactive and can produce mutations in DNA. For example, if cosmic radiation passes through the cell and ionizes a water molecule near a DNA molecule, the ions might react with the DNA and produce a mutation. Often, the cell can repair these mutations. If a number of mutations accumulate in a somatic cell, the result may be cell death or the induction of cancer. In germ cells, mutations that are not repaired are transmitted from generation to generation.

How Much Radiation Are We Exposed To?

A variety of units are used to measure radiation. These measurements are collectively called **dosimetry.** A dose of radiation can be measured in several ways: the amount a person is exposed to, the amount absorbed by the body, or the amount of damage caused. Most commonly, the dose is expressed as a **rem** (radiation equivalent man), the amount of radiation that causes the same damage as a standard amount of X rays. Because the amount of radiation to which people are exposed is very small, the dose is often expressed in **millirems** (1,000 millirems equals 1 rem). At doses in the millirem range, cells can repair most if not all the radiation damage. At doses of about 100 rem (100,000 millirems), cells begin to die, and radiation sickness results. At a dosage of 400 rem, about 50% of people will die within sixty days if they are not treated.

In the United States, the average person is exposed to 360 mrem/year, 82% of which is from natural sources (▶ Figure 11.3). Since most people are exposed to less than 1 rem/yr and will not get radiation sickness, what are the risks from radiation

■ **Background radiation** Radiation in the environment that contributes to radiation exposure.

■ **Dosimetry** The process of measuring radiation.

■ **Rem** The unit of radiation exposure used to measure radiation damage in humans. It is the amount of ionizing radiation that has the same effect as a standard amount of X rays.

■ **Millirem** Each rem is equal to 1000 millirems.

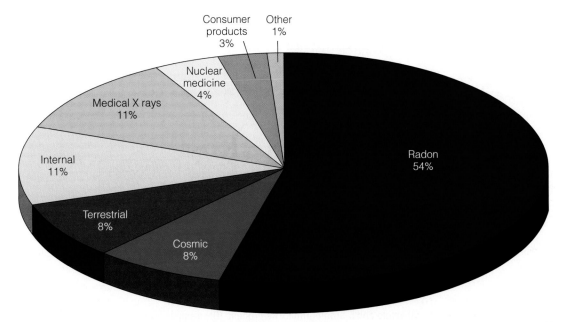

Sources of Radiation Exposure to the U.S. Population

Consumer products 3%
Other 1%
Nuclear medicine 4%
Medical X rays 11%
Internal 11%
Terrestrial 8%
Cosmic 8%
Radon 54%

▶ **FIGURE 11.3**   The sources of radiation received by individuals in the U.S. The average dose is 360 mrem, 82% of which is from background radiation.

exposure? At levels below 5 rem (5,000 millirems) the major risk is an increased susceptibility to cancer.

To understand how exposure to radiation increases your risk of getting cancer, some background information is necessary. In the U.S., the death rate from cancer is about 20%. If 10,000 people are chosen at random, about 20%, or 2000, will die from cancer. If each of these 10,000 people is exposed to a single radiation dose of 1,000 millirems (above the 360 mrem they receive each year), there would be about eight additional cancer deaths in this group over their lifetimes. However, the increase from 2,000 to 2,008 would not be detectable because of the fluctuations in the number of cancer cases from group to group. In other words, it is very difficult to estimate risks from low levels of radiation, and in most cases, the effects cannot be distinguished from the background levels. Overall, risk analysis suggests that the controlled use of radiation is a small risk compared to others in our daily lives.

Action of Chemical Mutagens

There are over 6 million chemical compounds now known, and almost 500,000 of these are used in manufacturing processes. Unfortunately, we know little or nothing about the mutagenic effects of most of them. Chemicals act as mutagens in several ways, and they are often classified by the type of damage they cause to DNA. Some chemicals cause nucleotide substitutions and frameshift mutations, or structurally change the bases in DNA, causing a base pair change after replication.

Base analogues A purine or pyrimidine that differs in chemical structure from those normally found in DNA or RNA.

Base Analogues Mutagenic chemicals that structurally resemble nucleotides and are incorporated during synthesis of DNA or RNA are called **base analogues**. The structure of the base analogue 5-bromouracil is similar to that of thymine (▶ Figure 11.4). Incorporation of 5-bromouracil into a DNA molecule in place of thymine generates

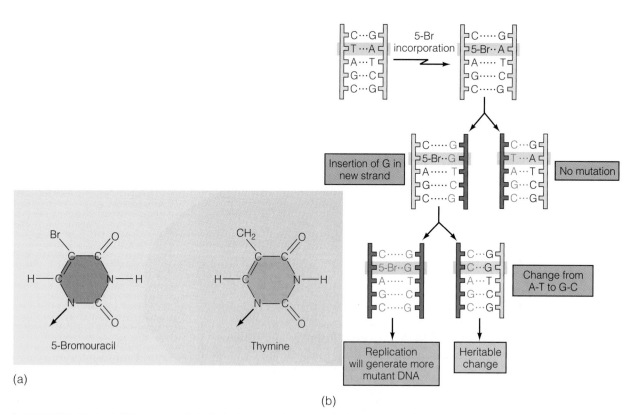

(a)

(b)

▶ **FIGURE 11.4** (a) The structural similarity between thymine and 5-bromouracil. The bromine (Br) group is about the same size as the CH_3 group in the thymine. (b) Rounds of DNA replication showing the incorporation of 5-Br and the base pair change it causes through several rounds of DNA replication. Boxes mark the beginning (T-A) base pair and the resulting (G-C) base pair.

Irradiated Food

During the past 40 years, research has demonstrated that radiation treatments can help preserve food and kill contaminating microorganisms. Irradiation prevents sprouting of root crops, such as potatoes, extends the shelf life of many fruits and vegetables, destroys bacteria and fungi in meat and fish, and kills insects and other pests in spices.

For irradiation, food is placed on a conveyor and moved to a sealed, heavily shielded chamber where it is exposed to radiation from a radioactive source (X rays or an electron beam). An operator who views the process on video camera delivers the dose. The food itself does not come in contact with the radioactive source, and the food is not made radioactive. Relatively low doses are used to inhibit sprouting of potatoes and to kill parasites in pork. Intermediate doses are used to retard spoilage in meat, poultry, and fish, and high doses can be used to sterilize foods, including meats. Worldwide, there are about 55 commercial facilities for food irradiation. The amount of food irradiated varies from country to country, ranging from a few tons of spices to hundreds of thousands of tons of grain.

Irradiated food has been used routinely by NASA to feed astronauts in space, and irradiated foods are sold in more than 20 countries, including the United States. The Food and Drug Administration (FDA) approved the first application for

food irradiation in 1964, and approval has been granted for the irradiation of spices, herbs, fruits and vegetables, and pork and chicken. All irradiated food sold in the United States must be labeled with an identifying logo (shown here).

Public concern about radiation has prevented the widespread sale of irradiated food in this country. Advocates point out that irradiation can eliminate the use of many chemical preservatives, lower food costs by preventing spoilage, and reduce the incidence of food-borne illnesses transmitted by *Salmonella* and *E. coli*. Those opposed to food irradiation argue that irradiation produces chemical changes in food and that the safety of these new chemicals has not been proven, although it should be noted that these same changes occur in foods preserved by other methods. Opponents also point out that treatment may select for radiation-resistant microorganisms.

---

a 5-Br/A base pair (Figure 11.4). In another round of replication, the 5-bromouridine may undergo a structural shift, and pair with guanine, creating a 5-Br/G base pair. After the next round of replication, the net result is that an A/T base pair is changed into a G/base pair. While most people will not come in contact with 5-bromouridine, other chemicals, such as caffeine are also base analogues. Fortunately, intensive investigation has shown that caffeine is rapidly broken down in the body and does not pose much of a threat as a mutagen.

Chemical Modification of Bases Some mutatgens change the chemical structure of bases in DNA molecules, resulting in a base pair change. Treatment of DNA with nitrous acid removes a chemical group from cytosine, converting it to uracil (❱ Figure 11.5). What was a G/C base pair is converted into a G/U pair. Uracil has the base pairing properties of thymine (T). In the next round of DNA replication, the U will pair with A, the G/U base pair is converted to an A/U pair. When A is used as a template for DNA replication, an A/T pair will be created, replacing the original G/C pair. Nitrates and nitrites used in the preservation of meats, fish and cheese are converted into nitrous acid in the body. Although studied extensively, the role of these dietary chemicals in mutagenesis has been difficult to assess.

Chemicals that Bind DNA Chemicals that bind directly to DNA generally result in mutations called frameshifts (described in a later section). These chemicals generally insert themselves into the DNA, distorting the double helix. This can cause a mistake during DNA replication, resulting in the addition or deletion of a base pair.

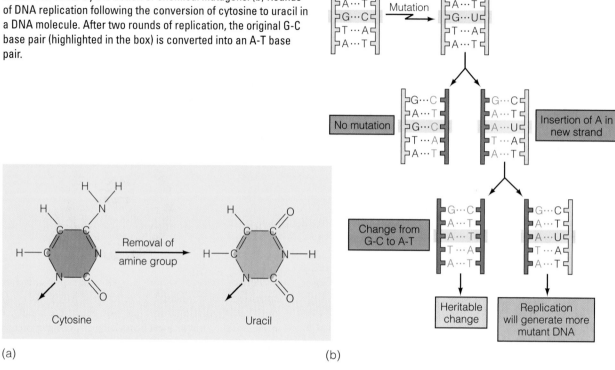

FIGURE 11.5 (a) The conversion of cytosine to uracil, which can occur by the action of chemical mutagens. (b) Rounds of DNA replication following the conversion of cytosine to uracil in a DNA molecule. After two rounds of replication, the original G-C base pair (highlighted in the box) is converted into an A-T base pair.

(a)

(b)

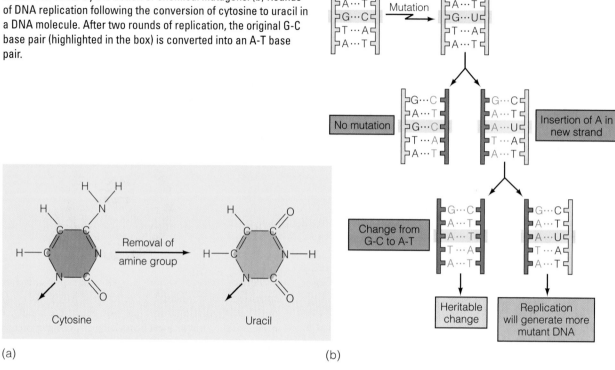

FIGURE 11.6 The molecular structure of acridine orange, an intercalating agent that inserts into the helical structure of DNA, distorting its shape. Replication in the distorted region can lead to the insertion or deletion of base pairs, producing a mutation.

Nucleotide substitutions Mutations that involve substitutions of one or more nucleotides in a DNA molecule.

These chemicals are called intercalating agents. The structure of one of these mutagens, acridine orange is shown in Figure 11.6. This molecule is about the same size as a purine-pyrimidine base pair, and wedges itself into DNA, distorting the shape of the double helix. When replication takes place in this distorted region, deletion or insertion of bases can take place, resulting in a frameshift mutation. Some components and breakdown products of pesticides are intercalating agents.

MUTATION AT THE MOLECULAR LEVEL: DNA AS A TARGET

At the molecular level, mutations can involve substitutions, insertions, or deletions of one or more nucleotides in a DNA molecule. Those mutations that involve an alteration in the sequence but not the number of nucleotides in a gene are called **nucleotide substitutions.** Generally, these involve one or a small number of nucleotides. A second type of mutation causes the *insertion* or *deletion* of one or more bases. Because codons are composed of three bases, changing the number of bases alters the sequence of all subsequent codons and results in large-scale changes in the amino acid sequence of the protein product of such mutated genes. We begin by examining the substitution of one nucleotide for another, and then consider mutations involving the addition or deletion of bases.

Nucleotide Substitutions

Several hundred variants of the alpha- and beta-globins that have single amino acid substitutions are known. These provide many well-studied examples of the effects of nucleotide substitutions on protein structure and function. Nucleotide substitutions in coding regions can have a number of outcomes, some of which are described in the following section. In this discussion, keep in mind that the term codon refers to the sequence of three nucleotides in mRNA that codes for an amino acid.

Missense mutation A mutation that causes the substitution of one amino acid for another in a protein.

Missense mutations are single nucleotide changes that cause the substitution of one amino acid for another in a protein. This substitution may or may not affect the function of the gene product and may or may not have phenotypic consequences. To illustrate this, let's consider amino acid number 6 in beta-globin (◗ Figure 11.7). A single nucleotide substitution in codon 6 from GAG (glu) to GUG (val) results in sickle cell anemia, a condition that has a potentially lethal phenotype. A condition known as HbC (hemoglobin C) has a nucleotide substitution at position 6 (GAG → AAG) that replaces glutamic acid with lysine and causes a mild set of clinical conditions. In a beta-globin variant called Hb Makassar, the codon at the 6th position is changed from GAG (glu) to GCG (ala), a substitution that causes no clinical symptoms and is regarded as harmless.

	1	2	3	4	5	6	7	8
Normal Hb A								
DNA	CAC	GTG	GAC	TGA	GGA	CTC	CTC	TTC
mRNA	GUG	CAC	CUG	ACU	CCU	GAG	GAG	AAG
Amino acid	val	his	leu	thr	pro	glu	glu	lys
Hb C								
DNA	CAC	GTG	GAC	TGA	GGA	TTC	CTC	TTC
mRNA	GUG	CAC	CUG	ACU	CCU	AAG	GAG	AAG
Amino acid	val	his	leu	thr	pro	lys	glu	lys
Sickle Hb S								
DNA	CAC	GTG	GAC	TGA	GGA	CAC	CTC	TTC
mRNA	GUG	CAC	CUG	ACU	CCU	GUG	GAG	AAG
Amino acid	val	his	leu	thr	pro	val	glu	lys

◗ **FIGURE 11.7** The DNA code word, mRNA codon, and the first eight amino acids of normal adult hemoglobin (Hb A), hemoglobin C (Hb C), and sickle cell hemoglobin (Hb S). A single nucleotide substitution in codon 6 is responsible for the changes in the two variant forms of hemoglobin.

In these examples, the resulting polypeptides differ only in the amino acid at position 6: Hb A has glutamic acid (glu), Hb S has valine (val), Hb C has lysine (lys), and Hb Makassar has alanine (ala). The sequence of the other 145 amino acids in the polypeptide is unchanged. In these three examples, single nucleotide changes in the 6th codon of the beta-globin gene result in phenotypes that range from harmless (Hb Makassar), to the mild clinical symptoms of Hb C, and to the serious and potentially life-threatening consequences of Hb S and sickle cell anemia.

Other nucleotide substitutions can produce proteins that are longer or shorter than normal. **Sense mutations** produce longer-than-normal proteins by changing a termination codon into one that codes for amino acids. Several hemoglobin variants with longer-than-normal globin molecules are shown in Table 11.2. In each case, the extended polypeptide chain can be explained by a single nucleotide substitution in the normal termination codon. In hemoglobin Constant Spring-1, the mRNA codon 142 is changed from UAA to CAA, and a glycine codon replaces a stop codon. Thirty more amino acids are inserted before another stop codon is reached.

Sense mutation A mutation that changes a termination codon into one that codes for an amino acid. Such mutations produce elongated proteins.

Nonsense mutations change codons that specify amino acids into one of the three termination codons: UAA, UAG, or UGA (see Table 9.3). This leads to the formation of shortened polypeptide chains. In the beta-globin variant McKees Rock, the last two amino acids are missing, and the protein is only 143 amino acids long. The change in codon 144 UAU (tyr) → UAA (termination) results in a beta chain that is shorter by two amino acids. This change has little or no effect on the function of the beta-globin molecule as a carrier of oxygen. However, some nucleotide substitutions

Nonsense mutation A mutation that changes an amino acid specifying codon to one of the three termination codons.

Table 11.2 **Alpha-Globins with Extended Chains Produced by Nucleotide Substitutions**

HB	Abnormal Chains
Constant Springs-1	gln (142) + 30 amino acids
Icaria	lys (142) + 30 amino acids
Seal Rock	glu (142) + 30 amino acids
Koya Dora	ser (142) + 30 amino acids

can produce more drastic changes in polypeptide length and more serious phenotypic changes.

Deletions and Insertions

Deletions and insertions can range from single nucleotides to the deletion or duplication of an entire gene. As more genes are analyzed at the molecular level, deletions and insertions are emerging as a major cause of genetic disorders and account for 5 to 10% of all known mutations. The insertion or deletion of nucleotides within the coding sequence of a gene causes **frameshift mutations**. Because codons consist of groups of three bases, adding or subtracting a base from a codon changes the coding sense of all subsequent codons. This changes the amino acid sequence of the protein encoded by the mutated gene. Suppose that a codon series reads as the following sentence:

THE FAT CAT ATE HIS HAT

An insertion in the second codon destroys the sense of the remaining message:

THE FAA TCA TAT EHI SHA T

Similarly, a deletion in the second codon can also generate an altered message:

THE FTC ATA TEH ISH AT

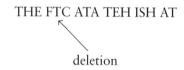

In nucleotide substitutions, usually only one amino acid in the protein is altered. In frameshift mutations, however, the addition or deletion of a single base can cause large-scale changes in the amino acid composition of the polypeptide chain and usually leads to a nonfunctional gene product. We will use a hemoglobin variant with an extended chain as a simple example of an altered gene product that results from a frameshift mutation. In this case, the frameshift occurs near the end of the gene and has a minimum impact on the function of the gene product.

In hemoglobin alpha, the mRNA codons for the last few amino acids are as follows:

Position number	138	139	140	141	TER
mRNA codon	UCC	AAA	UAC	CGU	UAA
Amino acid	ser	lys	tyr	arg	

In hemoglobin Wayne, a deletion in the last base of codon 139 produces a frameshift:

Position number	138	139	140	141	142	143	144	145	146	TER
mRNA Codon	UCC	AAU	ACC	GUU	AAG	CUG	GAG	CCU	CGG	UAG
Amino acid	ser	asn	thr	val	lys	leu	gln	pro	arg	

The deletion of a single base causes a shift in the codon reading frame so that the normal termination codon UAA adjacent to codon 141 is split into two codons, causing

Table 11.3 Mutations with Expanded Trinucleotide Repeats

Gene	Triplet Repeat	Normal Copy	Copy In Disease	MIM/OMIN Number
Spinal and bulbar muscular atrophy	CAG	12–34	40–62	313200
Spinocerebellar ataxia type 1	CAG	6–39	41–81	164400
Huntington disease	CAG	6–37	35–121	143100
Haw-River syndrome	CAG	7–34	54–70	140340
Machado-Joseph disease	CAG	13–36	68–79	109150
Fragile-X syndrome	CGG	5–52	230–72,000	309550
Myotomic dystrophy	CTG	5–37	50–72,000	160900
Friedreich ataxia	GAA	10–21	200–900	229300

new amino acids to be added until another stop codon (generated by the deletion) is reached. The result is an alpha chain variant that has 146 amino acid residues instead of 141.

Trinucleotide Repeats, Gene Expansion, and Mutation

Trinucleotide repeats are a recently discovered class of mutations associated with a number of genetic disorders. Trinucleotide repeats are a sequence of three nucleotides, repeated in tandem a variable number of times within a gene (Table 11.3). Mutations that involve an increase in the number of repeats are responsible for several genetic disorders. This expansion involves only one of the two alleles of a gene, and the phenomenon is called **allelic expansion**. The potential for expansion is a characteristic of a specific allele and occurs only within that allele. The discovery of allelic expansion in fragile-X syndrome has explained some aspects of the way this condition is inherited (see Chapter 6 for a review of fragile sites).

In fragile-X syndrome, the phenotype includes mental retardation. Up to 1% of all males institutionalized for mental retardation have this syndrome. Mothers of affected males are heterozygous carriers and pass the fragile-X chromosome to 50% of their offspring. In some cases, the phenotype has a low degree of penetrance in males. These males, who inherit the mutant allele but have a normal phenotype, are called transmitter males. Carrier mothers of transmitter males are phenotypically normal and have a low risk of bearing children who have fragile-X syndrome. Daughters of transmitter males, on the other hand, have a high risk of bearing affected children, leading to the type of pedigree shown in ▶ Figure 11.8.

The *FMR-1* gene at the fragile-X site has a repeated CGG sequence in the first exon. Normal individuals have 6 to 52 copies of this repeat. These normal alleles are stable and do not undergo expansion. Individuals who have more than 230 copies of this sequence have the clinical symptoms associated with the fragile-X syndrome. Those who have an intermediate number of copies (ranging from 60-200) are unaffected carriers. However, *FMR-1* alleles that carry an intermediate number of copies are called premutation alleles. Carriers of these alleles are unaffected, but their children and grandchildren are at high risk of being affected.

A surprising factor influencing the expansion of the fragile-X gene is the sex of the parent transmitting the expanded allele. When intermediate alleles are transmitted by males, the number of CGG repeats is more likely to remain constant or even decrease. When transmitted by females, the number of CGG repeats

■ **Trinucleotide repeats** A form of mutation associated with the expansion in copy number of a nucleotide triplet in or near a gene.

■ **Allelic expansion** Increase in gene size caused by an increase in the number of trinucleotide sequences.

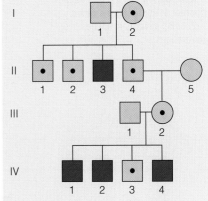

▶ **FIGURE 11.8** A pedigree that illustrates the inheritance of fragile-X syndrome. Mothers (I-2) of phenotypically normal but transmitting males (II-4) are phenotypically normal but have some offspring who have fragile-X syndrome (II-3). Daughters (III-2) of transmitting males are at high risk of having affected children. Allelic expansion of premutation alleles is more likely when inherited from a female. III-2 has inherited such an allele, which is likely to undergo expansion and affect her children.

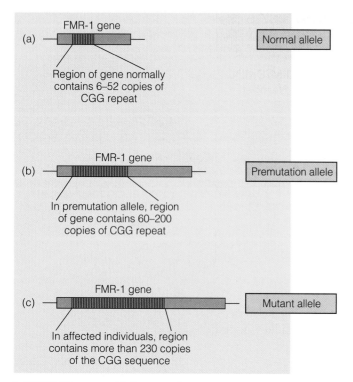

(a) FMR-1 gene — Normal allele
Region of gene normally contains 6–52 copies of CGG repeat

(b) FMR-1 gene — Premutation allele
In premutation allele, region of gene contains 60–200 copies of CGG repeat

(c) FMR-1 gene — Mutant allele
In affected individuals, region contains more than 230 copies of the CGG sequence

▶ **FIGURE 11.9** Allelic expansion in the FMR-1 gene at the fragile-X locus. (a) The gene normally contains 6-52 copies of a CGG trinucleotide repeat. (b) In premutation alleles, this region expands to include 60-200 copies of this repeat. (c) Affected individuals have more than 230 copies of the repeat. How expansion occurs and how it brings about the altered phenotype are still unknown.

■ **Anticipation** Onset of a genetic disorder at earlier ages and with increasing severity in successive generations.

is likely to increase, abolishing expression of the *FMR-1* gene (▶ Figure 11.9) and resulting in fragile-X syndrome.

More than a dozen genetic disorders are now known to be associated with expansion of trinucleotide repeats. Some of these are listed in Table 11.3. In myotonic dystrophy (DM), the mutation is caused by expansion of a CTG repeat adjacent to a structural gene. As with fragile-X syndrome, there is a progressive earlier onset of the disorder in succeeding generations and a correlation between the size of the expanded repeat, the age of onset, and the severity of symptoms. In contrast to the fragile-X syndrome, expansion of trinucleotide repeats in myotonic dystrophy and increased probability of clinical symptoms are more likely for male transmission than for female transmission.

Allelic Expansion and Anticipation

Five disorders (the first five in Table 11.3) associated with expansion of trinucleotide repeats have several similarities. All are progressive neurodegenerative disorders inherited as autosomal dominant traits. All have expanded CAG repeats, and all show a correlation between the increasing size of the repeat and earlier age of onset. In these disorders, mildly affected parents have more seriously affected offspring, who develop symptoms at an earlier age than the parents.

The appearance of more severe symptoms at earlier ages in succeeding generations, called **anticipation,** was first noted for myotonic dystrophy early in this century. Although carefully documented by clinicians, geneticists discounted the phenomenon of anticipation because genes were regarded as highly stable entities, which have only occasional mutations. In disorders showing anticipation, initial changes in the number of copies of a repeat sequence in or near a gene increase the chances that further changes in repeat number will occur, creating alleles with full mutations.

This finding means that the concept of mutation must now recognize the existence of unstable genomic regions that undergo these dynamic changes. In other words, when regions of the genome that contain trinucleotide repeats undergo an expansion, this event enhances the chance of further expansions, and the development of a disease phenotype. With the phenomenon of trinucleotide repeat expansion now well established, it is possible that other disorders that cannot be explained by sim-

Table 11.4 Rates of DNA Damage in a Mammalian Cell	
Damage	**Events/HR**
Depurination	580
Depyrimidation	29
Deamination of cytosine	8
Single-stranded breaks	2300
Single-stranded breaks after depurination	580
Methylation of guanine	130
Pyrimidinel (thymine) dimers in skin (noon Texas sun)	5×10^4
Single-stranded breaks from background ionizing radiation	10^{-4}

ple Mendelian inheritance (for example, those showing incomplete penetrance or variable expression) might also be explained by forms of genome instability.

DNA REPAIR MECHANISMS

Not every mutation results in a permanent alteration of the genome. Cells contain a number of enzyme systems that can repair damage to DNA. This repair function was first observed by cytologists, who noted that chromosome breaks, whether generated spontaneously or by exposure to chemicals or radiation, often rejoin with no apparent detrimental effects. Table 11.4 gives the estimated rate of spontaneous damage to DNA in a typical mammalian cell at 37°C (body temperature). If it were uncorrected, the accumulated damage would destroy much of the DNA in the cell. Fortunately, humans (and other organisms) have a number of highly efficient DNA repair systems (Table 11.5). However, because the rate of background damage is so high, it is easy to overload the repair systems. One type of damage, the formation of **thymine dimers**, is thought to be a major cause of cell death, mutation, and transformation to a cancerous condition.

DNA Repair Systems

One type of repair system can correct errors made during DNA replication. In humans, replication proceeds rapidly, with 10–20 nucleotides added each second to a DNA strand at each replication site. Because a little more than 3 billion nucleotides are copied in each round of cell division, it is little wonder that mistakes occur. On occasion, the wrong nucleotide is incorporated into the replicating strand, resulting in a potential spontaneous mutation. However, the enzyme DNA polymerase corrects many of these mistakes. In addition to directing DNA replication, the enzyme has a proofreading function. If an incorrect nucleotide is inserted by mistake, the enzyme can move backwards, removing nucleotides until the incorrect nucleotide has been eliminated. Then the enzyme resumes moving forward, continuing DNA replication.

Those few mistakes that elude the proofreading function of DNA polymerase remain as true spontaneous mutations.

Other biochemical systems recognize and repair damage to DNA that occurs during other phases of the cell cycle. These systems fall into several categories, each controlled by several genes. For example, exposure of DNA to ultraviolet light (from sunlight, tanning lamps, or other ultraviolet lamps) causes pairing of adjacent thymine molecules in the same DNA strand with each other, forming thymine dimers (❱ Figure 11.10). These dimers cause a regional distortion of the DNA molecule and can interfere with normal replication. These dimers can be corrected by a variety of DNA repair mechanisms.

Table 11.5
Maximum DNA Repair Rates in a Human Cell

Damage	Repairs/HR
Single-stranded breaks	2×10^5
Pyrimidine dimers	5×10^4
Guanine methylation	10^4–10^5

■ **Thymine dimer** A molecular lesion in which chemical bonds form between a pair of adjacent thymine bases in a DNA molecule.

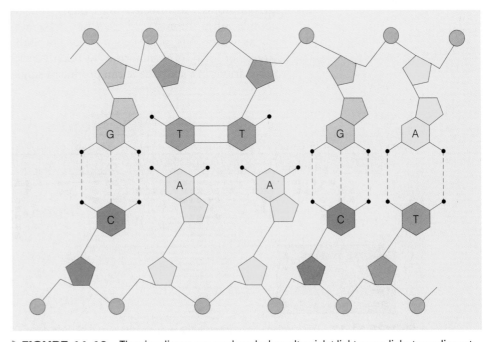

❱ **FIGURE 11.10** Thymine dimers are produced when ultraviolet light cross-links two adjacent thymine bases in the same strand of DNA. This structure causes a distortion in the DNA, and errors in replication are likely to occur unless corrected.

Genetic Disorders and DNA Repair

Because the mechanism of DNA repair is under genetic control, it too, is subject to mutation. Several genetic disorders, including xeroderma pigmentosum (MIM/OMIM 278700), are caused by mutations in genes that control DNA repair. Xeroderma pigmentosum (XP) is an autosomal recessive disorder which has a frequency of 1 in 250,000. Affected individuals are extremely sensitive to sunlight (which contains ultraviolet light). Even short exposure causes dry flaking skin, and pigmented spots that can develop into skin cancer (▶ Figure 11.11). Skin cancers of all kinds are about 1000 times more common in XP individuals, who develop these about 40 years sooner than the general population. Early death from cancer is the usual fate of XP individuals who do not take extraordinary measures to protect themselves from UV light.

Mutations in at least eight different genes can cause the symptoms of XP, and all are associated with the inability to repair damage to DNA caused by ultraviolet light. This disease illustrates a functional correlation between mutation and the development of cancer, a topic we consider in detail in Chapter 14.

Several other genetic disorders are characterized by unusual sensitivity to sunlight and/or to other forms of radiation due to defects in DNA repair. These include Fanconi anemia (MIM/OMIM 227650), ataxia telangiectasia (MIM/OMIM 208900), and Bloom syndrome (MIM/OMIM 210900). The genetic heterogeneity of these disorders indicate that DNA repair is a complex process that involves many different genes.

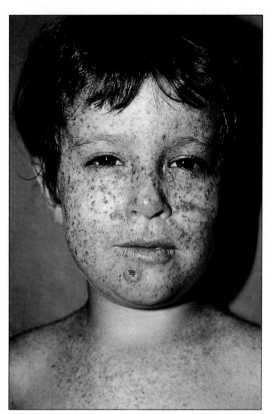

▶ **FIGURE 11.11** Child affected with xeroderma pigmentosum. Affected individuals cannot repair damage to DNA caused by ultraviolet light from the sun and other sources.

MUTATION, GENOTYPES, AND PHENOTYPES

Sickle cell anemia was the first genetic disorder to be analyzed at the molecular level. In this disorder, a nucleotide substitution in codon 6 changes the amino acid in the beta-globin polypeptide at position 6 and produces a distinctive set of clinical symptoms. All affected individuals and all heterozygotes have the same nucleotide substitution.

As it turns out, sickle cell anemia is probably an exception rather than the rule. Molecular analysis of mutations in other genes reveals that more often, a spectrum of mutations in a single gene can produce a single phenotype (▶ Figure 11.12). As the

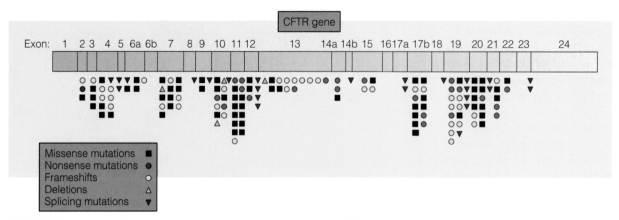

▶ **FIGURE 11.12** Distribution of mutations in the exons of the cystic fibrosis gene, CFTR. More than 500 different mutations have now been discovered. The mutations shown here include nucleotide substitutions, deletions, and frameshift mutations. Any of these mutations in the homozygous condition or in combination with each other (i.e., a compound heterozygote) results in the phenotype of cystic fibrosis.

molecular basis of mutations in a large number of genes becomes known, it is common to find that a genetic disorder can be caused by many different mutational events in a given gene.

MUTATION CAN CAUSE A RANGE OF PHENOTYPES

Cystic fibrosis provides a clear example of the variable nature of mutations that can occur in a single gene, all of which result in a single phenotype. More than 500 different mutations have been identified in the cystic fibrosis gene. These include single nucleotide substitutions, deletions of single amino acids, and larger deletions that involve one or more regions of the gene. In addition, there are frameshift mutations and splice-site mutations. Mutations are distributed in all regions of the CF gene (❱ Figure 11.13), strengthening the idea that any mutational event that interferes with expression of a gene produces an abnormal phenotype.

People who have cystic fibrosis have a wide range of clinical symptoms. The relationship between the genotype and phenotype has been investigated for a number of specific mutations in the *CFTR* gene. In some mutations, such as the Δ508 deletion (present in 70% of all cases of CF), the CFTR protein is produced but is not inserted into the plasma membrane. As a result, chloride ion transport is absent, and clinical symptoms are severe. In other mutations (Figure 11.13), the CFTR protein is produced and inserted into the membrane but is only partially functional. These mutations are associated with a milder form of cystic fibrosis. With over 500 different mutations known in the CF gene, it is possible for someone with CF to carry two different mutant alleles. This genetic variability contributes to the phenotypic variability seen in this disease.

❱ **FIGURE 11.13** Mutations in the *CFTR* gene differ in their phenotypic effects. When homozygous, mutations R117, R334, or R347 allow between 5 and 30% of normal activity for the gene product and produce only mild symptoms. These nucleotide substitution mutations are not common: Together they account for about 2% of all cases of cystic fibrosis. The most common mutation in European populations, Δ508, causes an amino acid deletion in a cytoplasmic region of the protein and accounts for 70% of all mutations in the *CF* gene. This mutation inactivates the CFTR protein and is associated with severe symptoms.

GENOMIC IMPRINTING: REVERSIBLE ALTERATIONS TO THE GENOME

Humans carry two copies of each gene, one received from the mother, the other from the father. Normally there is no difference in the expression of the two copies. But in certain genetic disorders, expression of genes may depend on whether they are inherited from the mother or the father. This differential expression is called **genomic imprinting.**

The first evidence for imprinting in mammals came from mice in which haploid germ cell nuclei were transplanted into eggs to produce zygotes containing two female or two male haploid genomes rather than one male and one female genome. Experimental embryos that have only a male genome develop abnormal embryonic structures but have normal placentas. Zygotes that have only a female genome develop normal embryonic structures and abnormal placentas. Both these conditions are lethal, leading to the conclusions that both a maternal and a paternal genome are required for normal development and that different sets of genes are inactivated during the formation of eggs and sperm.

■ **Genomic imprinting** Phenomenon in which the expression of a gene depends on whether it is inherited from the mother or the father. Also known as genetic or parental imprinting.

Genomic imprinting plays a role in several genetic disorders, including Prader-Willi syndrome (PWS) and Angelman syndrome (AS). Most cases of PWS, an autosomal recessive disorder characterized by obesity, uncontrolled appetite, and mental retardation, are associated with a small deletion in the long arm of chromosome 15 (see Figure 6.28). In about 40% of cases, however, no deletion can be detected. Using molecular markers, these nondeletion cases are found to have inherited both copies of chromosome 15 from the mother. This condition, where both copies of a given chromosome are inherited from a single parent, is called uniparental disomy. In this case, Prader-Willi syndrome is caused by the absence of a paternal copy of the chromosome or the presence of two maternal copies of chromosome 15. Interestingly, in cases of PWS that involve a deletion, the chromosome carrying the deleted region is always from the father, meaning that the only normal copy of the region comes from the mother.

Angelman syndrome is a genetic disorder characterized by severe mental retardation, uncontrollable puppetlike movements, and seizures of laughter. Cytogenetic analysis shows that about 50% of affected individuals have a small deletion in the long arm of chromosome 15 (the same region that is deleted in PWS). Studies using molecular markers indicate Angelman syndrome is a genetic mirror image of Prader-Willi syndrome. In Angelman syndrome, the chromosome carrying the deletion is always from the mother, leaving the affected individual with only the father's copy of the region. Many nondeletion cases of AS have inherited both copies of chromosome 15 from the father, another example of uniparental disomy. These findings indicate that both a paternal and maternal copy of chromosome 15 are required for normal development. The absence of a paternal copy results in PWS, and absence of a maternal copy results in AS.

Imprinting does not affect all regions of the genome but appears restricted to certain segments of chromosomes 4p, 8q, 17p, 18p, 18q, and 22q. Imprinting is not a mutation or permanent change in a gene or a chromosome region; what is affected is the expression of a gene, not the gene itself. Imprinting does not violate the Mendelian principles of segregation or independent assortment. Imprinting is not permanent. Remember that a chromosome received by a female from her father is transmitted as a maternal chromosome in the next generation (▶ Figure 11.14). In each generation, the previous imprinting is erased, and a new pattern of imprinting defines the chromosome as either paternal or maternal. Imprinting events are thought to take place during gamete formation and the effects are transmitted to all tissues of the offspring. Although imprinting is not strictly a mutational event, it does involve chemical modification of DNA in paternally and maternally derived chromosomes. The precise mechanism of imprinting remains unknown, but the process may play an important role in gene expression and may be part of the molecular explanation of the phenomenon of incomplete penetrance.

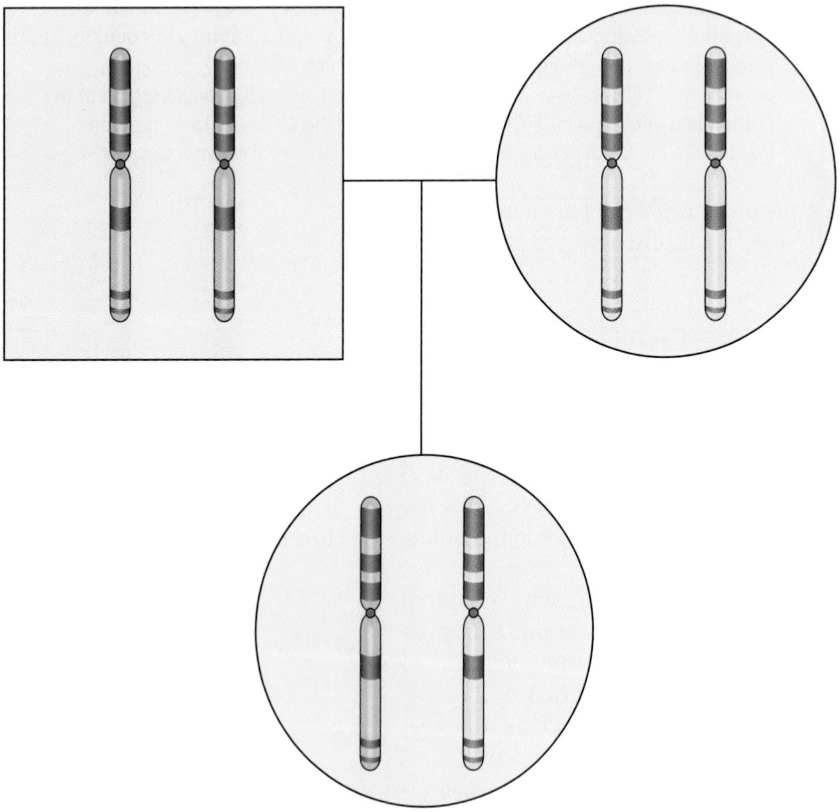

▶ **FIGURE 11.14** Females receive one copy of each chromosome from their fathers. These paternal chromosomes, imprinted as male chromosomes, must be reimprinted as female chromosomes before being transmitted to the next generation.

Case Studies

CASE 1

On April 26, 1986, one of the four reactors at the Chernobyl generating station in the Soviet Union melted down. It has been reported that the plant was running with disconnected safety measures. The result was fire, chaos, fear, a cloud of radioactive isotopes spreading across vast reaches of Eastern Europe, and the radioactive contamination of thousands of people.

Unfortunately, this human tragedy is not being investigated, as it should be, according to scientists who are trying to learn from it. Cancer prevention specialists claim that there is a lack of resources available to look at cancer and noncancer cases in the population that was exposed to the radiation from Chernobyl. Most of the information about radiation and health comes from studies of the aftermath of the bombings of Hiroshima and Nagasaki. However, these studies have examined the effects of a single, brief, and relatively intense dose. In contrast, Chernobyl's exposure is mainly chronic, low-level, and long-term. A better understanding of the health effects of low-level contamination would help in any future meltdowns.

There are many obstacles facing scientists who want to study the aftermath of Chernobyl, including these: (1) the dissolution of the Soviet Union split administrative, record-keeping and medical responsibilities among Belarus, Ukraine, and Russia; (2) a general decline in living standards has reduced the level of medical care in the area; (3) individual doses cannot be reconstructed accurately. In the atomic bomb studies, dose information was relatively accurate. But even there, dose estimates have been overhauled over time. The Chernobyl estimates have been complicated by the fact that some of the dose was external through exposure to radioactive dust, and part of it was internal, through eating contaminated food; and (4) little money for the kind of large studies that are needed for extracting the best data.

Here are some of the facts—as best we know them—about the Chernobyl meltdown. The accident released 1.85×10^{18} (1,850,000,000,000,000,000) international units of radioactive material. The releases contaminated an estimated 17 million people to some degree. The exact amount of exposure depended on location, wind, length of exposure, eating habits, and whether the person was a "liquidator." These unfortunate heroes were pressed into service in a crude cleanup effort after the accident. One hundred and thirty four people showed signs of acute radiation sickness immediately after the accident. Many of the 28 people who died from acute radiation sickness had skin lesions covering 50% or more of their bodies. After the fire, 135,000 people evacuated the area around the reactor, and 800,000 "liquidators" moved in to try to decontaminate the area. Approximately 17 to 45% of the "liquidators" received doses between 10 and 25 rads. (For comparison, in the United States, the annual dose permitted to the general public is 0.1 rads; nuclear workers are permitted 5 rads.)

Despite the confusion and uncertainty, some information is coming in from studies of Chernobyl's aftermath. The most compelling information is seen in the relationship between Chernobyl and thyroid cancer, particularly in children. According to the International Chernobyl Conference in April of 1996, the Chernobyl radiation exposures caused "a substantial increase in reported cases of thyroid cancer in Belarus, Ukraine, and some parts of Russia, especially in young children. This is thought to be due to exposure to radioiodine during the early phases of the accident in 1986. Up to the end of 1995, about 800 cases of thyroid cancer have been reported in children who were under age 15 at the time of diagnosis. To date, three of these thyroid cancer victims have died, and several thousand more cases of thyroid cancer are expected. Ironically, most of the thyroid cancers could have been prevented if people in the contaminated areas had taken iodine tablets immediately after the accident. Most iodine in the body goes to the thyroid gland; if enough normal iodine is available, only a small amount of radioactive isotope irradiates the little gland, whose hormones help regulate growth. More information on Chernobyl's effects will be learned as time passes. Unfortunately, it may be after many more deaths.

Summary

1. The study of mutations is an essential part of genetics. Mutation, the ultimate source of genetic variation, gives the geneticist a means of studying genetic phenomena. Without mutations it would be difficult to determine whether a trait is under genetic control at all and impossible to determine its mode of inheritance.

2. Mutations can be classified in a variety of ways using criteria such as morphology, biochemistry, and degrees of lethality.

3. As a class, dominant mutations are the easiest to detect because they are expressed in the heterozygous condition. Accurate pedigree information can often identify the individual in whom the mutation arose.

4. It is more difficult to determine the origin of sex-linked recessive mutations, but an examination of the male progeny is often informative. If the mutation in question is an autosomal recessive, it is almost impossible to identify the original mutant individual.

5. Studies of mutation rates in a variety of dominant and sex-linked recessive traits indicate that mutations in the human genome are rare events, that occur about once in every 1 million copies of a gene. The impact of mutation is diminished by several factors, including the redundant nature of the genetic code, the recessive nature of most mutations, and the lowered reproductive rate or early death associated with many genetic diseases.

6. Studies of the molecular basis of mutation have provided evidence for a direct link between gene, protein, and phenotype. Mutations can arise spontaneously as the result of an error in DNA replication or as the result of structural shifts in nucleotide bases. Environmental agents, including chemicals and radiation, also cause mutations. Frameshift mutations cause a change in the reading frame of codons, often resulting in dramatic alterations in the structure and function of polypeptide products.

7. Genomic imprinting alters expression of normal genes, depending on whether they are inherited maternally or paternally. Imprinting has been implicated in a number of disorders, including Prader-Willi and Angelman syndromes. Not all regions of the genome are affected, and only segments of chromosomes 4, 8, 17, 18, and 22 are known to be imprinted. Genes are not permanently altered by imprinting, but are reimprinted in each generation during gamete formation.

Questions and Problems

1. Replication involves a period of time during which DNA is particularly susceptible to the introduction of mutations. If nucleotides can be incorporated into DNA at a rate of 20 nucleotides/second and the human genome contains 3 billion nucleotides, how long would replication take? How is this time reduced so that replication can take place in a few hours?

2. Achondroplasia is an autosomal dominant form of dwarfism caused by a single gene mutation. Calculate the mutation rate of this gene given the following data: 10 achondroplastic births to unaffected parents in 245,000 births.

3. Why is it almost impossible to measure directly the mutation rates in autosomal recessive alleles?

4. Define the following terms:
   a. frameshift
   b. mutation rate
   c. trinucleotide repeats

5. What are the factors that influence mutation rates of human genes?

6. Define and compare the following types of nucleotide substitutions. Which is likely to cause the most dramatic mutant effect?
   a. missense mutation
   b. nonsense mutation
   c. sense mutation

7. If the coding region of a gene (the exons) contains 2100 base pairs of DNA, would a missense mutation cause a protein to be shorter, longer, or the same length as the normal 700 amino acid protein? What would be the effect of a nonsense mutation? A sense mutation?

8. A frameshift mutation causes:
   a. a nucleotide substitution
   b. a three base insertion
   c. a three base deletion
   d. a different amino acid sequence after the mutation
   e. the same amino sequence after the mutation

9. In the gene coding sequence shown below, which of the following events will produce a frame shift after the last mutational site?

normal mRNA: UCC AAA UAC CGU CGU UAA
              ser lys tyr arg arg stop

a. insertion of an A after the first codon
b. deletion of the second codon (AAA)
c. insertion of TA after the second codon and deletion of CG in the fourth codon
d. deletion of AC in the third codon.

10. Trinucleotide repeats cause serious neurodegenerative disorders such as Huntington disease, fragile-X syndrome and myotonic dystrophy. The process of anticipation causes the appearance of symptoms at earlier ages in succeeding generations. Describe the current theory on how anticipation works.

11. The cystic fibrosis gene encodes a chloride channel protein necessary for normal cellular functions. Let us assume that if at least 10% normal channels are present, the affected individual has mild symptoms of cystic fibrosis. Less than 10% normal channels produces severe symptoms. At least 50% of the channels must be expressed for the individual to be phenotypically normal. This gene has various mutant alleles:

allele	molecular defect	% functional channels	symptoms
CF100	deletion in exon	0%	severe
CF1	missense mutation in 5′ flanking region	25%	mild
CF2	nonsense mutation in exon	0%	severe
CF3	missense mutation in exon	5%	mild

Predict the percent of functional channels and severity of symptoms for the following genotypes:
a. heterozygous for CF100
b. homozygous for CF100
c. doubly heterozygous, with one copy of CF100 and one of CF3
d. doubly heterozygous, with one copy of CF1 and one copy of CF3

12. Familial retinoblastoma, a rare dominant autosomal defect, arose in a large family that had no prior history of the disease. Consider the following pedigree (the darkly colored symbols represent affected individuals):

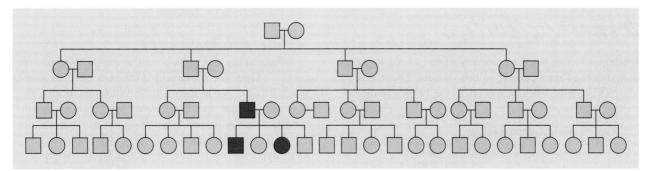

a. Circle the individual(s) in which the mutation most likely occurred.
b. Is this individual affected by the mutation? Justify your answer.
c. Assuming that the mutant allele is fully penetrant, what is the chance that an affected individual will have an affected child?

13. Achondroplasia is a rare dominant autosomal defect resulting in dwarfism. The unaffected brother of an individual with achondroplasia is seeking counsel on the likelihood of his being a carrier of the mutant allele. What is the probability that the unaffected client is carrying the achondroplasia allele?

14. Tay-Sachs disease is an autosomal recessive disease. Since affected individuals do not often survive to reproductive age, why has Tay-Sachs persisted in human population?

15. You are conducting an Ames test on a series of suspected mutagens. The following results are obtained:

dish 1	no mutagen	20 colonies
dish 2	substance A	50 colonies
dish 3	substance B	100 colonies
dish 4	substance C	20 colonies

a. Describe the Ames test
b. Describe the findings in dish 1
c. Rank the substances A, B, and C from most mutagenic to least mutagenic.

16. Our bodies are not defenseless against mutagens that alter our genomic DNA sequences. What mechanisms are used to repair DNA?

17. Even though it is well known that X rays cause mutations, they are routinely used to diagnose medical problems including potential tumors, broken bones, or dental cavities. Why is this done? What precautions need to be taken?

18. You are an expert witness called by the defense in a case where a former employee is suing an industrial company because his son was born with muscular dystrophy, an X-linked recessive disorder. The employee claims that he was exposed to mutagenic chemicals in the work place that caused his son's illness. His attorney argues that neither the employee, his wife or their parents have this genetic disorder, and therefore, the disease in the employee's son represents a new mutation.

   How would you analyze this case? What would you say to the jury to support or refute this man's case?

19. Bruce Ames and his colleagues have pointed out that while detailed toxicological analysis has been conducted on synthetic chemicals, almost no information is available about the mutagenic or carcinogenic effects of the toxins produced by plants as a natural defense against fungi, insects and animals predators. Tens of thousands of such compounds have been discovered, and he estimates that in the U.S. adults eat about 1.5 grams of these compounds each day, about 10,000 times higher than the levels of synthetic pesticides present in the diet. For example, cabbage contains 49 natural pesticides and metabolites, and only a few of these have been tested for their carcinogenic and mutagenic effects.

   a. With the introduction of new foods into the U.S. diet over the last two hundred years (mangoes, kiwi fruit, tomatoes, etc), has there been enough time for humans to evolve resistance to the mutagenic effects of toxins present in these foods?

   b. Since the natural pesticides present in plants constitute more than 99% of the toxins we eat, should diet planning, especially for vegetarians, take into account the doses of toxins present in the diet?

20. Two types of mutations discussed in this chapter are: (1) nucleotide changes and (2) unstable genome regions that undergo dynamic changes. Describe each type of mutation.

Internet Activities

The following activities use the resources of the World Wide Web to enhance the topics covered in this chapter. To investigate the topics described below, log on to the book's homepage at:

http://www.brookscole.com/biology

1. Review the various types of mutations and their consequences by completing the Gene Action/Mutation worksheet found at the Access Excellence Activities Site.

2. The Human Gene Mutation Database at the Institute of Medical Genetics at Cardiff, Wales is a resource which contains information about the mutations identified in human genes, including nucleotide substitutions, missense and nonsense mutations, splicing mutations and small insertions and deletions. The data provided includes the name and symbol for the gene, its chromosomal location, the mutant sequence codon number, and a reference to the paper that identified the mutation.

   Once at the site, search for the words "breast cancer". From the list provided for breast cancer genes, select *BRCA1* Scroll through the mutations, noting the type of mutation and its location. Are the mutations reported here scattered throughout the gene, or are they clustered in certain locations? Since this gene was identified only a few years ago, does this list seem comprehensive? How does this entry compare to the listing for the cystic fibrosis gene? What factors may account for these differences?

3. The National Toxicology Program of the U.S. Department of Health and Human Services oversees the screening of chemicals which may pose a "genotoxic" risk in the workplace. These chemicals are tested using *Salmonella typhimurium* reversion assays, *in vivo* cytogenetic assays and hepatocyte/DNA repair system assays. Read about these screening methods at the NCTR web site and determine why it is prudent to evaluate workplace chemicals using more than one of the assays described.

For Further Reading

Ames, B. (1974). Identifying environmental chemicals causing mutation and cancer. *Science 204:* 587–593.

Ames, B., Profet, M., & Gold, L. S. (1990). Dietary pesticides (99.00% all natural). *Proc. Nat. Acad. Sci. USA 87:* 7777–7781.

Ames, B., Profet, M., & Gold, L. S. (1990). Nature's chemicals and synthetic chemicals: Comparative toxicology. *Proc. Nat. Acad. Sci. USA 87:* 7782–7786.

Barnes, D. E., Lindahl, T., & Sedgwick, B. (1993). DNA repair. *Curr. Opinion Cell Biol. 5:* 424–433.

Brunner, H., Bruggenwirth, H., Nillesen, W., Jansen, G., Hamel, B., Hoppe, R., deDie, C., Howler, C., vanDost, B., Wieringa, B., Ropers, H., & Smeets, H. (1993). Influence of sex of the transmitting parent as well as of parental allele size on the CTG expansion in myotonic dystrophy (DM). *Am. J. Hum. Genet. 53:* 1016–1023.

Cohen, M. M., & Levy, H. P. (1989). Chromosome instability syndromes. *Adv. Hum. Genet. 18:* 43–149.

Deering, R. A. (1962). Ultraviolet radiation and nucleic acids. *Sci. Am. 207:* (December) 135–144.

Engel, E. (1993). Uniparental disomy revisited: The first twelve years. *Am. J. Med. Genet. 46:* 670–674.

Hanawalt, P. C., & Haynes, R. H. (1967). The repair of DNA. *Sci. Am. 216* (February): 36–43.

Harper, P., Harley, H., Reardon, W., & Shaw, D. (1992). Anticipation in myotonic dystrophy: New light on an old problem. *Am. J. Hum. Genet. 51:* 10–16.

Huntington's Disease Collaborative Research Group. (1993). A novel gene containing a trinucleotide repeat that is expanded and unstable on Huntington's disease chromosomes. *Cell 72:* 971–983.

Issa, J. P., & Baylin, S. B. (1996). Epigenetics and human disease. *Nat. Med. 2:* 281–282.

Ledbetter, D. H., & Engel, E. (1995). Uniparental disomy in humans: Development of an imprinting map and its implications for prenatal diagnosis. *Hum. Mol. Genet. 4* (Special No.): 1757–1764.

Muller, H., & Scott, R. (1992). Hereditary conditions in which the loss of heterozygosity may be important. *Mutation Res. 284:* 15–24.

McCann, J., Choi, E., Yamasaki, E., & Ames, B. (1975). Detection of carcinogens as mutagens in the Salmonella/microsome test: Assay of 300 chemicals. *Proc. Nat. Acad. Sci. USA 72:* 5135–5139.

Moss, T., & Sills, D. (1981). The Three Mile Island nuclear accident: Lessons and implications. *Ann. N.Y. Acad. Sci. 365:* entire issue.

Neel, J. V. (1995). New approaches to evaluating the genetic effects of the atomic bombs. *Am. J. Hum. Genet. 57:* 1275–1282.

Nicholls, R. (1993). Genomic imprinting and uniparental disomy in Angelman and Prader-Willi syndromes: A review. *Am. J. Med. Genet. 46:* 16–25.

Orr, H., Chung, M., Banfi, S., Kwiatkowski, T., Servadio, A., Beaudet, A., McCall, A., Duvick, A., Rnaum, L., & Zoghbi, H. (1993). Expansion of an unstable trinucleotide repeat in spinocerebellar ataxia type 1. *Nat. Genet. 4:* 221–226.

Sapienza, C. (1995). Genome imprinting: An overview. *Dev. Genet. 17:* 185–187.

Schwartz, E., & Surrey, S. (1986). Molecular biologic diagnosis of the hemoglobinopathies. *Hosp. Pract. 9:* 163–178.

Tsui, L-C. (1992). The spectrum of cystic fibrosis mutations. *Trends Genet. 8:* 392–398.

Warren, S. T. (1996). The expanding world of trinucleotide repeats. *Science 271:* 1423–1427.

Weeda, G., Hoeijmakers, J., & Bootsma, D. (1993). Genes controlling nucleotide excision repair in eukaryotic cells. *Bioessays 15:* 249–258.

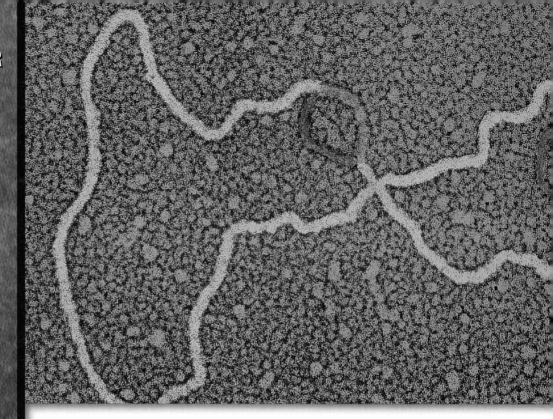

CHAPTER

12

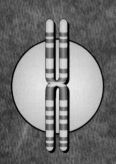

An Introduction to Cloning and Recombinant DNA

The story begins simply enough. A boy of Ghanaian descent, born in England and therefore a British citizen, moved to Ghana to live with his father, leaving behind his mother, two sisters, and a brother. Subsequently, he attempted to return to Britain to live with his mother and siblings, and the story becomes a little complicated. Immigration authorities suspected that the returning boy was an impostor and was either an unrelated child or possibly the nephew of the boy's mother. Acting on this belief, the authorities denied him residence. The boy's family sought assistance to help establish his identity and to allow him to live in the country of his birth. A series of medical tests was conducted, using genetic markers including ABO and other blood types and more sophisticated techniques, such as HLA testing (for details, see Chapter 15). The results indicated that the boy was closely related to the woman he claimed was his mother, but the tests could not tell whether she was the mother or an aunt.

The family's legal counsel then turned to Alec Jeffreys at the University of Leicester to see if the DNA fingerprinting technique developed in Jeffreys' research laboratory could help. To complicate the situation even more, neither the mother's sisters nor the boy's father was available for testing, and the mother was not sure about the boy's paternity. Jeffreys began by taking blood samples from the boy, his reputed siblings, and the woman who claimed to be his mother. DNA was extracted from the white blood cells in each sample and was treated with enzymes that cut the DNA at specific base sequences. The resulting fragments were separated by size. The pattern of these fragments, known as a DNA fingerprint, was analyzed to determine the boy's paternity and maternity. The results show that the boy has the same father as his brother and his sisters because they all share paternal DNA fragments (those not contributed by the mother) in common. The most important question was whether the boy and his "mother" were related.

Twenty-five fragments found in the woman's DNA fingerprint matched those in the boy's fingerprint, indicating that the two are very closely related and that in all probability the boy is the woman's child. (The chance that they are unrelated was calculated as 2×10^{-15}, or about one in a quadrillion.) Faced with this evidence, immigration authorities reversed their position and allowed the boy to enter the country.

DNA fingerprinting is but one example of the array of new techniques developed in recent years as a product of the ongoing revolution in genetic technology. This revolution began with the discovery of how to combine DNA from different organisms in a specific and directed way to create recombinant DNA molecules. In this and the next chapter, we examine how this revolution began, how recombinant DNA molecules are constructed, and how this technology is being used in areas of human genetics, biology, medicine, and agriculture.

WHAT ARE CLONES?

In some cases a fertilized egg divides into two cells, and these cells separate to form identical twins. Because they are derived from a common ancestor (in this case a fertilized egg or zygote), such twins are clones. **Clones** are molecules, cells, or individuals derived from a single ancestor. Methods for producing cloned organisms are not new; horticultural techniques for producing cloned fruit trees were used in ancient

■ **Clones** Genetically identical organisms, cells, or molecules all derived from a common ancestor.

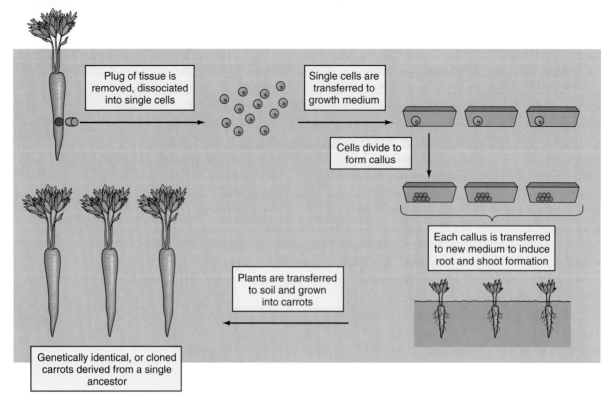

▶ FIGURE 12.1   Carrots can be cloned by removing a plug of tissue that is dissociated into single cells. These cells are placed in growth medium and divide to produce a mass of cells called a callus. Calluses are transferred to a new medium to induce the formation of roots and shoots. The developing plants are then transferred to soil, where they grow into genetically identical copies, or clones, of the original carrot.

▶ FIGURE 12.2   The cloned plant was grown from single cells removed from a parental plant.

times and are still used today. Techniques for producing cloned cells, single-celled organisms, and some animals have all been developed in this century. More recently it has become possible to clone DNA molecules.

The development of methods to produce identical copies of DNA molecules in large quantities has had a profound effect on genetic research and has generated applications of this technology across many disciplines, ranging from archaeology to environmental conservation. This chapter reviews the methods by which cloned DNA molecules are prepared and used in genetic research, forensics, medicine, agriculture, and the repair of defective human genes. Before we consider the basic techniques of DNA cloning, we will look briefly at recent advances in cloning plants and animals.

Plants Can Be Cloned from Single Cells

Domesticated plants and animals have been genetically manipulated for thousands of years by selective breeding. Organisms that have desirable characteristics were selected and bred together, and the offspring that had the best combinations of these characteristics were used for breeding the next generation. Although this method seems slow and unreliable, Charles Darwin noted that only a few generations of intense selection were required to produce many varieties of pigeons and minks.

In the 1950s Charles Steward demonstrated that individual carrot cells could be grown in the laboratory, using special nutrients. Under these conditions the cells would grow and divide to form a cell mass known as a callus (▶ Figure 12.1). When transferred to a different nutrient medium containing growth factors, the calluses grew into mature plants (▶ Figure 12.2). Using this method on a large scale, hundreds or thousands of clones could be produced from a single carrot. Variations on this method have been used to clone plants of several different species. One application of this method has been the development of a loblolly pine tree that is resistant to disease, grows rapidly, and has high wood content. Cells from such a tree were grown individually until they formed a callus and were converted into thousands of copies of the original tree. Using these clones, it is possible to plant a forest of genetically identical trees all of which mature simultaneously, allowing the forest to be harvested for pulpwood on a predictable schedule. This method offers a new approach to timber farming.

Animals Can Be Cloned by Several Methods

Cloning of domesticated animals such as cattle and sheep has moved from the research laboratory to the level of commercial enterprise in the last decade. The two main methods of cloning animals are embryo splitting and nuclear transfer. In embryo splitting, an egg is fertilized in a glass dish (*in vitro*) and allowed to develop into an embryo containing 8–16 cells. The cells of the embryo are then separated from one another under a microscope using very fine needles, and the individual cells are grown in the laboratory to form genetically identical embryos. These embryos can be transplanted into surrogate mothers for development. This method is an extension of the way that identical twins or triplets are produced naturally and can be used to clone any mammalian embryo, including human embryos.

Creating clones by nuclear transfer has the advantage of producing larger numbers of genetically identical embryos. The first successful cloning by nuclear transfer used nuclei from sheep embryos transferred into unfertilized sheep eggs. This method uses unfertilized eggs from which the nucleus has been removed with a fine glass pipette or a needle. These enucleated eggs are then fused with individual cells taken from a 16–32 cell embryo. In the first successful experiments, reported in 1986, cells from early stage sheep embryos were fused with unfertilized sheep eggs. The fused cells were grown into embryos before transplantation into surrogate mothers for development. Ideally, if the original embryo contains 16 or 32 cells, then 16 or 32 genetically identical offspring or clones can be produced (▶ Figure 12.3). As

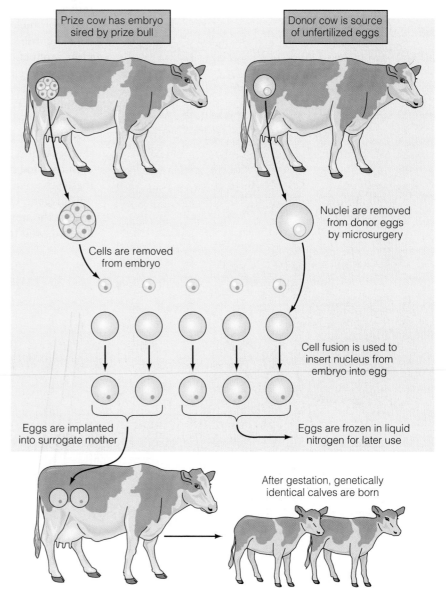

Prize cow has embryo sired by prize bull

Donor cow is source of unfertilized eggs

Cells are removed from embryo

Nuclei are removed from donor eggs by microsurgery

Cell fusion is used to insert nucleus from embryo into egg

Eggs are implanted into surrogate mother

Eggs are frozen in liquid nitrogen for later use

After gestation, genetically identical calves are born

▶ **FIGURE 12.3** Mammals, including cows, are cloned by cell fusion. This process has two stages. First, unfertilized eggs are removed from a donor, and the nucleus is removed from each egg by microsurgery. Second, the embryo to be cloned is recovered and separated into single cells. Embryo cells are fused with the donor eggs. Each egg then contains a genetically identical nucleus. These eggs can be frozen in liquid nitrogen for future use or implanted into the uterus of a surrogate mother to develop. All offspring of these eggs would be genetically identical copies, or clones, of the original embryo.

▶ **FIGURE 12.4** Dolly the sheep, with her surrogate mother.

the methods of nuclear transfer and of growing embryos improved, it was possible to do nuclear transfer with cells from older and older embryos. The cloning of Dolly, reported in 1997, was the first time that a nucleus from an adult cell (in this case from the udder) was successfully used to produce a cloned animal (▶ Figure 12.4). The cloning of Dolly was a significant advance because it overturned conclusions from earlier cloning experiments which indicated that as embryonic cells become more specialized, they are eventually unable to direct development when transferred into unfertilized eggs.

In 1998, a research team successfully cloned more than two dozen mice by directly injecting nuclei from adult cells rather than by cell fusion (▶ Figure 12.5). Using either cell fusion or nuclear transfer, it is possible to produce herds of cloned cattle or sheep that have superior wool, milk, or meat production.

The development of methods for cloning plants and animals represents a significant advance in agricultural technology. When coupled with methods of molecular screening for defective genes, gene transfer, and the prospect of gene surgery, these techniques present us with the possibility of transferring genes between species-and even creating new forms of life. With this background and these possibilities in mind, first we consider the methodology of gene cloning and recombinant DNA technology, and in the next chapter, explore the links between these methods and the cloning of plant and animals.

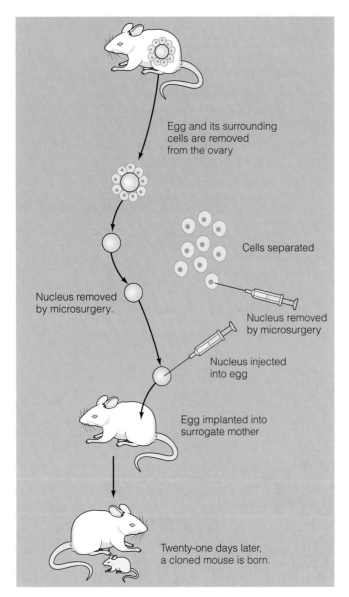

Egg and its surrounding cells are removed from the ovary

Cells separated

Nucleus removed by microsurgery.

Nucleus removed by microsurgery

Nucleus injected into egg

Egg implanted into surrogate mother

Twenty-one days later, a cloned mouse is born.

▶ **FIGURE 12.5** A research team based in Japan and Hawaii successfully cloned over two dozen mice using direct injection of adult cells into egg from which the nucleus was removed. This more direct method had a higher success rate (15% of transfers were successful).

CLONING GENES IS A MULTISTEP PROCESS

Cloning DNA molecules or segments of DNA molecules (genes) produces a large number of identical DNA molecules, all of which have a common ancestor. In this sense cloning DNA is similar to cloning whole plants or animals. The methods for cloning DNA molecules depend on genetic and biochemical discoveries made in the late 1960s and early 1970s. Collectively these techniques are often called **recombinant DNA technology**. A small number of techniques form the basis of recombinant DNA technology. These methods are used to

1. produce DNA fragments using enzymes that cut DNA at specific base sequences;
2. link these DNA fragments to self-replicating forms of DNA, called vectors, to create recombinant DNA molecules;
3. replicate the recombinant DNA molecule in a host organism to create hundreds or thousands of exact copies (clones) of the DNA segment inserted into the vector;
4. retrieve the cloned DNA insert in quantities large enough to allow further studies or modifications; and
5. produce and purify gene products encoded by the cloned DNA inserts.

■ **Recombinant DNA technology** A series of techniques in which DNA fragments are linked to self-replicating vectors to create recombinant DNA molecules, which are replicated in a host cell.

Restriction Enzymes Cut DNA at Specific Sites

The story of the way recombinant DNA technology was discovered illustrates well how basic research in one field often has an unexpected impact on unrelated areas. It might seem odd that mapping the gene for cystic fibrosis and the commercial production of human insulin by bacteria are a direct outgrowth of basic research into the way some bacteria can resist infection by viruses; but often, this is how science progresses. The discovery of bacterial enzymes that inactivate invading viruses by cutting viral DNA into fragments marks the beginning of genetic technology.

In the mid 1970s, Hamilton Smith and Daniel Nathans discovered a number of bacterial enzymes that attach to DNA molecules and cut both strands of DNA at sites of specific base sequences. The enzymes cut up the DNA of invading viruses, giving bacteria protection against viral infections. These enzymes, known as **restriction enzymes,** are a key component in recombinant DNA technology. The recognition and cutting site for an enzyme isolated from *Escherichia coli*, a bacterium that lives in the human intestine, is shown in ▶ Figure 12.6. This recognition and cutting sequence reads the same on either DNA strand (when read in the 5′ to 3′ direction). This recognition sequence is a **palindrome** because it reads the same on either strand of the DNA. Words like mom, pop, and radar are palindromes, as are phrases such as "Todd erases a red dot" or "live not on evil." This recognition and cutting sequence is that of the enzyme *Eco*RI. Some restriction enzymes, like *Eco*RI, create single-stranded tails when they cut DNA. These sticky ends can reassociate with each other or with other DNA molecules that have complementary (if necessary, review complementary base pairing in Chapter 8) tails to produce recombinant DNA molecules (▶ Figure 12.7). Restriction enzymes allow joining together DNA fragments from different organisms to form new

■ **Restriction enzymes** Enzymes that recognize a specific base sequence in a DNA molecule and cleave or nick the DNA at that site.

■ **Palindrome** A word, phrase, or sentence that reads identically in both directions. In DNA, a sequence of nucleotides that reads identically on both strands when read in the 5′ to 3′ direction.

▶ **FIGURE 12.6** The recognition and cutting sites for the enzyme *Eco*RI.

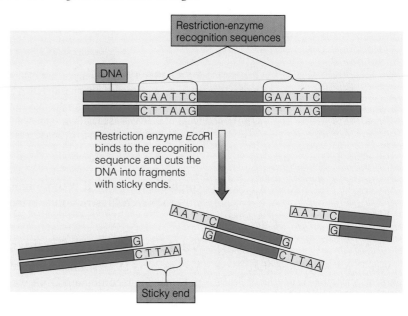

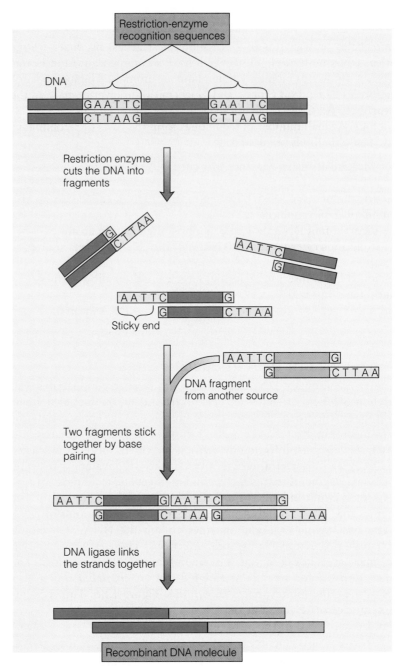

▶ **FIGURE 12.7** Recombinant DNA molecules can be created using DNA from two different sources, a restriction enzyme, and DNA ligase. DNA from each source is cut with the restriction enzyme, and the resulting fragments are mixed. In the mixture, the complementary ends of the two types of DNA will associate by forming hydrogen bonds. These linked fragments can be covalently joined by treatment with DNA ligase, creating recombinant DNA molecules.

combinations, called **recombinant DNA molecules.** The recognition and cutting sites for several restriction enzymes are shown in ▶ Figure 12.8.

Vectors Serve as Carriers of DNA

■ **Vector** A self-replicating DNA molecule that is used to transfer foreign DNA segments between host cells.

Vectors carry DNA segments into cells for replication (cloning). Many of the vectors initially used in recombinant DNA work were derived from self-replicating, circular DNA molecules, called plasmids, found in the cytoplasm of bacterial cells (▶ Figure 12.9).

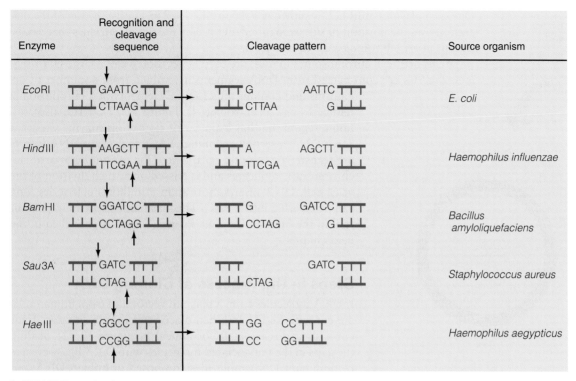

Enzyme	Recognition and cleavage sequence	Cleavage pattern		Source organism
EcoRI	GAATTC / CTTAAG	G / CTTAA	AATTC / G	E. coli
HindIII	AAGCTT / TTCGAA	A / TTCGA	AGCTT / A	Haemophilus influenzae
BamHI	GGATCC / CCTAGG	G / CCTAG	GATCC / G	Bacillus amyloliquefaciens
Sau3A	GATC / CTAG	/ CTAG	GATC /	Staphylococcus aureus
HaeIII	GGCC / CCGG	GG / CC	CC / GG	Haemophilus aegypticus

▶ **FIGURE 12.8**   Some common restriction enzymes and their cutting sites.

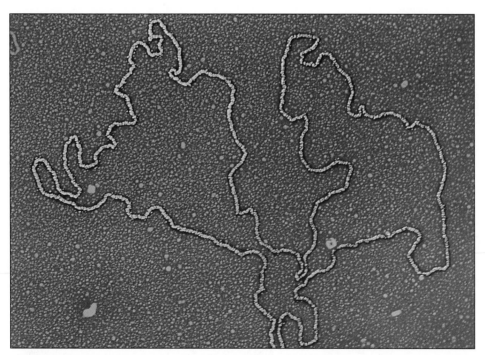

▶ **FIGURE 12.9**   A plasmid isolated from a bacterial cell. Such plasmids can be used as vectors for cloning DNA.

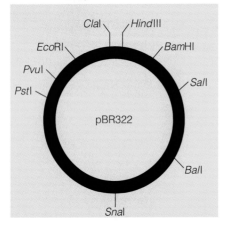

One such plasmid, pBR322, is shown in ▶ Figure 12.10. The middle section of the diagram shows the position of sites recognized and cut by restriction enzymes that can be used to insert DNA molecules to be cloned.

Yeast artificial chromosomes (YACs) are used as vectors to clone large DNA segments. YACs are constructed from DNA segments assembled from a number of organisms and contain the essential elements of a chromosome: telomeres and a centromere (▶ Figure 12.11). YACs also contain a region that has a cluster of restriction enzyme sites that carry DNA inserts for cloning. YACs that carry DNA inserts are transferred to yeast cells, where the recombinant artificial chromosome replicates and is passed on at each division of the yeast cell. DNA inserts more than a million nucleotides long can be inserted into a YAC. This has made them an important tool in the Human Genome Project, which is described in the next chapter.

Steps in the Process of Cloning DNA

If DNA molecules from a plasmid vector and from human cells are cut with EcoRI and then placed together in solution, their cut ends can reassociate, or anneal (▶ Figure 12.12). Sealing the gaps in the reassociated molecules with DNA ligase creates recombinant DNA molecules composed of human DNA and vector DNA. The recombinant vector is transferred into bacterial host cells, which are grown on a nutrient plate, where

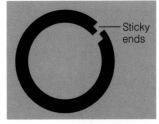

▶ **FIGURE 12.10** Plasmids, such as pBR322, contain restriction sites that can be used to carry DNA inserts for cloning.

■ **Yeast artificial chromosome (YAC)**
A cloning vector that has telomeres and a centromere that can accommodate large DNA inserts and that uses the eukaryote, yeast, as a host cell.

▶ **FIGURE 12.11** A yeast artificial chromosome used as a cloning vector. It contains two telomeres (TEL, structures at the end of all eukaryotic chromosomes), a centromere (CEN), and a region that has restriction-enzyme cutting sites that can be used to carry DNA inserts of more than one million (1Mb, a megabase) nucleotides.

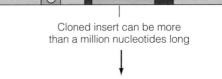

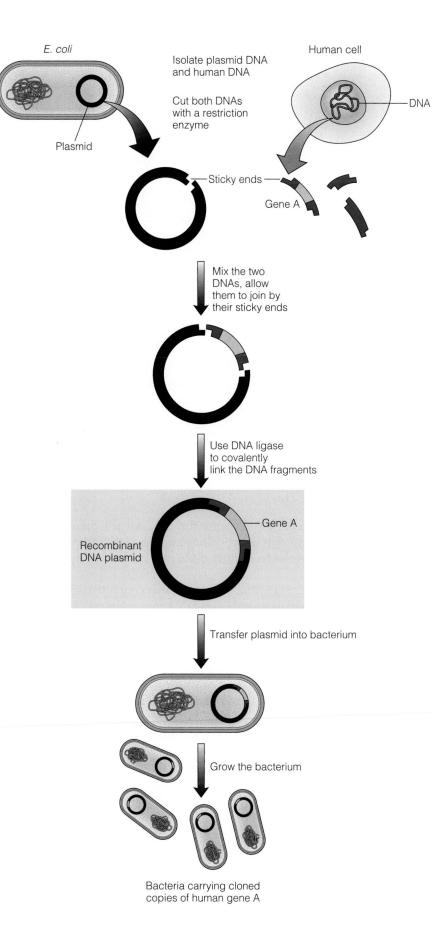

E. coli

Isolate plasmid DNA and human DNA

Cut both DNAs with a restriction enzyme

Human cell

DNA

Plasmid

Sticky ends

Gene A

Mix the two DNAs, allow them to join by their sticky ends

Use DNA ligase to covalently link the DNA fragments

Gene A

Recombinant DNA plasmid

Transfer plasmid into bacterium

Grow the bacterium

Bacteria carrying cloned copies of human gene A

▶ **FIGURE 12.12** A summary of the cloning process. Two types of DNA are isolated: vector DNA in the form of a plasmid and human DNA containing gene A. Both DNAs are cut with the same restriction enzyme. The cut DNAs are mixed and, after pairing of the sticky ends, are linked together by an enzyme (DNA ligase). The recombinant plasmid DNA is inserted into a bacterial host cell. At each bacterial cell division, the plasmid is replicated, producing many copies, or clones, of the DNA insert carrying the A gene.

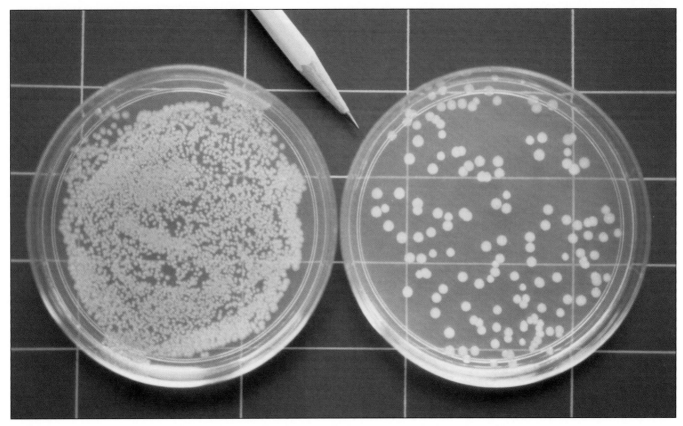

▶ **FIGURE 12.13** Colonies of bacteria on petri plates. Each colony is descended from a single cell. Therefore, each colony is a clone.

they form colonies (▶ Figure 12.13). Because the cells in each colony are derived from a single ancestral cell, all cells in the colony and the plasmids they contain are clones. The colonies are screened to identify those that have taken up recombinant plasmids.

Several methods can be used to identify colonies that contain plasmids that carry DNA inserts. One plasmid used in recombinant DNA research is pBR322, which carries two antibiotic resistance genes, one for tetracycline and one for ampicillin. If the DNA to be cloned is inserted into the gene for tetracycline resistance, this gene is inactivated (▶ Figure 12.14). Colonies formed from a cell that carry this recombinant vector do not grow in the presence of tetracycline but grow on plates that contain ampicillin. Host cells that have taken up a pBR322 vector with no insert grow on plates that contain ampicillin and on plates that contain tetracycline. Thus, colonies that grow on ampicillin but not on tetracyline carry vectors that have DNA inserts.

The steps involved in cloning DNA can be summarized as follows:

1. The DNA to be cloned is isolated and treated with restriction enzymes to produce segments that end in specific sequences.
2. These segments are linked to DNA molecules that serve as vectors, or carriers, and produce a recombinant DNA molecule.
3. Plasmids that carry DNA to be cloned are transferred into bacterial cells, where the recombinant plasmids replicate and produce many copies, or clones, of the inserted DNA.
4. After growth, the bacterial cells can be broken open, and the recombinant plasmids extracted.

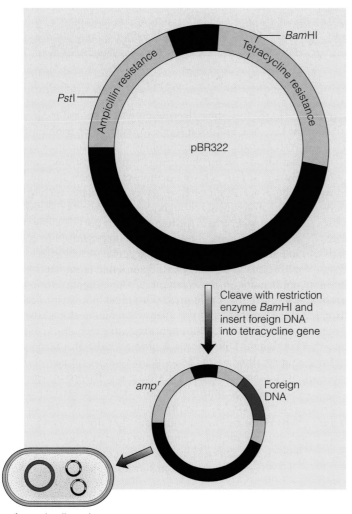

Cleave with restriction enzyme *Bam*HI and insert foreign DNA into tetracycline gene

amp^r

Foreign DNA

Transformed cells resistant to ampicillin, sensitive to tetracycline

▶ **FIGURE 12.14** The plasmid pBR322 carries two antibiotic resistance genes, one for tetracycline and one for ampicillin. If DNA is inserted into the tetracycline gene, bacterial host cells that take up this plasmid will be resistant to ampicillin, but not tetracycline. Cells that have not taken up a plasmid will be killed by either antibiotic.

The inserted human DNA can be released from the plasmid with the same restriction enzyme used in cloning. Cloning using YACs follows a similar procedure, except that yeast cells are used as hosts.

The cloned human DNA can then be used in further experiments or transferred to new vectors and host cells to synthesize the protein encoded by the cloned human DNA.

Scientists who work in the field of recombinant DNA technology were among the first to realize that there might be unrecognized dangers in using and releasing recombinant organisms. Accordingly, they called for a moratorium on all such work until issues related to the safety of genetic engineering could be discussed (see Concepts and Controversies: Asilomar–Scientists Get Involved). After years of discussion and experimentation, there is general agreement that such work poses little risk, but the episode demonstrates that scientists are concerned about the risks and the benefits of their work.

Asilomar: Scientists Get Involved

The first steps in creating recombinant DNA molecules were taken in 1973 and 1974. Scientists immediately realized that modifying the genetic information in *Escherichia coli,* a bacterium that lives in the human gut, could be potentially dangerous. A group of scientists asked the National Academy of Sciences to appoint a panel to assess the risks and the need to control recombinant DNA research. A second group published a letter in Science and in Nature, two leading scientific journals, calling for a moratorium on certain kinds of experiments until the potential hazards could be assessed. Shortly afterward, an international conference was held at Asilomar, California, to consider whether recombinant DNA technology poses any dangers and whether this form of research should be regulated. In 1975, a set of guidelines resulting from this conference was published under the direction of the National Institutes of Health (NIH), a government agency that sponsors biomedical research in the United States. In 1976, NIH published a new set of guidelines that prohibit certain kinds of experiments and dictated that other types of experiments were to be conducted only with appropriate containment to prevent the release of bacterial cells that carry recombinant DNA molecules. NIH also called for research to accurately assess the risks, if any, associated with recombinant DNA techniques.

In the meantime, legislation was proposed in Congress and at the state and local levels to regulate or prohibit the use of recombinant DNA technology. The federal legislation was withdrawn after exhaustive sessions of testimony and reports. By 1978, research had demonstrated that the common K12 laboratory strain of *E. coli* was much safer for use as a host cell than originally thought. Several projects showed that K12 could not survive in the human gut or outside a laboratory setting. Other work showed that recombinant DNA is produced in nature, and there are no detectable serious effects. In 1982, NIH issued a new set of guidelines that eliminated most of the constraints on recombinant DNA research. No experiments are currently prohibited. The most important lesson from these events is that the scientists who developed the methods were the first to call attention to the possible dangers of recombinant DNA research, and they did so based only on its potential for harm. There were no known cases of the release of recombinant DNA-carrying host cells into the environment. Scientists voluntarily shut down their research work until the situation could be properly and objectively assessed. Only when they reached a consensus that there was no danger did their work resume. Contrary to their portrayal in the popular media, scientists do care about the consequences of their work and do become involved in socially important issues.

CLONED LIBRARIES

Genetic library In recombinant DNA terminology, a collection of clones that contains all of the genetic information in an individual. Also known as a gene bank.

Because each cloned fragment of human DNA is small, relative to the size of the genome, many separate clones must be created to include a significant portion of the entire human genome. A collection of clones that contains all of the DNA sequences in an individual's genome is known as a **genetic library.** Cloned libraries can represent the entire genome of an individual, the DNA from a single chromosome, or the set of genes that are actively transcribed in a single cell type. In a genomic library, the number of clones required to cover the entire genome depends, of course, on the size of the genome and the number of segments created by the restriction enzyme. A human library composed of cloned fragments 1700 base pairs (1.7 kilobases) long would require about 8.1 million plasmids to contain the genetic information from one human cell. Vectors such as YACs can accept much larger DNA inserts and are now used in constructing clone libraries. A human genetic library can be contained in about 3,000 YACs that can be stored in a single test tube. In looking for a specific gene, it is easier to search through 3,000 clones rather than 8 million clones.

Using the techniques of molecular biology, any gene of interest can be recovered from a genetic library and used in experimental studies, in clinical applications, or commercially. Libraries from many organisms are now available, including bacteria, yeasts, crop plants, many endangered species, and humans.

FINDING A SPECIFIC CLONE IN A LIBRARY

A library of cloned DNA sequences can contain thousands or millions of different clones. One of the problems is finding a clone that contains a gene of interest. Most often, a specific clone is identified by using a labeled nucleic acid molecule called a **probe**. Probes can be radioactive-labeled DNA or RNA molecules that contain a base sequence complementary to all or part of the gene of interest. Other methods use chemical reactions, color or light to indicate the location of a specific clone. The ability of a probe to identify a gene of interest depends on the fact that under the right conditions, two single-stranded nucleic acid molecules that have a complementary base sequence will form a double-stranded hybrid molecule (Figure 12.15). Because the probe is labeled, the clone that carries a DNA insert that hybridizes with the probe can be identified.

Probes can come from a variety of sources including genes isolated from other species. Once a clone that carries part of the rat insulin gene was identified, it was used as a probe to search through a cloned library of human genes to find the human insulin gene. Or, if the gene being sought is expressed in certain cells, the mRNA from those cells can be used to make a probe. For example, hemoglobin is almost the only protein made in red blood cells. mRNA isolated and purified from these cells can be copied into a single-stranded DNA molecule, called cDNA. The cDNA can be used as a probe to recover the globin genes from a cloned library.

■ **Probe**   A labeled nucleic acid used to identify a complementary region in a clone or genome.

A REVOLUTION IN CLONING: THE POLYMERASE CHAIN REACTION

One advantage offered by the development of cloning techniques is the ability to produce large amounts of specific DNA sequences for various uses. Cloning is not the only way to produce a large number of copies of a specific DNA sequence. The **polymerase chain reaction (PCR)** technique, invented in 1986, has revolutionized and in some cases, replaced methods of recombinant DNA research.

■ **Polymerase chain reaction (PCR)**   A method for amplifying DNA segments that uses cycles of denaturation, annealing to primers, and DNA-polymerase directed DNA synthesis.

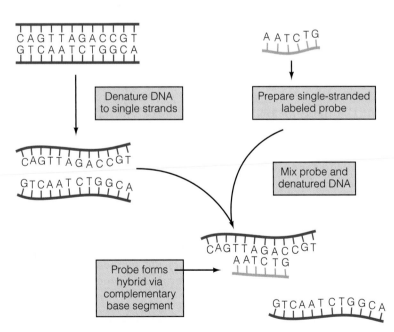

▶ **FIGURE 12.15**   A DNA probe is a single-stranded molecule labeled for identification in some way. Both chemical and radioactive labels are used. The single-stranded probe forms a double-stranded hybrid molecule that has complementary regions of the DNA being studied.

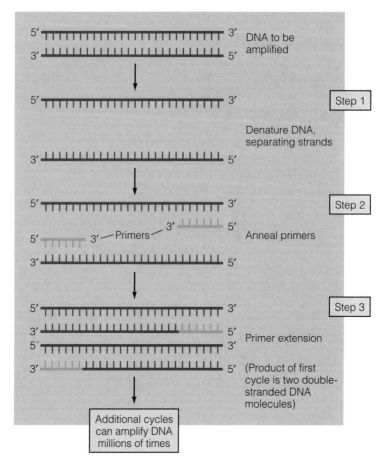

FIGURE 12.16   In step 1 of the polymerase chain reaction (PCR), the DNA to be amplified is denatured into single strands. In step 2, short single-stranded primers pair with the nucleotides adjacent to the region to be amplified. These primers are made synthetically and are complementary to the nucleotide sequence that flanks the region to be amplified. In step 3, enzymes and nucleotides for DNA synthesis are added, the primers are extended, forming a double-stranded DNA molecule, and the primers are incorporated into the newly synthesized strand. This series of three steps is called a cycle. In the second cycle, the double-stranded molecules produced in the first cycle are denatured to form single strands, which are templates for binding primers and another round of DNA synthesis. Repeated cycles can amplify the original DNA sequence by more than a million times.

PCR uses single-stranded DNA as a template to synthesize a complementary strand by the enzyme DNA polymerase, similar to the way DNA replication works in the cell nucleus. (See Chapter 8 for a discussion of DNA replication.) The following occur in the PCR reaction:

1. The DNA to be amplified is heated to break the hydrogen bonds that join the two polynucleotide strands, producing single strands (Figure 12.16).
2. Short nucleotide sequences that act as primers for DNA replication are added, and these primers bind to complementary regions on the single-stranded DNAs. Primers can be synthesized in the laboratory and are usually 20 to 30 nucleotides long.
3. The enzyme DNA polymerase begins at the primers and synthesizes a DNA strand complementary to the region between the primers, a process called primer extension.

This set of three steps is known as a PCR cycle. The entire cycle can be repeated by heating the mixture and separating the double-stranded DNAs into single strands, each of which can serve as a template. The end result is that the amount of DNA present doubles with each cycle. After n cycles, there is a 2^n increase in the amount of double-stranded DNA (Table 12.1).

Table 12.1
DNA Sequence Amplification by PCR

Cycle	Number of Copies
0	1
1	2
5	32
10	1024
15	32,768
20	1,048,576
25	33,544,432
30	1,073,741,820

▶ **FIGURE 12.17** Insects embedded in amber are the oldest organisms from which DNA has been extracted.

It is this power to amplify DNA that is one advantage of the PCR technique.

The DNA to be amplified does not have to be purified and can be present in minute amounts; even a single DNA molecule can serve as a template. DNA from many sources has been used as starting material for the PCR technique, including dried blood, hides from extinct animals such as the quagga (a zebra-like African animal exterminated by hunting in the late 19th century), single hairs, mummified remains, and fossils.

To date, the oldest DNA used in the PCR reaction has been extracted from insects preserved in amber for about 30 million years (▶ Figure 12.17). The DNA amplified from these samples will be used to study the evolution of these insects.

The PCR technique is also used in clinical diagnosis, forensic applications, and other areas, including conservation. Some of these applications are discussed in the next chapter.

ANALYZING CLONED SEQUENCES

Once a cloned sequence that represents all or a portion of a gene has been identified and selected from a library, it can be used to find and study regulatory sequences on adjacent chromosome regions, to investigate the internal organization of the gene, and to study its expression in cells and tissues. For these studies, geneticists use several techniques, some of which are outlined in the following sections.

Southern Blotting

Edward Southern developed the use of cloned DNA segments, that have been separated by electrophoresis, transferred to filters and screened by probes. This procedure, known as a **Southern blot,** has many applications. It is used to find differences in the organization of normal and mutant alleles, identify related genes in other genomes, and study the evolution of genes. To make a Southern blot, genomic DNA is extracted from the organism being studied, cut into fragments with one or more

■ **Southern blotting**   A method for transferring DNA fragments from a gel to a membrane filter, developed by Edward Southern for use in hybridization experiments.

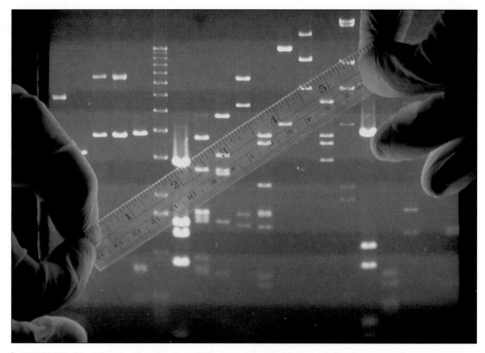

▶ **FIGURE 12.18** A gel stained to show the separation of restriction fragments by electrophoresis.

restriction enzymes, and placed in a gel made of agarose. The DNA fragments are then separated by gel electrophoresis, a technique in which an electric current is passed through the gel. DNA molecules are negatively charged, and the fragments migrate through the gel toward the positive pole. As they migrate through the gel, they separate by size, and the smaller fragments migrate faster. The gel is then stained and photographed (▶ Figure 12.18). The DNA in the gel is separated into single-stranded fragments and transferred to a sheet of DNA-binding material, usually nitrocellulose or a nylon derivative.

The transfer is made by placing the sheet of membrane on top of the gel and allowing buffer to flow through the gel and the nylon membrane by capillary action. Movement of the buffer transfers the DNA from the gel to the membrane. Placing the gel onto a thick sponge that acts as a wick does this. The sponge is partially immersed in a tray of buffer. A sheet of membrane filter is placed on top of the gel and covered by stacks of paper towels and a weight (▶ Figure 12.19). Capillary action draws the buffer up through the sponge, gel, and DNA-binding membrane into the paper towels. As the buffer passes through the gel, DNA fragments are transferred from the gel to the membrane, where they stick.

The membrane is placed in a heat-sealed food storage bag with the labeled, single-stranded probe that corresponds to the gene of interest. Only those single-stranded DNA fragments embedded in the membrane that are complementary to the base sequence of the probe will form hybrids. The unbound probe is washed away, and the hybridized fragments are visualized. For radioactive probes, a piece of X-ray film is placed next to the filter. The radioactivity in the probe exposes parts of the film. The film is developed, and a pattern of one or more bands is seen. These patterns are analyzed and compared with patterns from other experiments.

DNA Sequencing

It is possible to determine the exact order of nucleotides in a segment of DNA by **DNA sequencing** techniques. The most widely used method has been automated and uses a different-colored fluorescent probe for each of the four bases in DNA. A laser scanner then reads the results (▶ Figure 12.20). Banks of sequencing machines are used in large-scale sequencing projects such as the Human Genome Project.

■ **DNA sequencing** A technique for determining the nucleotide sequence of a fragment of DNA.

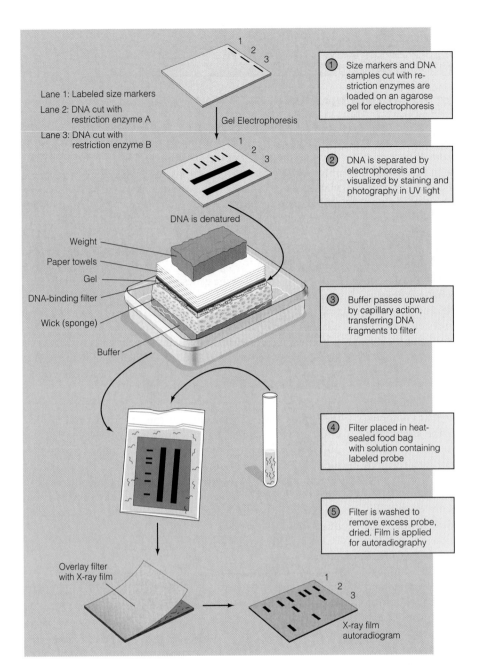

Lane 1: Labeled size markers

Lane 2: DNA cut with
restriction enzyme A

Lane 3: DNA cut with
restriction enzyme B

Gel Electrophoresis

DNA is denatured

Weight
Paper towels
Gel
DNA-binding filter
Wick (sponge)
Buffer

Overlay filter
with X-ray film

X-ray film
autoradiogram

① Size markers and DNA samples cut with restriction enzymes are loaded on an agarose gel for electrophoresis

② DNA is separated by electrophoresis and visualized by staining and photography in UV light

③ Buffer passes upward by capillary action, transferring DNA fragments to filter

④ Filter placed in heat-sealed food bag with solution containing labeled probe

⑤ Filter is washed to remove excess probe, dried. Film is applied for autoradiography

▶ **FIGURE 12.19** The Southern blotting technique. DNA is cut with a restriction enzyme, and the fragments are separated by gel electrophoresis. The DNA in the gel is denatured to single strands, and the gel is placed on a sponge that is partially immersed in buffer. The gel is covered with a DNA-binding membrane, layers of paper towels, and a weight. Capillary action draws the buffer up through the sponge, gel, DNA-binding membrane, and paper towels. This transfers the pattern of DNA fragments from the gel to the membrane. The membrane is placed in a heat-sealed food bag with the probe and a small amount of buffer. After hybrids have been allowed to form, the excess probe is washed away, and the regions of hybrid formation are visualized by overlaying the membrane with a piece of X-ray film. After development, regions of hybrid formation appear as bands on the film.

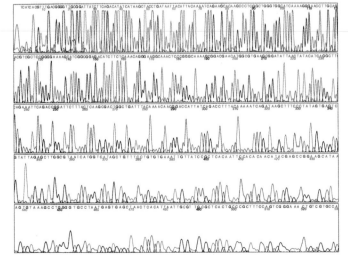

▶ **FIGURE 12.20** Using fluorescent dyes and robotic technology, DNA sequencing has been automated. The results are displayed in color, one for each of the four nucleotides in DNA. The results are read from left to right.

The Human Genome Project: Reading Our Own Genetic Blueprint

Francis Sellers Collins

I developed an early love for mathematics and science and, after learning about the elegant features of the structure of the atom and the chemical bond in a high school chemistry class, I became convinced that I wanted to be a chemist. Accordingly, I majored in chemistry at the University of Virginia, and immediately after graduation went into a Ph.D. program in physical chemistry at Yale, never glancing around to notice what other exciting areas of science I might be missing. Almost by accident, however, I was introduced to molecular biology. To my surprise, I learned that all of my biases about biology being chaotic, intellectually unsatisfying, and devoid of the principles that I prized so much in the physical sciences were really quite unjustified. In fact, it was clear that biology was poised on the threshold of a remarkable revolution, as it became possible to analyze the information content of DNA and the nature of the genetic code.

After a somewhat prolonged and tortuous route, including medical school, I resolved to pursue a career that combined clinical practice and research in medical genetics. Eventually, I got a "real job" as a junior faculty member at the University of Michigan, splitting my time between organizing a research program in human genetics and seeing patients who had genetic diseases. Often frustrated by the lack of information available about the fundamental basis of many genetic diseases, I was particularly drawn to those disorders where the action of a single mutant gene caused a great deal of sickness and distress, but where the cell biology had not been worked out sufficiently to allow direct identification of the responsible gene. Cystic fibrosis (CF) was a particularly puzzling and heart-breaking example of this situation. In collaboration with another researcher in Toronto, my research group began an intense effort to try to identify the CF gene by mapping it to the proper chromosome and then narrowing down the region where the gene must be located until finally the correct candidate was identified. This process, known as positional cloning, had never really succeeded for a problem of this complexity, so it was both gratifying and sobering to participate in such an effort. Many times we thought the problem was just too difficult and were tempted to give up. But finally the strategy worked. The successful cloning of the cystic fibrosis gene in 1989 made better diagnosis available and opened new doors to the design of therapies, including gene therapy. After this, similar efforts in which my research group participated yielded the genes for other puzzling human genetic disorders, including neurofibromatosis (often erroneously referred to as the Elephant Man disease) and Huntington disease.

At about the same time, an international effort to map and sequence all of the human DNA, known as the Human Genome Project, was getting under way. This would make it possible to find all of the genes responsible for human disease, and so I was happy to have the chance to participate by setting up a Human Genome Center at the University of Michigan. When the first U.S. director of the Human Genome Project, Dr. James Watson (the same Watson who with Francis Crick discovered the structure of DNA in 1953) resigned, I was asked to take on this role. Since 1993 I have been director of the National Center for Human Genome Research of the National Institutes of Health, which is the lead agency in the United States responsible for this ambitious and historic effort to map and sequence all of human DNA by the year 2005. So far the project is doing very well and is actually running ahead of schedule and under budget.

Although my own career path was far from focused, I can now look back on all the experiences that I had in science and medicine and see how they helped prepare me for overseeing this remarkable project.

FRANCIS SELLERS COLLINS has been the director of the National Center for Human Genome Research of the National Institutes of Health since 1993. In addition to numerous honors and awards, certifications, and society memberships, he is an associate editor of a number of scientific journals in the area of genetics. He received a Ph.D. from Yale University and an M.D. with honors from the University of North Carolina School of Medicine, Chapel Hill.

DNA sequencing provides information about the size, the organization of genes, and the nature and function of the gene products they encode. Analysis of mutant genes provides information about the way genes are altered by mutational events. Sequencing is also used to study the evolutionary history of genes and species-to-species variation.

Case Studies

CASE 1

Cloning an organism by nuclear transfer is not novel. The technique was first reported in frogs in 1952 and has been used widely since in amphibians to study early development. This work showed that the first few cell divisions after fertilization produce cells that are totipotent (i.e., they can develop into all of the cell types that make up the whole animal). As an embryo develops further, the cells lose this property, and the success of nuclear transfer rapidly declines. Nuclear transfer in mammals proved to be more difficult. The cloning of mice using nuclei from very early embryos was reported in 1977, but this work was not repeatable, and interest among biologist waned. In the 1980s several groups developed transgenic livestock and genetically modified pigs were considered as sources of organs for transplantation to human patients.

If animals can be derived from cells in culture, then it should be possible to make specific genetic modifications, including removing or substituting specific genes. This has been achieved in mice using embryonic stem (ES) cells, but to date no one has been successful in obtaining ES cells from cattle, sheep, or pigs. Dr. Ian Wilmut at the Roslin Institute in Edinburgh, Scotland, thought that nuclear transfer might provide an alternative. A major breakthrough came in 1995 when Wilmut and colleagues produced live lambs by nuclear transfer from cells from early embryos that had been cultured for several months in the laboratory.

So, how did they produce Dolly? Dolly was the first mammal cloned from a cell from an adult animal. She was derived from cells that had been taken from the udder of a 6-year old ewe and cultured for several weeks in the laboratory. Individual cells were then fused with unfertilized eggs from which the nuclei had been removed. Two hundred and seventy-seven of these reconstructed eggs—each now with a diploid nucleus from the adult animal—were cultured for six days in temporary recipients. Twenty-nine of the eggs that appeared to have developed normally to the blastocyst stage were implanted into surrogate Scottish Blackface ewes. One gave rise to a live lamb, Dolly, some 148 days later. Dolly was born on 5 July 1996.

CASE 2

We are a long way from being able to clone mammals with any degree of reliability. However, if this technique is perfected, the impact it may have on medical care may be enormous. For example, Polly, another sheep cloned from an embryo at the same laboratory where Dolly was cloned, has had a human gene inserted which means she is now producing milk with human Factor IX, a blood clotting agent needed by males who have hemophilia B, an X-linked condition. The cloning of human cells in the laboratory could similarly yield tissue, including blood cells and skin, for surgical or genetic repair or for skin grafting for burn victims. Some people cringe at the idea, but there could be potential benefits from cloning human beings. A possible application is in the treatment of sufferers from a rare inherited disorder of the mitochondria. This problem can cause blindness and epilepsy. By removing the nucleus—minus the defective mitochondria—from an embryo created by *in vitro* fertilization in the normal way and placing it in a donated egg stripped of its own nucleus, a cloned baby could be created that would be the genetic offspring of its parents, but without the disorder. Human therapeutic proteins could be developed in quantities larger than are available by current techniques. An example of this is currently being used in clinical trials for the treatment of cystic fibrosis.

During the past 20 years, transplantation of hearts and kidneys has become common. However, there is a shortage of suitable organs for transplant and many patients have to suffer prolonged dialysis and/or die as a result. Transgenic pigs are being developed as sources of organs to meet the shortfall. At present, these pigs contain an added human protein that is intended to prevent immediate rejection of the transplanted heart or kidney.

News of the existence of Dolly the sheep provoked intense worldwide interest, most of it about human cloning. Much of the early speculation was based on science fiction rather than true science. The news created such an uproar that President Clinton urged Congress to pass a five-year moratorium on research toward the cloning of humans. Ironically, nature, which follows its own set of rules, produces human clones every day—identical twins. Cloning techniques have some benefits, as outlined above, but the technology needs to be perfected before it is implemented, and policy for its uses needs to be established.

Summary

1. Cloning is the production of identical copies of molecules, cells, or organisms from a single ancestor. The development and refinement of methods for cloning higher plants and animals represents a significant advance in genetic technology that will speed up the process of improving crops and the production of domestic animals.

2. These developments have been paralleled by the discovery of methods to clone segments of DNA molecules. This technology is founded on the discovery that a series of enzymes, known as restriction endonucleases, recognize and cut DNA at specific nucleotide sequences. Linking DNA segments produced by restriction-enzyme treatment with vectors, such as plasmids or engineered viral chromosomes, produces recombinant DNA molecules.

3. Recombinant DNA molecules are transferred into host cells and cloned copies are produced as the host cells grow and divide. A variety of host cells can be used, but the most common is the bacterium, *E. coli*. The cloned DNA molecules can be recovered from the host cells and purified for further use.

4. A collection of cloned DNA sequences from one source is a library. The clones in the library are a resource for work on specific genes.

5. Clones for specific genes can be recovered from a library by using probes to screen the library.

6. Cloned sequences are characterized in several ways, including Southern blotting and DNA sequencing.

Questions and Problems

1. Nuclear transfer to clone cattle is done by the following technique:
   a. an eight cell embryo is divided into 2 4-cell embryos and implanted into a surrogate mother
   b. a sixteen cell embryo is divided into sixteen separate cells and these cells are allowed to form new 16-cell embryos and directly implanted into surrogate mothers
   c. a 2-cell embryo is divided into 2 separate cells and implanted into a surrogate mother
   d. a sixteen cell embryo is divided into sixteen separate cells and fused with enucleated eggs. The fused eggs are then implanted into surrogate mothers
   e. none of the above

2. Restriction enzymes:
   a. recognize specific nucleotide sequences in DNA.
   b. cut both strands of DNA.
   c. often produce single-stranded tails.
   d. do all of the above.
   e. do none of the above.

3. What is meant by the term recombinant DNA?
   a. DNA from bacteria and viruses
   b. DNA from different sources that are not normally found together
   c. DNA from restriction enzyme digestions
   d. DNA that can make RNA and proteins
   e. none of the above

4. Which enzyme is responsible for covalently linking DNA strands together?
   a. DNA polymerase
   b. DNA ligase
   c. *Eco*RI
   d. restriction enzymes
   e. RNA polymerase

5. A cloned library of an entire genome contains:
   a. the expressed genes in an organism.
   b. all the genes of an organism.
   c. only a representative selection of genes.
   d. a large number of alleles of each gene.
   e. none of the above.

6. The DNA sequence below contains a 6-base palindromic sequence that acts as a recognition and cutting site for a restriction enzyme. What is this sequence? Which enzyme will cut this sequence? Consult Figure 12.8
   CCGAGTAAGCTTAC
   GGCTCATTCGAATG

7. Briefly describe how to clone a segment of DNA. Start with cutting the DNA of interest and a vector molecule with a restriction enzyme.

8. You are given the task of preparing a cloned library from a human tissue culture cell line. What type of vector would you select for this library and why?

9. When cloning human DNA, why is it necessary to insert the DNA into a vector such as a bacterial plasmid?

10. Assume restriction enzyme sites as follows: E = *Eco*RI, H = *Hin*dIII and P = *Pst*I. What size bands would be present when the DNA shown below is cut with
    a. *Eco*RI
    b. *Hin*dIII and *Pst*I
    c. all three enzymes

2 kb	5 kb		3 kb	3.5 kb	4 kb		6 kb
———	E———	H———	P———	E———	H———

11. Name three of the many applications for Southern blotting.

12. You are running a PCR reaction to generate copies of a fragment of the cystic fibrosis (CF) gene. Beginning with two copies at the start, how much of an amplification of this fragment will be present after 6 cycles in the PCR machine?

13. A new gene has been isolated in mice that causes an inherited form of retinal degeneration. This disease leads to blindness in affected individuals. You are a researcher in a human vision lab and want to see if a similar gene exists in humans (genes with similar sequences usually mean similar functions). With your knowledge of the Southern blot procedure, how would you go about doing this? Start with the isolation of human DNA.

14. A cystic fibrosis (CF) mutation in a certain family involves a change in a single nucleotide. This mutation destroys an *Eco*RI restriction site normally found in this position.

    a. What experiments would you perform so that you can use this information in counseling individuals in this family about the likelihood that they are carriers of CF? Be precise in your description of the experiments. Assume that you find that a woman in this family is a carrier, and it happens that she is married to a man who is also heterozygous for CF, but for a different mutation in this gene.

    b. How would you counsel this couple about their risks of having a child with cystic fibrosis.

15. The steps in the polymerase chain reaction (PCR) are:

    a. break hydrogen bonds, anneal primers, synthesis using DNA polymerase

    b. restriction cutting, annealing primers, synthesis using DNA polymerase

    c. ligating, restriction cutting and transforming into bacteria

    d. DNA sequencing and restriction cutting

    e. transcription and translation

16. DNA studies are conducted using samples from a large family with a certain late onset disorder inherited in an autosomal dominant fashion. A DNA sample from each family member is digested with the restriction enzyme *Hae*III and the fragments are separated by size on an agarose gel. A Southern blot is then performed and the results are shown below, aligned with the family pedigree.

    a. Is there a relationship between the disorder and a band pattern in this family, excluding individual II-7 Is linkage shown? If so, what band pattern is linked to the disorder?

    b. How do you explain the genotype and phenotype of II-7?

    c. How would you use this information to counsel II-2, II-3 and II-6 about their risks of developing this disorder?

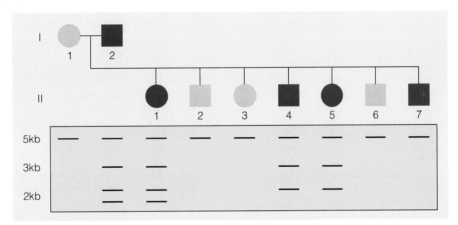

17. A base change (A to T) is the mutational event that creates the mutant sickle cell anemia allele of beta globin. This mutation destroys an *Mst*II restriction site that is normally present in the beta globin gene. This difference between the normal allele and the mutant allele can be detected by Southern blotting. Using a labeled beta globin gene as a probe, what differences would you expect to see for a Southern blot of the normal beta globin gene and the mutant sickle cell gene?

18. Dolly the sheep made headlines in 1997 when she was born. Why was her cloning so important, since cattle and sheep were already being cloned?

19. The cloning methods outlined in this chapter allow researchers to generate many copies of a gene they wish to study through the use of restriction enzymes, vectors, bacterial host cells, the creation of genetic libraries, and PCR. Once useful quantities of a gene are available by cloning, what kind of questions do you think should be asked about the gene? What kind of experiments can be done?

Internet Activities

The following activities use the resources of the World Wide Web to enhance the topics covered in this chapter. To investigate the topics described below, log on to the book's home-page at:

http://www.brookscole.com/biology

1. The DNA Learning Center of Cold Spring Harbor laboratory provides interactive animations of PCR, Southern blotting and cycle sequencing. Review these techniques by selecting "Resources" and "Biology Animation Library" at the site.

2. Public policy and laws concerning the use of genetic information are an important consideration as we acquire more knowledge of the human genome. Genetic discrimination in health insurance and employment is becoming a genuine concern for certain sectors of our society. Federal legislation is now being introduced to prevent misuse of this information. The National Center for Genome Resources has information available on genetic legislation on their home page.

3. The Roslin Institute, the location of Dolly's cloning, maintains a website on cloning and nuclear transfer. Access the background notes on cloning as well as abstracts on key papers from Roslin. How does this information compare with the accuracy of articles on Dolly that you may have read in the popular press?

Select one of the most recent pieces of legislation listed at this site and evaluate the bill. Do you think it has the ability to protect personal genetic information?

For Further Reading

Annas, G. J. (1998). Why we should ban human cloning. *N. Eng. J. Med.* 339: 122–125.

Arnheim, N., White, T., & Rainey, W. E. (1990). The application of PCR: Organismal and population biology. *Bioscience 40:* 174–182.

Drlica, K. (1992). *Understanding DNA and Gene Cloning: A Guide for the Curious,* 2nd ed. New York: Wiley.

Farello, A., Hillier, L., & Wilson, R. K. (1995). Genomic DNA sequencing methods. *Methods Cell Biol. 48:* 551–569.

Finch, J. L., Hope, R. M., & van Daal, A. (1996). Human sex determination using multiplex polymerase chain reaction. *Sci. Justice 36:* 93–95.

Heitman, J. (1993). On the origins, structure, and function of restriction-modification enzymes. *Genet. Eng. 15:* 57–108.

Hochmeister, M. M. (1995). DNA technology in forensic applications. *Mol. Aspects Med. 16:* 315–437.

Gasser, C. S. & Fraley, R. T. (1989). Genetic engineering of plants for crop improvement. *Science 244:* 1293–1299.

Jaenisch, R. (1989). Transgenic animals. *Science 240:* 1468–1474.

Lawn, R. M., & Vehar, G. (1986). The molecular genetics of hemophilia. *Sci. Am. 253:* 84–93.

Maniatis, T., Hardison, R., Lacy, E., Lauer, J., O'Connell, C., Quon, D., Sim, G., & Efstradiatis, A. (1978). The isolation of structural genes from libraries of eukaryotic DNA. *Cell 15:* 687–701.

Micklos, D. A., & Freyer, G. A. (1990). *DNA Science: A First Course in Recombinant DNA Technology.* Cold Spring Harbor, New York: Cold Spring Harbor Press.

Mullis, K. B. (1990). The unusual origin of the polymerase chain reaction. *Sci. Am. 262:* 56–65.

Nobel, D. (1995). Forensic PCR. Primed, amplified, and ready for court. *Anal. Chem. 67:* 613A–615A.

Pennisi, E. (1998). Cloned mice provide company for Dolly. *Science 281:* 495.

Sader, H. S., Hollis, R. J., & Pfaller, M. A. (1995). The use of molecular techniques in the epidemiology and control of infectious disease. *Clin. Lab. Med. 15:* 407–431.

Smith, A. E. (1995). Viral vectors in gene therapy. *Ann. Rev. Microbiol. 49:* 807–838.

Solter, D. (1998). Dolly is a clone—and no longer alone. *Nature 394:* 315–316.

Southern, E. M. (1975). Detection of specific sequences among DNA fragments separated by gel electrophoresis. *J. Mol. Biol. 98:* 503–517.

Watson, J. D., Gilman, M., Witkowski, J, & Zoller, M. (1992). *Recombinant DNA,* 2nd ed. New York: Scientific American Books.

Zilinskas, R. A. & Zimmerman, B. K. (1986). *The Gene Splicing Wars: Reflections on the Recombinant DNA Controversy.* New York: Macmillan.

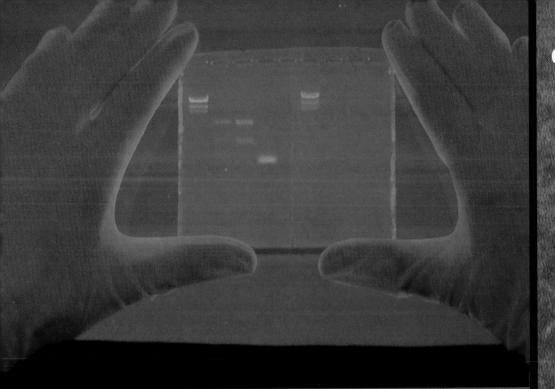

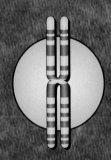

Applications of Recombinant DNA Technology

Bio-Pharming: Making Human Proteins in Animals

Genetic Disorders Can Be Corrected by Gene Therapy

New Plants and Animals Can Be Created by Gene Transfer

ETHICAL QUESTIONS ABOUT CLONING AND RECOMBINANT DNA

About 17–20 million years ago, a magnolia leaf fluttered into the cold water of a lake in what is now Idaho. Magnolias and other plants found in warm, humid, temperate climates flourished along the shores of the lake, which was formed when volcanic lava flowed into the streambed of a valley. The leaf sank into the cold, still water down to the oxygen-poor bottom and quickly became covered with mud. Because of the lack of oxygen, the leaf did not deteriorate; instead, over a period of millions of years, it formed a compression fossil as the mud turned to shale.

On a day millions of years later, researchers in the 20th century split the shale layers. The exposed fossil was still green, although its color faded rapidly in the oxygen-containing atmosphere. The researchers scraped the fossilized leaf into a mortar and ground it with dry ice to form a fine powder. From this powder, the DNA of the fossil leaf was extracted and purified. Using a recombinant DNA technique, the researchers isolated a chloroplast gene from the fossil leaf. This gene encodes the information for an enzyme, which in turn controls the production of a sugar that is an intermediate in photosynthesis.

The researchers determined the nucleotide sequence of this fossil gene and compared it with that of similar genes from present-day species of magnolias. This analysis indicated that although there are some differences in the sequence between the fossil gene and the contemporary gene, there has been little overall change as this gene has traveled through the intervening generations. The differences in nucleotide sequence between the fossil magnolia and present-day species were used to establish its relationship to the species we know today. The results indicate that the fossil species is closely related to one of the species of magnolia that grows in the eastern Unites States.

The ability to isolate and characterize a specific gene from a plant that lived about 17–20 million years ago and compare it to the gene from species alive today demonstrates the power of recombinant DNA technology. The data gathered in this experiment allow a direct measurement of the rate of mutation in an individual gene over evolutionary time and assist in establishing the evolutionary relationship between present species and their fossil ancestors.

Recombinant DNA technology is having a dramatic impact on many other areas, including molecular biology, pharmaceutical manufacturing, agriculture, diagnosis and treatment of genetic disorders, and criminal investigations. This revolution is just beginning, and the impact of recombinant DNA technology and the accompanying biotechnology industry will continue to spread and profoundly change the lives of many individuals. In this chapter, we discuss some of the changes this technology has brought about in medicine, genetics, and agriculture.

RECOMBINANT DNA TECHNIQUES HAVE REVOLUTIONIZED HUMAN GENE MAPPING

If we wish to know how many genetic disorders humans have and how to develop treatments for these diseases, we need to map all of the genes in the human genome and find out what these genes do. Genetic mapping is therefore one of the basic activities of human genetics, as it is in other areas of genetics. As discussed in Chapter 4, mapping human genes involves detecting linkage between genes (genetic evidence that two or more genes are on the same chromosome) and assigning these linked genes to individual chromosomes. Although genes can be assigned directly to the X chromosome by their unique pattern of inheritance, it is more difficult to map genes to individual autosomes.

Mapping of many genetic disorders is hampered because the nature of the mutant gene or its product is unknown. In some cases, the only information available is derived from family studies that have established the mode of inheritance as dominant or recessive and from population studies that estimate the frequency of the condition.

New methods of mapping human genes based on recombinant DNA techniques are now available. These methods have revolutionized mapping genes in humans and in organisms used in genetic research and are the foundation for the Human Genome Project. Mapping can now be done with no initial information about the nature of the mutant gene or the gene product. One method of mapping using recombinant DNA technology begins by identifying variations in the presence or absence of restriction-enzyme cutting sites. We all have slight variations in the nucleotide sequence of our genomes, mostly in stretches of DNA between genes or in the noncoding introns within genes. These nucleotide variations in turn create variations in cutting sites for restriction enzymes. These restriction-enzyme site variations are heritable and can be used as genetic markers. They can be assigned to specific chromosomes and chromosome regions, creating a genetic map of variable sites that cover the entire set of human chromosomes. Genetic disorders can then be mapped by showing linkage between these chromosome-specific markers and the disorder. In the following sections, we discuss these restriction site variations and show how they can be used in mapping human genetic disorders.

RFLPs Are Heritable Genetic Markers

Restriction enzymes work by recognizing specific nucleotide sequences in DNA and by cutting both strands of DNA at these sites. The nucleotide sequence in any long stretches of DNA contains a small degree of variation in the form of single nucleotide substitutions. This variation is not associated with any deleterious phenotype, and as mentioned above, occurs mostly in regions between genes. In some cases this variation in nucleotide sequence creates or destroys a cutting site for a restriction enzyme, altering the pattern of cuts made in the DNA. This creates a heritable difference in the length and number of DNA fragments generated by a given restriction enzyme.

This difference is known as a **restriction fragment length polymorphism,** or **RFLP.** RFLPs are heritable and are passed from generation to generation codominantly (review codominant inheritance in Chapter 3). In other words, RFLPs can serve as genetic markers, although the phenotype of an RFLP is not externally visible, as in albinism, or biochemically detectable, as with a blood type. Rather, the phenotype is present as differences in the number and/or size of DNA fragments. Thousands of RFLPs have been identified in humans, and many (often up to a hundred or more) of these markers have been assigned to each chromosome.

For RFLP analysis, a small blood sample (10 to 20 ml) is collected, and the white blood cells (leukocytes) are separated from the red blood cells (erythrocytes). DNA is extracted from the white cells and treated with a restriction enzyme. As a result, the DNA is cut into thousands of **DNA restriction fragments.** These fragments are placed on an agarose gel and separated by size using electrophoresis. The separated fragments are then Southern blotted, using an RFLP probe (⬤ Figure 13.1) which identifies a small number of specific fragments (those fragments that contain the sequence complementary to the probe).

If a mutational event has created or eliminated a restriction-enzyme cutting site, it can be detected as a change in the length of a DNA fragment, making it larger or smaller (see Figure 13.1). This variation in the cutting pattern is known as an RFLP. Using other methods, RFLPs can be mapped to specific chromosomes and to chromosome regions. Dozens of RFLP markers for each human chromosome are available.

▪ **Restriction fragment length polymorphism (RFLP)** Variations in the length of DNA fragments generated by a restriction endonuclease. Inherited codominantly, RFLPs are used as markers for specific chromosomes or genes.

▪ **DNA restriction fragment** A segment of a longer DNA molecule produced by the action of a restriction endonuclease.

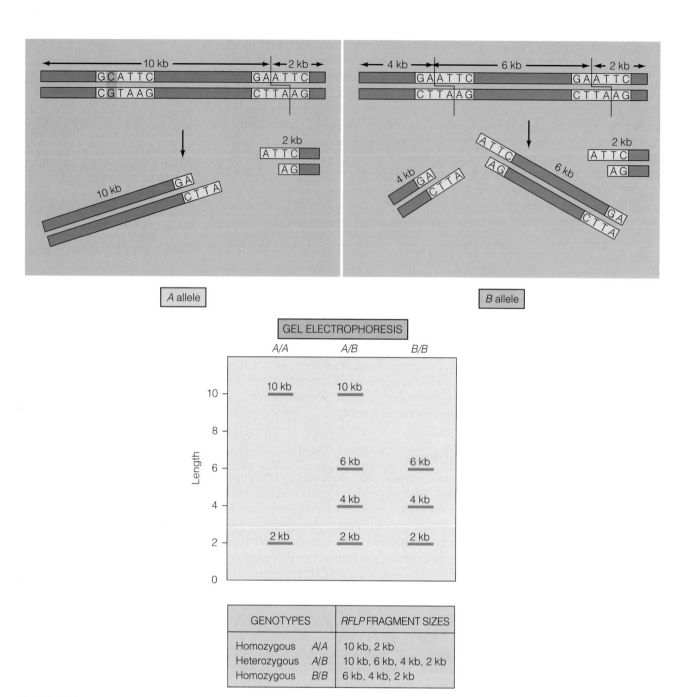

▶ **FIGURE 13.1** Restriction fragment length polymorphisms. The *A* and *B* alleles represent segments of DNA from homologous chromosomes. Arrows indicate recognition and cutting sites for restriction enzymes. Because of variation in nucleotide sequence (highlighted) in one region, a cutting site in *B* is missing in *A*. This variation produces differences in the number and length of DNA fragments in *A* and *B* when they are cut with a restriction enzyme. In *A*, two fragments, one of 10 kb (10,000 base pairs) and one of 2 kb are produced. In *B*, three fragments are produced: one of 6 kb, one of 4 kb, and one of 2 kb. Because these variations in cutting sites are inherited codominantly, there are three possible genotypes: *AA*, *AB*, and *BB*. The allele combination carried by any individual can be determined by restriction digestion of genomic DNA, followed by separation of the fragments by gel electrophoresis. The band patterns for the three combinations are shown as they would appear on a Southern blot.

Mapping Genes Using RFLPs

Two things are needed to assign a gene to a particular chromosome: a large, multi-generational family in which a genetic disorder is inherited and a collection of cloned sequences that detects RFLPs (at least one for each human chromosome). First, a pedigree is analyzed to determine the pattern of inheritance for the genetic disease and to identify affected individuals (❱ Figure 13.2). Then DNA from family members is tested to determine the pattern of inheritance for chromosome-specific RFLP markers, using markers for each chromosome, one at a time. If the genetic defect and a chromosome-specific RFLP marker are inherited together through several generations, then both the genetic disorder and the RFLP must be near each other on the same chromosome. When this happens, the genetic defect can be assigned to the chromosome represented by the RFLP marker.

Mapping the Gene for Cystic Fibrosis: RFLPs and Positional Cloning

The use of RFLPs in mapping human genes can be illustrated by considering the gene for cystic fibrosis (*CFTR*). Cystic fibrosis (CF) is an autosomal recessive condition

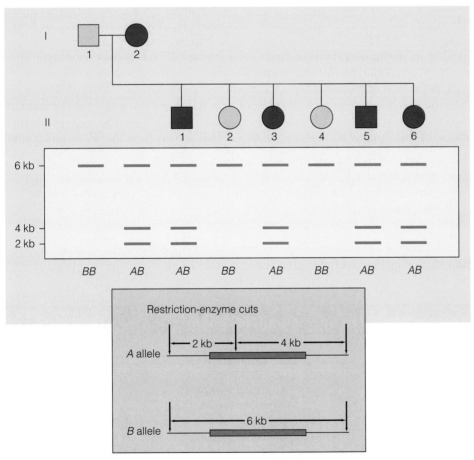

❱ **FIGURE 13.2**   Inheritance of an RFLP allele and a dominant trait. A family pedigree is positioned above a gel that shows the distribution of alleles *A* and *B* in members of the family. Members affected by a genetic disorder are shown by filled symbols. The unaffected father (I-1) is homozygous for allele *B,* and the affected mother (I-2) is heterozygous *AB*. Examination of the pedigree and the Southern blot indicates that affected children have inherited a maternal chromosome that carries allele *A* (II-1, II-3, II-5, and II-6), whereas unaffected children (II-2 and II-4) have inherited maternal and paternal chromosomes that carries allele *B*. If this pattern of inheritance is confirmed in large, multigenerational families, the mutant allele for the genetic disorder would be placed on the same chromosome as the *A* allele for the RFLP.

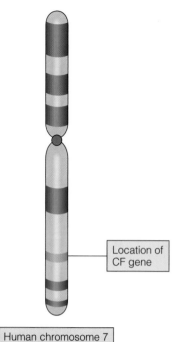

Location of
CF gene

Human chromosome 7

▶ **FIGURE 13.3** The gene for cystic fibrosis was mapped to the long arm of chromosome 7 by RFLP analysis.

■ **Positional cloning** Identification and cloning of a gene responsible for a genetic disorder that begins with no information about the gene or the function of the gene product.

that has an incidence of 1 in 2000 births among those of Northern European ancestry. CF homozygotes develop obstructive lung disease and infections that lead to premature death. The mapping of the CF gene was accomplished in several steps.

Large families that have a history of CF were studied to find linkage between CF and RFLP markers representing specific chromosomes. The result of these initial studies produced a list of chromosomes whose genes were not linked to markers. This list, called an exclusion map, indicates where the gene is not located. Finally, using other RFLP markers, it was discovered that the CF allele is linked to markers on chromosome 7.

In further work with RFLPs located in different regions of chromosome 7, researchers found that the gene for cystic fibrosis is located in a region near the tip of the long arm (▶ Figure 13.3). In the last stage, researchers cloned, recovered, and sequenced over 500,000 base pairs of DNA from the region. They identified the *CFTR* gene in this long stretch of DNA by finding a nucleotide sequence that could be converted into the amino acid sequence for an unknown protein. To confirm that this was the *CFTR* gene, this region of DNA was cloned from normal individuals and from CF patients, and the nucleotide sequences were compared. The sequence from CF patients differed from that from normal individuals. Most CF patients had a three-nucleotide deletion within the gene. In addition, the normal and mutant proteins were isolated and compared. The protein, known as CFTCR, is inserted in the cell membrane and functions in transporting ions (review the inheritance of CF in Chapter 4). In CF, the protein is missing or is defective in function.

The mapping, identification, and cloning of the *CFTR* gene was accomplished in about 5 years, beginning with no information about the gene product, the location of the gene, or the kind of mutational event that results in cystic fibrosis. This method of mapping and isolating human genes that begins with RFLP analysis is known as **positional cloning**. The successful mapping and cloning of the *CFTR* gene has several consequences. Studies of the structure and function of the CFTCR protein have the following benefits:

* They can lead to new and effective therapies for CF.
* Heterozygotes can be detected in the population (4% of the Caucasian population), and those couples at risk for bearing a child who has CF can be identified and counseled.
* They confirm the value of using recombinant DNA methods in gene mapping.

Over 100 other disease-causing genes have been mapped by RFLP analysis and positional cloning. Some of these are listed in Table 13.1.

Table 13.1	**Some of the Genes Identified by Positional Cloning**		
	mim/omim#		mim/omim#
Chromosome 4		Chromosome 17	
Huntington disease	143100	Breast cancer (*BRCA1*)	113705
		Neurofibromatosis (*NF1*)	162200
Chromosome 5			
Familial polyposis (APC)	175100	Chromosome 19	
		Myotonic dystrophy	160900
Chromosome 7			
Cystic fibrosis	219700	Chromosome 21	
		Amyotrophic lateral sclerosis	105400
Chromosome 11			
Wilms tumor	194070	X Chromosome	
Ataxia-telangiectasia	208900	Duchenne muscular dystrophy	310200
		Fragile-X syndrome	309550
Chromosome 13		Adrenoleukodystrophy	300100
Retinoblastoma	180200		
Chromosome 16			
Polycystic kidney disease	173900		

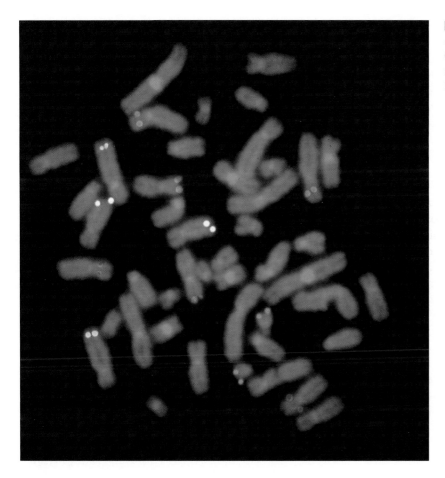

FIGURE 13.4 Genes mapped to homologous chromosomes by fluorescent *in situ* hybridization (FISH). In this example, several genes are being mapped at the same time, with different colors representing different genes.

Mapping at the Chromosomal Level

RFLP analysis and positional cloning are not the only new methods of gene mapping created by recombinant DNA technology. Another, more direct method is also available. For example, if a cloned gene has been recovered from a library, it can be mapped directly to its chromosomal locus. Recall from Chapter 6 that each member of the human chromosome set can be stained to produce a distinctive set of bands. For direct mapping, two things are needed. First, slides with mitotic chromosomes in metaphase are prepared (see Concepts and Controversies: Making a Karyotype in Chapter 6). Next, the cloned gene is labeled and used as a probe. The most commonly used labels are fluorescent dyes. In **fluoresence in situ hybridization (FISH)**, the probe is labeled with a fluorescent dye and hybridized to metaphase chromosomes that have been treated to partially unwind the chromosomes. The probe forms a hybrid that has complementary sequences on the chromosome (this is something like the hybrid that forms in a Southern blot). Unbound probe DNA is washed away, and the location of the hybridized probe is visualized using a microscope. A pair of bright fluorescent spots on homologous chromosomes seen through the microscope marks the location of the gene (Figure 13.4).

■ **Fluoresence in situ hybridization (FISH)**   A method of mapping DNA sequences to metaphase chromosomes by using probes labeled with fluorescent dyes.

PRENATAL AND PRESYMPTOMATIC TESTING FOR GENETIC DISORDERS

Using recombinant DNA techniques, it is possible to test prenatally for the presence of mutant genes. In addition, both children and adults can be tested for genetic disorders before symptoms appear (some genetic disorders appear only later in life). **Prenatal testing** can detect the presence of a genetic disorder in an embryo or fetus.

■ **Prenatal testing**   Testing to determine the presence of a genetic disorder in an embryo or fetus; commonly done by amniocentesis or chorionic villus sampling.

The use of recombinant DNA techniques extends the use of amniocentesis and chorionic villus sampling (CVS) in prenatal testing (see Chapter 6 for a discussion of amniocentesis and CVS). Rather than testing for chromosomal or biochemical abnormalities, linking of amniocentesis or CVS with recombinant DNA technology allows the direct testing of the genotype of an embryo or fetus. We will use the example of sickle cell anemia to illustrate how recombinant DNA methods are used in prenatal testing.

Prenatal Testing for Sickle Cell Anemia

As described in Chapter 4, sickle cell anemia is a recessive trait found in families whose ancestral origins are in areas of West Africa and the lowlands around the Mediterranean Sea. Beta-globin, the gene product defective in this condition, is not synthesized before birth, so the condition cannot be prenatally diagnosed by analyzing fetal hemoglobin.

The mutation that causes sickle cell anemia destroys a restriction-enzyme cutting site, changing the length and number of DNA fragments generated by a specific enzyme (▶ Figure 13.5). By collecting fetal cells using amniocentesis and analyzing restriction fragments from the fetal DNA, the genotypes of normal homozygous, heterozygous, and affected homozygous individuals can be visualized directly on a Southern blot. Prenatal diagnosis can be performed by amniocentesis after the 15th week of development or by chorionic villus sampling in the 8th week of development. At present, dozens of genetic disorders can be detected prenatally using recombinant DNA techniques, and the list is growing steadily.

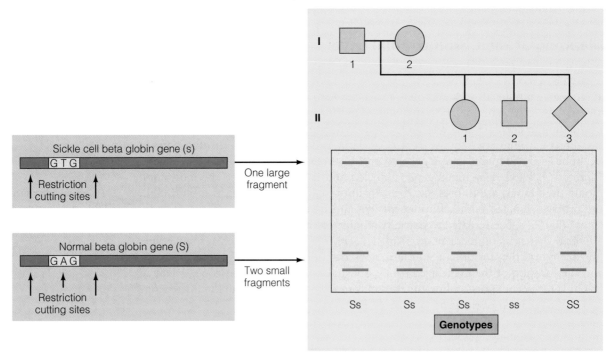

▶ **FIGURE 13.5** Prenatal diagnosis for sickle cell anemia by Southern blot analysis. In sickle cell anemia, the mutation destroys a cutting site for a restriction enzyme. As a result, the number and size of restriction enzyme fragments are altered. Because the mutant allele has a distinctive band pattern, the genotypes of family members can be read directly from the blot. The parents (I-1, I-2) are heterozygotes. The first child (II-1) is a heterozygous carrier, the second child (II-2) is affected with sickle cell anemia, and the unborn child (II-3) is homozygous for the normal alleles, will be unaffected, and will not be a carrier.

Preimplantation Testing

In **preimplantation testing,** human eggs are fertilized *in vitro* and allowed to develop in a culture dish for three days. At the end of this time, the fertilized egg has divided to form an early embryo. The embryo is then viewed through a microscope, and one of the 6 to 8 embryonic cells, called a blastomere, is removed by microdissection. The DNA in this single cell is extracted, amplified by PCR, and used to test for a genetic disorder in the embryo (Figure 13.6). To date, the method has been used successfully to test for the most common form of cystic fibrosis, muscular dystrophy, Lesch-Nyhan syndrome, and hemophilia.

Removing a single cell from an early stage embryo does not damage the embryo or alter its potential for development, and embryos not affected with a genetic disorder are implanted for development and birth. As the genes for more genetic disorders are identified and sequenced, it will be possible to test for an increasing number of these diseases.

Presymptomatic Testing for Genetic Disorders

Presymptomatic testing involves detecting genetic disorders that become apparent only later in life. The phenotypes of many dominant traits, including polycystic kidney disease (*PCKD*) are expressed in adulthood. Symptoms of *PCKD* usually appear between 35 and 50 years of age. This condition, inherited as a dominant trait, is characterized by the formation of cysts in the kidney that gradually destroy the organ. Treatment includes kidney dialysis and transplantation of a normal kidney, but many affected individuals die prematurely. Because *PCKD* is a dominant trait, anyone heterozygous or homozygous for the gene is affected. The gene for *PCKD* is located on chromosome 16, and markers can be used to determine which family members are most likely to develop *PCKD*. This testing can be done prenatally or at any age before (or after) the condition appears.

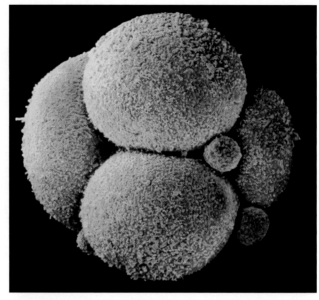

 FIGURE 13.6   In blastomeric testing, a single cell is surgically removed from an early embryo, and the DNA from this cell is used to test for a genetic disease.

 Preimplantation testing   Testing for a genetic disorder in an early embryo; testing is done by removing a single cell from the embryo.

 Presymptomatic testing   Genetic testing for adult-onset disorders; testing can be done at any time before symptoms appear.

DNA Chips and Genetic Testing

The recombinant DNA methods just described can test for a single genetic disorder or several alleles of a mutant gene. A new technology employing DNA chips (Figure 13.7) will allow screening for all known mutations in a disease gene and can be used to screen for mutations in thousands of genes at a time.

DNA chips employ the technology of the semiconductor industry and are made using the same equipment as computer chips. The DNA chips are made of glass squares about a half-inch on a side. The chip is divided into fields (smaller squares), each the width of a human hair. Each field contains linker molecules to which synthetic DNA probes of about 20 nucleotides are attached. In a row of fields, each probe differs by one nucleotide. Thus, for the first nucleotide in the probe, a set of four fields (one for each nucleotide) is needed. With a probe twenty nucleotides long, 160,000 fields (20^4) are needed to give all possible combinations of nucleotides. Current chips hold 400,000 fields, but chips with 1.6 million fields are now in development, and even larger chips are planned.

For testing, DNA is extracted from a blood sample and cut with one or more restriction enzymes. The resulting fragments are tagged with a fluorescent dye, melted into single strands, and pumped into the chip. Fragments whose nucleotide sequence matches the probe sequence exactly will bind, and those that do not match are

GeneChip® Probe Arrays

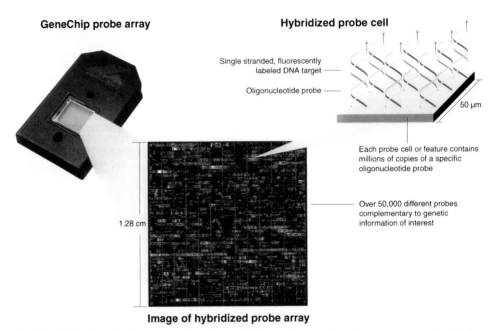

GeneChip probe array

Hybridized probe cell

Single stranded, fluorescently labeled DNA target ——

Oligonucleotide probe ——

50 μm

Each probe cell or feature contains millions of copies of a specific oligonucleotide probe

Over 50,000 different probes complementary to genetic information of interest

1.28 cm

Image of hybridized probe array

▶ **FIGURE 13.7** Probes are used to detect DNA mutations and monitor gene expression. Each chip holds between 50,000 and 500,000 probes. The nucleic acids to be analyzed are made single-stranded and tagged with fluorescent dyes, and hybridized to the probes on the chips. Unbound molecules are washed off the chip, and a laser scans the chip and detects the fluorescence that signals hybridization with a probe.

washed from the chip. A laser scanner reads the chip and detects the pattern of fluorescence, which is analyzed by a linked software program. The results are presented in the form of a nucleotide sequence for the DNA tested. This sequence can be compared with the normal sequence to detect any mutations.

Chips are now in use to screen for mutations in the *p53* gene, which is mutated in 60% of all cancers, and to screen for mutations in the *BRCA1* gene, which predisposes women to breast cancer. In the near future, chips will be loaded with gene sets for heart diseases, diabetes, Alzheimer's disease, and other genetically defined disease subtypes. Eventually, chips may be used to test for mutations in the entire human gene set, giving individuals a genetic profile of all the mutations they carry, and evaluating the risks they have for developing a genetic disorder or for transmitting a disorder to their children.

DNA FINGERPRINTING

As discussed earlier, the technique of RFLP analysis is widely used in mapping human genes. RFLP analysis is also used in many other applications, including forensics, ecology, archaeology, preservation of endangered species, and even dog breeding.

In addition to RFLPs, other nucleotide variations in the human genome can be used as genetic markers. In the 1980s, Alec Jeffreys and his colleagues at the University of Leicester discovered a form of nucleotide variation that depends on variability in the length of repetitive DNA clusters called minisatellites, which are located at specific chromosomal sites. Other nucleotide variations, including single nucleotide polymorphisms (SNPs), have been discovered in recent years. RFLPs, minisatellites,

SNPs, and other nucleotide variations are being used in many fields, including court cases, studies of human evolution, and tracing migratory patterns of humans across the world.

Minisatellites and VNTRs

Minisatellites are sequences ranging from 14 to 100 nucleotides long that are organized into clusters of tandem repeats. For example, the nucleotide sequence

CCTTCCCTTCCCTTCCCTTCCCTTCCCTTC

is made up of six tandemly repeated copies of the five-nucleotide sequence, CCTTC. Clusters of such repeats are scattered at many sites on all chromosomes. There are many different minisatellite sequences, and the number of repeats at each site is variable, ranging from 2 to more than 100. Each variant is an allele, and they are inherited codominantly. These minisatellites are called **variable-number tandem repeats (VNTRs)**. Because there are many different VNTR loci and because each locus can have dozens of alleles, heterozygosity is common.

When restriction enzymes are used to cut VNTR sequences in a DNA sample and the results are visualized by a Southern blot, a pattern of bands is produced. The term **DNA fingerprint** is used to describe this pattern (▶ Figure 13.8). As with a traditional fingerprint, the pattern is always the same for a given individual. In this case, it is a pattern of DNA fragments that remain constant no matter what tissue is used as a source of DNA, but the pattern varies from individual to individual. There is so much individual variation that if a number of different VNTRs are analyzed, each person's profile is unique (except, of course, for identical twins).

When coupled with PCR, DNA fingerprint analysis can be done using very small samples of DNA, obtained from a variety of sources, including single hairs, envelope flaps, cigarette butts, and the dried saliva on the back of a used postage stamp. DNA fingerprints can also be obtained from very old samples of DNA, increasing its usefulness in legal cases. In fact, DNA fingerprinting has been carried out on mummies that are more than 2400 years old.

■ **Variable-number tandem repeats (VNTRs)** Short nucleotide sequences, repeated in tandem, that are clustered at many sites in the genome. The number of repeats at homologous loci is variable.

■ **DNA fingerprint** A pattern of restriction fragments that is unique to an individual.

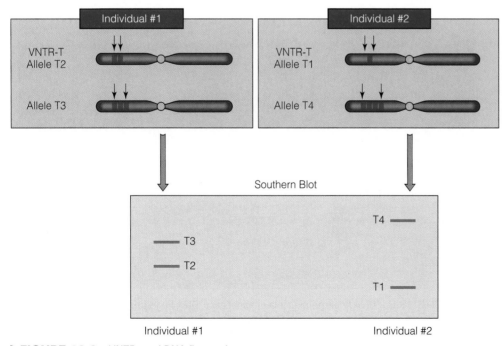

▶ **FIGURE 13.8** VNTRs and DNA fingerprints

Forensic Applications of DNA Fingerprinting

Forensics is the application of scientific knowledge to issues in civil and criminal law. In the U.S., forensic DNA fingerprinting was first used in 1987 in a criminal trial. Since then, it has been used in more than 50,000 criminal investigations and trials. At the present time, DNA fingerprinting is used in about 10,000 criminal cases every year, 75% of which involve criminal sexual assault. It is also used in about 200,000 civil cases each year for paternity testing. DNA analysis is performed in state and local police crime labs, private labs, and the FBI lab in Washington, D.C.

In criminal cases, DNA is extracted from biological material left at a crime scene. This can include blood, tissue, hair, skin fragments, and semen. The DNA is cut with restriction enzymes, and the resulting pattern of fragments is analyzed and compared with the DNA fingerprints of the victim and any suspects in the case. DNA fingerprinting was first used in England in 1986-87 to solve the rape and murder of two teenage girls. Over 4,000 men were DNA fingerprinted during the investigation. The results freed an innocent man who had been jailed for the crime and led to a confession from the killer. This case was described in the best-selling book *The Blooding,* written by Joseph Wambaugh. In the U.S., DNA fingerprinting first received widespread attention in the O.J. Simpson criminal trial.

The use of DNA fingerprints in legal cases requires care in collecting and handling samples, accurate identification of bands in the gel, methods for determining when bands match, and reliable calculations for the probability of matching bands from the evidence with those from the defendant.

In the U.S., about 12 VNTR probes located on different chromosomes are used in forensic DNA testing. Most of these have been patented by biotechnology companies and are available for purchase by law enforcement agencies. Each probe can be used to determine a fingerprint pattern for a specific VNTR. In a criminal case, at least four different probes are typically used to develop a suspect's DNA fingerprint profile.

If the suspect's DNA fingerprint profile does not match that of the evidence, then he or she can be excluded as the criminal (Figure 13.9). In DNA fingerprinting, exclusions account for about 30% of all results. On the other hand, if the suspect's pro-

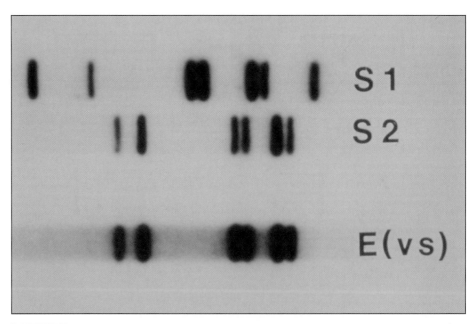

 FIGURE 13.9 DNA fingerprinting in a criminal case. DNA fingerprints for two suspects (S1 and S2) and a fingerprint from evidence [E(vs)] are shown. The profile of suspect 1 is different from that of the evidence, but the profile of suspect 2 matches that of the evidence band for band.

file matches that of the evidence, then there are two possibilities: The DNA finger-print profiles came from the same person, or the DNA profile came from someone else, who has the same pattern of DNA fingerprint bands (Figure 13.9).

Analysis of DNA fingerprints uses a combination of probability theory, statistics, and population genetics. These are used to estimate how frequently the profile of VNTR fingerprints in question might be found in an individual who is a member of a given population group or in the general population. This is done by first calculating the individual frequencies for each pair of VNTR alleles, using information from population studies. In a second step, the frequencies for each VNTR are multiplied together to produce the final estimate of frequency. The frequencies are multiplied together because the probes used represent VNTRs on different chromosomes and are inherited by independent assortment.

Multiplication of the frequencies of each VNTR locus typically results in a rare frequency, even when the individual frequencies are fairly common. The example in Table 13.2 shows how VNTR allele frequencies that individually are fairly common in the population can be combined to provide a profile that has a high degree of reliability. The allele detected at locus 1 has a frequency of 1 in 25 (4%) in the population, and the allele locus 2 has a frequency of 1 in 100 (1%). The combination of these two would be found in about 1 in every 2500 individuals. Combining the frequency of the first two loci with loci 3 and 4 gives a frequency of 1 in about 60 million. This means that only one person in a population of 60 million would be expected to have this particular combination of VNTR alleles. The frequency estimate does not prove the suspect guilty but is a factor that should be considered along with other facts in the case.

Other Applications of DNA Fingerprinting

DNA fingerprints have been used in many other applications, including a dispute over the bloodlines of a purebred dog and establishing the fate of Czar Nicholas II and other members of the Romanov family (see Concepts and Controversies: Death of a Czar). It has also been used to determine the degree of relatedness among members of endangered species, and on the frozen remains of a wooly mammoth recovered in Siberia to determine the evolutionary relationships between mammoths and modern-day elephants.

By far, the most common usage of DNA fingerprinting in the U.S. is in establishing paternity. Over 30% of the births in the U.S. are to single mothers, and over 200,000 paternity tests are performed each year, many using DNA fingerprints. These tests are performed to establish legal responsibility for financial support, and efforts by local, state and federal governments result in support payments of more than $5 billion per year.

Table 13.2	**Calculating a DNA Fingerprint Profile Frequency**	
Allele at Locus	**Frequency in Population**	**Combined Frequency**
1	1 in 25 (0.040)	—
2	1 in 100 (0.010)	1 in 2500
3	1 in 320 (0.0031)	1 in 806,000
4	1 in 75 (0.0133)	1 in 60,600,000

Death of a Czar

*C*zar Nicholas Romanov II of Russia was overthrown in the Bolshevik Revolution that began in 1917. He and the Empress Alexandra (granddaughter of Queen Victoria), their daughters Olga, Tatiana, Marie, Anastasia, and their son, Alexei (who had hemophilia), were taken prisoners. In July 1918, it was announced that the Czar had been executed, but for many years the fate of his family was unknown. In the 1920s, a Russian investigator, Nikolai Solokof, reported that the Czar, his wife and children, and four others were executed at Ekaterinburg, Russia on July 16, 1918, and their bodies were buried in a grave in the woods near the city. Other accounts indicated that at least one family member, Anastasia, escaped to live in Western Europe or the United States. Over the years, the mystery surrounding the family generated several books and movies.

In the late 1970s, two Russian amateur historians began investigating Sokolf's accounts, and after a painstaking search, nine skeletons were dug from a shallow grave at a site 20 miles from Ekaterinburg in July 1991. All of the skeletons bore marks and bullet wounds indicating violent death. Forensic experts examined the remains, and using computer-assisted facial reconstructions and other evidence, concluded that the remains were those of the Czar, the Czarina, and three of their five children. The remains of two children were missing: the son, Alexei and one daughter. DNA analysis was used to confirm the findings.

The investigators used a threefold strategy in the DNA analysis. DNA was extracted from bone fragments and used for sex testing, for DNA typing to establish family relationships, and for mitochondrial DNA testing to trace maternal relationships. The sex testing used a six-base pair difference in a gene carried on both the X and Y chromosome. The results indicated that the skeletons were of four males and five females, confirming the results of physical analysis. The family relationship was tested by using probes for five short tandem repeat (STR) sequences. These results indicate that skeletons 3–7 were a family group, 4 and 7 were parents, and 3, 5, and 6 were children. If the remains are those of the Romanovs, the combination of the sex tests and the STR tests establish that the remains of one of the princesses and the Tsarevitch, Alexei, are missing.

To determine whether the remains belonged to the Romanovs, mitochondrial DNA (mtDNA) testing was conducted. Because mtDNA is maternally inherited, living relatives of the Czarina, including Prince Philip, the husband of Queen Elizabeth of England, were included in the tests. This analysis shows an exact match between the remains of the Czarina, the three children, and living relatives. mtDNA from the Czar matched that of two living maternal relatives, confirming that the remains are those of the Czar, his wife, and three of his children. The fate of Alexei and one daughter remains unknown, but historical accounts suggest that their bodies were burned or buried separately. This study overcame several technical challenges but clearly established the identity of the skeletons as those of the Romanovs. The results are a significant application of genetic technology in solving a mystery of world history. The intrigues, mysteries, and the science surrounding the search for the Romanovs are told by Robert Massie in his book *The Romanovs: The Final Chapter*.

THE HUMAN GENOME PROJECT IS AN INTERNATIONAL EFFORT

The development of recombinant DNA technology and its use in gene mapping made it possible to consider mapping all of the genes in the human genome (recall that a genome is the set of genes carried by an individual). To coordinate this effort, the U.S. Congress established the Human Genome Project in 1990. The goal of this project is to map the location of the 50,000 to 100,000 genes in the human genome and to analyze the nucleotide sequences of these genes. The project has grown into an international effort coordinated by the Human Genome Organization (HUGO). The project is now scheduled for completion in 2003 (ahead of schedule) at a cost of about $3 billion (about $1 per nucleotide). Although not reflected in the name, the Human Genome Project includes work on other organisms that are used in genetic studies (Table 13.3). These include bacteria, yeast, a roundworm, the fruit fly, and the mouse. In addition, genome sequencing projects are underway for dozens of other species (in addition to those in the Human Genome Project), many of which cause disease. To date, the genomes for over two dozen species have been completely sequenced, and four or five more are to be completed by the end of 2000.

The Genome Project Has Several Steps

For the human genome, the major steps in the project are shown in ▶ Figure 13.10. Although the work is shown as a series of separate steps, work has progressed more or less simultaneously on all phases of the project in laboratories around the world.
   Briefly, the objectives are the following:

1. Development of high-resolution genetic maps for each human chromosome. Genetic maps are based on recombination frequencies and give the order and distance between genes (or other markers) along a chromosome. Using the recombinant DNA methods described earlier, geneticists generated maps (using about 3000 genes, RFLPs, and other markers) with markers spaced evenly along each chromosome. High-resolution maps are used to map new genes by testing for linkage to the markers already assigned to loci along the length of the chromosomes. High-resolution maps for all chromosomes are now available, and this phase of the project was completed in 1994.
2. Physical mapping of each chromosome. Physical maps show the order and distances between genes and markers, expressed in base pairs of DNA. This stage of the project is aimed at producing physical maps of each chromosome with markers spaced every 100,000 pairs (100 Kb) of DNA. High-resolution

Table 13.3   Organisms Included in the Human Genome Project

Organism	Genome Size (BP)	Estimated No. of Genes
Escherichia coli (bacteria)	4.2×10^6	3,237
Saccharomyces cerevisiae (yeast)	1.2×10^7	8,600
Caenorhabditis elegans (roundworm)	1.0×10^8	13,000
Arabidopsis thaliana (plant)	1.0×10^8	25,000
Drosophila melanogaster (fruit fly)	1.2×10^8	20,000
Mus musculus (mouse)	3×10^9	80,000
Homo sapiens (human)	3×10^9	80,000

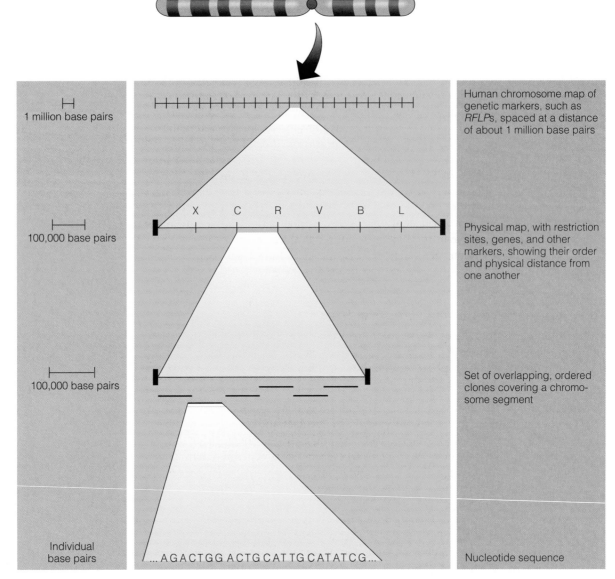

⊢─⊣ 1 million base pairs	Human chromosome map of genetic markers, such as *RFLP*s, spaced at a distance of about 1 million base pairs
⊢───⊣ 100,000 base pairs	Physical map, with restriction sites, genes, and other markers, showing their order and physical distance from one another
⊢───⊣ 100,000 base pairs	Set of overlapping, ordered clones covering a chromosome segment
Individual base pairs	Nucleotide sequence

In the physical map row, markers are labeled: X C R V B L

Nucleotide sequence: ...A G A C T G G A C T G C A T T G C A T A T C G...

▶ **FIGURE 13.10** The Human Genome Project is proceeding in four major steps. First, geneticists are producing genetic maps of each chromosome, and markers are spaced about 1 million base pairs apart. Next, high-resolution physical maps that have markers assigned to chromosomal sites about every 100,000 base pairs are prepared. Following that, overlapping clones that cover the regions between markers are being recovered. Finally, the nucleotide sequence of each clone is being determined, and the sequence that covers the entire genome is being assembled.

maps of almost all chromosomes are now available, and this phase of the project is now completed.

3. Cloning each chromosome. In the third stage, overlapping clones that cover the length of each chromosome will be developed. The problem is not in generating such clones but in identifying and arranging in order a collection of clones that overlap with each other and cover the entire chromosome. Dividing a chromosome into a number of segments and identifying clones that cover one segment at a time will accomplish this.

4. Sequencing the genome. The final stage will be determining the exact order of each nucleotide in each chromosome. This stage has already started, with the establishment of six centers in the U.S. to improve sequencing technology and to begin large-scale sequencing of the cloned DNA already available. About

6.5% of the genome is already sequenced, and the goal is to sequence most, if not all of the human genes by 2003.

The Human Genome Project: A Progress Report

By the end of 1998, The Human Genome Project successfully completed all goals set for the period from 1993–98. Sequencing of three of the five model organisms, *E. coli* (bacteria), *S. cerevisiae* (yeast), and the roundworm (*Caenorhabditis elegans*) are now completed. The genome of the fruit fly (*Drosophila melanogaster*) should be completed sometime in 2002, and the mouse genome (*Mus musculus*) is scheduled for completion in 2005.

High-resolution genetic maps for all human chromosomes are now complete, although some work is needed to refine the map in certain chromosome regions. The physical map is essentially complete. About 30,000 markers are required to produce a map that has markers at 1 million base pair (1 Mb) intervals. By 1998, some 52,000 markers were mapped, although some regions need filling in. Altogether, about 30,000 genes have been identified and placed on the physical map.

The ultimate goal of the project, to determine the nucleotide sequence of the 50,000 to 80,000 genes in the human genome, is now underway, and about 6.5% of the sequence is already available. The current goal is to finish one-third of the sequence by 2001 and prepare a working draft of the remainder of the sequence. The combined working draft and finished sequence are expected to cover slightly more than 90% of the genome with an accuracy of at least 99%. The sequence of the entire gene set is to be completed by 2003. Of this, about 60% is expected to come from the U.S. Human Genome Project, and the rest will be contributed by international and private efforts.

What have we learned so far from the genome project? Many significant findings have already resulted from the genome project. First, completing the genome sequence of model organisms allows geneticists to turn their attention from identifying and mapping genes to understanding their functions. Because many cellular processes are the same in bacteria, yeast, and humans, this change in focus will speed and enhance our understanding of basic cellular mechanisms. In turn, this understanding will translate into progress in diagnosing and treating of human diseases.

Important findings in the human genome are changing our concepts of genetic processes, including mutation and gene function. A new mechanism of mutation, trinucleotide expansion (discussed in Chapter 11), has been identified. This mechanism is important in genetic diseases that affect the nervous system, and understanding why this is so will play an important role in treating these disorders. Another unexpected finding is that mutations in a single gene can give rise to different genetic disorders, depending on how the gene is affected. For example, the gene *RET* (MIM/OMIM 164761) encodes a protein that is a cell-surface receptor engaged in transferring signals across the plasma membrane. Depending on the type and location of mutations within this gene, four distinct genetic disorders can result: two types of multiple endocrine neoplasia (MIM/OMIM 171400 and MIM/OMIM 162300), familial medullary thyroid carcinoma (MIM/OMIM 155240), and Hirschprung's disease (MIM/OMIM 142623). Another significant discovery is that some mutations that affect DNA repair can destabilize distant regions of the genome and make them susceptible to more mutations, often resulting in cancer (these mutations are discussed in Chapter 14). These and similar discoveries have already had an impact on diagnosing, treating, and genetic counseling in several groups of genetic disorders.

The Human Genome Project has raised a number of related legal, ethical, and moral issues, including the use of and access to information about an individual's genetic status or predisposition. Within the project, a program has been established to identify and discuss such issues. This program, called ELSI (Ethical, Legal, and Social Implications), draws on the expertise of scientists, lawyers, ethicists, philosophers, and others to consider how the information generated by the Human Genome Project affects individuals and society. It also examines the possible uses of information and techniques generated by the project and develops public policy options to ensure that the information is used for the benefit of individuals and society. We discuss the ethical implications of cloning and recombinant DNA research in a later section.

GENE TRANSFER TECHNOLOGY HAS MANY APPLICATIONS

Recombinant DNA technology was originally used to transfer foreign genes into bacterial cells as a way of cloning large amounts of a specific gene for further study. Later, methods were developed for producing eukaryotic proteins in the host bacterial cells. Gene transfer is not limited to using bacteria as host cells; it is also possible to transfer genes between higher organisms. In this section, we review some of the current applications of gene transfer technology.

The Rise of Biotechnology

One of the first commercial applications of recombinant DNA technology was the production of gene products (in the form of proteins) that could be used in the therapeutic treatment of human disease. Human insulin, derived from a cloned gene inserted into bacteria was first marketed in 1992 and is widely used to control diabetes. Previously, such proteins were collected from animals, pooled blood samples, or even human cadavers. In some cases, these proteins can be produced only in limited amounts, and their use can pose serious and in some cases, potentially fatal risks. For example, individuals afflicted with hemophilia, an X-linked recessive disorder, cannot manufacture a clotting factor and suffer episodes of uncontrollable bleeding. In the past, the missing clotting factor was extracted from pooled blood donations. However, before blood testing for HIV became routine, it is estimated that up to 90% of hemophiliacs treated with clotting factor prepared in this way were infected with the human immunodeficiency virus (HIV) associated with the development of AIDS (acquired immunodeficiency syndrome). The use of recombinant DNA technology to manufacture specific proteins such as insulin, clotting factors, and alpha-1-antitrypsin ensures a controlled supply of a product free from contamination by disease-causing agents. Now, clotting factors and dozens of other proteins are manufactured by using recombinant DNA techniques and are free of contaminating agents. Some of these are listed in Table 13.4.

Proteins made by recombinant DNA techniques are not limited to treating human disease and have other applications. When your cat is inoculated against feline leukemia virus, it receives a vaccine produced by recombinant DNA technology; and farmers routinely use recombinant-DNA-derived vaccines to prevent diseases in hogs.

Bio-Pharming: Making Human Proteins in Animals

At first, recombinant proteins were produced in bacterial hosts. Although bacterial hosts can be genetically modified to produce human (and other eukaryotic) proteins, the bacteria cannot process and chemically modify these proteins to convert them into biologically active forms. In addition, some eukaryotic proteins do not fold into the proper three-dimensional form in bacterial cells and are inactive. To overcome these problems, a new generation of eukaryotic hosts is now in use. This new generation includes farm animals as sources of human proteins of pharmaceutical importance.

In humans, an enzyme deficiency is associated with heritable forms of emphysema (MIM/OMIM 107400), a progressive and fatal respiratory disorder. The defective enzyme is alpha-1-antitrypsin, which can be produced for treating emphysema by gene transfer technology. The human alpha-1-antitrypsin gene has been cloned into a vector at a site adjacent to a DNA sequence that regulates expression of milk proteins. Vectors that carry this gene were microinjected into fertilized sheep

Table 13.4 Some Products Made by Recombinant DNA Technology

Product	Use
Atrial natriuvetic factor	Treatment for hypertension, heart failure
Bovine growth hormone	Improve milk production in dairy cows
Cellulase	Break down cellulose in animal feed
Colony stimulating factor	Treatment for leukemia
Epidermal growth factor	Treatment of burns, improve survival of skin grafts
Erythropoietin	Treatment for anemia
Hepatitis B vaccine	Prevent infection by hepatitis B virus
Human insulin	Treatment for diabetes
Human growth hormone	Treatment for some forms of dwarfism, other growth defects
Interferons (alpha, gamma)	Treatment for cancer, viral infections
Interleukin-2	Treatment for cancer
Superoxide dismutase	Improve survival of tissue transplants
Tissue plasminogen activator	Treatment of heart attacks

eggs, and the eggs were implanted into female sheep (▶ Figure 13.11). The resulting transgenic sheep developed normally, and the females produce milk that contains up to one-third of a pound of alpha-1-antitrypsin per gallon of milk. This recombinant protein is now in clinical testing, and soon a small herd of sheep will supply the world's need for this protein.

The use of such proteins can be controversial, however. When introduced, there was considerable disagreement about the use of recombinant-DNA-derived bovine

▶ **FIGURE 13.11** Transgenic sheep that carry the human gene for alpha-1-antitrypsin secrete the gene product into their milk. The protein can be extracted, purified, and used for treating emphysema.

growth hormone to boost milk production in dairy cows. The use of this hormone was approved by the U.S. Food and Drug Administration in February 1994 and is now used on about 15% of the dairy cows in the United States. On the one hand, use of the hormone, known as BST, benefits farmers by increasing milk production by about 10%, and results in lowered fixed costs. On the other hand, opponents charge that the technology is unnecessary, that cows given this hormone are at greater risk of disease, and that the risks of long-term use of this protein are unknown. Nevertheless, it is estimated that by the year 2000, 70% of the nation's milk supply will be produced with BST.

Genetic Disorders Can Be Corrected by Gene Therapy

Recombinant DNA technology has made it possible to treat human genetic disorders by transferring normal copies of genes into cells that carry defective copies. A number of methods for transferring cloned genes into human cells are in use, including viral vectors, chemical methods that aid transfer of DNA across the cell membrane, and physical methods such as microinjection or fusion of cells with vesicles that carry cloned DNA sequences.

Viral vectors, especially viruses known as retroviruses, are most often used for gene therapy. The viruses have been genetically modified, and some viral genes are removed, allowing a human gene to be inserted (▶ Figure 13.12). Human gene therapy began in 1990, when a human gene for the enzyme adenosine deaminase (ADA) was inserted into a retrovirus and then transferred into the white blood cells of a young girl, who

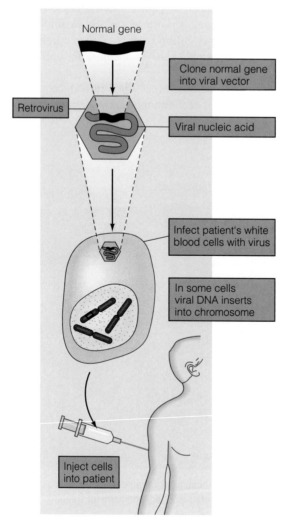

Normal gene

Clone normal gene into viral vector

Retrovirus

Viral nucleic acid

Infect patient's white blood cells with virus

In some cells viral DNA inserts into chromosome

Inject cells into patient

▶ **FIGURE 13.12** The present method of gene therapy uses a virus as a vector to insert a normal copy of a gene into the white blood cells of a patient who has a genetic disorder. The normal gene becomes active, and the cells are reinserted into the affected individual, curing the genetic disorder. Because white blood cells die after a few months, the procedure has to be repeated regularly. In the future, it is hoped that transferring a normal gene into the mitotically active cells of the bone marrow will make gene therapy a one-time procedure.

was suffering from severe combined immunodeficiency (SCID) (MIM/OMIM 102700). Affected individuals have no functional immune system and are prone to infections, many of which can be fatal. The normal *ADA* gene, inserted into her white blood cells, encodes an enzyme that allows cells of the immune system to mature properly. As a result, she now has a functional immune system and is leading a normal life.

Gene therapy is now being used in more than 100 clinical trials for a wide range of purposes, including the treatment of genetic disorders, cancer and HIV infection. Although some successes have been recorded in treating cancer and HIV infection, gene therapy has been less successful in treating genetic disorders. One problem has been inconsistent results. After three years, the first child treated with gene therapy for SCID had a new *ADA* gene in more than 50% of her circulating white blood cells and had an immune response. A second child, who began treatment a short time later, has the normal gene in only 0.1% to 1% of her cells, and has not developed an immune system. In 11 gene therapy trials for cystic fibrosis, only about 5% of the target cells express the normal CF gene, and in many recipients, no expression can be detected. Similar results have plagued other trials. Most of these problems have been traced to inefficient vectors. Retroviral vectors can only insert genes into cells that are active in growth and division. In addition, there is no way to direct where the transferred gene inserts into the host cell DNA, and as a result, many transferred genes are not expressed. A new generation of viral vectors is now under development, and other methods of gene transfer are being investigated.

New Plants and Animals Can Be Created by Gene Transfer

Gene transfer technology has been used for some time to improve crop plants and farm animals. Corn is a major cereal crop in the United States and has an annual market value of about $22 billion. Corn is used as an animal feed, a source of sweeteners for foods and beverages, and to produce ethanol, which may grow in importance as an alternative to gasoline.

Cloned genes for a desired trait are blasted into cultured corn cells with a "gene gun" that forces the genes into the cell (▶ Figure 13.13). The cells are cultured to form

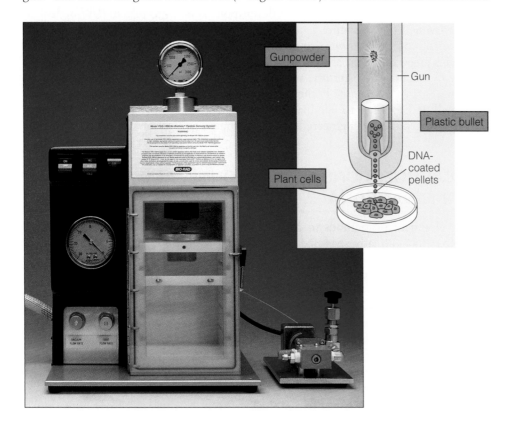

▶ **FIGURE 13.13** (a) A gene gun is used to transfer genes into plant cells. (b) A plastic bullet in the gene gun is used to drive DNA-coated pellets into cells.

▶ **FIGURE 13.14** This transgenic pig carries human growth hormone genes.

a tissue mass that will grow into a corn plant. Transgenic (carrying a foreign gene) strains of corn that carry a gene for herbicide resistance have completed field trials and are now on the market. Other strains of corn now carry genes for resistance to insect pests. Field testing of genetically modified strains of other crop plants, including potatoes, tomatoes, and cotton, are underway, and these will reach the market in a few years. Manipulating organisms by genetic engineering achieves the same goals as selective breeding but is more efficient and faster than traditional methods.

Genetic engineering of crop plants is being accompanied by parallel developments in genetic alterations of domestic animals. Gene transfer technology has been used to transfer human and cow growth hormone genes into pigs to develop leaner, faster growing hogs (▶ Figure 13.14). The genes were transferred by injection into newly fertilized eggs that were implanted into a foster mother. Although the transgenic pigs grow faster on a high-protein diet, they have a number of problems including ulcers, arthritis, sterility, and premature death resulting from excessive production of growth hormone in their tissues. Further work on regulating transferred genes will be necessary to eliminate these problems and pave the way for the appearance of genetically engineered pork in the meat case at the supermarket.

ETHICAL QUESTIONS ABOUT CLONING AND RECOMBINANT DNA

The development of methods to clone plants and animals and recombinant DNA technology have already spawned the biotechnology industry, a multibillion dollar segment of our economy, and are now beginning to revolutionize biomedical research and the diagnosis and treatment of human diseases. Unlike many other technologies, the use of these techniques raises serious social and ethical issues that need to be identified, discussed, and resolved. A program within the Human Genome Project has been established to deal with these issues. This program, called the Ethical, Legal, and Social Implications (ELSI) project brings together individuals from biology, social sciences, history, law, ethics, and philosophy to explore issues and lay out public policy alternatives.

There are many issues to be resolved, and the technology and information are developing much faster than policy and legislation. Because we can do some things does not mean we should do those things without a consensus from society. For example, should we permit human cloning as part of programs in assisted reproduction? Should we allow gene transfer to germ cells, altering the genetic makeup of future generation without their consent? Should gene therapy be used as a form of "improving" individuals rather than curing a genetic disease? Should very short children receive recombinant-DNA-produced growth hormone to make them average in height? If it is acceptable to treat abnormally short children, is it ethical to treat children of normal height to enhance their chances of becoming professional basketball or volleyball players?

The current guidelines of the U.S. Food and Drug Administration do not require identifying labels on food produced by recombinant DNA technology. Should such food be labeled, and if so, why? It can be argued that food products have been genetically manipulated for thousands of years and that gene transfer is simply an extension of past practices. It can also be claimed that consumers have a need and a right to know that gene transfer has altered food products.

The legal, moral, and ethical implications of cloning and recombinant DNA technology need to be considered carefully. As outlined earlier, some of these issues are being addressed by the Human Genome Project; but in other areas, discussion, education, and policy formation lag far behind the technology. As citizens, it is our responsibility to become informed about these issues and to participate in formulating policy about the use of this technology.

Case Studies

CASE 1

Todd and Shelly Z. were referred for genetic counseling because of advanced maternal age (40 years) in their current pregnancy. While obtaining the family history, the counselor learned that during their first pregnancy, in 1990, they had elected to have an amniocentesis for prenatal diagnosis of cytogenetic abnormalities because Shelly was 36 years old at the time. During that pregnancy, Shelly and Todd reported that they were also concerned that Todd's family history of cystic fibrosis (CF) increased their risk for having an affected child. Todd's only sister had CF, and she had severe respiratory complications.

The genetic counseling and testing was performed at an outside institution, and the couple had not brought copies of the report with them. They did state that they had completed studies to determine their CF carrier status and that Todd was found to be a CF carrier, but Shelly's results were negative. The couple was no longer concerned about their risk of having a child with CF based on these results. To support their belief, they had a healthy 5-year-old son who had a negative sweat test at the age of 4 months. The counselor explained the need to review the records and scheduled a follow-up appointment.

Review of the test report from 1990 indicated that the only CF mutation tested at that time was the delta F508 mutation. The reports confirmed that Shelly is not a carrier of this mutation and that Todd is. Shelly's report from 1990 indicated that her CF risk status had been reduced from 1 in 25 to about 1 in 300. More information has been learned about the different mutations in the CF gene since the last time they received genetic counseling.

The counselor conveyed the information about recent advances in CF testing to the couple, and Shelly decided to have her blood drawn for CF mutational analysis with an expanded panel of mutations. Her results showed that she is a carrier for the W1282X CF mutation. The family was given a 25% risk for CF for each of their pregnancies based on their combined molecular test results. They proceeded with the amniocentesis because of the risks associated with advanced maternal age and requested fetal DNA analysis for CF mutations. The fetal results were positive for both parental mutations, indicating that the fetus had a greater than 99% chance of being affected with CF.

CASE 2

Can DNA fingerprinting uniquely identify the source of a sample? Because any two human genomes differ at about 3 million sites, no two persons (except identical twins) have the same DNA sequence. Unique identification with DNA typing is therefore possible if enough sites of variation are examined. However, the DNA typing systems used today examine only a few sites of variation and have only limited resolution for measuring the variability at each site. There is a chance that two persons might have DNA patterns (i.e., genetic types) that match at the small number of sites examined. Nonetheless, even with today's technology, which uses 3–5 loci, a match between two DNA patterns can be considered strong evidence that the two samples came from the same source. How is DNA fingerprinting currently being used?

1. Paternity and Maternity Testing: Because a person inherits his or her VNTRs (variable number of tandem repeats) from his or her parents, VNTR patterns can be used to establish paternity and maternity. The patterns are so specific that a parental VNTR pattern can be reconstructed even if only the children's VNTR patterns are known (the more children produced, the more reliable the reconstruction). Parent-child VNTR pattern analysis has been used to solve standard father-identification cases and more complicated cases of confirming legal nationality and, in instances of adoption, biological parenthood.

2. Criminal Identification and Forensics: DNA isolated from blood, hair, skin cells, or other genetic evidence left at the scene of a crime can be compared, through VNTR patterns, with the DNA of a criminal suspect to determine guilt or innocence. VNTR patterns are also useful in establishing the identity of a homicide victim, either from DNA found as evidence or from the body itself. The O.J. Simpson trial used DNA evidence in its investigation.

3. Personal Identification: The notion of using DNA fingerprints as a sort of genetic bar code to identify individuals has been discussed, but this is not likely to happen anytime in the foreseeable future. The technology required to isolate, keep on file, and analyze millions of very specific VNTR patterns is both expensive and impractical. Social security numbers, picture IDs, and other more mundane methods are much more likely to remain the ways to establish personal identification. However, the FBI and other police agencies, and the military, are enthusiastic proponents of DNA databanks. The British police keep DNA files on suspected Irish terrorists. The practice has civil-liberties implications because police can go on "fishing expeditions" with DNA readouts.

Summary

1. Recombinant DNA techniques can be used in a variety of applications, including gene mapping, forensic applications such as DNA fingerprinting, prenatal diagnosis, gene therapy, and the production of human gene products for therapeutic uses.

2. Gene mapping has been greatly enhanced by using recombinant DNA techniques. Genes can be mapped by several techniques, including positional cloning and direct mapping to chromosomes.

3. An ever-growing list of genetic disorders can be diagnosed by RFLP analysis. Mutant genes and heterozygous carrier status can be detected in adults, children, fetuses, embryos, and even gametes.

4. The Human Genome Project uses recombinant DNA technology to map all of the estimated 50,000 to 100,000 genes carried in human cells. The goal of the project is to determine the sequence of the 3 billion base pairs of human DNA.

5. The first clinical trials involving the insertion of genes into humans have already started, and although some technical barriers remain, gene therapy will probably become a more common form of medical treatment within the next decade.

6. Genetic engineering is also being used to transfer genes into crop plants, conferring resistance to herbicides, insect pests, and plant diseases. Gene transfers to alter the protein, carbohydrate, and oil content of cereal crops will dramatically alter farming practices in the next century. Experiments to produce genetically altered farm animals used in meat and milk production are underway, and the products from these experiments should reach consumers within the next decade.

Questions and Problems

1. VNTRs are:
   a. used for DNA fingerprinting
   b. repeated sequences present in the human genome
   c. highly variable in copy number
   d. all of the above
   e. none of the above

2. Gene therapy involves:
   a. the introduction of recombinant proteins into individuals
   b. cloning human genes into plants
   c. the introduction of a normal gene into an individual carrying a mutant copy
   d. DNA fingerprinting
   e. none of the above

3. Which of the following is not a goal of the Human Genome Project?
   a. construction of high resolution genetic maps
   b. sequencing the entire genome
   c. curing all genetic diseases
   d. generating a physical map of each chromosome
   e. obtaining a set of overlapping clones covering the genome

4. RFLP sites are useful as genetic markers for linkage and mapping studies as well as disease diagnosis. Keeping in mind that RFLPs behave as genes, what factors are important in selecting RFLPs for use in such studies?

5. In the examples given in this chapter (Figure 13.2), the RFLP is always inherited with (linked to) the genetic disorder. In practice, however, it is sometimes observed that in some progeny the RFLP is not linked to the genetic disorder, reducing its effectiveness in detecting carrier heterozygotes. How do you explain this observation, and what circumstances are likely to affect the frequency with which this nonlinkage occurs?

6. In selecting target cells to receive a transferred gene in gene therapy, what factors must be taken into account?

7. At what stages in life are prenatal testing, preimplantation testing and presymptomatic testing done?

8. What do the letters RFLP stand for? Explain this term word by word.

9. A paternity test is conducted using PCR to analyze an RFLP that consistently produces a unique DNA fragment pattern from a single chromosome. Examining the results of the Southern blot below, which male(s) can be excluded as the father of the child? Which male(s) could be the father of the child?

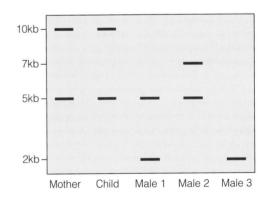

10. What are the steps involved in the procedures currently used for gene therapy?

11. DNA fingerprinting has been used in criminal identification, paternity cases, establishing blood lines of pure bred dogs, identifying endangered species, and studying extinct animals. Can you think of other uses for DNA fingerprinting?

12. What are some of the potential dangers of recombinant DNA technology? How should these potential problems be investigated to determine if they are in fact dangers?

13. A crime is committed and the only piece of evidence the police are able to gather is a small bloodstain. The forensic scientist at the crime lab is able to:
    a. extract DNA from the blood.
    b. cut the DNA is cut with a restriction enzyme
    c. separate the fragments by electrophoresis.
    d. transfer the DNA from a gel to a membrane and probe with radioactive DNA

    Probe 1 is used to visualize the pattern of bands. The forensic scientist compares the band pattern in the evidence (E) with the patterns from the suspects (S1, S2). The first probe is removed, and the membrane is hybridized using another probe (Probe 2), and the band patterns are compared. This process is repeated for Probe 3 and Probe 4.

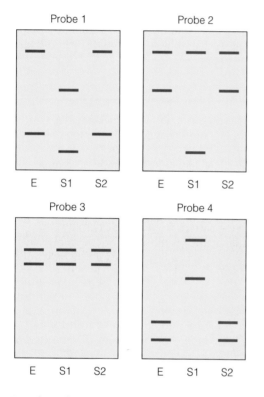

a. Based on the results of this testing, can either of the suspects be excluded as the one who committed the crime?
b. If so, which one? Why?
c. Is the pattern from the evidence consistent with the band pattern of one of the suspects? Which one?

14. Fluorescent in situ hybridization (FISH) is a technique that uses a fluorescently-labeled probe to bind to metaphase chromosomes. The results are then viewed using a microscope. What kind of probe would you suggest to detect Down syndrome?

15. You are serving on a jury in a murder case in a large city. The prosecutor has just stated that DNA fingerprinting shows that the suspect must have committed the crime. He says that two VNTR probes were used instead of four for cost reasons, but that the results are just as accurate. The first probe detected a locus that has a population frequency of 1 in 100 (1%), and the second probe has a population frequency of 1 in 500 (0.2%).
    a. what is the combined frequency of these alleles in the population?
    b. does this point to the suspect as the perpetrator of the crime?
    c. is there any other evidence you want from the DNA lab?
    d. do you think that most people on juries understand basic probabilities or do you think that DNA evidence can be used to trick jurors into reaching false conclusions?

16. The prospect of using gene therapy to alleviate genetic conditions is still a vision of the future. Gene therapy for adenosine deaminase deficiency has proven to be quite promising but many obstacles remain to be overcome. Currently, the correction of human genetic defects is done using retroviruses as vectors. For this purpose, viral genes are removed from the retroviral genome, creating a vector capable of transferring human structural genes into sites on human chromosomes within target tissue cells. Do you see any potential problems with inserting pieces of a retroviral genome into humans? If so, are there ways to combat or prevent these problems?

17. Genetic testing using DNA markers for specifically identified mutations allows the diagnosis of many adult onset disorders years or decades before symptoms appear. Presymptomatic testing to determine whether individuals are at risk for certain conditions such as familial hypercholesterolemia, Huntington disease or cancer, raises many ethical, legal and social issues.
    a. What are the possible psychological and medical effects of a positive test for a condition that causes a premature death?
    b. Is it appropriate to offer such testing before we have effective measures to treat or cure the condition?
    c. Does this testing raise the possibility of discrimination against those who test positive for a fatal disorder? How? Should the law require that such individuals be given life and health insurance?

Internet Activities

The following activities use the resources of the World Wide Web to enhance the topics covered in this chapter. To investigate the topics covered below, log on to the book's homepage at:

http://www.brookscole.com/biology

1. The Union of Concerned Scientists sponsors a web site which provides information and a forum to discuss biotechnology and the marketing of genetically engineered products. At the site, read about the current uses of biotechnology in agriculture and formulate an opinion regarding the development and use of these products. You may also choose to become involved in the online discussion at the Activists Corner of the site.

2. RFLP analysis can be used to "fingerprint" human DNA samples. This evidence can then provide proof for either paternity challenges or questions of criminal guilt. Learn more about this technique and practice interpreting RFLP data in a web DNA forensics activity.

3. The safety, efficacy, and ethic concerns surrounding gene therapy have been debated within and outside of the biomedical community since the first gene therapy protocol for ADA was carried out. Unfortunately, few protocols have lived up to the optimism and early promises attached to gene therapy. Review the gene therapy debate sponsored by Nature Genetics on the HMS Beagle homepage. (You are required to register to use this internet resource)

For Further Reading

Anderson, W. F. (1992). Human gene therapy. *Science 256*: 808–813.

Caskey, C. (1993). Presymptomatic diagnosis: A first step toward genetic health care. *Science 262*: 48–49.

Collins, F., & Galas, D. (1993). A new five-year plan for the U.S. Human Genome Project. *Science 262*: 43–46.

Crystal, R. G. (1995). Transfer of genes to humans: Early lessons and obstacles to success. *Science 270*: 404–410.

deWachter, M. (1993). Ethical aspects of human germ-line therapy. *Bioethics 7*: 166–177.

Friedman, T. (1989). Progress toward human gene therapy. *Science 244*: 1275–1281.

Gilbert, W., & Villa-Komaroff, L. (1980). Useful proteins from recombinant bacteria. *Sci. Am. 242* (April): 74–97.

Guyer, M., & Collins, F. S. (1995). How is the Human Genome Project doing, and what have we learned so far? *Proc. Nat. Acad. Sci. USA 92*: 10841–10848.

Harris, J. (1992). *Wonderwoman and Superman: The Ethics of Human Biotechnology.* New York: Oxford University Press.

Hubbard, R., & Wald, E. (1993). *Exploding the Gene Myth.* Boston: Beacon Press.

Kantoff, P. W., Freeman, S. M., & Anderson, W. F. (1988). Prospects for gene therapy for immuno-deficiency diseases. *Ann. Rev. Immunol. 6*: 581–594.

Kevles, D., & Hood, L. (1992). *The Code of Codes: Scientific and Social Issues in the Human Genome Project.* Cambridge, MA: Harvard University Press.

Kohn, D. B., Anderson, W. F., & Blaese, R. M. (1989). Gene therapy for genetic diseases. *Cancer Invest. 7*: 179–192.

Lapham, E. V., Kozma, C., & Weiss, J. O. (1996). Genetic discrimination: Perspectives of consumers. *Science 274*: 621–624.

Marshall, E. (1995). Gene therapy's growing pains. *Science 269*: 1050–1055.

Marshall, E. (1996). The genome program's conscience. *Science 274*: 488–490.

Marx, J. L. (1987). Assessing the risks of microbial release. *Science 237*: 1413–1417.

Morsy, M., Mitani, K., & Clemens, P. (1993). Progress toward human gene therapy. *JAMA 270*: 2338–2340.

Neufeld, P. J., & Colman, N. (1990). When science takes the witness stand. *Sci. Am. 262* (May): 46–53.

Nichols, E. K. (1988). *Human Gene Therapy.* Cambridge, MA: Harvard University Press.

Olson, M. V. (1993). The human genome project. *Proc. Nat. Acad. Sci. USA 90*: 4338–4344.

Porteus, D., & Alton, E. (1993). Cystic fibrosis: Prospects for therapy. *Bioessays 15*: 485–486.

Pursel, V. G., Pinkert, C. A., Miller, K. F., Bolt, D. J., Campbell, R. G., Palmiter, R. D., Brinster, R. L., & Hammer, R. E. (1989). Genetic engineering of livestock. *Science 244*: 1281–1288.

Roberts, L. (1989). Ethical questions haunt new genetic technologies. *Science 243*: 1134–1135.

Tacket, C. O., Mason, H. S., Losonsky, G., Clements, J. D., Levine, M. M. and Arntzen, C. J. (1998). Immunogenicity in humans of a recombinant bacterial antigen delivered in a transgenic potato. *Nature Medicine 4*: 607–609.

Travis, J. (1998). Another human genome project. *Science News 153*: 334–335.

Watson, F. (1993). Human gene therapy—progress on all fronts. *Trends Biotechnol. 11*: 114–117.

Weatherall, D. J. (1989). Gene therapy: Getting there slowly. *Br. Med. J. 298*: 691–693.

White, R., & Lalouel, J. M. (1988). Chromosome mapping with DNA markers. *Sci. Am. 258* (February): 40–47.

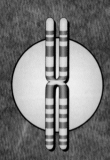

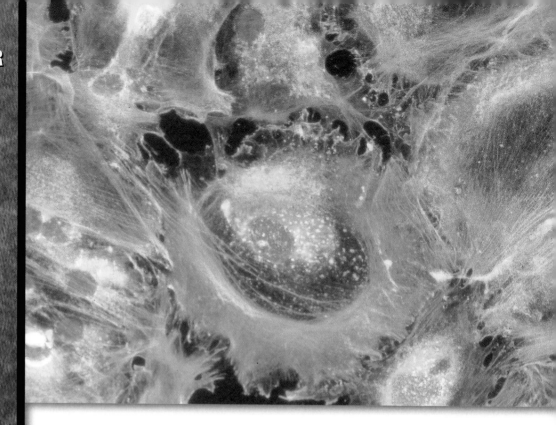

Genes and Cancer

Chapter Outline

*A*s you sit in a classroom, a crowded stadium, or a concert hall, look around. According to the American Cancer Society, about one in three of the people you see will develop cancer at some point in their life, and about one in four will die from cancer. Each year about 500,000 people die of cancer, a rate of about 1 death per minute, and almost 1.5 million new cases of cancer are diagnosed annually in the United States (Table 14.1). Currently more than 10 million individuals are receiving medical treatment for cancer in U.S. hospitals and medical centers.

Cancer is a complex group of diseases that affect many different cells and tissues in the body. It is characterized by uncontrolled growth and division of cells and by the ability of these cells to spread, or metastasize, to other sites within the body. Unchecked, the growth and metastasis result in death, making cancer a devastating and feared disease. Improvements in medical care have reduced deaths from infectious disease and have led to increases in life span, but these benefits have also helped make cancer a major cause of illness and death in our society. Because the risk of many cancers is age-related and because more Americans are living longer, they are at greater risk of developing cancer.

Although our society seems preoccupied with cancer, the idea that it is a disease of modern civilization is not accurate. Ancient Egyptian and Indian manuscripts reveal that cancer was recognized as a life-threatening disease more than 3500 years ago. The Greek physician Galen (A.D. 131–201) provided clear descriptions of cancers and referred to them as *karkinos* or *karkinomas*, terms that translate into the Latin word *cancer*.

The link between cancer and mutation was forged early in this century by Theodore Boveri, who proposed that normal cells mutate into malignant cells because of changes in chromosome constitution. Four lines of evidence support the idea that cancer has a genetic origin:

1. More than 50 forms of cancer are known to be inherited to one degree or another.

Table 14.1   Estimated New Cases of Cancer in the United States, 1996

Site	Number of New Cases
Skin	>700,000
Lung	171,500
Colon-rectum	132,500
Breast (female)	180,300
Prostate	184,500
Urinary System	86,300
Uterus	49,800
Pancreas	29,000
Ovary	25,400

2. It has been shown that most carcinogens are also mutagens.
3. Some viruses carry mutant genes, known as oncogenes, that promote and maintain the growth of a tumor.
4. As Boveri proposed, specific chromosomal changes are found in particular forms of cancer, especially leukemia.

Mutation is a common feature of all cancers. In most cases, these mutations take place in somatic tissue, and the mutant alleles are not passed on to offspring. In other cases, the mutations take place in germ cells and are passed on to succeeding generations. Cancer, then, is a genetic disorder that acts at the cellular level.

Because mutation is the ultimate cause of cancer, and because there is always a background rate of spontaneous mutation, there will always be a baseline rate of cancer. The environment (ultraviolet light, chemicals, and viruses) and behavior (diet, smoking) can also play a significant role in cancer risks by increasing the rate of mutation.

In this chapter we examine the relationship between genes and cancer and describe the role of tumor suppressor genes and oncogenes in causing and supporting malignant transformation in cells. We discuss the relationship between leukemia and chromosomal aberrations, and finally, we analyze the interaction between cancer and the environment to explain how a multitude of factors may initiate the multistep process required for developing malignant growth.

CANCERS ARE MALIGNANT TUMORS

Before we consider cancer, let's make a distinction between tumors and cancer. Tumors are abnormal growths of tissue. Benign tumors are self-contained, noncancerous growths that do not spread to other tissues and are not invasive. Benign tumors, such as cysts, usually cause problems by increasing in size until they interfere with the function of neighboring organs.

Cancers are malignant tumors that have several characteristics. They are clonal in origin; cancers arise from a single cell (usually in somatic tissue). Cancers develop through a series of genetic alterations that result in more aggressive growth with each mutation. Third, cancers are invasive and metastatic. Cells can detach from the primary tumor and move to other sites in the body where new malignant tumors are formed. The property of metastasis is conferred as a result of mutational changes in the cell. In the following sections we consider the genetic changes that take place within single cells that lead to the formation of malignant tumors.

MUTATIONS IN SPECIFIC GENES CAN PREDISPOSE TO CANCER

Families that have high rates of specific cancers have been known for hundreds of years, but in most cases no clear-cut pattern of inheritance can be identified. How is it, then, that some families have a rate of cancer that is much higher than average? Many explanations have been offered, including multiple gene inheritance, environmental agents, or chance alone (Table 14.2).

Recent advances in cancer research have now linked mutations in specific genes to cellular events that form malignant tumors. Cancer is now viewed as a disease that results from a small number of independent mutations, that can take place over a long period of time. Although the exact number of mutations varies for different cancers, two mutations may be the minimum number needed to cause cancer (▶ Figure 14.1).

Table 14.2 Factors in Cancer	
Genetic	**Nongenetic**
Single gene mutations	Environmental carcinogens
Chromosome aberrations	Abnormal hormone levels
	Diet
	Chance

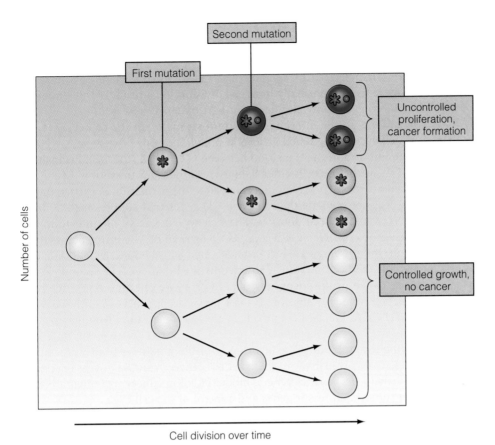

▶ **FIGURE 14.1** Over time, cells acquire mutations. Two independently acquired mutations in the same cell may be enough to cause uncontrolled cell growth and cancer.

Table 14.3 Heritable Predispositions to Cancer		
Disorder	**Chromosome**	**MIM/OMIM Number**
Early-onset familial breast cancer	17q	113705
Familial adenomatous polyposis	5q	175100
Hereditary nonpolyposis colorectal cancer	2p	120435
Li-Fraumeni syndrome	17p	151623
Multiple endocrine neoplasia type 1	11q	131100
Multiple endocrine neoplasia type 2	10	171400
Neurofibromatous type 1	17q	162200
Neurofibromatous type 2	22q	101000
Retinoblastoma	13	180200
Von Hippel-Lindau disease	3p	193300
Wilms tumor	11p	194070

In the majority of cases, these mutations accumulate randomly over a period of years, which is why age is a leading risk factor for many cancers.

Some individuals inherit a predisposition to cancer. In these cases, the first mutation is present in the germ cells and all the other cells of the body (Table 14.3). Additional mutations accumulate in somatic cells spontaneously or by exposure to environmental agents that cause genetic damage. The result can be cancer. In some cases, individuals who carry an allele that predisposes them to cancer have a 100,000-fold increase in the risk of developing cancer. But not all individuals who inherit the first mutation develop cancer. If no other mutations take place, then no malignant tumor develop.

TUMOR SUPPRESSOR GENES, ONCOGENES, AND THE CELL CYCLE

Cancer cells are characterized by uncontrolled cell division. These cells have by-passed control systems that normally limit cell division. As a result of this discovery, studies of the cell cycle are now an important part of cancer research.

Recall that events in the interphase and mitosis make up the cell cycle (Figure 14.2). The cell cycle is regulated at several points. For our discussion, two regulatory points are important: the transition between G2 and M (the G2/M transition) and a point in G1 just before cells enter S, known as the G1/S transition. Mutations in genes that regulate these control points can lead to the formation of cancer. In general, two types of genes regulate the cell cycle: (1) genes that suppress cell division; and (2) genes that stimulate cell division. The first type is known as **tumor suppressor genes**. These genes can act at either the G1/S or the G2/M control points to inhibit cell division. If these genes are lost or inactivated by mutation, cell division cannot be switched off, and cells divide in an uncontrolled manner. The second class of regulatory genes, called **proto-oncogenes,** acts to promote cell division. When these genes are active, cells undergo division. If these genes become permanently activated or overproduce their products, uncontrolled cell division results. Mutant forms of proto-oncogenes are called **oncogenes.**

The following examples describe how mutations in a number of different tumor suppressor genes are involved in the development of cancer. After we discuss oncogenes and their involvement in cancer, a genetic model of colon cancer will integrate what is known about tumor suppressor genes and control of the cell cycle.

Retinoblastoma Is Caused by Mutation of a Tumor Suppressor Gene

Retinoblastoma (*RB1*) (MIM/OMIM 180200) is a cancer of the eye that affects the light-sensitive retina. It occurs in 1 in 14,000 to 1 in 20,000 births. Although it may be present in infancy, it is most often diagnosed between the ages of 1 and 3 years. Two forms of retinoblastoma can be distinguished. One form, hereditary retinoblas-

■ **Tumor suppressor gene**   A gene that normally functions to suppress cell division.

■ **Proto-oncogene**   A gene that normally functions to control cell division and may become a cancer gene (oncogene) by mutation.

■ **Oncogene**   A gene that induces or continues uncontrolled cell proliferation.

■ **Retinoblastoma**   A malignant tumor of the eye that arises in retinoblasts (embryonic retinal cells that disappear at about two years of age). Mature retinal cells do not transform into tumors, so this is a tumor that usually occurs in children. It is associated with a deletion on the long arm of chromosome 13.

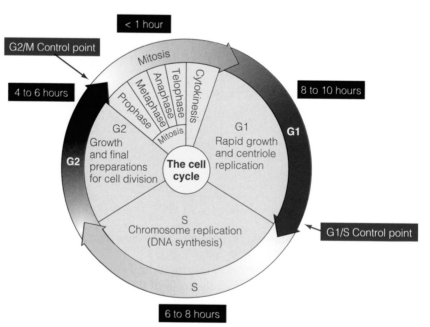

▶ **FIGURE 14.2**   The cell cycle consists of two parts: the interphase and mitosis. The interphase is divided into three stages G1, S, and G2, which make up the major part of the cycle. Mitosis involves partitioning cytoplasm and replicated chromosomes to daughter cells.

toma (that accounts for 40% of all cases), is an autosomal dominant trait in which a susceptibility to retinoblastoma is inherited. In families that have this trait, offspring have a 50% chance of receiving the mutant *RB* gene, and 90% of these individuals will develop retinoblastoma, usually in both eyes. In addition, those who carry the mutation are at high risk of developing other cancers, especially osteosarcoma and fibrosarcoma. The second form (60% of all cases) is sporadic retinoblastoma. Affected individuals have not inherited any mutant alleles of the retinoblastoma gene. Instead, sometime after birth, they acquire the mutations that result in tumor formation. In these cases, tumors usually develop in only one eye and patients are not at high risk for other cancers (▶ Figure 14.3).

Alfred Knudson and his colleagues proposed a two-step model for retinoblastoma that explains both forms of the disease. According to this model, retinoblastoma develops when two mutant copies of the *RB* gene are present in a single retinal cell. This model predicts that retinoblastoma can arise in two ways. If a child inherits a mutated *RB* allele, all cells of the body, including those in the retina, carry this mutation. If the normal copy of the *RB* gene in a retinal cell becomes mutated, the child will develop retinoblastoma (▶ Figure 14.4). Because the retinal cells already carry one mutant gene, this form of the disease is more likely to involve both eyes (bilateral cases) and occur at an earlier age than the sporadic form.

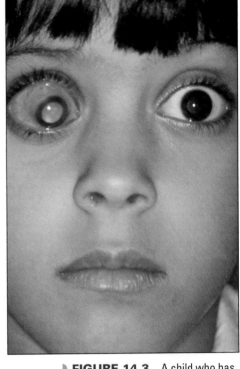

▶ **FIGURE 14.3** A child who has a retinoblastoma in one eye.

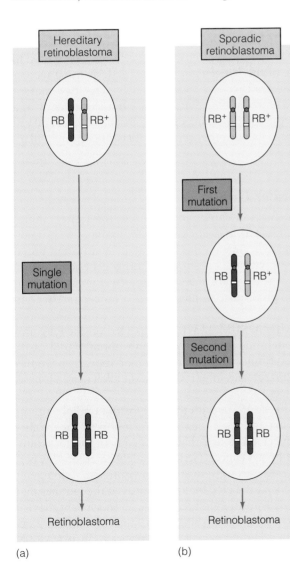

(a) (b)

▶ **FIGURE 14.4** A model of retinoblastoma. In hereditary cases, one mutation is inherited. (a) A single mutation in the normal allele of the *RB* gene will cause retinoblastoma. (b) Sporadic retinoblastoma requires two independent mutations in the *RB* gene in a single cell.

If a child begins life with two normal alleles of the *RB* gene, then both copies of the *RB* gene must become mutated in a single retinal cell for a tumor to develop. Because the chance of having two *RB* mutations occur in the same cell is low, sporadic cases are more likely to involve only one eye (unilateral cases), and because more time is required to acquire two separate mutations, this form of retinoblastoma arises later in childhood.

Surveys of patients who have retinoblastoma have confirmed these predictions. Pedigree analysis indicates that most cases that involve two eyes are inherited, whereas the vast majority of unilateral cases are sporadic and noninherited. In addition, most cases that involve only one eye occur at a later age (although still in childhood) than bilateral cases.

The *RB* gene is located on chromosome 13 at 13q14 (◗ Figure 14.5), and it encodes a protein (pRB) that is confined to the nucleus. The protein is found in cells of the retina and in almost all other cell types and tissues of the body. The pRB is present at all stages of the cell cycle, but its *activity* is regulated synchronously with the cell cycle. The protein acts as a molecular switch that controls progression through the cell cycle. During G1, if the protein is active, it stops the cell from moving into S and prevents cell division. On the other hand, if the pRB protein is inactive, it allows the cell to pass through the S phase, through G2, and on to mitosis. If both copies of the *RB* gene are deleted or become mutated in a retinal cell, the absent or defective pRB cannot regulate cell division, and the cell begins to divide in an uncontrolled manner, forming a tumor.

What happens if both copies of the *RB* gene become mutated in a cell type other than retinal cells? If this happens in bone cells, the result is a cancer known as osteosarcoma. In fact, cultured osteosarcoma cells have been used to provide evidence for the role of the *RB* gene in cancer. Analysis using recombinant DNA techniques indicates that osteosarcoma cells carry two mutant copies of the *RB* gene and produce no retinoblastoma protein. If a cloned copy of a normal *RB* gene is transferred into the osteosarcoma cells, pRB is produced, and cell division stops. This finding reinforces the idea that the retinoblastoma protein plays a central role in regulating the cell cycle.

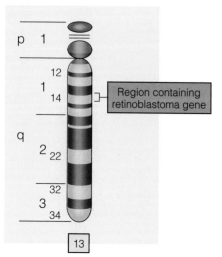

◗ **FIGURE 14.5** A diagram of chromosome 13 that shows the locus of the retinoblastoma gene.

The Search for Breast Cancer Genes

Breast cancer is the most common form of cancer in women in the United States. Each year, more than 44,000 women die from breast cancer, and more than 180,000 new cases are diagnosed. Although environmental factors may be involved in breast cancer, geneticists struggled for years with the question, Is there a genetic predisposition to breast cancer? After more than 20 years of work, the answer is yes. The mutant form of two different genes can predispose women to breast cancer and ovarian cancer.

One of these genes, *BRCA1* (MIM/OMIM 113705), is located on the long arm of chromosome 17 at q21.1. Although it is responsible for only about 5% of cases, this gene is responsible for the susceptibility to breast cancer that appears in women under 40 years of age. About 1 in 200 females inherit the mutant allele, and of these, about 90% will develop breast cancer.

The search for this gene began in the 1970s with the analysis of familial patterns of breast cancer by Mary Claire King and her colleagues. They looked for families that had a clear history of breast cancer and found that about 15% of the 1500 families surveyed had multiple cases of breast cancer. Genetic models based on this pattern suggested that about 5% of these cases were related to an autosomal dominant pattern of inheritance. The models also predicted that two-thirds of the families that had multiple cases had no genetic predisposition to breast cancer. This meant that for further research on the gene, it was impossible to single out families that had a genetic predisposition from families that had no genetic predisposition. Instead of being discouraged, King and her colleagues decided on a brute force approach. They began testing as many of the families as possible, knowing that finding linkage between breast cancer and a genetic marker was a long shot.

The search began by testing multiple-case families to find linkage between protein markers and breast cancer. In the 1980s, as recombinant DNA techniques became widely used, King and her colleagues switched to DNA markers and the PCR technique for screening and began the search all over again. Finally, in 1990, after testing hundreds of families using hundreds of markers, they found linkage. The 183rd marker they used, from a VNTR locus on chromosome 17 called D17S74, was tested on family members from 23 pedigrees that had a history of breast cancer. This marker was inherited along with cases of breast cancer and was clearly linked to the disease. Other laboratories quickly confirmed their results and found that this marker was also linked to familial cases of ovarian cancer.

The investigators formed an international consortium, and using DNA from members of 214 families that had multiple cases of either breast or ovarian cancer, used RFLP markers to narrow the search for the gene to a small region on the long arm of chromosome 17 (▶ Figure 14.6). The *BRCA1* gene was identified and cloned in 1994. Mutant alleles of this gene cause about half of all hereditary forms of breast cancer.

A second gene for predisposition to breast cancer, *BRCA2* (MIM/OMIM 600185), was discovered in 1995. Mutant alleles of this gene are inherited in an autosomal dominant manner. It maps to region q12-13 of chromosome 13 and may be responsible for the majority of inherited cases that are not caused by a mutation in *BRCA1*. Together, mutations in *BRCA1* and *BRCA2* account for a large majority of inherited cases of breast cancer but are not involved in sporadic cases of breast or ovarian cancers, which make up 80 percent of all cases.

Many questions about the normal functions of both genes remain unanswered and about the way mutations in these genes lead to breast cancer. It is known that gene expression for *BRCA1* and *BRCA2* peaks at the G1/S boundary in rapidly dividing cells. The proteins encoded by both genes bind to another protein, called RAD51, which is involved in repairing double-stranded breaks in DNA molecules.

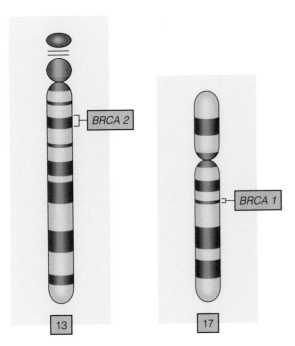

▶ **FIGURE 14.6** The chromosome locations for *BRCA1* and *BRCA2*. Together these genes account for the majority of heritable cases of breast cancer.

ONCOGENES AND CANCER

Oncogenes are mutant alleles of genes that normally stimulate or maintain cell division. In mutant form, the genes induce or maintain uncontrolled cell division associated with cancer. The existence of oncogenes was first discovered by work on a virus that causes cancer in chickens.

Rous Sarcoma Virus and the Discovery of Oncogenes

The role of viruses in cancer began with Peyton Rous, who studied a malignant tumor, known as a **sarcoma,** found in chickens. In 1911 Rous found that cell-free extracts from sarcomas would cause tumor formation when injected into healthy chickens. Decades later, it was shown that the extract contains a virus, known as the Rous sarcoma virus (RSV) (▶ Figure 14.7). At the time, Rous's work was criticized by those who claimed that his extract was not cell-free and that he was simply transferring cancer cells that grew in the injected chickens to form tumors. As a result of these criticisms, Rous gradually abandoned the project. Several decades later, when RSV was identified as the cause of the tumors, it became one of the most widely studied animal tumor viruses. In belated recognition of his pioneering work on viral tumors, Rous was awarded the Nobel Prize in 1966 (at the age of 85).

Viruses can be grouped into two classes: DNA viruses or RNA viruses, depending on the nature of the genetic material they carry. Cancer-causing viruses are found in both groups (Table 14.4). DNA tumor viruses include SV40 (simian virus 40), first identified in monkey tumors, and polyoma, a virus that produces tumors in several species of animals, including mice, hamsters, and rats. Some RNA viruses are known as **retroviruses** because the single-stranded RNA viral genome must return to the form of a double-stranded DNA molecule before it can replicate. Tumor-causing retroviruses have been found in the chicken, mouse, rat, hamster, and other species. The Rous sarcoma virus is an example of a tumor-producing retrovirus. The results of work with RSV and similar viruses have been important in understanding the origins of human cancer.

■ **Sarcoma** A cancer of connective tissue. One type of sarcoma in chickens is associated with the retrovirus known as the Rous sarcoma virus.

■ **Retrovirus** Viruses that use RNA as a genetic material. During the viral life cycle, the RNA is reverse-transcribed into DNA. The name retrovirus symbolizes this backward order of transcription.

Table 14.4
Viruses and Cancer

DNA Cancer Viruses	RNA Cancer Viruses
SV40	Rous sarcoma virus
Polyoma	Mouse mammary tumor virus
Adenovirus	

▶ **FIGURE 14.7** The Rous sarcoma virus as seen in the transmission electron microscope.

The ability of the Rous sarcoma virus to cause tumor formation in chickens is due to the presence of a single gene (▶ Figure 14.8). Because this gene is associated with the ability of the virus to cause cancer, it is called an oncogene. The discovery of the oncogene in RSV was a central and important event in cancer research because it indicated that cancer could be caused by changes in a small number of genes. Over the years, other types of RNA tumor viruses have been discovered, and it was shown that many forms of animal cancers, including mouse leukemia, cat leukemia, and mouse breast cancer are caused by retroviruses. Like RSV, some of these RNA tumor viruses carry a single gene that causes cancer. These genes are of several different types, but all have the common property of being oncogenic. More than 20 different oncogenes have been identified by their presence in retroviruses, and altogether, more than 50 oncogenes have been identified (Table 14.5). Oncogenes are named for the

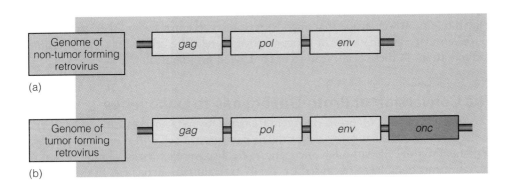

▶ **FIGURE 14.8** Organization of the genome in a nontransforming strain of Rous sarcoma virus. (a) The three genes carried by the virus enable the virus to infect cells and replicate, but not cause tumors. (b) In a tumor-producing strain, the genome contains an extra gene, an oncogene. This gene is derived from the host cell in a previous infection and confers upon the virus the ability to produce uncontrolled growth and tumor formation in another host cell.

Table 14.5 Retroviral Oncogenes and Human Proto-Oncogenes

Viral Oncogene	Associated Tumor	Human Proto-Oncogene	Human Chromosome
v-*src*	Sarcoma	c-*src*	20
v-*fps*/v-*fes*	Sarcoma	c-*fps*/c-*fes*	15
v-*yes*	Sarcoma	c-*yes*	?
v-*ros*	Sarcoma	?	?
v-*ski*	?	?	?
v-*myc*	Carcinoma, sarcoma, myelocytoma	c-*myc*	8
v-*erb*-A	?	c-*erb*-A	17
v-*erb*-B	Erythroleukemia, sarcoma	c-*erb*-B	7
v-*myb*	Myeloblastic leukemia	c-*myb*	6
v-*rel*	Lymphatic leukemia	?	?
v-*mos*	Sarcoma	c-*mos*	8
v-*abl*	B-Cell lymphoma	c-*abl*	9
v-*fos*	Sarcoma	c-*fos*	14
v-*raf*	?	c-*raf*-1	3
v-Haras/v-*bas*	Sarcoma, erythroleukemia	c-Ha-*ras*-1	11
		c-Ha-*ras*-2	X
v-Ki-*ras*	Sarcoma, erythroleukemia	c-Ki-*ras*-1	6
		c-Ki-*ras*-2	12
v-*fms*	Sarcoma	c-*fms*	5
v-*sis*	Sarcoma	c-*sis*	22

virus in which they were discovered. In Rous sarcoma virus, the gene is known as *v-src,* the oncogene in avian erythroblastosis virus is known as *v-erb,* and so forth.

Not all retroviruses carry oncogenes, so where do oncogenes come from? They could be viral in origin or might represent genes captured from animal cells that the viruses infect. Researchers have shown that oncogenes carried by retroviruses are acquired from animal cells during viral infection. The normal, cellular versions of these genes are called proto-oncogenes, or cellular oncogenes (*c-onc*). Proto-oncogenes are nonmutant genes that are present in all cells. These genes have the potential to cause cancer if they are mutated or if their usual pattern of expression is altered.

But what is the usual function of such genes? For the most part, proto-oncogenes are associated with cell growth, cell division, and cell differentiation, and their gene products function at many sites within the cell (Table 14.6). For example, the DNA sequence of the *sis* oncogene closely matches that of a proto-oncogene that encodes a growth factor called PDGF. Similarly, the DNA sequence of the *erb*-B oncogene is related to a proto-oncogene that encodes a cellular receptor for another growth factor called EGF. The majority of proto-oncogenes regulate cell growth and division, and mutations in these genes lead to uncontrolled proliferation and cancer.

The Conversion of Proto-Oncogenes to Oncogenes

The mutant version of proto-oncogenes carried by retroviruses are called *v-onc* genes. Retroviruses that carry a *v-onc* gene can infect a cell and transform it into a malignant tumor cell. Although oncogenes were discovered in viruses, only a few rare human forms of cancer are caused by virally transmitted oncogenes. In most cases, the conversion from proto-oncogene to oncogene takes place in the nucleus of a somatic cell without the action of a virus.

What is the difference between a proto-oncogene in a normal cell and a mutant version of that gene (an oncogene) in a cancer cell? Many differences are possible, including mutations that produce an altered gene product and those that cause underproduction or overproduction of the normal gene product. In fact, all of these kinds of mutations have been found in human oncogenes or in their adjacent regulatory regions.

The normal *ras* gene encodes a protein 189 amino acids long that is located on the cytoplasmic side of the plasma membrane. The protein can cycle between an active and an inactive state. In its active form, the ras protein transfers growth-promoting signals that have come across the cell membrane. Analysis of 12 different mutant forms of the *ras* gene that were isolated from human tumors reveals that in each case a single nucleotide change is the only difference between the mutant oncogene found in tumor cells and the proto-oncogene found in normal cells. In all 12 cases examined, a nucleotide substitution leads to an amino acid substitution in the

Table 14.6	Cellular Localization of Oncogene (*c-onc* and *v-onc*) Gene Products	
Gene	**Location of c-*onc* Proteins**	**Location of v-*onc* Proteins**
abl	Nucleus	Cytoplasm
erb-B	Plasma membrane	Plasma membrane and Golgi
fos	Cytoplasm	Cytoplasm and membranes
myc	Nucleus	Nucleus
ras	Membranes	Membranes
src	Membranes	Membranes

encoded protein. The mutant gene and the mutant oncoprotein are found only in the tumor cells and not in the normal tissue of the patient. Somewhat surprisingly, the amino acid substitution in all 12 mutant genes occurred at one of two places in the protein: amino acid 12 or amino acid 61 (▶ Figure 14.9). Work using X-ray analysis of protein crystals has shown that changing the amino acid glycine at position 12 disrupts the structure of the protein and prevents it from folding into its normal shape. The mutant ras protein cannot cycle between the inactive and active state and is locked in the active state. As a result, an affected cell escapes from growth control and becomes cancerous.

Knowledge about the molecular organization of oncogenes and their products is being used to develop new methods to diagnose and treat cancer. For example, if oncogene proteins are present in the blood, antibodies that bind to these proteins can be used to detect cancer at a very early stage. Tests are already available for a mutant protein released into the bloodstream by breast cancer cells.

New strategies for treatment may also be derived from knowledge about oncogenes. In some cases, multiple copies of oncogenes are present in lung cancer, breast cancer, and cervical cancer. The number of copies of these oncogenes can be detected by using recombinant DNA techniques, and these tumors receive more aggressive treatment and therapy. Laboratory studies of cultured cells indicate that if an oncogene is switched off, the cell becomes nonmalignant. Further work on the mechanism of gene switch-off may result in development of a new class of anticancer drugs directed against the regulatory regions of specific genes. Alternatively, because the protein products of oncogenes have been characterized, it may be possible to block or reduce the action of such proteins, causing cancer cells to become quiescent.

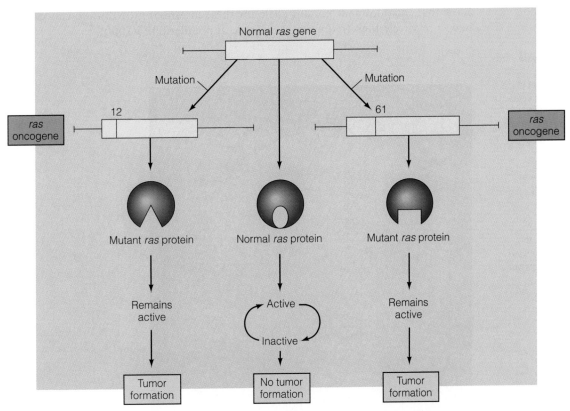

▶ **FIGURE 14.9** The *ras* proto-oncogene is a normal component of the genome and encodes a gene product that receives and transduces signals needed for cell growth. Mutations at amino acid positions 12 or 61 cause the formation of an oncoprotein that promotes the formation of a cancer cell.

A GENETIC MODEL FOR CANCER: COLON CANCER

As discussed earlier in the chapter, cancer is regarded as a multistep process that involves mutations in a number of specific genes. In retinoblastoma, only two mutational steps are required to convert a normal cell into a cancerous one. In other cases, a half dozen or more mutations are required to initiate formation of a cancer cell. Cancer of the colon is one of those latter types. This form of cancer has been studied to examine the order of mutations necessary for tumor formation and was selected for a number of reasons. First, colon/rectal cancer is one of the most common forms of cancer in the U.S. (Table 14.7). Second, tumors develop from preexisting benign tumors, and tumors at all stages of development are available for study. Third, two forms of genetic predisposition to colon cancer are known: an autosomal dominant trait, **familial adenomatous polyposis** (*FAP*) (MIM/OMIM 175100), and **hereditary nonpolyposis colon cancer** (*HNPCC*) (MIM/OMIM 120435 and 120436). *FAP* accounts for only about 1% of all cases of colon cancer but has been useful in deriving the main features of a genetic model for colon cancer that is described below. *HNPCC* accounts for about 15% of all cases and is associated with a form of genetic instability, described in the next section.

To clarify the role of inheritance in colorectal cancer, Randall Burt and his colleagues studied a large pedigree that has more than 5000 members covering six generations. This family contains clusters of siblings and relatives who have colon cancer. Like many other families, this one shows no definite pattern of inheritance for the cancer. However, as part of this study, about 200 family members were examined for intestinal growths (▶ Figure 14.10). These benign tumors, known as **polyps**, usually precede or accompany colon cancer and are regarded as precursors of malignancy. When intestinal growths and colon cancer are considered *together* as a single phenotype, an autosomal dominant pattern of inheritance is clear. The results also show that the dominant mutant allele for these intestinal growths and cancer may have a relatively high frequency in the general population (3/1000).

Table 14.7 Colon and Rectal Cancer in the United States

Estimated new cases, 1999	
Colon	94,700
Rectum	34,700
Total	129,400
Mortality (estimated deaths, 1999)	
Colon	47,900
Rectum	8,700
Total	56,600
5-year survival rate (early detection)	
Colon	91%
Rectum	85%

■ **Familial adenomatous polyposis (*FAP*)** A dominant condition associated with the development of growths known as polyps in the colon. These polyps often develop into malignant growths and cause cancer of the colon and/or rectum.

■ **Hereditary nonpolyposis colon cancer (*HNPCC*)** A form of colon cancer associated with genomic instability of microsatellite DNA sequences.

■ **Polyp** A growth attached to the substrate by a small stalk. Commonly found in the nose, rectum, and uterus.

▶ **FIGURE 14.10** Polyps in the colon are a precursor to colon cancer. If a cell in one of these polyps acquires enough mutations, it will transform into a cancer cell and form a tumor.

By analyzing the mutations in tumors at various stages of development, from benign to cancerous, researchers have defined the number and order of steps that change a normal intestinal cell into a malignant cell. The genetic model for colon cancer, shown in Figure 14.11, has several features. First, the development of colon/rectal cancer requires five to seven mutations. If fewer mutations are present, benign growths or intermediate stages of malignant tumor formation result. Second, analysis of many cases suggests that the order of mutations usually follows that shown in the figure, indicating that *both* the number and order of mutations are important in tumor formation. A dominant mutation in the *APC* gene on chromosome 5 is the first step in the model. In a normal homozygote, no growth of polyps occurs. Heterozygotes who inherit a mutant copy of *APC* develop hundreds or thousands of benign tumors in the colon and rectum. In spontaneous cases, mutation of a single copy of *APC* takes place in a single cell. This cell then divides to form a benign tumor. In either case, the benign tumors are made up of clones of cells, each of which carries a mutant *APC* gene. The formation of tumors caused by a single copy of the mutant gene suggests that *APC* is a tumor suppressor gene. Mutation in *APC* is not enough to cause cancer, and mutations in other genes are required to cause the transition from benign tumor to colon cancer.

Intermediate stages between benign growth and colon cancer carry an intermediate number of mutations (Figure 14.11). Mutation of a single copy of the K-*ras* gene causes a benign tumor to progress to form an adenoma, an intermediate stage tumor that has fingerlike projections. To progress further, *both* alleles of the downstream genes shown in the figure must be mutated. The 18q region contains a number of genes involved in colon cancer (all of them are tumor suppressor genes), including *DCC*, *DPC4*, and *JV18-1*. Mutation in both alleles of one of these genes results in the formation of late stage adenomas. In the last stage, mutations that involve the loss or inactivation of both alleles of the *p53* gene on chromosome 17 cause the late stage adenoma to become cancerous and form colon cancer. Mutations in the *p53* gene are pivotal in the formation of other cancers, including lung, breast, and brain cancer. The *p53* gene is a tumor suppressor gene that is active in regulating the passage of cells from the G1 to the S phase of the cell cycle.

In sum, the genetic model for colon cancer requires a series of mutations that accumulate over time in a single cell. Each mutation confers a slight growth advantage on the cell, first allowing it to proliferate and form a benign tumor, then enlarging the tumor through a series of stages. Eventually, one cell accumulates enough mutations to escape from cell cycle control to form a malignant tumor. Later, additional mutations allow tumor cells to become metastatic and break away to form tumors at remote sites.

Recombinant DNA techniques have been used to identify other cases in which cancer involves a number of mutations at specific chromosomal sites, often on different chromosomes (Table 14.8).

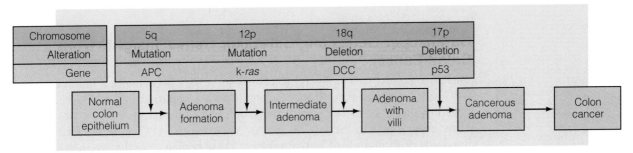

 FIGURE 14.11 A model for colon cancer. In this multiple-step model, the first mutation occurs in the *APC* gene, leading to the formation of polyps. Subsequent mutations in genes on chromosomes 12, 17, and 18 cause the transformation of the polyp into a tumor. In this model, the sum of the changes is more important than the order in which the changes take place.

Table 14.8	Number of Mutations Associated with Specific Forms of Cancer	
Cancer	**Chromosomal Sites of Mutations**	**Minimal Number of Mutations Required**
Retinoblastoma	13q14	2
Wilms tumor	11p13	2
Colon cancer	5q, 12p, 17p, 18q	5 to 7
Small-cell lung cancer	3p, 11p, 13q, 17p	10 to 15

GENOMIC CHANGES AND CANCER

Most cancers are caused by mutations in two or more genes that accumulate over time. If one of these mutations is inherited, fewer mutations are required to cause cancer, resulting in a genetic predisposition to cancer. But if the mutation rate is low (as we have seen in Chapter 11), how do the multiple mutations needed for cancer formation accumulate in a single cell? As it turns out, work on a second form of colon cancer has provided a partial answer to this question.

DNA Repair Mutations Can Result in Genomic Instability

There are two forms of hereditary nonpolyposis colon cancer (*HPNCC1* and *HNPCC2*). *HNPCC1* has been mapped to chromosome 2 (2p16), and *HNPCC2* has been mapped to chromosome 3 (3p23-21.3). This form of cancer may be one of the most common human hereditary disorders, and affects about 1 in 200 individuals. Mutations of either of these two genes can destabilize the genome, generating a cascade of mutations in microsatellite sequences (review microsatellite sequences in Chapter 13) located on many chromosomes. It has been estimated that cells from *HNPCC* tumors can carry more than 100,000 mutations in microsatellite sequences scattered throughout the genome.

Because of these findings, it has been proposed that the *HNPCC1* and *HNPCC2* genes are *neither* tumor suppressor genes *nor* oncogenes but represent a new class of genes, which when mutated lead to genomic instability and cancer. Normally, proteins encoded by the genes at these loci repair errors made during DNA replication. When these genes are inactivated by mutation, the mutation rate of microsatellite repeats increases by at least 100-fold. It is thought that this genomic instability promotes mutations in other genes, including tumor suppressor and oncogenes, leading to cancer.

Gatekeeper Genes and Caretaker Genes: Insights from Colon Cancer

There are at least two pathways to colon cancer; one begins with a mutation in the *APC* gene, and the other (*HNPCC*) begins with a mutation in a DNA repair gene. Together, these two mechanisms provide insight into the nature of the genes that can cause normal cells to become cancerous. Mutations in *APC* cause the formation of hundreds or thousands of benign tumors. These benign growths progress slowly to malignancy by accumulating mutations in other genes. Because there are thousands of benign growths, there is a high risk that at least one benign tumor will progress to form colon cancer. In *HNPCC*, benign growths accumulate slowly during the aging process. However, mutations in these benign tumors accumulate at a rate two to

Table 14.9		Human Genetic Disorders Associated with Chromosome Instability and Cancer Susceptibility		
Disorder	**Inheritance**	**Chromosome Damage**	**Cancer Susceptibility**	**Hypersensitivity**
Ataxia telangiectasia	Autosomal recessive	Translocations on 7, 14	Lymphoid, others	X rays
Bloom syndrome	Autosomal recessive	Breaks, translocations	Lymphoid, others	Sunlight
Fanconi anemia	Autosomal recessive	Breaks, translocations	Leukemia	X rays
Xeroderma pigmentosum	Autosomal recessive	Breaks	Skin	Sunlight

three times faster than in normal cells, making it almost certain that at least one benign growth will progress to colon cancer.

These different pathways to colon cancer led to the idea that two different types of genes are involved in cancer: **gatekeeper genes** and **caretaker genes**. Gatekeeper genes inhibit cell growth and control the cell cycle. Mutation in these genes opens the gate to uncontrolled cell division. Tumor suppressor genes are one type of gatekeeper gene. Many types of oncogenes are also gatekeeper genes.

Caretaker genes, on the other hand, encode proteins that maintain the integrity of the genome, e.g., DNA repair genes. Diseases associated with mutations in genes that encode DNA repair proteins are listed in Table 14.9. These proteins normally repair damage to DNA, caused by mistakes in DNA replication, or damage caused by environmental agents (ultraviolet light, for example). Mutation of a caretaker gene does not directly lead to tumor formation but creates a state of genomic instability that increases the mutation rate of all genes, including gatekeeper genes.

This insight about gene types and genomic instability may also explain why many forms of cancer become associated with chromosomal instability and aneuploidy.

■ **Gatekeeper genes** Genes that regulate cell growth and passage through the cell cycle, e.g., tumor suppressor genes.

■ **Caretaker genes** Genes that help maintain the integrity of the genome, e.g., DNA repair genes.

CHROMOSOME CHANGES AND CANCER

Changes in chromosome structure and number are a common feature of cancer cells. In some cases, the relationship between a single chromosome change and the development of cancer is not clear. For example, Down syndrome is caused by the presence of an extra copy of chromosome 21. This quantitative change in genetic information is not associated with any known gene mutation. In addition to defects in cardiac structure and in the immune system, children with Down syndrome are 18 to 20 times more likely to develop leukemia than children in the general population. How extra copies of genes on chromosome 21 predispose to cancer is not yet known, but it may be an effect of increasing the dosage of some proto-oncogenes. In a limited number of cases, more direct information is available about the way chromosomal rearrangements are associated with developing and/or maintaining the cancerous condition.

Chromosome Rearrangements and Leukemias

The connection between chromosome rearrangements and cancer is perhaps most clearly seen in leukemias. In these cancers (which involve uncontrolled division of the white blood cells), specific chromosome changes are well-defined and diagnostic (Table 14.10).

The first and one of the best studied examples of the link between cancer and a chromosomal aberration is a translocation between chromosome 9 and chromosome

Table 14.10	Chromosomal Translocation Associated with Human Cancers
Chromosomal Translocation	**Cancer**
t(9;22)	Chronic myelogenous leukemia (Philadelphia chromosome)
t(15;17)	Acute promyelocytic leukemia
t(11;19)	Acute monocytic leukemia, acute myelomonocytic leukemia
t(1;9)	Pre-B-cell leukemia
t(8;14),t(8;22),t(2;8)	Burkitt's lymphoma, acute lymphocytic leukemia of the B-cell type
t(8;21)	Acute myelogenous leukemia, acute myeloblastic leukemia
t(11;14)	Chronic lymphocytic leukemia, diffuse lymphoma, multiple myeloma
t(4;18)	Follicular lymphoma
t(4;11)	Acute lymphocytic leukemia
t(11;14)(p13;q13)	Acute lymphocytic leukemia

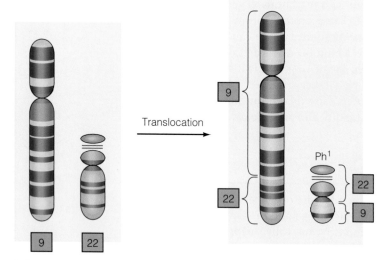

▶ **FIGURE 14.12** A reciprocal translocation between chromosomes 9 and 22 results in the formation of a chromosome involved in chronic myelogenous leukemia (CML).

■ **Philadelphia chromosome** An abnormal chromosome produced by an exchange of portions of the long arms of chromosomes 9 and 22.

22 found in chronic myelogenous leukemia (CML)(▶ Figure 14.12). Originally called the **Philadelphia chromosome** (the city in which it was first found), this discovery by Janet Rowley of the University of Chicago was the first example of a chromosome translocation accompanying a human disease.

Now, it is known that other cancers, including acute myeloblastic leukemia, Burkitt's lymphoma, and multiple myeloma, are associated with specific translocations (Table 14.10). The chromosomal defects are usually deletions of a specific band or a reciprocal exchange of chromosome parts between different chromosomes. The finding that certain forms of cancer are consistently associated with specific chromosomal abnormalities suggests that these aberrations are related to the development of the cancer. There is strong evidence that chromosome rearrangements are not by-products of malignancy, but are important steps in the development of certain cancers. The genetic and molecular basis for this role is now becoming clear as the field of cancer cytogenetics merges with the molecular biology of oncogenes.

Translocations, Hybrid Genes, and Leukemias

As oncogenes were identified, cytogeneticists systematically mapped the chromosomal locations of the normal versions of these genes (the proto-oncogenes). It soon became clear that many of these genes are located at or very close to the break points of chromosomal translocations involved with specific forms of leukemia. In fact, chromosome breaks can alter the expression of proto-oncogenes and bring about the development of cancer.

Chronic myelogenous leukemia (CML) is associated with the fusion of two genes at the break points of the 9; 22 translocation characteristic of this disease. The Abel-

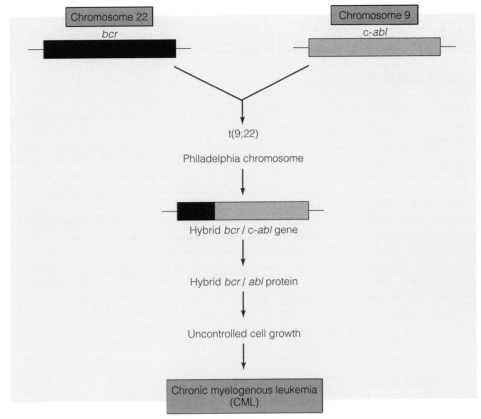

FIGURE 14.13 Gene fusion associated with the 9;22 translocation in chronic myelogenous leukemia (CML). The *bcr* gene on chromosome 22 is fused to the *c-abl* proto-oncogene on chromosome 9. The hybrid gene is transcribed, and the resulting fusion protein stimulates cell division in white blood cells. Overproduction of these cells results in CML.

son oncogene *c-abl* maps at the break point on chromosome 9, and the *bcr* gene maps at the break point on chromosome 22. The translocation produces a hybrid gene that has *bcr* sequences at the beginning of the gene and most of the normal *c-abl* sequences at the end of the gene (Figure 14.13). This hybrid gene is transcribed, and the message is translated to produce a fusion protein in which the *bcr*-encoded amino acids activate the amino acids in the region encoded by the *c-abl* gene. The hybrid protein switches on cell division in white blood cells, resulting in CML.

Other genes at translocation break points have been identified, and in these cases, hybrid genes are formed by the translocation. The activity of these hybrid genes (often involving a proto-oncogene) cause the cell to undergo a malignant transformation.

CANCER AND THE ENVIRONMENT

The relationship between environmental factors and cancer has been studied for more than 50 years. During that time, the development of more sophisticated data gathering and analysis has provided solid evidence for the relationship between the environment and cancer. **Epidemiology** is the study of factors that control the presence or absence of a disease. It is an indirect and inferential science that provides correlations between factors and the existence of a disease, such as cancer. These correlations provide working hypotheses that must be confirmed in laboratory experiments on animal models and then in carefully controlled clinical trials with humans (see Concepts and Controversies: Epidemiology, Asbestos and Cancer).

Epidemiology The study of the factors that control the presence, absence, or frequency of a disease.

Epidemiology, Asbestos, and Cancer

Asbestos is a fibrous material known since ancient times. The emperor Charlemagne is said to have had a tablecloth made of asbestos that was thrown into the flames after meals and emerged from the flames unburned, to the amazement of his guests. In more modern times, asbestos has been widely used in manufactured goods and is present in our homes, schools, and automobiles. It has been used in brake linings, ceiling tiles, wallboard, textiles, ironing boards, and kitchen gloves. Almost everyone in the United States has been exposed to asbestos in one form or another. To determine the possible role of asbestos in cancer, an epidemiological study compared the cause of death in a group of asbestos workers with an age- and sex-matched group of individuals selected from the general population. Some data from this study are shown in the following table.

A total of 444 deaths were recorded in the asbestos workers, and 301 deaths occurred in the control group. In determining whether a disease such as cancer is linked to asbestos, the number of cancer deaths in asbestos workers is divided by the number of cancer deaths in a similar-sized sample of the control population. If the number of deaths is the same, the ratio is about 1.0. If the number of cancer deaths among asbestos workers is greater than the population at large, the ratio is higher. The results show that the ratio of cancer deaths is 3.86, and the ratio of deaths from lung cancer is 7.62.

This circumstantial evidence was tested in a laboratory on rats and mice. Animals were exposed to various amounts of asbestos fibers and monitored for the development of cancer. A control group was not exposed to asbestos. Cancers in the control group were used as a baseline measurement of cancer rates among the experimental animals. These experiments supported the link between asbestos exposure and cancer, leading to government standards for maximum permissible exposure to asbestos fibers.

Number of Deaths, 1943–1973

Cause of Death	Expected	Observed	Ratio of Obs/Exp
Cancer, all sites	51	198	3.86
Lung cancer	12	89	7.62
Pleural mesothelioma	0	10	
Peritoneal mesothelioma	0	25	
Stomach cancer	5	18	3.53
Colon-rectal cancer	8	22	2.93
Asbestosis	0	37	
All other causes	249	209	0.84
Total deaths, all cases	301	444	1.48

Source: Selicoff, I. J., & Hammond, E. C. (1975). Major risk factors in environmental cancer. In Fraumeni, J. F. (ed.), Persons at High Risk of Cancer. New York: Academic Press, pp. 467–483.

Epidemiology and Links to Environmental Factors

Typically, an epidemiological study of cancer measures the incidence of cancer in several different populations. If statistically significant differences are found, further studies seek to identify factors that correlate with these differences. The results of many such studies illustrate that there are widespread geographic variations in cancer cases and mortality that are presumably correlated with environmental factors (Table 14.11). Many cases of cancer in the United States are related to our physical surroundings, personal behavior, or both. Estimates indicate that at least 50% of all cancer can be attributed to environmental factors.

	All Sites	
Country	**Male**	**Female**
United States	165.3 (27)*	111.1 (18)
Australia	158.5 (28)	100.2 (20)
Austria	171.6 (20)	105.6 (16)
Denmark	178.7 (17)	138.1 (1)
Germany	177.3 (18)	108.2 (11)
Hungary	258.7 (1)	135.2 (2)
Japan	149.8 (32)	75.2 (43)
Latvia	206.1 (6)	98.7 (23)
Mauritius	85.4 (47)	63.8 (46)
Mexico	81.6 (48)	77.6 (41)
Poland	204.2 (8)	107.6 (13)
Romania	140.2 (36)	84.5 (38)
Slovenia	203.9 (9)	108.0 (12)
Switzerland	167.2 (24)	96.5 (26)
Trinidad, Tobaco	120.0 (42)	91.4 (31)
United Kingdom	179.1 (16)	124.6 (5)

Table 14.11 **Age-Adjusted Cancer Death Rates per 100,000 Population, 1990–1993**

*Rank among 48 countries surveyed.

Occupational Hazards and Cancer Risk

A relationship between occupation and cancer was first noted in 1775 by the English physician Percival Potts, who recorded that London chimney sweeps had a high rate of scrotal cancer, presumably caused by exposure to soot and coal tars. Coal tars and more than 500,000 other chemicals are in commercial and industrial use in the United States. Fewer than 1% of these have been adequately tested for their ability to cause cancer. Occupational exposure to some chemicals is known to cause cancer, but because of the long lag between the time of exposure and the onset of cancer, identification of people at risk is often difficult. One study estimates that occupational exposure to only a few materials that are used in industrial processes may account for 18% to 38% of all cancer cases in the next few decades. Among these materials is vinyl chloride.

Vinyl chloride is used in the manufacture of many plastic items, and more than 8 billion pounds is produced each year. Vinyl chloride is used to make polyvinyl chloride (PVC), which in turn is used to manufacture products as diverse as floor tile, bottles, food wrap, and pipes. In laboratory experiments, vinyl chloride causes a rare form of liver cancer in rats at a dose of 50 parts per million (ppm). At the time these animal experiments were performed (1970), the permissible occupational level of exposure to vinyl chloride workers was 500 ppm. Medical surveys of workers in vinyl chloride plants showed that several were diagnosed as having the same rare form of liver cancer as the rats exposed to vinyl chloride. In 1974, permissible exposure levels were reduced to 50 ppm, and in 1975 they were reduced to 1 ppm. The Food and Drug Administration also banned the use of PVC in beverage containers. Although workers in vinyl chloride plants receive the highest levels of exposure, finished plastics always contain entrapped vinyl chloride gas in amounts that can be significant. The effects of this exposure on the general population are yet to be assessed.

Guest Essay: Research and Applications

A Journey through Science

Bruce Ames

As a youngster, I was always interested in biology and chemistry and read the books left lying around the house by my father, who was chairman of a high school chemistry department and later supervisor of science for all of the New York City public schools. During the summers, my sisters and I explored the natural world and collected animals at our family's summer cabin on a lake in the Adirondack Mountains. Throughout my childhood, I read voraciously and would come back from the library with a whole stack of books. I attended the Bronx High School of Science, where I conducted my first scientific experiments, studying the effect of plant hormones on the growth of tomato root tips. Motivated by my experiences with research, I enrolled at Cornell University to study chemistry and biology. After graduating, I headed west to study at the California Institute of Technology. In the early 1950s, there was an exceptional group of faculty and students at Cal Tech, many of whom were learning about genes by studying biochemical genetics. I joined the laboratory of Herschel K. Mitchell and began using mutant strains to work out the steps used by the bread mold *Neurospora* to make the amino acid, histidine.

After completing my studies at Cal Tech, I moved to the National Institutes of Health at Bethesda, Maryland, to do research on gene regulation, using mutant strains of the bacterium *Salmonella*. By 1964, I was married and had two children, a daughter, Sofia, and a son, Matteo. Sometime in that same year, I happened to read the list of ingredients on a box of potato chips and began to think about all the new synthetic chemicals being used, and wondered if they might cause genetic damage to human cells.

To test the ability of chemicals to cause mutation, I devised a simple test, using strains of *Salmonella*. Chemicals that cause mutations in bacteria may cause mutation in human genes. This test showed that more than 80% of cancer-causing substances cause mutations. Now, the *Salmonella* test, widely known as the Ames test, is used by more than 3000 laboratories as a first step in identifying chemicals that might cause cancer.

In 1968, I joined the faculty at the University of California at Berkeley, where my students and I developed genetically engineered bacterial strains that can be used to identify what types of DNA changes are caused by mutagens. My colleague Lois Gold and I have assembled a database on the results of animal cancer tests. I have also studied the mechanisms of aging and cancer.

The accumulated evidence indicates that there is no epidemic of cancer caused by synthetic chemicals. In fact, pollution accounts for less than 1% of human cancer. Several factors, including tobacco and diet, have been identified as the major contributors to cancer in the U.S. The use of tobacco contributes to about one-third of all cases of cancer, and the quarter of the population that has the lowest dietary intake of fruits and vegetables compared to the highest quarter, has twice the cancer rate for most types of cancer. Hormonal factors contribute to most breast cancer cases. Decreases in physical activity and recreational exposure to the sun have also contributed importantly to increases in some cancers. These results strongly suggest that a large portion of cancer deaths can be avoided by using knowledge at hand to modify lifestyles.

Bruce Ames is a professor of biochemistry and molecular biology at the University of California, Berkeley. He has been the international leader in the field of mutagenesis and genetic toxicology for more than 20 years. His work has had a major impact on, and changed the direction of, basic and applied research on mutation, cancer, and aging. He earned his B.A. from Cornell University and his Ph.D. from the California Institute of Technology.

Environmental Factors and Cancer

The American Cancer Society estimates that 85% of the lung cancer cases among men and 75% of the cases among women are related to smoking. Smoking produces cancers of the oral cavity, larynx, esophagus, and lungs and accounts for 30% of all cancer deaths. Most of these cancers have very low survival rates. Lung cancer, for example, has a 5-year survival rate of 13%. Cancer risks associated with tobacco are not limited to smoking; the use of snuff or chewing tobacco carries a 50-fold increased risk of oral cancer.

About 800,000 new cases of skin cancer are reported in the United States every year, almost all of them related to ultraviolet exposure from sunlight or tanning lamps. The incidence of skin cancer is increasing rapidly in the population, presumably as a result of an increase in outdoor recreation. Skin color is a trait controlled by several genes and is associated with a continuous variation in phenotype. In humans, traits controlled by several genes typically produce a wide range of phenotypes. Epidemiological surveys show that lightly pigmented people are at much higher risk for skin cancer than heavily pigmented individuals. This supports the idea that genetic characteristics can affect the susceptibility of individuals or subpopulations to environmental agents that cause a specific form of cancer.

During the past decade, epidemiological and laboratory studies have increasingly focused on the role of diet and nutrition in the development of cancer. Many investigators believe that 30% to 40% of all cancers are related to diet. Colon cancer occurs with much higher frequency in North America than in Asia, Africa, or other parts of the world. When people migrate from a low risk-area to a high-risk area, their rate of colon cancer rises to that of the native residents, suggesting that environmental factors, including diet, may play a role in the development of colon cancer. The health of more than 88,000 registered nurses across the United States has been monitored since 1980. In one 6-year period, 150 cases of colon cancer occurred in this group. Analysis of dietary intake indicates that the risk of colon cancer is positively associated with the intake of animal fat. Nurses who ate beef, pork, or lamb daily had a 2.5-fold increased risk of colon cancer compared to those who ate such foods less than once a month. This correlation suggests that animal fat may be an environmental risk factor for colon cancer. This correlation has been challenged by a more recent assessment, indicating that more work remains to be done.

A correlation does not mean a cause-and-effect relationship. Other dietary factors, including total calories consumed, protein intake, and trace elements, might be associated with colon cancer. In the meantime, the National Academy of Sciences has issued cautionary diet guidelines. These include the reduction of fat intake to 20% or less of total calories and increases in the intake of high-fiber foods such as whole grains and vegetables rich in vitamins A and C.

Behavior and Cancer

Most research indicates that bad lifestyle choices contribute heavily to the burden of cancer cases and may contribute to 50% of all cancer deaths. How can you reduce your cancer risks? There are three simple steps: stay out of the sun (or use protective sunscreens), follow a low-fat diet that is high in fiber, and don't use tobacco.

Case Studies

CASE 1

Julie was a 23-year-old who was examined in the family cancer risk assessment clinic because she was concerned about her risk of developing breast cancer on the basis of her family history. Julie had been going to the local breast center because of fibrocystic breast disease and had requested prophylactic mastectomies at her last mammogram. Julie came to the counseling session with a maternal aunt who had prophylactic mastectomies several years earlier. The genetic counseling session explored the reasons that had prompted Julie's request for surgery and reviewed her family history. Three of Julie's six maternal aunts were reported to have had breast cancer in their late thirties or early forties. One maternal first cousin had also been diagnosed with breast cancer in her thirties. Julie's mother was in her fifties and had no history of cancer.

Discussion about the family revealed that Julie and her aunt believed that the only way not be diagnosed with cancer was to have their breasts removed. Julie went on to say that her mother had prophylactic mastectomies at age 34 and the aunt attending the session had prophylactic mastectomies last year, when she turned 40. The aunt told stories of caring for her dying sisters who were much older than she.

She also stated that "It would be the same as a death sentence not to do this." The women in Julie's family believed that the only way to avoid this disease was to have their healthy breasts removed before it could occur. This led the remaining aunts to seek surgery, and now Julie was being encouraged to take this step.

The genetic counselor explained the possibility that one of the maternal aunts could take a genetic test to determine if an altered gene is contributing to the development of breast cancer in the family. The counselor explained that if a mutation is present in her aunt, Julie could be tested for this same mutation. If it is present, prophylactic mastectomies might be a reasonable option. However, if Julie did not inherit the mutation, her risk for breast cancer would be the same as the general population risk of 10%, and prophylactic mastectomies would be unnecessary. One of Julie's aunts with cancer was tested, and it was found that she carries a mutation that increases the risk of breast cancer. Julie and her mother were tested for this same mutation and neither of them had the mutation. Julie continues to be followed by the local breast center for her fibrocystic breast disease but has decided not to have prophylactic mastectomies.

CASE 2

Mike was referred for genetic counseling because he was concerned about his extensive family history of colon cancer. His family history, outlined below, is highly suggestive of an inherited form of colon cancer, known as *HNPCC* (hereditary nonpolyposis colon cancer). This is an autosomal dominant condition and individuals who inherited the altered gene have a 75% chance of developing colon cancer by age 65. Mike was extensively counseled about the inheritance of this condition, the associated cancers, and the possibility of genetic testing (on an affected family member). Mike's aunt elected to be tested for one of the genes that may be altered in this condition, and it was found that she has an altered *MSH2* gene. Other family members are in the process of being tested for this mutation.

Summary

1. The primary risk factor for cancer is age. Heritable predisposition to cancer usually show a dominant pattern of inheritance. In some cases, such as retinoblastoma, a gene for cancer susceptibility is inherited, and at least one additional mutation is required to bring about malignancy.

2. Cancers that result from a heritable predisposition have been used to develop models in which cancer is a multistep process, requiring a minimum of two separate mutations to occur within a cell to produce cancer. According to this model, cancer cells are clonal descendants from a single mutant cell.

3. The study of two classes of genes, tumor suppressor genes and oncogenes, has established the relationship between cancer, the regulation of cell growth and division, and the cell cycle.

4. The discovery of tumor suppressor genes that normally act to inhibit cell division has provided insight into the regulation of the cell cycle. These gene products act at control points in the cell cycle at G1/S or G2/M. Deletion or inactivation of these products causes cells to continuously divide.

5. The discovery of an oncogene in the Rous sarcoma virus led to the identification of other oncogenes and their normal cellular counterparts, proto-oncogenes. Proto-oncogenes may regulate cell growth and are converted to oncogenes by mutations that produce a defective gene product. Human tumors express at least two oncogenes and occasionally as many as seven such genes. Because all human tumors show alterations in oncogenes, it is possible that a limited number of molecular pathways lead to the development of cancer.

6. Other human disorders, including Down syndrome, are associated with high rates of cancer. This predisposition may result from the presence of an initial mutation or genetic imbalance that moves cells closer to a cancerous state.

7. It is now apparent that many cancers are environmentally induced. Occupational exposure to minerals and chemicals poses a cancer risk to workers in a number of industries. The widespread dissemination of these materials poses an undefined but potentially large risk to the general population. Social behavior contributes to about 50% of all cancer cases in the United States, most, if not all of which are preventable.

Questions and Problems

1. Benign tumors:
   a. are non-cancerous growths that do not spread to other tissues
   b. do not contain mutations
   c. are malignant and clonal in origin
   d. metastasize to other tissues
   e. none of the above

2. Evidence that cancer has a genetic origin includes:
   a. most carcinogens are not mutagens
   b. the discovery of oncogenes, that promote tumor growth
   c. cancer does not spread to other tissues
   d. tumors become benign
   e. the environment plays no role in cancer

3. Cancer is now viewed as a disease that develops in stages. What are the stages that this statement is referring to?
   a. non-malignant tumors become malignant
   b. proto-oncogenes become tumor suppressor genes
   c. younger people get cancer more than older people
   d. small numbers of individual mutational events that can be separated by long periods of time
   e. none of the above

4. Metastasis refers to the process in which:
   a. tumor cells die
   b. tumor cells detach and move to secondary sites
   c. cancer does not spread to other tissues
   d. tumors become benign
   e. cancer can be cured

5. It is often the case that a predisposition to certain forms of cancer is inherited. Examples are familial retinoblastoma and colon cancer. What does it mean to have inherited an increased probability to acquire a certain form of cancer? What subsequent event(s) must occur?

6. Distinguish between dominant inheritance and recessive inheritance in retinoblastoma.

7. Describe the likelihood of developing bilateral (both eyes affected) retinoblastoma in the inherited versus the sporadic form of the disease.

8. Parents of a one-year old son are concerned that their son may be susceptible to retinoblastoma because the father's brother had bilateral retinoblastoma. Both parents are normal (no retinoblastoma). They have their son tested for the *RB* gene, and find to their surprise that he has inherited a mutant allele. The father is then tested and also found to carry a mutant allele.
   a. Why didn't the father develop retinoblastoma?
   b. What is the chance that the couple will have another child carrying the mutant allele?
   c. Is there a benefit to knowing their son may develop retinoblastoma?

9. Describe how the following terms relate to cancer:
   a. benign tumors
   b. malignant tumors
   c. metastasis

10. A proto-oncogene is a gene that:
    a. normally causes cancer
    b. normally suppresses tumor formation
    c. normally functions to promote cell division
    d. is involved in forming only benign tumors
    e. is only expressed in blood cells

11. Theodore Boveri predicted that malignancies would often be associated with chromosomal mutation. What lines of evidence substantiate this prediction?

12. Can you postulate a reason or reasons why children with Down syndrome are 20 times more likely to develop leukemia than children in the general population?

13. What is the difference between an oncogene and a proto-oncogene?

14. Viral oncogenes are often transduced copies of normal cellular genes-they become oncogenic when their gene product or its regulation is subtly altered. What are the roles of cellular proto-oncogenes, and how is this role consistent with their implication in oncogenesis?

15. What oncogene has been implicated in Burkitt lymphoma? What mutational event is typically associated with this tumor?

16. You are in charge of a new gene therapy clinic. Two cases have been referred to you for review and possible therapy.
    Case 1. A mutation in the promoter of an oncogene causes the gene to make too much of its normal product, a receptor protein that promotes cell division. The uncontrolled cell division has caused cancer.
    Case 2. A mutation in an exon of a tumor suppressor gene makes this gene nonfunctional. The product of this gene normally suppresses cell division. The mutant gene cannot suppress cell division which has led to cancer. Given the current state of knowledge, in which case is gene therapy a viable option? Why?

17. What are some factors that epidemiologists have associated with a relatively high risk of developing cancer?

18. Which of the following mutations will result in cancer?
    a. homozygous recessive point mutations in a tumor suppressor gene coding for a non-functional protein
    b. dominant mutation in a tumor suppressor gene in which the protein product is over-expressed
    c. homozygous recessive mutation in which there is a deletion in the coding region of a proto-oncogene, leaving it non-functional
    d. dominant mutation in a proto-oncogene in which the protein product is over-expressed

19. In DNA repair, what are caretaker genes? How do they work?

20. The search for the *BRCA1* breast cancer gene discussed in this chapter was widely publicized in the media (see for example, *Newsweek*, Dec. 6, 1993). Describe the steps taken by Mary Clare King and her colleagues to clone this gene. How long did this process take?

21. Distinguish between a carcinogen and a mutagen.

22. Smoking cigarettes has been shown to be associated with the development of lung cancer. However, a direct correlation between how many cigarettes one smokes and the onset of lung cancer does not exist. A heavy smoker may not get lung cancer, while a light smoker may get the disease. Explain why this may be.

23. Discuss the relevance of epidemiologic and experimental evidence in recent governmental decisions to regulate exposure to asbestos in the environment.

24. The following is a description of a family that has a history of inherited breast cancer.

    Betty (grandmother) does not carry the gene. Don, her husband, does. Don's mother and sister had breast cancer. One of Betty and Don's daughters (Sarah) has breast cancer, the other (Karen) does not. Sarah's daughters are in their 30s. Dawn, 33, has breast cancer, Debbie, 31, does not. Debbie is wondering if she will get the disease because she looks like her mother. Dawn is wondering if her two-year old daughter (Nicole) will get the disease.
    a. Draw a pedigree, indicate affected individuals and identify all individuals.
    b. What is the most likely mode of inheritance of this trait?
    c. What is Don's genotype and phenotype?
    d. What is the genotype of the unaffected women (Betty and Karen)?

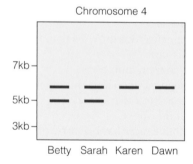

Chromosome 4

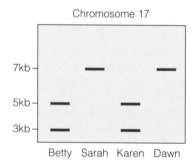

Chromosome 17

e. An RFLP marker has been found that maps very close to the gene. Given the following RFLP data for chromosomes 4 and 17, which chromosome does this gene map to?
f. Using the same RFLP marker, Debbie and Nicole were tested. The results are shown below. Based on

their genotypes, are either of them at increased risk for breast cancer?

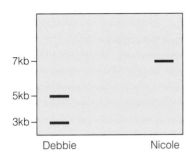

25. Studies have shown that there are significant differences in cancer rates among different ethnic groups. For example, the Japanese have very high rates of colon cancer, but very low rates of breast cancer. It has also been demonstrated that members of low-risk ethnic groups move to high-risk areas, their cancer risks rise to that of the high-risk area. For example, Japanese who live in the United States, where the risk of breast cancer is high, have higher rates of breast cancer than Japanese who live in Japan. What are some of the possible explanations for this interesting phenomenon? What factors may explain why the Japanese have higher rates of colon cancer than other ethnic groups?

Internet Activities

The following activities use the resources of the World Wide Web to enhance the topics covered in this chapter. To investigate the topics described below, log on to the book's home page at:

http://www.brookscole.com/biology

1. Oncolink is the University of Pennsylvania's resource for cancer information. By using the menus at this site you can supplement the coverage in this chapter on cancers with a genetic component. Use the Oncolink to access the most recent developments in cancer research.

2. Breast cancer screening followed by prophylactic mastectomy has now become an option for some women with strong family histories of breast cancer. What would you do if you were found to carry the gene that would predispose you to this type of cancer? Access the Breast Cancer and Genetic Screening page of Lawrence Berkeley National Laboratory's ELSI project and read about breast cancer. Answer the questions under "What would you do?"

3. The MD Anderson Oncolog is a site maintained by the University of Texas M.D. Anderson Cancer Center. Use this site for information on cancer patient care, diagnostic advances and therapies.

4. Several cancer case studies are presented at the Genetics of Cancer web site. Information provided includes family histories, pedigrees, and personal and ethical questions. Select one of the cases, evaluate it, and answer the questions posed.

For Further Reading

Alberts, D. S., Lipkin, M., & Levin, B. (1996). Genetic screening for colorectal cancer and intervention. *Int. J. Cancer 69:* 62–63.

Ames, B., Gold, L. S., & Willett, W. C. (1995). The causes and prevention of cancer. *Proc. Nat. Acad. Sci. USA 92:* 5258–5265.

Benedict, W. F., Xu, H. J., Hu, S. X., & Takahashi, R. (1990). Role of the retinoblastoma gene in the initiation and progression of human cancer. *J. Clin. Invest. 85:* 988–993.

Bodmer, W. F., Bailey, C. J., Bodmer, J., Bussey, H., Ellis, A., Gorman, P., Lucibello, F., Murday, V., Rider, S., Scambler, P., Sheer, D., Solomon, E., & Spurr, N. (1987). Localization of the gene for familial adenomatous polyposis on chromosome 5. *Nature 328:* 614–616.

Burt, R., Bishop, D. T., Cannon, L. A., Dowdle, M. A., Lee, R. G., & Skolnick, M. H. (1985). Dominant inheritance of adenomatous polyps and colorectal cancer. *N. Engl. J. Med. 312:* 1540–1544.

Call, K. M., Glaser, T., Ito, C. Y., Buckler, A. J., Pelletier, J., Haber, D. A., Rose, E. A., Kral, A., Yeger, H., Lewis, W. H., Jones, C., & Housman, D. E. (1990). Isolation and characterization of a zinc finger polypeptide gene at the human chromosome 11 Wilms tumor locus. *Cell 61:* 509–520.

Cancer Facts and Figures. (1999). New York: American Cancer Society.

Committee on Diet, Nutrition and Cancer, Assembly of Life Sciences, National Research Council. (1982). *Diet, Nutrition, and Cancer.* Washington, D.C.: National Academy Press.

Corbett, T. H. (1977). *Cancer and Chemicals.* Chicago: Nelson-Hall.

Croce, C., & Klein, G. (1985). Chromosome translocations and human cancer. *Sci. Am. 252* (March): 54–60.

Fearon, E. R., & Vogelstein, B. (1990). A genetic model for colorectal tumorigenesis. *Cell 61:* 759–767.

Hamm, R. D. (1990). Occupational cancer in the oncogene era. *Br. J. Ind. Med. 47:* 217–220.

Kinzler, K. W. and Vogelstein, B. (1997). Gatekeepers and caretakers. *Nature 386:* 761–763.

Lee, W-H., Bookstein, R., & Lee, E. (1988). Studies on the human retinoblastoma susceptibility gene. *J. Cell. Biochem. 38:* 213–227.

Maitland, N., Brown, K., Poirier, V., Shaw, A., and Williams, J. (1989). Molecular and cellular biology of Wilms tumor. *Anticancer Res. 9:* 1417–1426.

Malkin, D., & Knoppers, B. M. (1996). Genetic predisposition to cancer—issues to consider. *Semin. Cancer Biol. 7:* 49–53.

Marshall, E. (1993). Search for a killer: Focus shifts from fat to hormones. *Science 259:* 618–621.

Perrera F. P. (1996). Molecular epidemiology: Insights into cancer susceptibility, risk assessment and prevention. *J. Natl. Cancer Inst. 88:* 496–509.

Rous, P. (1911). Transmission of a malignant new growth by means of a cell-free filtrate. *JAMA 56:* 1981.

Ruoslahti, E. (1996). How cancer spreads. *Sci. Am. 275* (Sept): 72–77.

Russell, P. (1998). Checkpoints on the road to mitosis. *Trends Biochem. Sci. 10:* 399–402.

Stahl, A., Levy, N., Wadzynska, T., Sussan, J. M., Jourdan-Fonta, D., & Saracco, J. B. (1994). The genetics of retinoblastoma. *Ann. Genet.* 37: 172–178.

Sugimura, T. (1990). Cancer prevention: Underlying principles and practical proposals. *Basic Life Sci. 52:* 225–232.

Vasen, H. F. (1994). What is hereditary nonpolyposis colon cancer (HNPCC)? *Anticancer Res. 14* (4B): 1613–1615.

Vogelstein, B., Fearson, E. R., Kern, S., Hamilton, S., Preisinger, A., Nakamura, Y., & White, R. (1989). Allelotypes of colorectal carcinomas. *Science 244:* 207–211.

Weinberg, R. A. (1985). The action of oncogenes in the cytoplasm and nucleus. *Science 203:* 770–776.

Weinberg, R. A. (1996). How cancer arises. *Sci. Am. 275* (Sept): 62–70.

Yunis, J. J. (1983). The chromosomal basis of human neoplasia. *Science 221:* 227–236.

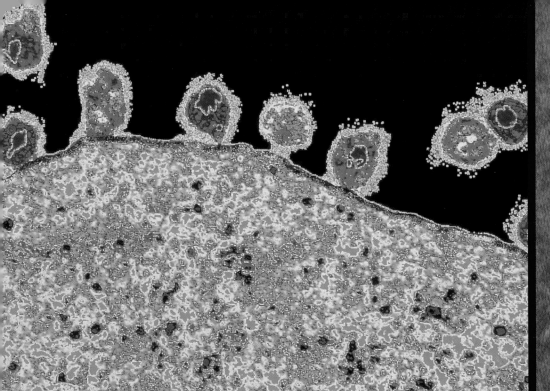

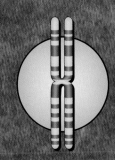

Genetics of the Immune System

Approximately 2500 years ago, a mysterious plague swept through Athens, Greece, killing thousands of residents. The historian Thucydides wrote an account of the epidemic and noted that those who recovered from the disease could care for the sick without becoming ill a second time. Physicians in ancient China observed that people who contracted smallpox became resistant to the disease. Records from eighth-century China indicate that partially successful attempts were made to transfer resistance to uninfected individuals by injecting them with fluid obtained from smallpox victims. The first safe and successful method of transferring resistance to a disease was developed about a thousand years later by Edward Jenner, an English physician. He noted that people who developed cowpox are resistant to smallpox infections. To see whether he could transfer resistance to smallpox, Jenner inoculated a boy with dried scrapings from a cowpox patient and followed the inoculation a few weeks later with an injection from a smallpox patient. The boy did not develop smallpox, and the method, called vaccination (from vaccus, *the Latin word for "cow"), became an effective tool in controlling this disease. The World Health Organization led a worldwide vaccination campaign against smallpox and eradicated the disease by 1980. Because the virus that causes smallpox cannot reproduce outside the body of an infected individual, the disease has disappeared. The only remaining samples of the virus are stored in research facilities in the United States and Russia.*

Vaccines are used to prevent a number of diseases, including diphtheria, whooping cough, measles, and hepatitis B. But the study of the body's resistance to disease is a young and developing science, and new diseases still arise to remind us of our limited knowledge. In the early 1970s, women began to develop a condition known as toxic shock syndrome. By 1980, several thousand cases and more than 25 deaths had been reported. Careful investigation determined that certain brands of highly absorbent tampons led to vaginal infections with certain strains of Staphylococcus aureus. *These bacteria secrete a toxin that produces the symptoms of fever, rash, low blood pressure, and in some cases, death.*

Other new diseases remain a mystery. In 1981, an infectious disease known as acquired immunodeficiency syndrome (AIDS) appeared in the United States. Affected individuals exhibit rare forms of cancer, pneumonia, and other infections. In all cases, there is a complete and irreversible breakdown of the immune system, associated with infection by the human immunodeficiency virus (HIV). This breakdown allows the development of infections, one or more of which proves fatal.

THE IMMUNE SYSTEM DEFENDS THE BODY AGAINST INFECTION

Infections caused by viruses, bacteria, and fungi remain a health problem in industrialized nations and are a major cause of death in developing nations. The responses of the body to infection include a nonspecific response and a specific set of responses known as the immune reaction.

Nonspecific responses are designed (1) to block the entry of disease-causing agents into the body, and (2) to block the spread of infectious agents if they get into the body. The immune system has two types of responses to invading agents: antibody-mediated immunity and cell-mediated immunity. Each of these is highly specific and consists of two stages, a primary response and a later-developing secondary response. In addition, the immune system is responsible for the success or failure of blood transfusions and organ transplants. In this chapter, we examine the cells of the immune system and how they are mobilized to mount an immune response.

We also consider how the immune system is involved in determining blood groups, and mother-fetus incompatibility. The immune system also plays a role in organ transplants, and in determining risk factors for a wide range of diseases. Finally, we describe a number of disorders of the immune system, including how AIDS acts to cripple the immune response of infected individuals.

THE INFLAMMATORY RESPONSE IS A GENERAL REACTION

The skin is a barrier that prevents infectious agents, such as viruses and bacteria, from entering the body. The surface of the skin is home to bacteria, fungi, and even mites, but they cannot penetrate the protective layers of dead skin cells to cause infection. Glands in the skin secrete acidic oils that inhibit bacterial growth. In addition, sweat, tears, and saliva contain enzymes that break down the outer walls of many bacteria, thus helping to prevent infection.

The mucus membranes that line the respiratory, digestive, urinary, and reproductive tracts secrete mucus that forms a barrier against infection. Cells in the respiratory system have cilia that sweep out bacteria trapped in the mucus. These physical barriers form the first line of defense against infection by disease-causing agents.

Nonspecific Responses Are Activated by the Inflammatory Reaction

If microorganisms penetrate the skin or the cells that line the respiratory, digestive, or urinary systems, an inflammatory reaction takes place (▶ Figure 15.1). Cells infected by microorganisms release chemical signals, including histamine. These chemicals increase capillary blood flow in the affected area. (The action of these chemicals is responsible for the heat and redness that develop around a cut or scrape.) The increased heat creates an unfavorable environment for the growth of microorganisms, mobilizes white blood cells, and raises the metabolic rate in nearby cells. These reactions promote

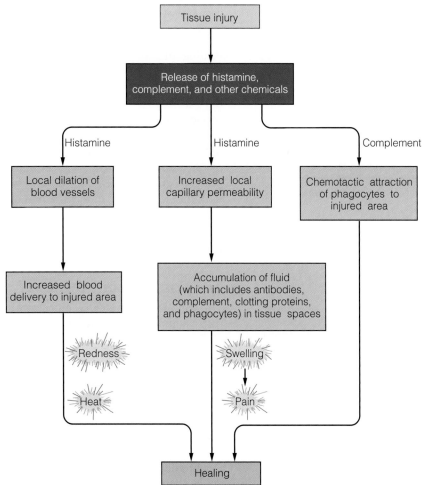

▶ **FIGURE 15.1** Tissue injury or infection generates an inflammatory response. The responses generate four signs of inflammation: redness, swelling, heat, and pain.

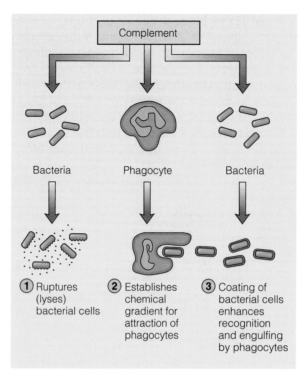

FIGURE 15.2 The complement system has several actions, all associated with defending the body against infection.

1 Ruptures (lyses) bacterial cells

2 Establishes chemical gradient for attraction of phagocytes

3 Coating of bacterial cells enhances recognition and engulfing by phagocytes

Bacteria   Phagocyte   Bacteria

Complement

healing. In addition, white blood cells migrate to the area in response to the chemical signals to engulf and destroy the invading microorganisms.

If infection persists, the capillary beds in the infected area become leaky, allowing plasma to flow into the injured tissue, causing it to swell. Clotting factors in the plasma trigger a cascade of small blood clots which seal off the injured area, preventing the escape of invading organisms. Finally, the area becomes the target for another type of white blood cell, the **monocyte,** which cleans up dead viruses, bacteria, or fungi and disposes of dead cells and debris.

This chain of events, beginning with the release of chemical signals and ending with the cleanup by monocytes, makes up the **inflammatory response.** This response is an active defense mechanism that the body employs to resist infection. This localized, limited reaction is usually enough to stop the spread of infection. If this system fails, another more powerful system—the immune response—is called into action.

The Complement System Kills Microorganisms Directly

The **complement system** is a chemical defense system that kills microorganisms directly, supplements the inflammatory response, and works with the immune response (Figure 15.2). The name of the system is derived from the way it complements the action of the immune system. Complement proteins are synthesized in the liver and circulate in the bloodstream as inactive precursors. When activated, the first protein in the series (C1) activates the second (C2), and so forth, in a cascade of activation. The final five components (C5–C9) form a large, cylindrical multiprotein complex, called the membrane-attack complex (MAC). The MAC embeds itself in the plasma membrane of an invading microorganism, creating a pore (Figure 15.3). Water flows through the pore in response to an osmotic gradient, bursting the cell.

In addition to directly destroying microorganisms, some complement proteins guide phagocytes to the site of microbial invasion. Other components aid the immune response by binding to the outer surface of microorganisms and marking them for destruction.

■ Monocytes White blood cells that clean up viruses, bacteria, and fungi and dispose of dead cells and debris at the end of the inflammatory response.

■ Inflammatory response The body's reaction to invading microorganisms.

■ Complement system A chemical defense system that kills microorganisms directly, supplements the inflammatory response, and works with (complements) the immune system.

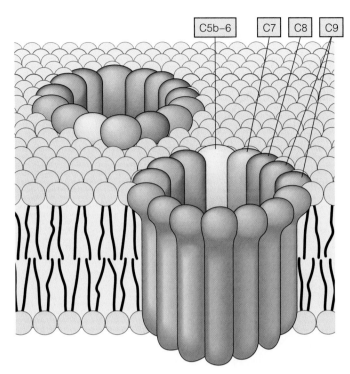

C5b–6 C7 C8 C9

▶ **FIGURE 15.3** The membrane attack complex (MAC). The complement proteins insert into and become part of the plasma membrane of an invading microorganism. Once in the membrane, the complement proteins form a pore, causing water to flow into the cell and burst the cell.

THE IMMUNE RESPONSE IS A SPECIFIC DEFENSE AGAINST INFECTION

The immune system generates a response that neutralizes and/or destroys viruses, bacteria, fungi, and cancer cells. The immune system is more effective than the non-specific defense system and has a memory component that remembers previous encounters with infectious agents. Immunological memory allows a rapid, massive response to a second exposure to a foreign substance.

An Overview of the Immune Response

Immunity results from the production of proteins called **antibodies.** Antibodies bind to foreign molecules and microorganisms and can inactivate them in several ways. Molecules produced by microorganisms that initiate antibody production are called **antigens** (antibody generators). Most antigens are proteins or proteins combined with polysaccharides, but *any* molecule, regardless of its source, that causes antibody production is an antigen.

The immune response is mediated by white blood cells called **lymphocytes.** These cells arise in the bone marrow by mitotic division of **stem cells** (▶ Figure 15.4). One type of lymphocyte is formed when daughter cells migrate from the bone marrow to the thymus gland, where they are irreversibly programmed to become **T cells.** Mature T cells circulate in the blood and become associated with lymph nodes and the spleen. **B cells** mature in the bone marrow and move directly to the circulatory system and the lymph system. B cells are genetically programmed to produce antibodies. Each B cell produces only one type of antibody.

The immune system has two parts, **antibody-mediated immunity,** regulated by B cell antibody production, and **cell-mediated immunity,** controlled by T cells (Table 15.1). Antibody-mediated reactions defend the body against invading viruses and bacteria. Cell-mediated immunity attacks cells of the body that are infected

■ **Antibody** A class of proteins produced by B cells that binds to foreign molecules (antigens) and inactivates them.

■ **Antigens** Molecules carried or produced by microorganisms that initiate antibody production.

■ **Lymphocytes** White blood cells that arise in bone marrow and mediate the immune response.

■ **Stem cells** Cells in bone marrow that produce lymphocytes by mitotic division.

■ **T cells** A type of lymphocyte that undergoes maturation in the thymus and mediates cellular immunity.

■ **B cells** A type of lymphocyte that matures in the bone marrow and mediates antibody-directed immunity.

■ **Antibody-mediated immunity** Immune reaction that protects primarily against invading viruses and bacteria by antibodies produced by plasma cells.

■ **Cell-mediated immunity** Immune reaction mediated by T cells that is directed against body cells that have been infected by viruses or bacteria.

FIGURE 15.4 Immature lymphocytes (lymphoblasts) produced in bone marrow that travel to the thymus for maturation become T cells. Those that remain and mature in bone marrow become B cells. Once mature, T and B lymphocytes migrate through the body in the circulatory system as part of the immune system.

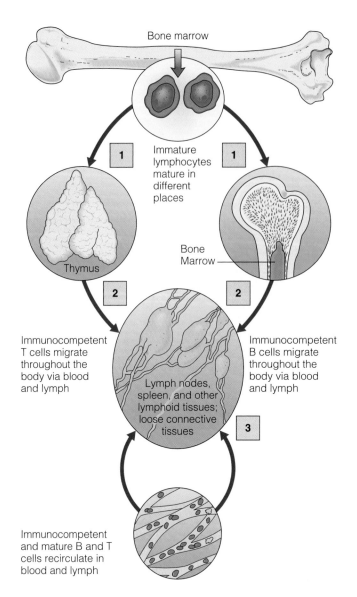

Bone marrow

1 Immature lymphocytes mature in different places **1**

Thymus

Bone Marrow

2 Immunocompetent T cells migrate throughout the body via blood and lymph

2 Immunocompetent B cells migrate throughout the body via blood and lymph

Lymph nodes, spleen, and other lymphoid tissues; loose connective tissues

3

Immunocompetent and mature B and T cells recirculate in blood and lymph

Table 15.1	Comparison of Antibody-Mediated and Cell-Mediated Immunity	
Antibody-Mediated	**Cell-Mediated**
Principal cellular agent is the B cell. B cell responds to bacteria, bacterial toxins, and some viruses.	Principal cellular agent is the T cell. T cells respond to cancer cells, virally infected cells, single-celled fungi, parasites, and foreign cells in an organ transplant.
When activated, B cells form memory cells and plasma cells, which produce antibodies to these antigens.	When activated, T cells differentiate into memory cells, cytotoxic cells, suppressor cells, and helper cells; cytotoxic T cells attack the antigen directly.

by viruses and bacteria. T cells also protect against infection by parasites, fungi, and protozoans. One group of T cells can also kill cells of the body if they become cancerous.

The Antibody-Mediated Immune Response Involves Several Stages

The antibody-mediated immune response involves several stages: antigen detection, activation of helper T cells, and antibody production by B cells. A specific cell type of the immune system controls each of these steps. White blood cells, called **macrophages,** continuously wander through the circulatory system and the spaces between cells searching for foreign (nonself) antigenic molecules, viruses, or microorganisms. When a macrophage encounters an antigen, the macrophage engulfs it, internalizes it (▶ Figure 15.5), and destroys it with enzymes. Small fragments of the antigen move to the outer surface of the macrophage's plasma membrane.

■ **Macrophages** Large white blood cells derived from monocytes that engulf antigens and present them to T cells, activating the immune response.

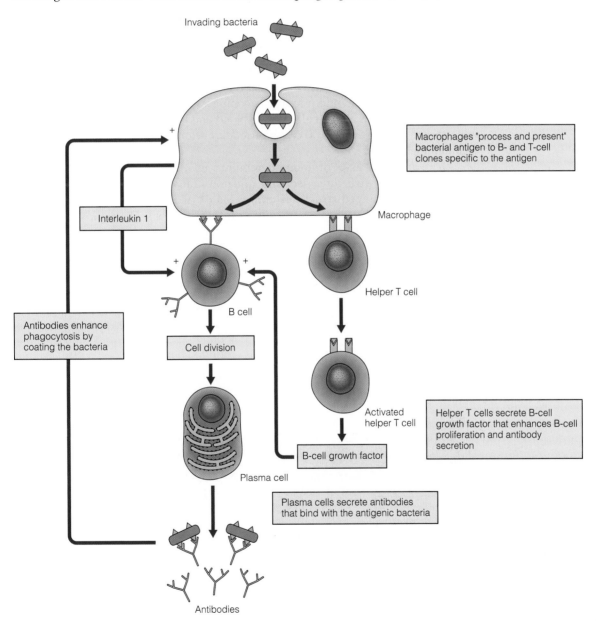

▶ **FIGURE 15.5** Macrophages engulf invading microorganisms and process their antigens for presentation to the helper T cells of the immune system. The activated T cell, in turn, stimulates division of the B cell, which produces antibodies against the antigen presented by the macrophage. Daughter B cells become differentiated as plasma cells for the production of large quantities of antibody. Macrophages can also directly stimulate antibody production by interaction with a B cell.

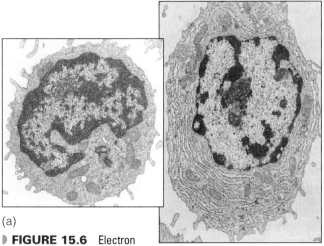

(a)

(b)

▶ **FIGURE 15.6** Electron micrographs of (a) a mature, unactivated B cell, and (b) a differentiated plasma cell (an activated B cell).

As the macrophage moves about, it may encounter a lymphocyte called a **helper T cell.** Surface receptors on the T cell make contact with the antigen fragment on the macrophage. This contact activates the T cell. The activated T cell in turn identifies and activates B cells that are synthesizing an antibody against the antigen encountered by the T cell. The activated B cells divide and form two types of daughter cells. The first type is **plasma cells,** which synthesize and secrete 2,000 to 20,000 antibody molecules *per second* into the bloodstream (▶ Figure 15.6). A second type of cell, called a **B-memory cell,** also forms at this time. Plasma cells live only a few days, but the memory cells have a life span of months or even years. These memory cells are part of the immune memory system and are described in a later section.

■ **Helper T cells**   A type of lymphocyte that stimulates the production of antibodies by B cells when an antigen is present.

■ **Plasma cells**   Cells produced by mitotic division of B cells that synthesize and secrete antibodies.

■ **B-memory cells**   Long-lived B cells produced after exposure to an antigen that play an important role in secondary immunity.

■ **Immunoglobulins**   The five classes of proteins to which antibodies belong.

Antibodies Are Molecular Weapons against Antigens

Antibodies are Y-shaped protein molecules that bind to specific antigens in a lock-and-key manner to form an antigen-antibody complex (▶ Figure 15.7). Antibodies are secreted by plasma cells and circulate in the blood and lymph. Some types of antibodies are attached to the surface of B cells. Antibodies belong to a class of proteins known as **immunoglobulins.**

There are five classes of immunoglobulins: IgG, IgA, IgM, IgD, and IgE. Each class has a unique structure, size, and function (Table 15.2). In general, antibody molecules are Y-shaped structures composed of two identical long polypeptides (H chains) and two identical short polypeptides (L chains). The chains are held together by chemical bonds.

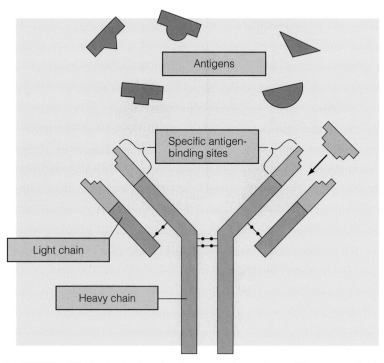

▶ **FIGURE 15.7** Antibody molecules are made up of two different proteins (an H chain and an L chain). The molecule is Y-shaped and forms a specific antigen-binding site at the ends.

Table 15.2 Types and Functions of the Immunoglobulins

Class	Location and Function
IgD	Present on surface of many B cells, but function uncertain; may be a surface receptor for B cells; plays a role in activating B cells.
IgM	Found on surface of B cells and in plasma; acts as a B-cell surface receptor for antigen, secreted early in primary response; powerful agglutinating agent.
IgG	Most abundant immunoglobulin in the blood plasma; produced during primary and secondary response; can pass through the placenta, entering fetal bloodstream, thus providing protection to fetus.
IgA	Produced by plasma cells in the digestive, respiratory, and urinary systems, where it protects the surface linings by preventing attachment of bacteria to surfaces of epithelial cells; also present in tears and breast milk; protects lining of digestive, respiratory, and urinary systems.
IgE	Produced by plasma cells in skin, tonsils, and the digestive and respiratory systems, overproduction is responsible for allergic reactions, including hay fever and asthma.

The structure of antibodies is related to their functions: (1) recognize and bind antigens, and (2) inactivate the antigen. At one end of the antibody molecule is an antigen-combining site formed by the ends of the H and L chains. This unique site recognizes and binds to a site on the antigen, called the **antigenic determinant**. The formation of an antigen-antibody complex leads to the destruction of an antigen in several ways (▶ Figure 15.8).

■ **Antigenic determinant** The site on an antigen to which an antibody binds, forming an antigen-antibody complex.

Antibody Genes Undergo Recombination in B Cells

Humans can produce billions of different antibody molecules, each of which can bind to a different antigen. Because there are billions of such combinations, it is impossible that each antibody molecule is encoded directly in the genome; there simply is not enough DNA in the human genome to encode hundreds of million antibodies.

The vast number of different antibodies are produced as a result of genetic recombination in three clusters of antibody genes: the H-chain genes on chromosome 14, the kappa L genes on chromosome 2, and the lambda light genes on chromosome 22. This recombination takes place during B-cell maturation before antibody genes are transcribed and before antibody production begins. In each antibody gene cluster, DNA segments that encode various portions of the H and L chains undergo recombination so that each mature B cell encodes, synthesizes, and secretes only one type of antibody. These rearranged antibody genes are stable and are passed to all daughter B cells.

Is there enough genetic recombination to produce hundreds of millions or billions of different antibodies? The DNA sequences that encode H- and L-chain components have been identified, isolated, and studied in detail. The random recombination of all H-chain components can generate about 30,000 different H chains. Similar random recombination can generate about 3600 different light chains. The combination of all possible H chains (30,000) with all possible L chains (3600) gives 108 million possible antibodies (30,000 × 3600) from a few hundred components. Other events that occur during B-cell maturation expand the number of possible H and L chains, allowing production of billions of possible antibody combinations from a few hundred segments at three loci.

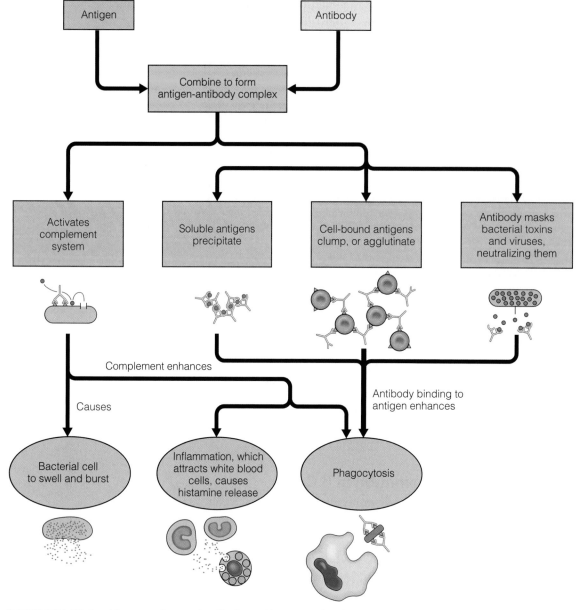

FIGURE 15.8 After an antigen and antibody combine, several pathways lead to the destruction of the antigen.

T Cells Mediate the Cellular Immune Response

There are several types of T cells in the immune system (Table 15.3). Helper T cells, described earlier, activate B cells to produce antibodies. **Suppressor T cells** slow down and stop the immune response and act as an "off" switch for the immune system. A third type, the **killer T cells**, finds and destroys cells of the body that are infected with a virus, bacteria or other infectious agents (▶ Figure 15.9). If a cell in the body is infected with a virus, viral proteins will appear on its surface. These foreign antigens are recognized by receptors on the surface of a killer T cell. The T cell attaches to the infected cell and secretes a protein that punches holes in the plasma membrane of the infected cell. The cytoplasmic contents of the infected cell leak out through these holes, and the infected cell dies and is removed by phagocytes. Cytotoxic T cells also bind to and kill cells of transplanted organs if they recognize them as for-

■ **Suppressor T cells**   T cells that slow or stop the immune response of B cells and other T cells.

■ **Killer T cells**   T cells that destroy body cells infected by viruses or bacteria. These cells can also directly attack viruses, bacteria, cancer cells, and cells of transplanted organs.

Table 15.3　Summary of T Cells

Cell Type	Action
Killer T cells	Destroy body cells infected by viruses and attack and kill bacteria, fungi, parasites, and cancer cells.
Helper T cells	Produce a growth factor that stimulates B-cell proliferation and differentiation and also stimulates antibody production by plasma cells; enhance activity of cytotoxic T cells.
Suppressor T cells	May inhibit immune reaction by decreasing B- and T-cell activity and B- and T-cell division.
Memory T cells	Remain in body awaiting reintroduction of antigen, when they proliferate and differentiate into cytotoxic T cells, helper T cells, suppressor T cells, and additional memory cells.

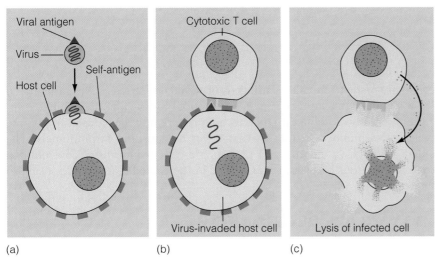

(a)　　　　　　　　(b)　　　　　　　　(c)

▶ **FIGURE 15.9**　(a) Virally-infected cells display viral antigens on their surface. (b) Cytotoxic T cells can recognize virally-infected cells of the body. (c) These T cells can bind to the infected cell and release chemicals that lead to the destruction of the infected cell before the virus begins to replicate.

eign. The nonspecific and specific reactions of the immune system are summarized in Table 15.4.

The Immune System Has a Memory Function

As described in the opening section of the chapter, ancient writers observed that exposure to certain diseases made people resistant to second infections by the same disease. B and T memory cells that are produced in the first infection control this resistance, called **secondary immunity.** A second exposure to the same antigen results in an immediate, large-scale production of antibodies and killer T cells. Because of the presence of the memory cells, the second reaction is faster, more massive, and lasts longer than the primary immune response.

The secondary immune response is the reason that we can be vaccinated against infectious diseases. A **vaccine** stimulates the production of memory cells against a disease-causing agent. A vaccine is really a weakened, disease-causing antigen, given orally or by injection, that provokes a primary immune response and the production of memory cells. Often, a second dose is administered to elicit a secondary response that raises, or "boosts," the number of memory cells (which is why such shots are called booster shots).

▪ **Secondary immunity**　Resistance to an antigen the second time it appears, due to T and B memory cells. This second response is faster, larger, and lasts longer than the first.

▪ **Vaccine**　A preparation containing dead or weakened pathogens that elicit an immune response when injected into the body.

Table 15.4 Nonspecific and Specific Immune Responses to Bacterial Invasion

Nonspecific Immune Mechanisms	Specific Immune Mechanisms
INFLAMMATION Engulfment of invading bacteria by resident tissue macrophages Histamine-induced vascular responses to increase blood flow to area, bringing in additional immune cells Walling off of invaded area by fibrin clot Migration of neutrophils and monocytes/macrophages to the area to engulf and destroy foreign invaders and to remover cellular debris Secretion by phagocytic cells of chemical mediators, which enhance both nonspecific and specific immune responses NONSPECIFIC ACTIVATION OF THE COMPLEMENT SYSTEM Formation of hole-punching membrane attack complex that lyses bacterial cells Enhancement of many steps of inflammation	Processing and presenting of bacterial antigen by macrophages Proliferation and differentiation of activated B-cell clone into plasma cells and memory cells Secretion by plasma cells of customized antibodies, which specifically bind to invading bacteria Enhancement by helper T cells, which have been activated by the same bacterial antigen processed and presented to them by macrophages Binding of antibodies to invading bacteria and activation of mechanisms that lead to their destruction Activation of lethal complement system Stimulation of killer cells, which directly lyse bacteria Persistence of memory cells capable of responding more rapidly and more forcefully should the same bacteria be encountered again

Vaccines are made from killed pathogens or weakened strains (called attenuated strains) that stimulate the immune system but cannot produce life-threatening symptoms of the disease. Recombinant DNA techniques are now being used to prepare vaccines against a number of diseases that affect humans and farm animals. The irony is that although vaccination is a proven and effective means of controlling a wide range of infectious diseases, fewer people in the United States are choosing to be vaccinated.

BLOOD TYPES ARE DETERMINED BY CELL-SURFACE ANTIGENS

Antigens on the surface of blood cells determine compatibility in blood transfusions. There are about 30 known antigens on blood cells; each of these constitutes a blood group or **blood type.** In transfusions, certain critical antigens of the donor and recipient must be identical. If transfused red blood cells have a nonmatching antigen on their surface, the recipient's immune system will produce antibodies against this antigen, causing the transfused cells to clump together. The clumped blood cells block circulation in capillaries and other small blood vessels, with severe and often fatal results. In transfusions, two blood groups are of major significance: the ABO system and the Rh blood group.

■ **Blood type** One of the classes into which blood can be separated based on the presence or absence of certain antigens.

Table 15.5 Summary of ABO Blood Types

Blood Type	Antigens on Plasma Membranes of RBCs	Antibodies in Blood	Safe to Transfuse To	Safe to Transfuse From
A	A	Anti-B	A, AB	A, O
B	B	Anti-A	B, AB	B, O
AB	A + B	none	AB	A, B, AB, O
O	–	Anti-A Anti-B	A, B, AB, O	

ABO Blood Typing Allows Safe Blood Transfusions

ABO blood types are determined by a gene *I* (for isoagglutinin) that encodes an enzyme that alters a cell-surface protein. This gene has three alleles, I^A I^B, and I^O, often written as *A, B,* and *O*. The *A* and *B* alleles each produce a slightly different version of a gene product, and *O* produces no gene product. Individuals who have type A blood carry the *A* antigen on their red blood cells, so they do not produce antibodies against this cell surface marker. However, people with type A blood have antibodies against the antigen encoded by the *B* allele (Table 15.5). Those who have type B blood carry the *B* antigen on their red cells and have antibodies against the *A* antigen. If you have type AB blood, both antigens are present on the surface of red cells, and no antibodies against *A* or *B* are made. In those who have type O blood, neither antigen is present, but antibodies against both the *A* and *B* antigen are present in the blood.

In transfusions, AB individuals have no serum antibodies against *A* or *B* and can receive blood of any type. Type O individuals have neither red cell antigen and can donate blood to anyone, even though their plasma contains antibodies against *A* and *B;* after transfusion, the concentration of these antibodies is too low to cause problems.

When transfusions are made between incompatible blood types, several problems arise. ▶ Figure 15.10 shows the cascade of reactions that follows transfusing someone who is type A with type B blood. Antibodies to the B antigen are in the blood of the recipient. These bind to the transfused red blood cells, causing them to clump together and burst. The clumped cells restrict blood flow in capillaries, reducing oxygen delivery. Lysis of red blood cells releases large amounts of hemoglobin into the blood. The hemoglobin forms deposits in the kidneys that block the tubules of the kidney and often cause kidney failure.

Rh Blood Types Can Cause Immune Reactions between Mother and Fetus

The Rh blood group (named for the rhesus monkey, in which it was discovered) consists of those who can make the *Rh* antigen (*Rh positive*, Rh⁺) and those who cannot make the antigen (*Rh negative,* Rh⁻).

The Rh blood group is a major concern when incompatibility exists between mother and fetus, a condition known as **hemolytic disease of the newborn** or **HDN**. This occurs most often when the mother is Rh⁻ and the fetus is Rh⁺ (▶ Figure 15.11). If the mother is Rh⁻, and Rh⁺ blood from the fetus enters the maternal circulation, the mother's immune system makes antibodies against the Rh antigen. This mingling of Rh⁺ blood with that of the Rh⁻ mother commonly occurs during birth, so the first Rh⁺ positive child is often unaffected. But, as a result, the maternal circulation now contains antibodies against the Rh antigen, and a subsequent Rh⁺ fetus evokes a response from the maternal immune system. Massive amounts of antibodies

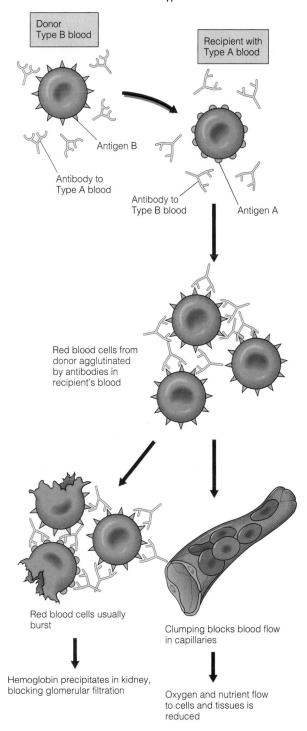

▶ **FIGURE 15.10** A transfusion reaction that results when type B blood is transfused into a recipient who has type A blood.

Donor Type B blood

Recipient with Type A blood

Antigen B

Antibody to Type A blood

Antibody to Type B blood

Antigen A

Red blood cells from donor agglutinated by antibodies in recipient's blood

Red blood cells usually burst

Clumping blocks blood flow in capillaries

Hemoglobin precipitates in kidney, blocking glomerular filtration

Oxygen and nutrient flow to cells and tissues is reduced

■ **Hemolytic disease of the newborn (HDN)** A condition of immunological incompatibility between mother and fetus that occurs when the mother is Rh⁻ and the fetus is Rh+.

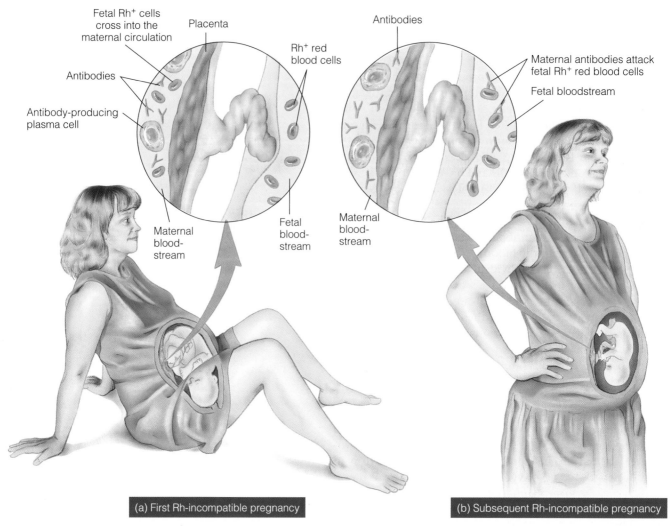

Fetal Rh⁺ cells cross into the maternal circulation

Placenta

Antibodies

Rh⁺ red blood cells

Antibodies

Antibody-producing plasma cell

Maternal antibodies attack fetal Rh⁺ red blood cells

Fetal bloodstream

Maternal bloodstream

Fetal bloodstream

Maternal bloodstream

(a) First Rh-incompatible pregnancy

(b) Subsequent Rh-incompatible pregnancy

▶ **FIGURE 15.11** The Rh factor and pregnancy. (a) Rh-positive cells from the fetus can enter the maternal circulation at birth. If the mother is Rh-negative, she produces antibodies against the Rh factor. (b) In a subsequent pregnancy, if the fetus is Rh-positive, the maternal antibodies cross into the fetal circulation and destroy fetal red blood cells, producing hemolytic disease of the newborn (HDN).

cross the placenta in late stages of the pregnancy and destroy the red blood cells of the fetus (Figure 15.11), resulting in HDN.

To prevent HDN, Rh^- mothers are given an Rh-antibody preparation during the first pregnancy in which the fetus is Rh^+ and in all subsequent pregnancies involving an Rh^+ fetus. The injected Rh-antibodies move through the maternal circulatory system and destroy any Rh^+ fetal cells that may have entered the mother's circulation. To be effective, this antibody must be administered before the maternal immune system can make its own antibodies against the Rh antigen.

ORGAN TRANSPLANTS MUST BE IMMUNOLOGICALLY MATCHED

The success of organ transplants and skin grafts depends on matching histocompatibility antigens between the donor and recipient. These antigens are proteins on the surface of all cells in the body. In humans, antigens produced by a cluster of genes on chromosome 6, known as the major histocompatibility complex (MHC).

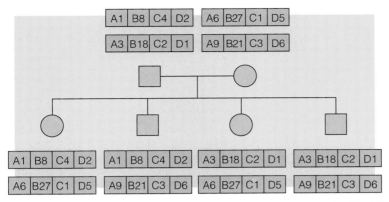

▶ **FIGURE 15.12** The transmission of HLA haplotypes. In this simplified diagram, each haplotype contains four genes, each of which encodes a different antigen.

The HLA genes of this complex play a critical role in the outcome of transplants. The HLA complex consists of several gene clusters. One group, called class I, consists of *HLA-A, HLA-B,* and *HLA-C.* Adjacent to this is a cluster called class II, which consists of *HLA-DR, HLA-DQ,* and *HLA-DP.* A large number of alleles have been identified for each HLA gene, making millions of allele combinations possible. The array of HLA alleles on a given copy of chromosome 6 is known as a **haplotype.** Because each of us carries two copies of chromosome 6, we each have two HLA haplotypes (▶ Figure 15.12).

Because of the large number of allele combinations possible, it is rare that any two individuals have a perfect HLA match. The exceptions are identical twins, who are HLA matched, and siblings, who have a 25% chance of being matched. In the example in Figure 15.12, each child receives one haplotype from each parent. As a result, four new haplotype combinations are represented in the children. (Thus, siblings have a one in four chance of having the same haplotypes.)

Successful Transplants Depend on HLA Matching

Successful transplantation of organs and tissues depends largely on matching HLA haplotypes between donor and recipient. Because there is such a large number of HLA alleles, the best chance for a match is usually between related individuals, with identical twins having a 100% match. The order of preference for organ and tissue donors among relatives is identical twin > sibling > parent > unrelated donor. Among unrelated donors and recipients, the chances for a successful match are only 1 in 100,000 to 1 in 200,000. Because the frequency of HLA alleles differs widely among racial and ethnic groups, matches across racial and ethnic lines are often more difficult. When HLA types are matched, the survival of transplanted organs is dramatically improved. Survival rates for matched and unmatched kidney transplants over a 4-year period are shown in ▶ Figure 15.13.

Animal-Human Organ Transplants

In the United States, about 18,000 organs are transplanted each year, but about 40,000 qualified patients are on waiting lists. Each year, almost 3000 people on the waiting list die before receiving an organ transplant, and another 100,000 die even *before* they are placed on the waiting list. Although the demand for organ transplants is rising, the number of donated organs is growing very slowly. Experts estimate that more than 50,000 lives would be saved each year if enough organs were available.

■ **Haplotype** A cluster of closely linked genes or markers that are inherited together. In the immune system, the HLA alleles on chromosome 6 are a haplotype.

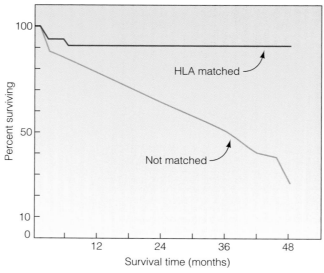

▶ FIGURE 15.13 The outcome of kidney transplants with (upper curve) and without (lower curve) HLA matching.

One way to increase the supply of organs is to use animal donors for transplants. Animal-human transplants (called **xenotransplants**) have been attempted many times, but with little success. Two important biological problems are related to xenotransplants: (1) complement-mediated rejection and (2) T-cell mediated rejection. In the first, cell-surface proteins (gene products of the HLA system) act as antigens, and these antigens are very different across species. When an animal organ (say, from a pig) is transplanted into a human, the cell-surface proteins on the pig organ are so different that they trigger an immediate and massive immune response, known as **hyperacute rejection**. This reaction is mediated by the complement system, and usually within hours, the transplanted organ is destroyed. The second problem is the same as in transplants between two humans: the difficulty involved in suppressing the cellular immune (T-cell mediated) rejection of the transplant.

To overcome the first problem, several research groups have isolated and cloned human genes that suppress the hyperacute rejection. These cloned genes were injected into fertilized pig eggs, and the resulting transgenic pigs carry human-recognition antigens on their cell surfaces. Organs from these transgenic pigs should appear as human in origin to the recipient's immune system, thus preventing a hyperacute rejection. Transplants from genetically engineered pigs to monkey hosts have been successful, but the ultimate step will be an organ transplant from a transgenic pig to a human. However, even if hyperacute rejection can be suppressed, using pig organs will still cause problems with T-cell-mediated rejection of the transplant. Because transplants from pig donors to humans occur across species, the tendency toward rejection may be stronger and require the lifelong use of immunosuppressive drugs. These powerful drugs may be toxic when taken over a period of years or may weaken the immune system, paving the way for continuing rounds of infections. To deal with this problem, it may be necessary to transplant bone marrow from the donor pig to the human recipient. The resulting pig-human immune system (called a chimeric immune system) would recognize the pig organ as "self" and still retain normal human immunity. As far-fetched as this may sound, animal experiments using this approach have been successful in preventing rejection more than two years after transplantation without the use of immunosuppressive drugs.

As recently as 1993, the possibility of animal-human organ transplants seemed like a remote possibility, more suited to science fiction than to medical fact. However, the advances described here make it likely that cross-species transplants will be attempted in the next few years. Although the final combination of transgenic

Table 15.6 HLA Alleles and Disease

Disease	HLA Allele	Frequency in Patients %	Frequency in General Population %	Relative Risk Factor
Ankylosing spondylitis	B27	>90	8	>100
Congenital adrenal hyperplasia	B47	9	0.6	15
Goodpasture syndrome	DR2	88	32	16
Juvenile rheumatoid arthritis	DR5	50	16	5
Multiple sclerosis	DR2	59	26	4
Pernicious anemia	DR5	25	6	5
Psoriasis	B17	38	8	6
Reiter's syndrome	B27	75	8	50
Rheumatoid arthritis	DR4	70	28	6
Systemic lupus erythematosis (SLE)	DR3	50	25	3

donors, immunosuppressive drugs, and immune tolerance remains to be worked out, guidelines for xenotransplantation are in place, and the use of animal organ donors will probably become common in the near future.

The HLA System and Disease Associations

In studying the distribution of HLA alleles in the population, researchers have discovered a relationship between certain HLA alleles and specific diseases. For example, more than 90% of the individuals who have the connective tissue disease **ankylosing spondylitis** (MIM/OMIM 106300) carry the HLA-B27 allele. This disease is a chronic inflammatory condition associated with joints and the spine that leads to fusion of the joints between vertebrae. The available evidence suggests that B27 class I antigens are involved in promoting the joint inflammation characteristic of this disease. Other diseases also show an association with certain HLA alleles; some of these are listed in Table 15.6.

Several hypotheses have been proposed to explain the relationship between HLA alleles and disease, but so far none has proven valid. It appears that most of these disorders result from a combination of HLA and non-HLA effects with environmental activation, possibly by infection. For example, ankylosing spondylitis may be triggered in B27 patients by an infection, and the bacterium *Klebsiella* has been implicated in the onset of this disease.

Ankylosing spondylitis An autoimmune disease that produces an arthritic condition of the spine; associated with HLA allele B27.

DISORDERS OF THE IMMUNE SYSTEM

We are able to resist most infectious diseases because we have an immune system. Unfortunately, failures in the immune system can result in abnormal or even absent immune responses. The consequences of these failures can range from mild inconvenience to systemic failure and death. In this section, we briefly catalog some ways in which the immune system can fail.

Overreaction in the Immune System Causes Allergies

Allergies result when the immune system overreacts to the presence of weak antigens that do not provoke an immune response in most people (❱ Figure 15.14). These

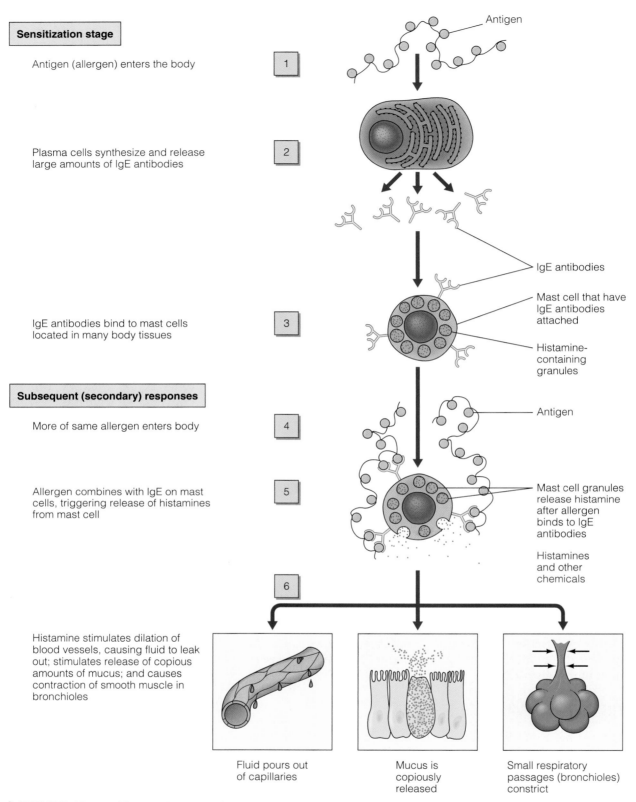

Antigen (allergen) enters the body 1

Plasma cells synthesize and release large amounts of IgE antibodies 2

IgE antibodies bind to mast cells located in many body tissues 3

Subsequent (secondary) responses

More of same allergen enters body 4

Allergen combines with IgE on mast cells, triggering release of histamines from mast cell 5

Histamine stimulates dilation of blood vessels, causing fluid to leak out; stimulates release of copious amounts of mucus; and causes contraction of smooth muscle in bronchioles 6

Antigen

IgE antibodies

Mast cell that have IgE antibodies attached

Histamine-containing granules

Antigen

Mast cell granules release histamine after allergen binds to IgE antibodies

Histamines and other chemicals

Fluid pours out of capillaries

Mucus is copiously released

Small respiratory passages (bronchioles) constrict

▶ **FIGURE 15.14** The steps in an allergic reaction.

weak antigens, called **allergens**, include a wide range of substances: house dust, pollen, cat dander, certain foods, and even such medicines as penicillin. It is estimated that up to 10% of the U.S. population suffer from at least one allergy. Typically, allergic reactions develop after a first exposure to an allergen. The allergen causes B cells to make antibodies that belong to the IgE class of immunoglobulins. These antibodies attach to mast cells in tissues including those of the nose and respiratory system.

In a subsequent exposure to the same antigen, the allergen binds to the IgE, the mast cells release histamine, which triggers an inflammatory response, resulting in fluid accumulation, tissue swelling, and mucus secretion.

This reaction is severe in some individuals, and histamine is released into the circulatory system, causing a life-threatening decrease in blood pressure and constriction of airways in the lungs. This reaction, called **anaphylaxis** or anaphylactic shock, most often occurs after exposure to the antibiotic penicillin or to the venom in bee or wasp stings. Prompt treatment of anaphylaxis with antihistamines, epinephrine, and steroids can reverse the reaction. As the name suggests, antihistamines block the action of histamine. Epinephrine opens the airways and constricts blood vessels, raising blood pressure. Steroids, such as prednisone, inhibit the inflammatory response. Some people who have a history of severe reaction to insect stings carry these drugs with them in a kit.

Autoimmune Reactions Cause the Immune System to Attack the Body

One of the most elegant properties of the immune system is its capacity to distinguish self from nonself. During development, the immune system "learns" not to react against cells of the body. In some disorders, this immune tolerance breaks down, and the immune system attacks and kills cells and tissues in the body. Juvenile diabetes, also known as insulin-dependent diabetes (*IDDM*) (MIM/OMIM 222100), is an autoimmune disease. Clusters of cells in the pancreas normally produce insulin, a hormone that lowers blood sugar levels. In IDDM, the immune system attacks and kills the insulin-producing cells, resulting in diabetes and the need for insulin injections to control blood sugar levels.

Other forms of autoimmunity, such as systemic lupus erythematosus (*SLE*) (MIM/OMIM 152700), are directed against many of the major organ systems in the body instead of against just one cell type. Some of these autoimmune disorders are listed in Table 15.7.

Genetic Disorders Can Impair the Immune System

In 1952 Ogden Bruton, a physician at Walter Reed Army Hospital, first described an immunodeficiency disease. He examined a young boy who had suffered at least 20 serious infections in the preceding five years. Blood tests indicated that the child had no antibodies. Other patients were discovered who had similar characteristics. Affected individuals were usually boys who were highly susceptible to bacterial infections. In all of these cases, either the B cells were completely absent, or the B cells were immature and unable to produce antibodies. Without functional B cells, no antibodies can be produced. On the other hand, these individuals usually have nearly normal levels of T cells. In other words, antibody-mediated immunity is absent, but cellular immunity is normal. This X-linked disease, called **X-linked agammaglobulinemia** (*XLA*) (MIM/OMIM 300300), usually appears five to six months after birth when maternal antibodies transferred to the fetus during pregnancy disappear and the infant's B-cell population normally begins to produce antibodies. Patients who have this syndrome are highly susceptible to pneumonia and streptococcal infections and pass from one life-threatening infection to another.

■ **Allergens** Antigens that provoke an immune response.

■ **Anaphylaxis** A severe allergic response in which histamine is released into the circulatory system.

Table 15.7 Some Autoimmune Diseases

Autoimmune Diseases
Addison's disease
Autoimmune hemolytic anemia
Diabetes mellitus— insulin-dependent
Graves' disease
Membranous glomerulonephritis
Multiple sclerosis
Myasthenia gravis
Polymyositis
Rheumatoid arthritis
Scleroderma
Sjögren's syndrome
Systemic lupus erythematosus

■ **X-linked agammaglobulinemia (XLA)** A rare, sex-linked, recessive trait characterized by the total absence of immunoglobulins and B cells.

Why Bee Stings Can Be Fatal

*I*n most cases an allergy is an inconvenience, not a life-threatening condition. Typically, the response to contact with an allergen involves localized itching, swelling, and reddening of the skin, often in the form of hives. Inhaled allergens usually result in a localized response accompanied by itching and watery eyes, runny nose, and constricted bronchial tubes. The most common form of allergy involving inhaled allergens is hay fever. In the United States, more than 20 million people are affected.

In response to injected allergen (such as venom from a bee sting) or certain drugs (such as penicillin), the reaction can be systemic rather than localized and result in anaphylactic shock. Anaphylaxis is usually a two-stage process. In the first stage, say, a bee sting, the venom enters the body, and IgE antibodies are made in response, causing sensitization to future stings. It is not clear why some people become sensitized and others do not. If a person has become sensitized, a second exposure to bee venom can provoke a life-threatening allergic reaction. When a bee stings a sensitized individual, the allergen in the venom enters the body and binds to immunoglobulin E antibodies made in response to previous exposure. Within 1 to 15 minutes after exposure, the IgE antibodies activate mast cells. The stimulated mast cells release large amounts of histamines and chemotactic factors, which attract other white blood cells as part of the inflammatory response.

In addition, the mast cells release prostaglandins and the slow-reacting substance of anaphylaxis (SRS-A). SRS-A is more than 100 times more powerful than histamine and prostaglandins in eliciting an allergic reaction and intensifies the response to the allergen.

Release of SRS-A, histamine, and other factors into the bloodstream causes a systemic reaction. In response, the bronchial tubes constrict, closing the airways, and fluids pass from the tissues into the lungs, making breathing difficult. Blood vessels dilate, dropping blood pressure, and plasma escapes into the tissues, causing shock. Heart arrhythmias and cardiac shock can develop and cause death within one to two minutes after the onset of symptoms. Immediate treatment with epinephrine, antihistamines, and steroids is effective in reversing the symptoms.

In the case of hypersensitivity to insect stings, desensitization can be used as immunotherapy. This consists of injecting small quantities of the allergen, gradually increasing the dose. Exposure over time to small doses of the allergen causes the production of IgG antibodies against the allergen rather than IgE antibodies. When desensitization is complete, subsequent exposure to the allergen causes the more numerous IgG antibodies to bind to the mast cells, preventing the binding of IgE antibodies and the resulting anaphylactic reaction.

▓ **Severe combined immunodeficiency disease (SCID)** A genetic disorder in which affected individuals have no immune response; both the cell-mediated and antibody-mediated responses are missing.

Individuals who have *XLA* lack mature B cells, but they have normal populations of B-cell precursors, indicating that the defect is in the maturation of B cells. The *XLA* gene was mapped to Xq21.3–Xq22 and encodes an enzyme that transmits signals from the external environment into the cytoplasm. Chemical signals from outside the cell help trigger B-cell maturation, and the gene product that is defective in *XLA* plays a critical role in this signaling process. Understanding the role of this protein in B-cell development may permit the use of gene therapy to treat this disorder.

A rare genetic disorder causes a complete absence of *both* the antibody-mediated and cell-mediated immune responses. This condition is called **severe combined immunodeficiency (SCID)** (MIM/OMIM 102700, 600802, and others). Affected individuals have recurring and severe infections and usually die at an early age from seemingly minor infections. One of the oldest known survivors of this condition was David, the "boy in the bubble," who died at 12 years of age after being isolated in a sterile plastic bubble for all but the last 15 days of his life (▶ Figure 15.15).

One form of *SCID* is associated with a deficiency of an enzyme known as adenosine deaminase (*ADA*). A small group of children affected with *ADA-deficient SCID* (MIM/OMIM 102700) is currently undergoing gene therapy to give them a normal copy of the gene. The genetically modified white blood cells are replaced into their circulatory systems by transfusion. Expression of the normal *ADA* gene stimulates

▶ **FIGURE 15.15** David, the "boy in the bubble," the oldest known survivor of severe combined immunodeficiency disease.

the development of functional T and B cells and at least partially restores a functional immune system. The recombinant DNA techniques used in gene therapy are reviewed in Chapter 13.

AIDS Attacks the Immune System

The immunodeficiency disorder that is currently receiving the most attention is **acquired immunodeficiency syndrome (AIDS).** AIDS is a collection of disorders that develop as a result of infection with a retrovirus known as the human immunodeficiency virus (HIV) (▶ Figure 15.16). HIV consists of three components: (1) a protein coat that encloses (2) RNA molecules (the genetic material) and (3) an enzyme called reverse transcriptase. The entire viral particle is enclosed in a coat derived from the plasma membrane of a T cell. The virus selectively infects and kills T4 helper cells of the immune system. Inside a cell, reverse transcriptase transcribes the RNA into a DNA molecule, and the viral DNA is inserted into a human chromosome, where it can remain for months or years.

Later, when the infected T cell is called upon to participate in an immune response, the viral genes are activated. Viral RNA and proteins are made, and new viral particles are formed. These bud off the surface of the T cell, rupturing and killing the

▶ **FIGURE 15.16** Particles of HIV.

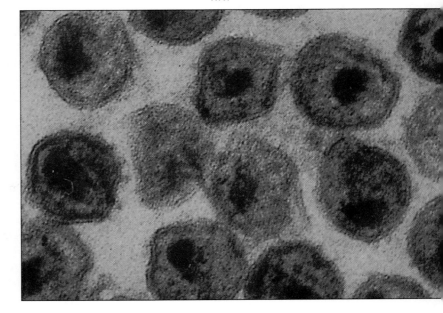

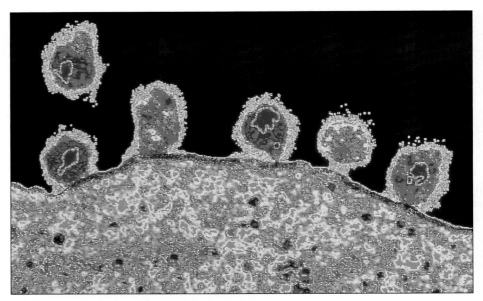

▶ **FIGURE 15.17** A colored electron micrograph showing particles of HIV budding off the surface of a helper T-lymphocyte.

cell and setting off a new round of T-cell infection (▶ Figure 15.17). Gradually, over the course of HIV infection, the number of helper T4 cells decreases. (Recall that these cells act as the master "on" switch for the immune system.) As the T4 cell population falls, the ability to mount an immune response decreases. The result is increased susceptibility to infection and increased risk of certain forms of cancer. The eventual outcome is premature death brought about by any of a number of diseases that overwhelm the body and its compromised immune system.

HIV is transmitted from infected to uninfected individuals through body fluids, including blood, semen, vaginal secretions, and breast milk. The virus cannot live for more than 1 to 2 hours outside the body and cannot be transmitted by food, water, or casual contact.

Guest Essay: Research and Applications

Medicine—A Scientific Safari

M. Michael Glovsky

*W*hat motivates a person to become a physician or a scientist? Chance, the desire for adventure, and the opportunity to discover the unknown are important ingredients. Let me explain.

I grew up in a small town in central Massachusetts and attended Tufts University. Chemistry was a favorite subject. I learned that it was possible to question why certain chemical reactions occurred and that several different approaches could be used to solve problems. My desire to work in a rewarding and stimulating profession was also important. I then entered Tufts Medical School. Most medical students are overwhelmed by the amount of knowledge that is available. I was not an exception. Yet, after the initial shock, coupled with long hours of study, I adapted to the need to absorb the most important information in anatomy, physiology, and biochemistry. During my second year, I became intrigued by problems in immunology. How is the fetus able to tolerate the foreign environment of the mother's uterus? Why do habitual abortions occur? I had an opportunity to explore these two questions in the laboratory of Dr. Wadi Bardawell, an obstetrical pathologist. In the final two years of medical school, patient contact and the reality of the consequences of disease were opportunities to integrate knowledge gained in the basic sciences.

After graduation from medical school, I spent almost four years learning about complement and immunoglobulins. I was fortunate to have as mentors Dr. Elmer Becker at the Walter Reed Army Institute of Research and Hugh Fudenberg, M.D., at the University of California, San Francisco. My initial research addressed what chemicals could block the inflammatory reactions in the skin when complement and immunoglobulins interact.

Because of my interest in immunology, I became an allergist. Can chemicals derived from the structure of IgE, the allergy antibody, interfere in the allergic response? We were fortunate to have as collaborators Drs. Bergitt Helm and Hannah Gould from King's College, London. They provided recombinant proteins synthesized from the known structure of the heavy chain regions of the IgE myeloma proteins. Together, we were able to show that the IgE recombinant fragment (301–376) could block ragweed allergens and grass allergen reactions in human skin and in blood basophils. Further studies were performed to pinpoint the binding site of the allergy antibody, IgE, that binds to human basophils.

More recently we have studied substances in air pollution that cause the nose to run and the bronchial tubes to constrict. We are exploring whether latex, an ingredient in natural rubber, produces allergenic proteins that are important in asthma. Together with Dr. Ann Miguel and Dr. Glenn Cass at Caltech, we have found latex allergen in air samples and roadside dust samples. Latex allergen, present in natural rubber gloves, balloons, and tire dust, is one of the most potent allergens of the last 10 years. We are studying whether latex allergy is important in the increasing symptoms of asthma in the last decade.

Medical science has evolved to provide sophisticated tools, especially in molecular biology, to address important questions and seek relevant answers. Biological science is a safari through the wilderness. Every path leads to adventure and exploration with both frustration and rewards. But the trail is always interesting and provocative.

M. Michael Glovsky is director of the Asthma and Allergy Center at Huntington Memorial Hospital in Pasadena, California. He received a B.S. degree from Tufts University in 1957 and an M.D. degree from Tufts Medical School in 1962.

Case Studies

CASE 1

Mary and John Smith went for genetic counseling because John did not believe that their newborn son was his. Mary and John wanted to test their blood types to help rule out the possibility that someone other than John is the father of this baby. The counselor explained that ABO blood typing could give some preliminary indications about possible paternity. Its use is limited, however, because there are only four possible ABO blood types, and the vast majority of people in any population have only two of those types (A and O). This means that a man may have a blood type consistent with paternity and still not be the father of the tested child. Because the allele for type O can be masked by the genes for A or B, inheritance of blood type can be unclear. More modern genetic tests, such as DNA typing, would lead to a more reliable conclusion regarding paternity.

Mary's blood was tested and identified as type O. John's blood was tested and identified as type O. Based on these two parental combinations of blood types, the only possible blood type that their son could be is type O. The baby was tested and his blood type is, indeed, type O.

And the Father is

If the Mother is		A	B	AB	O	The Child must be
	A	A or O	A, B, AB or O	A, B, or AB	A or O	
	B	A, B, AB, or O	B or O	A, B, or AB	B or O	
	AB	A, B, or AB	A, B, or AB	A, B, or AB	A or B	
	O	A or O	B or O	A or B	O	

CASE 2

The Jones were referred to a clinical geneticist because their six-month-old daughter was failing to grow adequately and was having recurrent infections. The geneticist took a detailed family history (which was uninformative) and medical history of their daughter. He discovered that their daughter had a history of several ear infections against which antibiotics had no effect, she had a difficult time gaining weight (failure to thrive), and had an extensive history of yeast infection (thrush) in her mouth. The geneticist did a simple blood test to check their daughter's white blood count and determined that she had severe combined immunodeficiency (SCID).

The geneticist explained that SCID is an immune deficiency that causes a marked susceptibility to infections. The defining characteristic is usually a severe defect in both the T- and B-lymphocyte systems. This usually results in one or more infections within the first few months of life. These infections are usually serious and may even be life threatening. They may include pneumonia, meningitis, or bloodstream infections. Based on the family history, it is possible that their daughter inherited an altered gene from each of them and therefore was homozygous for the gene that causes SCID. Each time the Jones had a child there is a 25% chance that the child will have SCID. Prenatal testing is available to determine whether the developing fetus has SCID.

Summary

1. The immune system protects the body against infection by a graded series of responses that attack and inactivate foreign molecules and organisms. The lowest level called innate immunity, involves a nonspecific, local, inflammatory response.

2. The immune system has two components: antibody-mediated immunity, which is regulated by B cells and antibody production, and cell-mediated immunity, which is controlled by T cells.

3. The primary function of antibody-mediated reactions is to defend the body against invading viruses and bacteria. Cell-mediated immunity is directed against cells of the body that have been infected by agents such as viruses and bacteria.

4. The presence or absence of certain antigens on the surface of blood cells is the basis of blood transfusions and blood types. Two blood groups are of major significance: the ABO system and the Rh blood group.

5. Matching ABO blood types is important in blood transfusions. In some cases, mother-fetus incompatibility in the Rh system can cause maternal antigens to destroy red blood cells of the fetus, resulting in hemolytic disease in the newborn.

6. The success of organ transplants and skin grafts depends on matching histocompatibility antigens found on the surface of all cells in the body. In humans, the antigens produced by a group of genes on chromosome 6 (known as the MHC complex) play a critical role in the outcome of transplants.

7. Allergies are the result of immunological hypersensitivity to weak antigens that do not provoke an immune response in most people. These weak antigens, known as allergens, include a wide range of substances: house dust, pollen, cat hair, certain foods, and even medicines, such as penicillin.

8. Acquired immunodeficiency syndrome (AIDS) is a collection of disorders that develop as a result of infection with a retrovirus known as the human immunodeficiency virus (HIV). The virus selectively infects and kills the T4 helper cells of the immune system.

Questions and Problems

1. Identify the components of cellular immunity, and define their roles in the immune response.

2. Compare the general inflammatory response, the complement system, and the specific immune response.

3. Compare the roles of the three types of T cells: helper cells, suppressor cells and killer T cells.

4. Distinguish between antibody-mediated and cell-mediated immunity. What components are involved in each?

5. Describe the rationale for vaccines as a form of preventive medicine.

6. Antibodies have long been a subject for study because of the staggering variety of antigens to which they are capable of responding. What other element of the immune system is equally adept at antigen recognition?

7. The molecular weight of IgG is 150,000 kd. What are the molecular weights of each individual subunit assuming that the two heavy chains are equivalent and the two light chains are equivalent? Also assume that the molecular weights of the light chains are half the molecular weight of the heavy chains.

8. Name the class of molecules that includes antibodies and the five groups that make up this class.

9. a. How is Rh incompatibility involved in hemolytic disease of the newborn? Is the mother Rh^+ or Rh^-? Is the fetus Rh^+ or Rh^-?

   b. Why is a second child that is Rh^+ more susceptible to attack from the mother's immune system?

10. It is often helpful to draw a complicated pathway in the form of a flow chart in order to visualize the multiple steps and the ways in which the steps are connected to each other. Draw the antibody-mediated immune response pathway that acts in response to an invading virus.

11. AIDS is an immunodeficiency syndrome. In the flow chart you drew above, describe where AIDS sufferers are deficient. Why can't our immune systems fight off this disease?

12. Describe the genetic basis of antibody diversity.

13. In the human HLA system there are 23 HLA-A alleles, 47 for HLA-B, 8 for HLA-C, 14 for HLA-DR, 3 for HLA-DQ, and 6 for HLA-DP. How many different human HLA genotypes are possible?

14. What mode of inheritance has been observed for the HLA system in humans?

15. Why can someone with blood type AB receive blood of any type? Why can a blood type O individual donate blood to anyone?

16. What is more important to match during blood transfusions: the antibodies of the donor or the antigens of the donor/recipient?

17. The following data were presented to a court during a paternity suit: (1) the infant is a universal donor for blood transfusions; (2) the mother bears antibodies against the B antigen only; (3) the alleged father is a universal recipient in blood transfusions. Can you identify the ABO genotypes of the three individuals? Can the court draw any conclusions?

18. A man has the genotype $I^A I^A$ and his wife is $I^B I^B$. If their son needed an emergency blood transfusion, would either parent be able to be a donor? Why or why not?

19. In cystic fibrosis gene therapy, scientists propose the use of viral vectors to deliver normal genes to cells in the lungs. What immunological risks are involved in this procedure?

20. A burn victim receives a skin graft from her brother; however; her body rejects the graft a few weeks later. The procedure is attempted again, but this time the graft is rejected in a few days. Explain why the graft was rejected the first time and why it was rejected faster the second time.

21. In the near future, pig organs may be used for organ transplants. How are researchers attempting to prevent rejection of the pig organs by human recipients?

22. An individual has an immunodeficiency that prevents helper T cells from recognizing the surface antigens presented by macrophages. As a result, the helper T cells are not activated, and they, in turn, fail to activate the appropriate B cells. At this point, is it certain that the viral infection will continue unchecked?

23. A couple have the following HLA genotypes:
Male A12, B36, C7, DR14, DQ2, DP5/A9, B27, C3, DR14, DQ3, DP1
Female A8, B15, C2, DR5, DQ2, DP3/A2, B20, C3, DR8, DQ1, DP3
What is the probability that an offspring will suffer from ankylosing spondylitis if:
   a. The child is male and has HLA-A9?
   b. The child is female and has HLA-A9?
   c. The child is male and has HLA-DP1?
   d. The child is male and has HLA-C7?

24. A patient of yours has just undergone shoulder surgery and is now experiencing kidney failure for no apparent reason. You check his chart and find that his blood is type B, but he has been mistakenly transfused with type A. Describe why he is experiencing kidney failure.

25. Assume a single gene having alleles that show complete dominance relationships at the phenotypic level controls the Rh character. An Rh^+ father and an Rh^- mother have 8 boys and 8 girls, all Rh^+.
   a. What are the Rh genotypes of the parents?
   b. Should they have been concerned about hemolytic disease of the newborn?

26. Why are allergens called "weak" antigens?

27. Antihistamines are used as anti-allergy drugs. How do these drugs work to relieve allergy symptoms?

28. A young boy, who has suffered from over a dozen viral and bacterial infections in the past 2 years, comes to your office for an examination. You examine the boy and determine by testing that he has no circulating antibodies. What syndrome does he have and what are its characteristics? What component of the two-part immune system is nonfunctional?

29. Researchers have been having a difficult time developing a vaccine against a certain pathogenic virus due to the lack of a weakened strain. They turn to you because of your wide knowledge of recombinant DNA technology and the immune system. How could you vaccinate someone against the virus using a cloned gene from the virus that encodes a cell-surface protein?

30. A couple has a young child who needs a bone marrow transplant. They propose that pre-implantation screening be done on several embryos fertilized *in vitro* to find a match for their child. What do they need to match in this transplant procedure? They propose that the matching embryo will be transplanted to the mother's uterus, and will serve as a bone marrow donor when old enough. What are the ethical issues involved in this proposal?

Internet Activities

The following activities use the resources of the World Wide Web to enhance the topics covered in this chapter. To investigate the topics described below, log on to the book's home page at:

   http://www.brookscole.com/biology

1. Because the normal functioning of the immune response in humans requires the delicate interplay of B cells, T cells, and phagocytic cells, as well as the action of several immunoglobulin classes and cytokines, it is easy to conclude that several genetic disorders are recognized as being related to an absent or abnormal immune function. Access the National Institute of Allergy and Infectious Disease fact sheet on Primary Immune Deficiencies. Which of the described disorders have a demonstrated genetic link and which have no described clear pattern of inheritance?

2. The Journal of Immunology is available on-line, and includes a "Cutting Edge" section dealing with the latest advances in immunology. Browse this section for information on immunogenetics. Use the site to see how lymphocytes produce antibodies (at the Cells Alive site). You may also use this site as a personal resource by posting a question to be answered by immunologists, on any topic presented in this chapter or on something you read somewhere else.

3. Adenosine deaminase deficiency is the most common cause of severe combined immunodeficiency (SCID) affecting both humoral and cell-mediated immune responses. Much information is known about this particular genetic disorder. Use the Pyrimidine and Purine Data Base page of the European Society for the Study of Purine and Pyrimidine Metabolism in Man to find details on the biochemistry, clinical presentation, molecular biology, diagnosis, and treatment of this disorder.

For Further Reading

Allan, J. S. (1996). Xenotransplantation at a crossroads: Prevention versus progress. *Nat. Med. 2:* 18–21.

Arnett, F. (1986). HLA genes and predisposition to rheumatic diseases. *Hosp. Pract. 20:* 89–100.

Bach, F. H., Winkler, H., Wrighton, C. J., Robson, S. C., Stuhlmeier, K., & Ferran, C. (1996). Xenotransplantation: A possible solution to the shortage of donor organs. *Transplant. Proc. 28:* 416–417.

Barre-Sinoussi, F. (1996). HIV as the cause of AIDS. *Lancet 348:* 31–35.

Bosma, M. J. (1989). The SCID mutation: Occurrence and effect. *Curr. Topics Microbiol. Immunol. 152:* 3–9.

Britton, C. (1993). HIV infection. *Neurol. Clin. 11:* 605–624.

Buckley, R. (1992). Immunodeficiency diseases. *JAMA 268:* 2797–2806.

Burrows, P., & Cooper, M. (1993). B-cell development in man. *Curr. Opinion Immunol. 5:* 201–206.

Dierich, M. P., Stoiber, H., & Clivio, A. (1996). A complementary AIDS vaccine. *Nat. Med. 2:* 153–155.

Hors, J., & Gony, J. (1986). HLA and disease. *Adv. Nephrol. 15:* 329–351.

Iseki, M., & Heiner, D. (1993). Immunodeficiency disorders. *Pediatr. Rev. 14:* 226–236.

Just, J. J. (1995). Genetic predisposition to HIV-1 infection and acquired immunodeficiency syndrome: A review of the literature examining association with HLA. *Hum. Immunol. 44:* 156–159.

Kotzin, B., Leung, D., Kappler, J., & Marrack, P. (1993). Superantigens and their potential role in human disease. *Adv. Immunol. 54:* 99–166.

Lokki, M. L., & Colten, H. R. (1995). Genetic deficiency of complement. *Ann. Med. 27:* 451–459.

McDivitt, H. (1986). The molecular basis of autoimmunity. *Clin. Res. 34:* 163–175.

Morgan, B. P. (1996). Intervention in the complement system: A therapeutic strategy in inflammation. *Biochem. Soc. Trans. 24:* 224–229.

Puck, J. (1993). X-linked immunodeficiencies. *Adv. Hum. Genet. 21:* 107–144.

Quinn, T. C. (1996). Global burden of the HIV pandemic. *Lancet 348:* 99–106.

Rapaport, F. T. (1995). Skin transplantation—experimental basis for the study of human histocompatibility. *Transplant. Proc. 27:* 2205–2210.

Tomlinson, I. P., & Bodmer, W. F. (1995). The HLA system and the analysis of multifactorial genetic disease. *Trends Genet. 11:* 493–498.

Tonegawa, S. (1985). The molecules of the immune system. *Sci. Am. 253* (October): 123–131.

Vetri, D., Vorechovsky, I., Sideras, P., Holland, J., Davies, A., Flinter, F., Hammarstrom, L., Kinnon, C., et al. (1993). The gene involved in X-linked agammaglobulinaemia is a member of the src family of protein-tyrosine kinases. *Nature 361:* 226–233.

Vladutiu, A. (1993). The severe combined immunodeficient (SCID) mouse as a model for the study of autoimmune diseases. *Clin. Exp. Immunol. 93:* 1–8.

Yamamoto, F. (1995). Molecular genetics of the ABO histo-blood group system. *Vox Sang. 69:* 1–7.

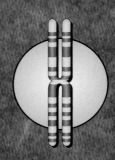

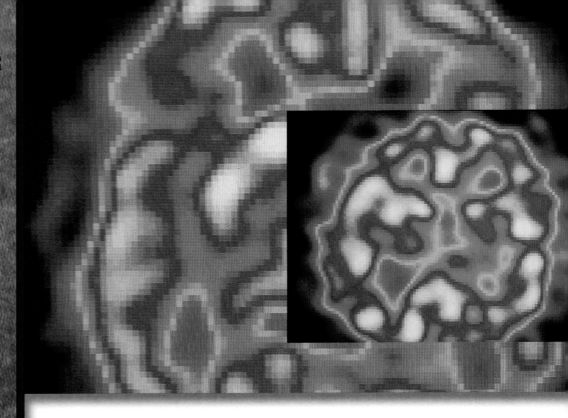

Genetics of Behavior

Chapter Outline

*S*ometime in the late 1850s, an 8-year-old boy and his father, a physician, drove along a road through the woods on eastern Long Island. They met two women walking along the road, and this chance encounter had a profound effect on the young boy. Years later, he would recall that meeting:

> I recall it as vividly as though it had occurred but yesterday. It made a most enduring impression upon my boyish mind, an impression every detail of which I recall today, an impression which was the very first impulse to my choosing chorea as my virgin contribution to medical lore. Driving with my father through a wooded road leading from East Hampton to Amagansett, we suddenly came upon two women, mother and daughter, both tall, thin, almost cadaverous, both bowing, twisting, grimacing. I stared in wonderment, almost in fear. What could it mean? My father paused to speak to them and we passed on.

The young boy, George Huntington, went on to study medicine at Columbia University. In 1872, a year after graduation, he published an account of this disorder, which became known as Huntington's chorea, or as it is called today, Huntington disease. In the paper, his summary of the condition is a model of brevity and clarity:

> There are three characteristics of this disorder: 1. Its hereditary nature, 2. A tendency to insanity and suicide, 3. Its manifesting itself as a grave disease only in adult life.

Huntington described the pattern of inheritance of the disorder in a way that is consistent with an autosomal dominant trait. Unfortunately, having described the condition, Huntington turned his attention in other directions, and little more was accomplished in the next 88 years until recombinant DNA techniques were used to map the locus for this disease to the short arm of chromosome 4.

In several ways, conditions such as Huntington disease (HD) are an ideal model for genetic disorders that affect behavior. The pattern of autosomal dominant inheritance is well defined and clear-cut. The gene for HD was one of the first to be mapped using restriction fragment length polymorphisms (RFLPs), which demonstrated the power of molecular techniques in analyzing human behavior. More recently, positional cloning, one of the newer genetic techniques of recombinant DNA technology, was used to isolate the gene. Lastly, the molecular basis of mutation in the HD gene represents a new class of mutations that affect the nervous system. From all of this, it would seem that by following the methods used in studying Huntington disease, researchers could identify, map, and isolate many genes affecting human behavior.

Unfortunately, searches for single genes (like the HD gene) that control behavior may have only limited success. One of the most difficult problems in *human behavioral genetics is defining the phenotype. Although Huntington disease has a well-defined phenotype and progression, many behavioral traits do not. In addition, the phenotype of some conditions, such as schizophrenia (well defined as a medical condition), can be genetically heterogeneous and may actually include several genetic disorders, each of which has a similar phenotype. Finally, many behavioral traits are multifactorial; the phenotype is determined by several genes and environmental interactions, and no single gene has a major effect.*

To understand the issues in behavior genetics and how decisions about phenotypic definitions, genetic models of behavior, and methods influence both the speed and the outcome of this research, this chapter begins with a discussion of the genetic models and methods used in studying human behavior. Then we briefly consider animal models, for which single-gene effects on behavior have been well documented. We discuss single genes that affect human behavior through their effect on the nervous system and then consider more complex traits and those that have the greatest social impact. The chapter ends with a summary of the current state of human behavior genetics and the ethical, legal, and social implications of this research.

MODELS, METHODS, AND PHENOTYPES IN STUDYING BEHAVIOR

The results of pedigree analysis, family studies, adoption, and twin studies suggest that aspects of human behavior are under genetic control. However, most behavioral characteristics that have genetic components are likely to be determined by several genes, by interaction with other genes, and are influenced by environmental components. In fact, most behaviors are not inherited as simple Mendelian traits, demonstrating the need for genetic models that can explain observed patterns of inheritance. To a large extent, the model of inheritance proposed for a trait determines the methods used to analyze its pattern of inheritance and the techniques that can be pursued in mapping and isolating the gene or genes responsible for the trait's characteristic phenotype (discussed below).

Genetic Models of Inheritance and Behavior

Several models for genetic effects on behavior have been proposed (Table 16.1). The simplest model is a single gene, dominant or recessive, which affects a well-defined behavior. In fact, several human behavioral traits—including Huntington disease, Lesch-Nyhan syndrome, fragile-X syndrome, and others—can be described by such a model (◗ Figure 16.1). Multiple-gene models are also possible. The simplest of these is a polygenic additive model in which two or more genes contribute equally in an additive manner to the phenotype. This model has been proposed (along with oth-

Table 16.1	*Models for Genetic Analysis of Behavior*
Model	**Description**
Single gene	One gene controls a defined behavior
Polygenic trait	Additive model that has two or more genes
	One or more major genes with other genes contributing to phenotype
Multiple genes	Interaction of alleles at different loci generates a unique phenotype

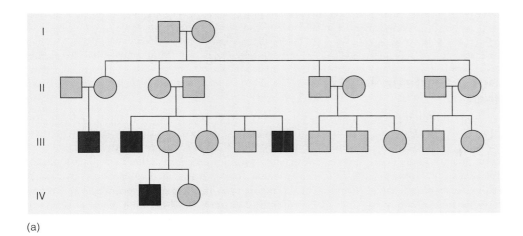

(a)

(b)

▶ **FIGURE 16.1** Many genetic disorders inherited as sex-linked recessive traits (a) or as autosomal dominant traits (b) have behavioral phenotypes.

ers) to explain schizophrenia (the inheritance of additive polygenic traits was considered in Chapter 5). In a variation of this model, one or more genes might have a major effect, and other genes make smaller contributions to the phenotype. Still another multigene model involves a phenomenon called epistasis, a form of gene interaction in which an allele of one gene masks the expression of a second gene's allele. This form of gene interaction has been well documented in experimental genetics, although it has not yet been invoked to explain a human behavior trait.

In each of these models, the environment can make significant contributions, and the study of behavior must take this into account (see Concepts and Controversies: Is Going to Medical School A Genetic Trait?). When this happens, geneticists must use methods that combine a genetic basis and a way to assess the role of the environment in the phenotype.

Methods of Studying Behavior Genetics

For the most part, the methods used to study the inheritance of behavior follow the pattern for other human traits. If the model involves a single gene, pedigree analysis and linkage studies, including the use of RFLP markers and other methods of recombinant DNA technology, are the most appropriate methods. However, because many behavior traits are polygenic, twin studies play a prominent role in human behavior genetics. Concordance and heritability values based on twin studies have established

Is Going to Medical School a Genetic Trait?

*I*t is clear that many behavioral traits exhibit a familial pattern of inheritance. This observation, along with twin studies and adoption studies, indicates that there is a genetic component to many complex behavioral disorders, such as manic depression and schizophrenia. In most cases, these phenotypes are not inherited as simple Mendelian traits. Researchers are then faced with selecting a model to describe how a behavioral trait is inherited. Using this model, further choices select the methods used in genetic analysis of the trait. A common strategy is to find a family in which the behavior appears to be inherited as a recessive or an incompletely penetrant dominant trait controlled by a single gene. One or more molecular markers (such as RFLPs) are used in linkage analysis to identify the chromosome that carries the gene that controls the trait.

If researchers are looking for a single gene when the trait is controlled by two or more genes or by genes that interact with environmental factors, then the work may produce negative results, even though preliminary findings can be encouraging. Reports of loci for manic depression on chromosome 11 and the X chromosome were based on single-gene models, and after initial successes, it was found that these reports were flawed. Overall, regions on 14 chromosomes have been proposed as candidates for genes that control manic depression, but none have been substantiated.

To illustrate some of the pitfalls associated with model selection in behavior genetics, one study selected attendance at medical school as a behavioral phenotype and attempted to determine if the distribution of this trait in families is consistent with a genetic model. This study surveyed 249 first- and second-year medical students. Thirteen percent of these students had first-degree relatives who had also attended medical school, compared with 0.22% relatives of individuals selected from the general population. Thus, the overall risk factor among first-degree relatives for medical school attendance was 61 times higher than in the general population, indicating a strong familial pattern. To determine whether this trait was inherited in a Mendelian manner, researchers used standard statistical analysis, which supported familial inheritance and rejected the model of no inheritance. Analysis of the pedigrees most strongly supported a simple recessive mode of inheritance, although other models including polygenic inheritance were not excluded. Using a further set of statistical tests, the researchers concluded that the recessive mode of inheritance was just at the border of statistical acceptance.

Similar results are often found in studies of other behavioral traits, and it is usually argued that another, larger study would confirm the results, in this case, that attendance at medical school is a recessive Mendelian trait. Although it is true that genetic factors may partly determine whether one will attend medical school, it is unlikely that a single recessive gene controls this decision, although that conclusion is supported by this family study and segregation analysis of the results.

The authors of this study were not serious in their claims that a decision to attend medical school is a genetic trait, nor was it their intention to cast doubts on the methods used in the genetic analysis of behavior. Rather, it was intended to point out the folly of accepting simple explanations for complex behavioral traits.

that there is a genetic component to mental illnesses (manic depression and schizophrenia), and to behavioral traits (sexual preference and alcoholism). The results of such studies should be interpreted with caution because they are subject to the limitations inherent in interpreting heritability (see Chapter 5) and often use small sample sizes, where minor variations can have a disproportionately large effect on the outcome.

To overcome these problems, geneticists are extending and adapting twin studies as a genetic tool to study behavior. One innovation is the study of the children of twins, which is being used to confirm the existence of genes predisposing to a certain behavior. Twin studies are also being coupled to recombinant DNA techniques to search for behavior genes, and this combination may prove to be a powerful method for identifying such genes.

Phenotypes: How Is Behavior Defined?

A second fundamental problem that has limited progress in human behavior genetics is choosing a consistent phenotype as the basis for study. The phenotypic definition must be precise enough to distinguish the behavior from other, similar behaviors and from the behavior of the control group but not so narrow that it excludes some variable parts of a behavior pattern. Recall that gene mapping uses the phenotype as a guide, and beginning with the most accurate phenotype is of primary importance.

For some mental illnesses, clinical definitions are provided by guidelines such as the *Diagnostic and Statistical Manual of Mental Disorders* of the American Psychiatric Association. For other behaviors, the phenotypes are often poorly defined and may not reflect the underlying biochemical and molecular basis of the behavior. For example, alcoholism can be defined as the development of characteristic deviant behaviors associated with excessive consumption of alcohol. Is this definition explicit enough to be useful as a phenotype in genetic analysis? Is there too much room for interpreting what is deviant behavior or what is excessive consumption? As we will see, whether the behavioral phenotype is narrowly or broadly defined can affect the outcome of the genetic analysis and even the model of inheritance for the trait.

ANIMAL MODELS: THE SEARCH FOR BEHAVIOR GENES

One way to begin studying the genetics of human behavior is to ask whether behavior genetics can be studied in model systems. Several approaches have been used in experimental genetics to study behavior in animals. In one method, two closely related species or two strains of the same species are studied to detect variations in behavioral phenotypes. Genetic crosses are used to establish whether these variations in behavior are inherited and if so, to determine the pattern of inheritance.

In another approach, individuals who exhibit variant behavior are isolated from a population and interbred to establish a strain that has a distinct behavioral pattern. More recently, the effects of single genes on animal behavior have been studied. In some cases, these studies have led to isolating and cloning genes that affect behavior. In the following sections, we describe some examples of behavior genetic studies of experimental organisms.

Open-Field Behavior in Mice

Beginning in the mid-1930s, the emotional and exploratory behaviors of mice were tested by studying open-field behavior. When mice are introduced into a brightly lit environment, some mice actively explore the new area, whereas others are apprehensive, do not move about, and have elevated rates of urination and defecation (▶ Figure 16.2). This behavior pattern is under genetic control, since strains exhibiting both types of behavior have been established.

To test the genetic components of this behavior, beginning in the 1960s, studies used an enclosed, illuminated box whose floor is marked into squares. Exploration is tested by counting the mouse's movements in different squares, and emotion is quantified by counting the number of defecations. The BALB/cJ strain, homozygous for a recessive albino allele, shows low exploratory behavior and is highly emotional. The C57BL/6j strain has normal pigmentation, is active in exploration, and shows low levels of emotional behavior.

If these two strains are crossed and the offspring are interbred, each generation beyond the F1 includes both albino and normally pigmented mice. When tested for open-field behavior, pigmented mice behaved like the C57 parental line, showing active exploration and low levels of emotional behavior. The albino mice behaved like the BALB parental line and showed low exploratory activity and high levels of emotional behavior, indicating that the albino gene affects behavior and pigmentation. The overall results show that open-field behavior is a polygenic trait.

Learning in *Drosophila*

The fruit fly, *Drosophila*, offers several advantages to study behavior, including the existence of behavioral mutants (Table 16.2). To learn, flies are presented with a series of odors, one of which is accompanied by an electric shock. Flies learn to avoid

▶ **FIGURE 16.2** An open-field trial. Movements are automatically recorded as the rat moves across the open field.

Table 16.2	**Some Behavior Mutants of Drosophila**	
Category	**Mutation**	**Phenotype**
Learning	*dunce*	Cannot learn conditioned response.
	turnip	Impaired in learning conditioned response.
	rutabaga	Impaired in several types of learning and memory.
Sexual Behavior	*fruitless*	Males court each other.
	savoir-faire	Males unsuccessful in courtship.
	coitus-interruptus	Males stop copulation prematurely.
Motor Behavior	*flightless*	Lacks coordination for flying.
	sluggish	Moves slowly.
	wings up	Holds wings perpendicular to body.

the odor associated with the shock. Mutant screens have identified a number of genes that influence this learning ability, including the mutants *dunce, turnip,* and *rutabaga.*

These single-gene mutants exert their effect through an intracellular signaling system that involves a molecule called cyclic AMP (cAMP). Inside the cells of the nervous system, cAMP sets off a cascade of biochemical reactions that controls gene transcription and the responses associated with learning. Cyclic AMP is produced by the enzyme adenyl cyclase (▶ Figure 16.3); the normal *rutabaga* gene encodes this enzyme. In the mutant, there is no adenyl cyclase, and no production of cAMP. The *turnip* gene encodes a protein that activates adenyl cyclase, and the *dunce* gene controls the pathway by which adenyl cyclase is recycled. The clustering of these muta-

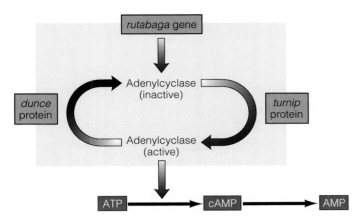

▶ **FIGURE 16.3** The metabolic pathway that involves production of cyclic AMP (cAMP) is involved in learning. In this pathway, ATP is converted into cAMP by the enzyme adenyl cyclase. cAMP is active in signal transduction and then is converted into AMP. In *Drosophila*, the *rutabaga* gene encodes adenyl cyclase. The enzyme is inactive when first produced, and is activated by the action of a protein encoded by the *turnip* gene. When no longer needed, a gene product encoded by the *dunce* gene inactivates the enzyme. When any of these genes are mutant in *Drosophila*, flies have difficulty in learning.

tions in the cAMP pathways provides strong evidence for the involvement of cAMP in learning. Experiments in other organisms support this finding and imply that some aspects of learning and memory in humans may involve cyclic nucleotides.

SINGLE-GENE EFFECTS ON HUMAN BEHAVIOR

In Chapter 10, we discussed the role of genes in metabolism. Mutations that disrupt metabolic pathways or interfere with the synthesis of essential gene products can influence the function of cells and in turn produce an altered phenotype. If the affected cells are part of the nervous system, alterations in behavior may be part of the phenotype. In fact, some genetic disorders do affect cells in the nervous system and in turn affect behavior. In PKU, for example, brain cells are damaged by excess levels of phenylalanine, causing mental retardation and other behavioral deficits.

In this section we discuss several single-gene defects that have specific effects on the development, structure, and/or function of the nervous system, and that consequently, affect behavior. Following this we discuss more complex interactions between the genotype and behavior, where the number and functional roles of genes are not well understood and where effects on the nervous system may be more subtle.

Lesch-Nyhan Syndrome Is a Disorder of Nucleic Acid Metabolism

Lesch-Nyhan syndrome (MIM/OMIM 308000) is a rare X-linked recessive trait that affects 1 in 100,000 males and is associated with kidney failure, spastic movements, mental retardation, high levels of uric acid in the blood and urine, and a strong tendency for self-mutilation. Hemizygous males are unaffected at birth, but delays in development appear at about 3 months. Later, writhing and uncontrollable spastic movements become apparent, but the most striking feature of this disease is the compulsive self-mutilation that usually appears between 2 and 4 years of age. Unless restrained, affected individuals bite off pieces of their fingers, lips, and cheeks. No affected females have been described, and most affected males die by the age of 20 years.

Lesch and Nyhan proposed that this condition is caused by an inborn error in purine metabolism (purines are bases found in DNA and RNA; review the structure of nucleic acids in Chapter 8). In 1967, J. E. Seegmiller and his colleagues demonstrated that the disease is caused by a lack of the enzyme hypoxanthine-guanine phosphoribosyltransferase (HGPRT) (▶ Figure 16.4). Normally, excess nucleic acids are degraded into nucleotides during cell metabolism, and the bases are stripped off. A large fraction of the purine bases are reused, and end up in newly synthesized DNA and RNA molecules. In the absence of HGPRT, the purine bases hypoxanthine and guanine are converted to uric acid, which build up in the blood to levels up to six times normal.

■ **Lesch-Nyhan syndrome** An X-linked recessive condition associated with a defect in purine metabolism that results in mental retardation, spastic movements, and self-mutilation.

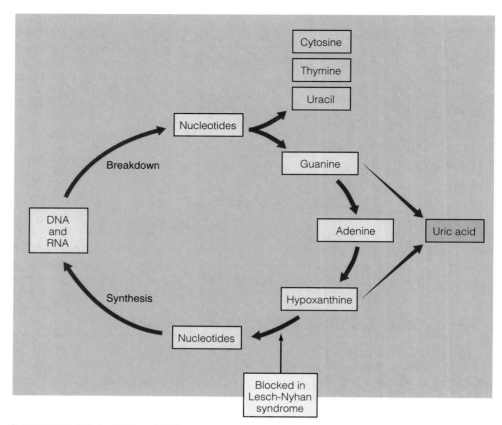

FIGURE 16.4 DNA and RNA are metabolically degraded into nucleotides. The bases are removed from the nucleotides by the action of enzymes. Most of the guanine and hypoxanthine (derived from adenine) are reused in making new nucleotides. Excess amounts are converted into uric acid, which is excreted in the urine. In the Lesch-Nyhan syndrome, guanine and hypoxanthine cannot be reused and are converted into uric acid. The buildup of uric acid is toxic to the nervous system and produces the symptoms of the Lesch-Nyhan syndrome.

Lesch-Nyhan syndrome affects the function of the brain and nervous systems, but no detectable changes occur in the structure of cells in the nervous system. Normally, only low levels of new brain cells are produced in children older than 2 years of age, and recent work has suggested that the progressive neurological degeneration in the Lesch-Nyhan syndrome may result from the gradual accumulation of damage to DNA in brain cells. How the lack of the HGPRT enzyme brings about DNA damage and the resulting phenotype is not known.

Charcot-Marie-Tooth Disease Affects Nerve Fibers

This disorder of the peripheral nervous system is characterized by the onset of weakness and atrophy in the muscles of the lower leg and later, may involve arm muscles (teeth are not involved; H.H. Tooth described the condition in 1886). Loss of sensation in the feet and hands may also occur. Affected nerve fibers lose their myelin sheath (▶ Figure 16.5), and the speed of nerve impulse conduction is greatly reduced. One form of **Charcot-Marie-Tooth** (MIM/OMIM 118200) **disease** (*CMT-1*) is inherited as an autosomal dominant disorder; mutations at any of three loci are associated with this form of CMT. One of these has been mapped to chromosome 17, one to chromosome 1, and a third gene has yet to be localized. Individuals who have *CMT-1* begin to show loss of control over foot movement during childhood and develop muscle wasting and loss of sensation in the lower legs. Muscle atrophy and loss

■ **Charcot-Marie-Tooth disease** A heritable form of progressive muscle weakness and atrophy. One form, CMT-1, can result from mutation at any of three loci.

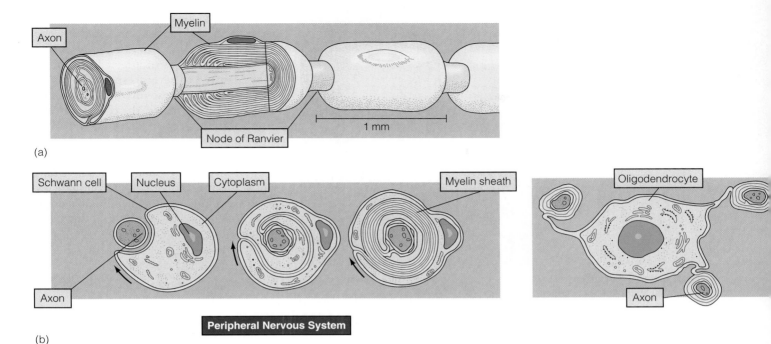

(a)

(b)

Peripheral Nervous System

(c)

▶ **FIGURE 16.5** (a) A sheath surrounds myelinated nerve fibers at regular intervals. Each nonmyelinated interval is called the node of Ranvier. (b) In the peripheral nervous system, a Schwann cell that wraps itself around a single nerve fiber forms the myelin sheath. In *CMT-1*, a gene called the peripheral myelin gene (PMP-22) is mutant, producing a defective myelin sheath that breaks down, slowing nerve impulses. (c) A transmission electron micrograph of a myelinated nerve fiber viewed in cross-section.

of sensation in hands appears later. Walking becomes difficult, and leg braces are often needed.

The *CMT-1* locus on chromosome 17 encodes a membrane-associated protein that becomes part of the myelin sheath around nerve cells in the peripheral nervous system. Altered expression of this protein impairs peripheral nerve function, which in turn leads to muscle atrophy. The gene product encoded by the locus on chromosome 1 is also a myelin-associated protein, and presumably, mutation in this gene causes muscle atrophy indirectly through its effects on the myelin sheath.

Huntington Disease Is an Adult Onset Disorder

This autosomal dominant disorder is usually first expressed in mid-adult life as involuntary muscular movements and jerky motions of the arms, legs, and torso. As **Huntington disease** (MIM/OMIM 143100) progresses, personality changes, agitated behavior, and dementia occur. Most affected individuals die within 10 to 15 years after the onset of symptoms.

The gene for Huntington disease maps to the tip of the short arm of chromosome 4 (4p16.3). The mutant form of the gene is associated with increases in the number of CAG triplet repeats. Normal individuals have from 6 to 37 copies of this repeat, whereas 35–121 copies are found in affected individuals. A number of other neurodegenerative

Huntington disease An autosomal dominant disorder associated with progressive neural degeneration and dementia. Adult onset is followed by death 10–15 years after symptoms appear.

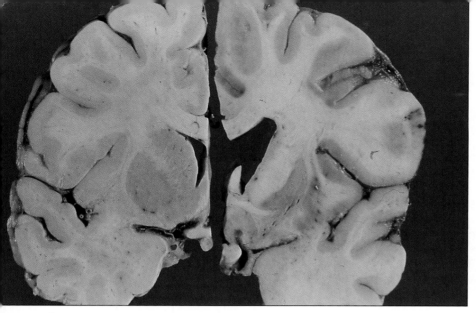

diseases are associated with expansion of repeated nucleotide triplets (discussed in Chapter 11).

Brain autopsies of affected individuals show damage to several, specific brain regions, including those involved with motor activity. In many cases, the cells in affected regions are altered in shape or destroyed (▶ Figure 16.6). Although the gene has been isolated, the gene product has not yet been functionally characterized. Brains from affected individuals accumulate excessive amounts of a neurotoxic chemical, quinolonic acid. In Huntington disease, it appears likely that abnormal metabolism leads to destruction of brain cells, which in turn affects behavior.

▶ **FIGURE 16.6** At left is a slice from a normal brain. At right is a slice from a Huntington patient's brain. The central region (caudate nucleus) is missing.

SINGLE GENES, AGGRESSIVE BEHAVIOR, AND BRAIN METABOLISM

In 1993, a new form of X-linked mild mental retardation was identified in a large European family. All affected males showed forms of aggressive and often violent behavior. Gene mapping and biochemical studies indicate that this condition has a direct link between a single-gene defect and a phenotype characterized by aggressive and/or violent behavior (▶ Figure 16.7). In particular, eight males who have mild or borderline mental retardation showed a characteristic pattern of aggressive and often violent behavior triggered by anger, fear, or frustration. The behavioral phenotype varied widely in levels of violence and time but included acts of attempted rape, arson, stabbings, and exhibitionism.

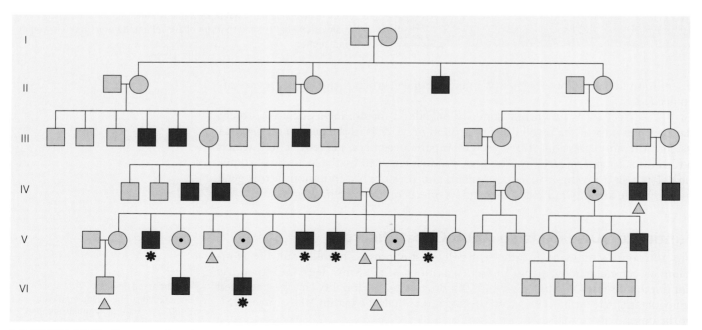

▶ **FIGURE 16.7** Cosegregation of mental retardation, aggressive behavior, and a mutation in the monoamine oxidase type A (*MAOA*) gene. Affected males are indicated by the filled symbols. Symbols marked with an asterisk represent males known to carry a mutation of the *MAOA* gene; those marked with a triangle are known to carry the normal allele. Symbols marked with a dot inside represent females known to be heterozygous carriers.

Mapping a Gene for Aggression

Using RFLP markers, the gene for this behavior was mapped to the short arm of the X chromosome in region Xp11.23-11.4. One of the genes in this region encodes an enzyme called monoamine oxidase (MAOA) that breaks down a neurotransmitter (Table 16.3). Neurotransmitters are chemical signals that carry nerve impulses across synapses in the brain and nervous system (Figure 16.8). Failure to break down these chemical signals rapidly can disrupt the normal function of the nervous system.

The urine of the eight affected individuals contained abnormal levels of chemicals made by monoamine oxidase type A (MAOA). The researchers concluded that the eight affected males carried a mutation in the gene that encodes this enzyme, and that lack of MAOA activity is associated with their behavior pattern.

A follow-up study analyzed the *MAOA* gene (MIM/OMIM 309850) in five of the eight affected individuals and showed that all five carry a mutation that results in a nonfunctional gene product. This mutation was also found in two known female heterozygotes and is not present in any of the unaffected males in this pedigree. Lack

Table 16.3
Some Common Neurotransmitters

Acetylcholine

Dopamine

Norepinephrine

Epinephrine

Serotonin

Histamine

Glycine

Glutamate

Gamma-Aminobutyric acid (GABA)

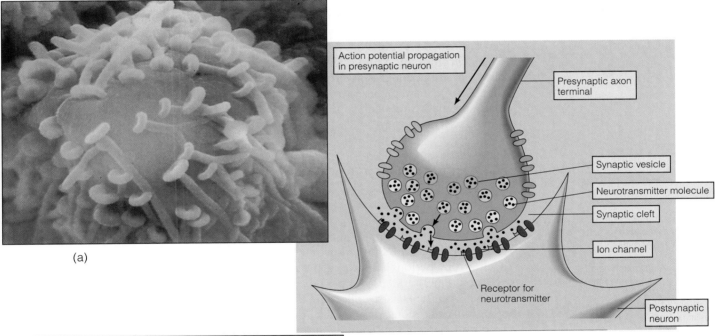

(a)

(b)

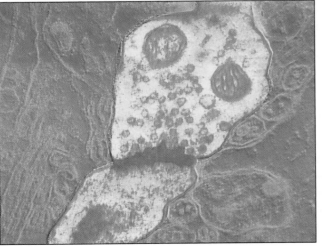

(c)

 FIGURE 16.8 The synapse and synaptic transmission. (a) A scanning electron micrograph of the endings (called terminal boutons) of an axon in contact with another cell. (b) In this diagram, a nerve impulse that arrives at the end of an axon triggers the release of a chemical neurotransmitter from storage in synaptic vesicles. The neurotransmitters diffuse across the synapse and bind to receptors on the membrane of the cell on the other side of the synapse (postsynaptic neuron), where they trigger another nerve impulse. (c) A transmission electron micrograph of a synapse. The cell above the synapse contains synaptic vesicles that have stored neurotransmitters.

of MAOA activity in affected males prevents normal breakdown of certain neuro-transmitters that is reflected in elevated levels of toxic compounds in the urine.

Because of the difficulty in relating a phenotype such as aggression (for example, what exactly constitutes aggression?) to a specific genotype, further work is needed to determine whether mutations of *MAOA* are associated with altered behavior in other families and in animal model systems. In addition, the interaction of this disorder with external factors such as diet, drugs, and environmental stress remain to be established. However, the identification of a specific mutation associated with this behavior pattern is an important discovery and suggests that biochemical or pharmacological treatment for this disorder may be possible.

Problems with Single-Gene Models

Although using recombinant DNA markers in linkage studies as a way of identifying single genes that affect behavior has been successful in the case of Huntington disease and the other disorders just described, in other cases it has led to erroneous results. In a 1987 linkage study using DNA markers, a gene for manic depression was mapped to a region of chromosome 11. Later, individuals from the study group who did not carry the suspect copy of chromosome 11 developed manic depression, indicating the lack of linkage between the markers on chromosome 11 and the gene or genes for manic depression. Similarly, a report on the linkage between DNA markers on chromosome 5 and schizophrenia was found to be coincidental, or at best applied only to a small, isolated population. These early failures to find single genes that control these disorders have led to the reevaluation of single-gene models for many behavior traits and to the development of alternative models, as described in the next section.

THE GENETICS OF MOOD DISORDERS AND SCHIZOPHRENIA

Mood disorders, also known as affective disorders, are psychological conditions in which there are profound emotional disturbances. **Moods** are defined as sustained emotions; **affects** are short-term expressions of emotion. Affective disorders are characterized by periods of prolonged depression (**unipolar disorder**) or by cycles of depression that alternate with periods of elation (**bipolar disorder**).

Schizophrenia is a collection of mental disorders characterized by psychotic symptoms, delusions, thought disorders, and hallucinations, often called the schizoid spectrum. Schizoid individuals suffer from disordered thinking, inappropriate emotional responses, and social deterioration. Mood disorders and schizophrenia are complex, often difficult to diagnose, have genetic components, and are common disorders. Heredity is regarded as a predisposing factor in both types of conditions, but the mode or modes of inheritance are unclear, and the role of social and environmental factors are unknown. Nonetheless, genetic components of these conditions are emerging, and despite recent setbacks in identifying single genes, progress is being made in forming genetic models of these disorders.

Mood Disorders: Unipolar and Bipolar Illnesses

The lifetime risk for a clinically identifiable mood disorder in the U.S. population is 8 to 9%. Depression (unipolar illness), is the most common of these disorders. It is more common in females (about a 2:1 ratio), usually begins in the fourth or fifth decade of life, and is often protracted or recurring. Depression has several characteristics, including weight loss, insomnia, poor concentration, irritability and anxiety, and lack of interest in surrounding events.

■ **Mood disorders** A group of behavior disorders associated with manic and/or depressive syndromes.

■ **Mood** A sustained emotion that influences perception of the world.

■ **Affect** Pertaining to emotion or feelings.

■ **Unipolar disorder** An emotional disorder characterized by prolonged periods of deep depression.

■ **Bipolar disorder** An emotional disorder characterized by mood swings that vary between manic activity and depression.

■ **Schizophrenia** A behavioral disorder characterized by disordered thought processes and withdrawal from reality. Genetic and environmental factors are involved in this disease.

About 1% of the U.S. population suffers from bipolar illness. The age of onset is during adolescence or the second and third decades of life, and males and females are at equal risk for this condition. Manic activity is characterized by hyperactivity, acceleration of thought processes, low attention span, creativity, and feelings of elation or power.

The link between mood disorders and genetics is derived from twin studies that show concordance of 57% for monozygotic (MZ) twins and 14% for dizygotic (DZ) twins. The evidence for genetic factors is stronger for bipolar disorders than for unipolar disorders. As discussed earlier, several attempts have been made to map genes for bipolar illnesses by using a single-gene model but were unsuccessful. The failure to find linkage between genetic markers and single genes for manic depression does not undermine the role of genes in bipolar illness but means that new strategies of linkage analysis are required to identify the genes involved. A new strategy, called an association study, is being used in a worldwide effort to screen all human chromosomes for genes that control bipolar illness. Association studies use DNA markers but follow the inheritance of the marker and the disorder (bipolar illness in this case) in unrelated individuals affected with the disorder rather than following the trait in large families. The idea is to identify portions of the genome that are more common in affected individuals than in those without the trait. Association studies have identified regions on chromosomes 6, 13, 15, and 18 as candidates that may contain genes for manic depression. Now that these regions have been identified, they are being closely studied to search for genes responsible for bipolar illness.

In addition to its elusive genetic nature, bipolar illness remains a fascinating behavioral disturbance because of its close association with creativity (see Concepts and Controversies: The Link between Madness and Genius). Many great artists, authors, and poets have been afflicted with manic depressive illness (◗ Figure 16.9). Studies of the nature of creativity have shown that the thought patterns of the creative mind parallel those of the manic stage of bipolar illness. In her book *Touched with Fire*, Kay Jamison explores the relationship among genetics, neuroscience, and

◗ **FIGURE 16.9** Virginia Woolf, the author and poet, was affected by manic depression. Like others, she often commented on the relationship between creativity and her illness.

the lives and temperaments of creative individuals, including Byron, Van Gogh, Poe, and Virginia Woolf.

Schizophrenia Has a Complex Phenotype

Schizophrenia is a relatively common mental illness that affects about 1% of the population (about 2.5 million people in the United States are affected). The disorder usually appears in late adolescence or early adult life. It has been estimated that half of all hospitalized, mentally ill, and mentally retarded individuals are schizophrenic.

Schizophrenia is a disorder of the thought processes rather than of mood. Diagnosis is often difficult, and there is notable disagreement on the definition of schizophrenia because it has no single distinguishing feature and causes no characteristic brain pathology. Some features of the disorder include

- psychotic symptoms, including delusions of persecution
- disorders of thought; loss of the ability to use logic in reasoning
- perceptual disorders, including auditory hallucinations (hearing voices)
- behavioral changes, ranging from mannerisms of gait and movement to violent attacks on others
- withdrawal from reality and inability to participate in normal activities

Several models proposed for schizophrenia generally fall into two groups: models in which biological factors (including genetics) play a major role and environmental factors are secondary; and conversely, models in which environmental factors are primary and biological factors are secondary. Some evidence points to metabolic differences in the brains of schizophrenics compared to those of normal individuals (❱ Figure 16.10), but it is unclear whether these differences are genetic. Overall,

❱ **FIGURE 16.10** Brain metabolism in a set of monozygotic quadruplets, all of whom suffer from varying degrees of schizophrenia. These scans of glucose utilization by brain cells are visualized by positron emission tomography (PET scan). They show low metabolic rates in the frontal lobe (top of each scan) compared to nonschizophrenics (top left image). The frontal lobe is where cognitive ability resides.

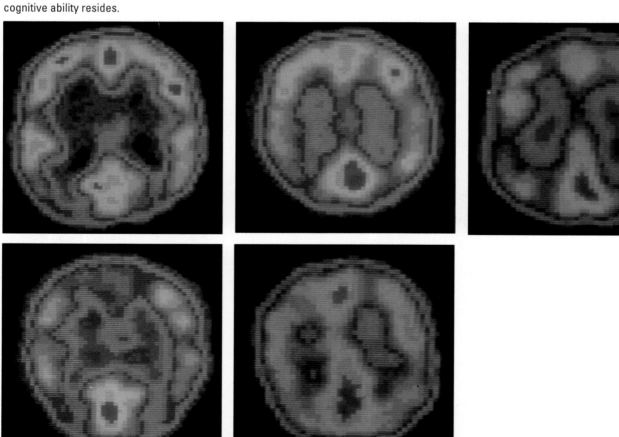

Concepts and Controversies

The Link between Madness and Genius

*A*s long ago as the 4th century B.C. Aristotle observed that talented philosophers, poets, and artists tend to have mental problems. This idea has been incorporated into the popular wisdom of our culture, as illustrated by statements such as "There is a thin line between genius and madness." In recent years, a substantial amount of evidence has accumulated that distinguished artists, poets, authors, and composers suffer from mood disorders, particularly depression and manic depression, at rates 10 to 30 times more frequent than the general population. Based on these results, some researchers have proposed that there is a link between bipolar disorder and creativity. Several books, including *Touched with Fire* (Kay Jamison), *The Price of Greatness* (Arnold Ludwig), and *The Broken Brain* (Nancy Andreasen), have explored this link.

As described by Jamison, the poet Alfred Lord Tennyson and members of his family were affected by mental instability, ranging from unstable moods to insanity. Alfred was affected by lifelong bouts of depression, as were two of his brothers, his sister, father, two uncles, an aunt, and his grandfather. Along with manic depression went a consuming passion for poetry and verse. Although Alfred is the best known member of his family, his brothers published volumes of poetry and each won prestigious awards for translating ancient Greek poems and epics. Alfred's siblings and his aunt also wrote verse, and one of his nieces became a poet and playwright, illustrating the link between creativity and bipolar illness.

We now know that different regions of the brain are affected in the depressive and in manic stages of bipolar illness. Manic depressives have a unique pattern of metabolism and blood flow in the prefrontal cortex, the part of the brain associated with intellect. It is thought that nerve cell connections ("wiring") may be different in certain regions of the brains of manic depressives and that the transition from mania to depression may stimulate mental activity and creativity.

---

however, the best evidence supports the role of genetics as a primary factor in schizophrenia, but environmental factors are needed for full expression.

Examining risk factors for relatives of schizophrenics (▶ Figure 16.11) reveals the influence of genotype on schizophrenia. Overall, relatives of affected individuals have a 15% chance of developing the disorder (as opposed to 1% among unrelated individuals). Using a narrow definition of schizophrenia, the concordance value for MZ twins is 46% versus 14% for DZ twins. MZ twins raised apart show the same level of concordance as MZ twins raised together. If a broader definition of the phenotype is used, one which combines schizophrenia and borderline or schizoid personalities, the concordance for MZ twins approaches 100%, and the risk for siblings, parents, and offspring of schizophrenics is about 45%. This strongly supports the role of genes in this disorder.

The mode of inheritance of schizophrenia (or susceptibility to this illness) is unknown. In the past, pedigree and linkage studies have suggested loci on the X chromosome and a number of autosomes as sites of genes contributing to this condition. Although all the linkage studies have been contested and contradicted by other studies, several tentative conclusions can be drawn about the genetics of schizophrenia, as discussed below.

▶ **FIGURE 16.11** The lifetime risk for schizophrenia varies with the degree of the relationship to an affected individual. The observed risks are more compatible with a multifactorial mode of transmission than a single-gene or polygenic mode of inheritance.

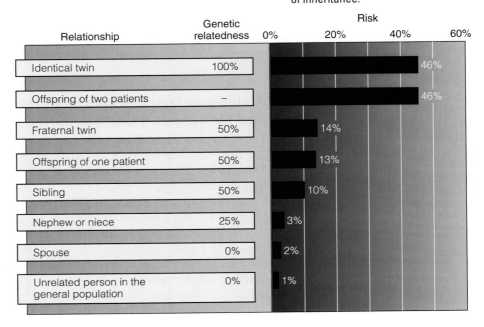

Relationship	Genetic relatedness	Risk
Identical twin	100%	46%
Offspring of two patients	—	46%
Fraternal twin	50%	14%
Offspring of one patient	50%	13%
Sibling	50%	10%
Nephew or niece	25%	3%
Spouse	0%	2%
Unrelated person in the general population	0%	1%

Genetic models indicate that a polygenic model with a single major gene making most of the contribution is consistent with the results from family studies as well as with other observations, including concordance in MZ twins and incidence of the disorder in the general population.

Familial linkage studies appear to be more valuable than association studies for identifying genes related to schizophrenia. A group of research laboratories has formed the Schizophrenia Collaborative Linkage Group to conduct genomewide scans for major genes that contribute to schizophrenia and bipolar illness. In one study, 452 microsatellite markers covering all regions of the genome were used to study individuals in 54 pedigrees, each of which have multiple cases of schizophrenia. To date, linkages between schizophrenia and loci on chromosomes 3, 8, 13, and 22 have been identified.

GENETICS AND SOCIAL BEHAVIOR

Although single genes control many genetic disorders that have behavioral effects, these behavioral alterations are often secondary to a biochemical abnormality. Human geneticists have long been interested in aspects of human behavior that take place in a social context, that is, behavior that results from the interaction between and among individuals. This behavior is often complex, and evidence indicates that such behavior involves multifactorial inheritance. Several traits that have different levels of social behavior are discussed in the following sections.

Tourette Syndrome Affects Speech and Behavior

■ **Tourette syndrome** A behavioral disorder characterized by motor and vocal tics and inappropriate language. Genetic components are suggested by family studies that show increased risk for relatives of affected individuals.

Tourette syndrome *(GTS)* (MIM/OMIM 137580) is characterized by motor and behavioral disorders. About 10% of affected individuals have a family history of the condition. Males are affected more frequently than females (3:1), and onset is usually between 2 and 14 years of age. *GTS* is characterized by episodes of motor and vocal tics that can progress to more complex behaviors involving a series of grunts and barking noises. The vocal tics include outbursts of profane and vulgar language and parrotlike repetition of words spoken by others. Because of variable expression, the incidence of the condition is unknown, but it has been suggested that the disorder may be very common.

Family studies indicate that relatives of affected individuals are at significantly greater risk for GTS than relatives of unaffected controls. Linkage studies of more than 1000 GTS families have been performed, and it has been proposed that a model of autosomal dominant inheritance that shows incomplete penetrance and variable expression is most compatible with the linkage results.

Alzheimer Disease Has Genetic and Nongenetic Components

■ **Alzheimer disease** A heterogeneous condition associated with the development of brain lesions, personality changes, and degeneration of intellect. Genetic forms are associated with loci on chromosomes 14, 19, and 21.

The symptoms of **Alzheimer disease** *(AD)* begin with loss of memory and a progressive dementia that involves disturbances of speech, motor activity, and recognition. There is an ongoing degeneration of personality and intellect, and eventually, affected individuals are unable to care for themselves.

Brain lesions (Figure 16.12) accompany these behavioral changes. The lesions are formed from a protein fragment called amyloid beta-protein, which accumulates outside cells in aggregates known as senile plaques. These plaques cause the degeneration and death of nearby neurons, affecting selected regions of the brain (Figure 16.13). Formation of senile plaques is not specific to *AD;* almost everyone who lives beyond the age of 80 will have such lesions. The difference between normal aging of the brain and Alzheimer disease appears to be the number of such plaques (greatly increased in *AD*) and the rate of accumulation (earlier and faster in *AD*).

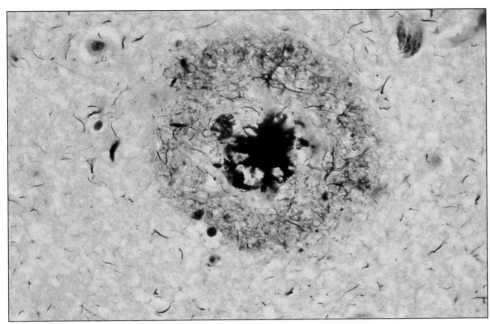

▶ **FIGURE 16.12**　A lesion called a plaque in the brain of someone with Alzheimer disease. A ring of degenerating nerve cells surrounds the deposit of protein.

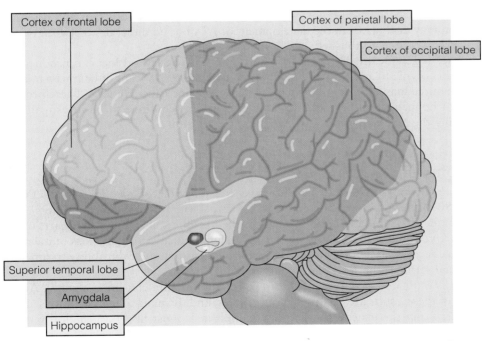

▶ **FIGURE 16.13**　Location of brain lesions in Alzheimer disease. Plaques are most heavily concentrated in the amygdala and hippocampus. These brain regions are part of the limbic system, a region of the brain involved in controlling memory.

It is estimated that Alzheimer disease affects 10% of the U.S. population over the age of 65 and 50% of those over the age of 80 and that the cost of treatment and care for those affected is more than $80 billion.

The genetics of Alzheimer disease is complex. Less than 50% of all cases can be traced to genetic causes, indicating that the environment plays a significant role in this disorder. We first examine the genetic evidence and then discuss some of the proposed environmental factors associated with this disorder.

The gene that encodes the amyloid beta-protein (known as APP) is located on the long arm of chromosome 21, and mutations of this gene are responsible for an early-onset form of *AD,* called *AD1* (MIM/OMIM 104300), which is inherited as an autosomal dominant trait. A second form, called *AD2* (MIM/OMIM 107741), is associated with an allele (*APOE*4*) of the *APOE* gene on chromosome 19. A third form, *AD3,* is caused by a mutation in a membrane protein encoded by a gene on chromosome 14 (MIM/OMIM 104311), and *AD4* (MIM/OMIM 600759) is caused by a mutation in a gene on chromosome 1 that also encodes a membrane protein. There is evidence for *AD* genes on other chromosomes, and mitochondrial DNA polymorphisms may also play a role in susceptibility to this disease.

It is not clear whether all cases of *AD* have a genetic explanation, because other genes are yet to be discovered, or whether environmental factors are important in the development of this behavioral disorder. In an analysis of 232 families in which *AD* is segregating, it was concluded that *AD* is multifactorial, indicating an important role for other genetic and nongenetic factors. Among the environmental factors that may play a role in the development of *AD,* two have received the most attention. Aluminum is found deposited in the brain lesions associated with *AD.* Thus, it has been suggested that aluminum intake and differences in the uptake, transport, and metabolism of this metal may play a role in the development of *AD.* More recent work has cast doubt on the role of aluminum in *AD,* but the question is still open.

One of the proteins in the brain lesions of *AD* is similar to that found in infectious proteins, called **prions** (MIM/OMIM 176640). Prions cause neurodegenerative diseases in domesticated animals, including Mad Cow disease, and are associated with several human neurological diseases, including Creutzfeldt-Jakob disease (MIM/OMIM 123400). It has been proposed that exposure to prions and prion infection may be a factor in the development of *AD,* but no direct link between prions and *AD* has been established.

AD is genetically heterogeneous and mutations at several loci can produce the *AD* phenotype. In addition, the fact that many cases of *AD* cannot be traced to a genetic source may indicate that there is more than one cause for *AD.* Therefore, the role of nongenetic influences and their mechanisms continue to be investigated in defining the risk factors for this debilitating condition.

■ Prion An infectious protein that is not a virus but is the cause of several disorders, including Creutzfeldt-Jakob syndrome.

Alcoholism Has Several Components

As a behavioral disorder, the excessive consumption of alcohol (MIM/OMIM 103780) has two important components. First, consumption of large amounts of alcohol can cause cell and tissue damage in the nervous system and other parts of the body. Over time, this results in altered behavior, hallucinations and loss of memory. These effects are secondary to the behavior patterns that alter the ability to function in social settings, the workplace, and the home.

It is estimated that 75% of the adult U.S. population consumes alcohol. About 10% of these adults are classified as alcoholics, and the male to female ratio is about 4 to 1. From the genetic standpoint, alcoholism is most likely a genetically influenced, multifactorial (genetic and environmental) disorder. The role of genetic factors in alcoholism is indicated by a number of findings:

- In mice, experiments indicate that alcohol preference can be selected for; some strains of mice choose 75% alcohol over water, whereas genetically different strains shun all alcohol.
- There is a 25% to 50% risk of alcoholism in sons and brothers of alcoholic men.
- There is a 55% concordance for alcoholism in MZ twins, and a 28% rate in same-sex DZ twins.
- Sons adopted by alcoholic men show a rate of alcoholism more like that of their biological fathers.

The nature of the genetic influence on alcoholism and its site or sites of action are unknown. Segregation analysis in families that have alcoholic members has produced evidence against the Mendelian inheritance of a single major gene and for multifactorial inheritance involving several genes. Other researchers have advanced a single-gene model and reported an association between an allele of a gene that encodes a neurotransmitter receptor protein (called D2) and alcoholism. This evidence is based on the finding that in brain tissue, the *A1* allele of the *D2* gene was found in 69% of the samples from severe alcoholics but in only 20% of the samples from nonalcoholics, implying that the *A1* allele is involved in alcoholism.

Subsequent linkage and association studies on the *A1* allele have not supported the idea that this allele is involved in alcoholism. Taken together, the available studies have failed to show any relationship between abnormal neurotransmitter metabolism or receptor function and alcoholic behavior.

The search for genetic factors in alcoholism illustrates the problem of selecting the proper genetic model to analyze for behavioral traits. Segregation and linkage studies indicate that there is no single gene for alcoholism. If a multifactorial model involving a number of genes, each with a small additive effect, is invoked, the problem becomes more complicated. How do you prove or disprove that a given gene contributes, say, 10% to the behavioral phenotype? At present, the only method would involve studying thousands of individuals to find such effects.

Sexual Orientation Is a Multifactorial Trait

In sexual behavior, most humans are heterosexual and prefer the opposite sex, but a fraction of the population is homosexual and prefers sexual activity with members of the same sex. These variations in sexual behavior have been recorded since ancient times, but biological models for these behaviors have been proposed only recently.

Twin studies and adoption studies have investigated the role of genetics in sexual orientation. In a twin study that employed 56 MZ twins, 54 DZ twins, and 57 genetically unrelated adopted brothers, concordance for homosexuality was 52% for MZ twins, 22% for DZ twins, and 11% for the unrelated adopted sibs. Overall heritability ranged from 31% to 74%. Another study investigating homosexual behavior in women employed 115 twin pairs and 32 genetically unrelated adopted sisters. In this study, heritability ranged from 27% to 76%.

The results from these and other studies indicate that homosexual behavior has a strong genetic component. These studies have been challenged on the grounds that the results can be affected by the phrasing of the interview questions, by the methods used to recruit participants, and that the phenotype is self-described. But the average heritability estimates from these studies parallel those from the Minnesota Twin Project, a long-term study of MZ twins separated at birth and reared apart. Further twin studies are needed to determine whether the heritability values are accurate. If confirmed, the studies to date indicate that homosexual behavior is a multifactorial trait that involves several genes and unidentified environmental components.

Work using RFLP markers to study male homosexual behavior has found linkage between one subtype of homosexuality and markers on the long arm of the X chromosome (▶ Figure 16.14). In this study, a two-step approach was employed. First, family histories were collected from 114 homosexual males. Pedigree analysis was performed on 76 randomly-selected individuals from this group, using interviews with male relatives to ascertain sexual preference. The results indicated the possibility of X-linked inheritance (MIM/OMIM 306995).

A further pedigree analysis was conducted using 38 families in which there were two homosexual brothers, based on the idea that this might show a stronger trend for X-linked inheritance since there were

▶ **FIGURE 16.14** Region of the X chromosome found by linkage analysis to be associated with one form of male homosexual behavior.

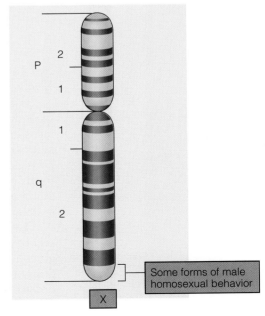

Some forms of male homosexual behavior

X

Neurogenetics: From Mutants to Molecules

Barry S. Ganetzky

*F*rom the time I was young, I was fascinated by living things and enjoyed reading books on natural history. But I really had no idea how biologists earn a living. I just knew that I was about 100 years too late to become a naturalist. So I planned on studying chemistry, although it did not give me the same sheer pleasure as biology. In introductory biology I became interested in genetics and molecular biology and realized that biology was what I wanted to immerse myself in. I have never regretted that choice.

In my junior year, I signed up to do an honor's research project. My mentor was a new young professor (a certain Michael R. Cummings) who at the time was studying egg development in *Drosophila* (fruit flies). This was my first exposure to research, and using my imagination to discover answers to some of nature's secrets was the most exciting thing I had ever done. Although my project was to last only 10 weeks, I remained for the next two years.

I pursued a Ph.D. in genetics at the University of Washington with the late Larry Sandler, who was so intellectually gifted that I knew instinctively no one could provide me with better graduate training.

As a postdoctoral fellow with Seymour Benzer, I was interested in the molecular basis of the signaling mechanisms in neurons. It was known that ion channel proteins play key roles in nerve impulses, but little was known about their molecular structure or how they worked. I isolated mutations that were defective in neuronal signaling to identify the genes that encode ion channels.

The trick was to find the right mutations. I began by screening for mutants that became paralyzed when exposed to elevated temperatures. In a stroke of luck, one of the first paralytic mutations I found caused a complete block of action potentials. This mutation led us to identify other mutants with neuronal defects. After taking a faculty position at the University of Wisconsin, my colleagues and I succeeded in cloning these genes. We now have the largest collection of mutations that affect ion channels in any organism, and they are providing us with new insights into the molecular basis of neuronal activity. Because one of the human genes we identified is similar to a *Drosophila* gene, it turns out to be defective in a heritable form of cardiac arrhythmia. Identification of the affected gene opens the way to identifying individuals at risk. It is gratifying to know that work pursued primarily because it was interesting and fun is also important and useful.

BARRY S. GANETZKY received a B.S. in biology from the University of Illinois-Chicago and a Ph.D. in genetics from the University of Washington. He has been a faculty member in the Laboratory of Genetics at the University of Wisconsin, Madison, since 1979, where he is the Steenbock Professor of Biological Sciences. After spending more than half his life working with fruit flies, he still derives great pleasure from discovering new mutations that have interesting and unusual phenotypes.

two homosexual sibs in the same family. The results show a stronger trend toward X-linked inheritance and an absence of paternal transmission (▶ Figure 16.15).

Using information about traits and relatedness from the pedigree analysis, the second part of the study employed DNA markers to determine whether an X-linked locus or loci were associated with male homosexual behavior in the 38 families that had two homosexual brothers. Linkage was detected to RFLP markers from the distal region of the long arm in the Xq28 region. These markers were present in two-thirds of the 32 pairs of homosexual brothers and in about one-fourth of the heterosexual brothers. More work is needed to confirm the linkage relationship and to search the region for a locus or loci that affects sexual orientation.

Two important factors related to this study need to be mentioned. First, seven sets of homosexual brothers did not coinherit all of the markers in the Xq28 region, and about one-fourth of all heterosexual brothers inherited the markers, but did not display homosexual behavior. This indicates that genetic heterogeneity or nongenetic factors are significant in this behavioral variation. Second, the study cannot estimate what fraction of homosexual behavior might be related to the

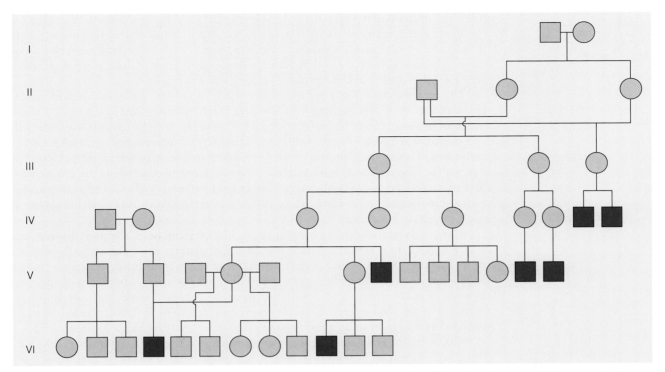

▶ FIGURE 16.15 Pedigree showing X-linked transmission of male homosexuality. Affected males, some of whom are brothers, are indicated by the filled symbols.

Xq28 region or whether this region influences lesbian sexual behavior. In spite of its preliminary nature and (to some) its controversial conclusions, this work applies genomewide screening with molecular markers to study the role of genes in one type of male sexual behavior and represents a model for future studies in this area of behavior genetics.

SUMMING UP: THE CURRENT STATUS OF HUMAN BEHAVIOR GENETICS

In reviewing the current status of human behavior genetics, several elements are apparent. Almost all studies of complex human behavior have provided only indirect and correlative evidence for the role of specific genes. Segregation studies and heritability estimates indicate that most behaviors are complex traits that involve several genes. Searches for single-gene effects have proven unsuccessful to date, and initial reports of single genes that control bipolar illness, schizophrenia, and alcoholism have been retracted, or remain unconfirmed.

The multifactorial nature of behavioral traits means that methods for identifying genes with small, incremental effects must be developed. Although twin and adoption studies have been valuable in behavior genetics, these studies typically involve small numbers of individuals. For example, fewer than 300 pairs of MZ twins raised apart have been identified worldwide. Where traits involve multiple genes, confirmation results can require detailed examination of thousands of individuals in hundreds of families. This process is necessarily slow and labor-intensive. Perhaps newer approaches, such as association studies in combination with new and quantitative ways of defining phenotypes, can be used to dissect the genetic components of a behavioral phenotype.

The limited evidence currently available indicates that the environment plays a significant role in behavior. As confirmation of the role of genes in behavior becomes available, investigations into the role of environmental factors cannot be neglected. The history of human behavior genetics in the eugenics movement of the early part of this century provides a lesson in the consequences of overemphasizing the role of genetics in behavior. Attempts to provide single-gene explanations for complex behavior inhibited the growth of human genetics as a discipline.

The increasing evidence for the involvement of specific genes in controlling human behavior has implications for society at large. As discussed in Chapter 13, the Human Genome Project has raised questions about the way genetic information will be disseminated and used, who will have access to this information, and under what conditions. These same concerns need to be addressed for genes that affect behavior. If genes for alcoholism or homosexuality can be identified, will this information be used to predict an individual's future behavioral patterns? Who should have access to this information, and what can be done to prevent health care discrimination in employment or insurance?

Many behavioral phenotypes, such as Huntington or Alzheimer disease, are clearly regarded as abnormal. Few would argue against the development of treatments for intervening in and perhaps preventing these conditions. When do behavior phenotypes move from being abnormal to being variants? If there is a connection between bipolar illness and creativity, to what extent should this condition be treated? If genes that influence sexual orientation are identified, will this behavior be regarded as a variant or as a condition that should be treated and/or prevented?

Although research can provide information about the biological factors that play a role in determining human behavior, it cannot provide answers to questions of social policy. Social policy and laws have to be formulated by using information from research.

Case Studies

CASE 1

Rachel asked to see a genetic counselor because she was concerned about her risk of developing schizophrenia. Her mother and maternal grandmother both had schizophrenia and had to be institutionalized for most of their adult life. Rachel's three maternal aunts are all in their sixties and have not shown any signs of this disease. Rachel's father is alive and healthy, and his family history does not suggest any behavioral or genetic conditions. The genetic counselor discussed the multifactorial nature of schizophrenia and explained that there have been many candidate genes identified that may be mutated in individuals who have this condition. However, a genetic test is not available for presymptomatic testing of at-risk persons. The counselor explained that based on Rachel's family history and her relatedness to the individuals who have schizophrenia, her risk of developing it is approximately 13%. If an altered gene is in the family and her mother passed it to Rachel, she would have a 50% chance of inheriting this gene.

CASE 2

A genetic counselor was called to the pediatric ward of the hospital for a genetics consultation. Her patient, an 8-year-old boy, was having a "temper tantrum" and was biting his own fingers and toes. The nurse called after she noticed in the boy's chart that he had been previously followed by a clinical geneticist at another institution. The counselor carefully reviewed the boy's chart and noted a history of growth retardation and self-mutilation since the age of 3. His movements were very "jerky," and he was banging his head against the bedpost. The nurses were having a very difficult time controlling him. The counselor immediately recognized these symptoms as part of a genetic syndrome known as Lesch-Nyhan syndrome.

Lesch-Nyhan syndrome is an X-linked recessive condition (Xq26) that is due to a mutation in the hypoxanthine phosphoribosyltransferase gene. This rare condition affects about 1 in 100,000 males. The onset usually occurs between the ages of 3 to 6 months. Prenatal testing is available for families who have this condition. There is currently no treatment for Lesch-Nyhan, and most affected individuals die by the second decade of life.

Summary

1. Many forms of behavior represent complex phenotypes based on multifactorial inheritance. Single genes that affect behavior do so as a consequence of their effect on the development, structure, and function of the nervous system.

2. The methods used to study inheritance of behavior encompass both classical methods of linkage and pedigree analysis, newer methods of recombinant DNA analysis, and new combinations of techniques, such as twin studies combined with molecular methods. Refined definitions of behavior phenotypes are also being used in the genetic analysis of behavior.

3. Results from work on experimental animals indicate that behavior is under genetic control and have provided estimates of heritability. The molecular basis of single-gene effects in some forms of behavior has been identified and provides useful models to study gene action and behavior.

4. Several single-gene effects on human behavior are known. Most of these affect the development, structure, or function of the nervous system and consequently affect behavior.

5. Bipolar illness and schizophrenia are common behavior disorders, each affecting about 1% of the population. Simple models of single-gene inheritance for these disorders have not been supported by extensive studies of affected families, and more complex forms of multifactorial inheritance seem likely.

6. Other multifactorial traits that affect behavior include Tourette syndrome, Alzheimer disease, alcoholism, and sexual orientation.

7. Twin studies combined with molecular markers have identified a region of the X chromosome that may affect one form of homosexual behavior. If this study can be confirmed and the gene or genes identified, this combination of methods may be useful as a model for future work to identify genes that affect behavior.

Questions and Problems

1. When studying the genetics of behavior, what are the major differences in the methods used for single gene versus polygenic traits?

2. In human behavior genetics, why is it important for the trait under study to be defined accurately?

3. What are the advantages of using *Drosophila* for the study of behavior genetics? Can this organism serve as a model for human behavior genetics? Why or why not?

4. A premature stop codon leads to the altered expression of a membrane-associated protein found in cells that form the myelin sheath in the peripheral nervous system. What phenotype results and what disease is this phenotype linked to?

5. In a long-term study of over 100 pairs of MZ and DZ twins separated shortly after birth and reared apart, one of the conclusions was that "general intelligence or IQ is strongly affected by genetic factors." The study concluded that about 70% of the variation in IQ is due to genetic variability (review the concept of heritability in Chapter 5). Discuss this conclusion, and include in your answer the relationship between IQ and intelligence, and to what extent these conclusions can be generalized. In evaluating their conclusion, what would you like to know about the twins?

6. One of the models for behavioral traits in humans involves a form of interaction known as epistasis. In a simplified example involving two genes, the expression of one gene affects the expression of the other. How might this interaction work, and what patterns of inheritance might be shown?

7. Perfect pitch is the ability to name a note when it is sounded. In a study of this behavior, perfect pitch was found to predominate in females (24 out of 35 in one group). In one group of 7 families, 2 individuals had perfect pitch. In 2 of these families, the affected individuals included a parent and a child. In another group of 3 families, 3 or more members (up to 5) had perfect pitch, and in all 3 families, 2 generations were involved. Given this information, what if any conclusions can you draw as to whether this behavioral trait might be genetic? How would you test your conclusion? What further evidence would be needed to confirm your conclusion?

8. Opposite to perfect pitch is tune deafness, the inability to identify musical notes. In one study, a bimodal distribution in populations was found, with frequent segregation in families and sib pairs. The author of the study concluded that the trait might be dominant. In a family study, segregation analysis suggested an autosomal dominant inheritance of tune deafness with imperfect penetrance. One of the pedigrees is presented below. On the basis of the results, do you agree with this conclusion? Could perfect pitch and tune deafness be alleles of a gene for musical ability?

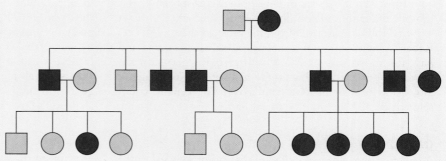

9. A woman diagnosed with Alzheimer disease wants to know the probability of her children inheriting this disorder. Explain to her the complications of determining heritability for this disease.

10. You are a researcher studying manic depression. Your RFLP data shows linkage between a marker on chromosome 7 and manic depression. Later in the study, you find that a number of individuals lack this RFLP marker, but still develop the disease. Does this mean that manic depression is not genetic?

11. A pedigree analysis was performed on the family of a man with schizophrenia. Based on the known concordance statistics, would his monozygotic twin be at high risk for the disease? Would the twin's risk decrease if he were raised in a different environment than his schizophrenic brother?

12. In schizophrenia, a region on chromosome 6 has been linked to this disease, but researchers have not yet found a gene. Explain this linkage and why linkage does not necessarily locate a gene.

13. If you discover a single-gene mutation that leads to an altered behavioral phenotype, what would you suspect about the nature of the mutant gene and the focus of its action?

14. The two main affective disorders are manic depression and schizophrenia. What are the essential differences and similarities between these disorders?

15. What evidence does not support the idea that alcoholism is genetic?
    a. some strains of mice select alcohol over water 75% of the time, while others shun alcohol
    b. the concordance value for MZ twins is 55% and for DZ twins it is 28%
    c. sons of adopted alcoholic men have a rate of alcoholism more like their adoptive fathers
    d. there is a 20-25% risk of alcoholism in sons of alcoholic men
    e. none of the above

16. There could be significant personal, social, and even legal effects if sexual orientation is proven to have genetic components. What are some of these implications?

17. In July of 1996, The Independent, a popular newspaper published in London, England reported a study conducted by Dr. Aikarakudy Alias, a psychiatrist, who has been working on the relationship between body hair and intelligence for 22 years. Dr. Alias told the 8th Congress of the Association of Europeans Psychiatrists that hairy chests are more likely to be found among the most intelligent and highly educated than in the general population. According to this new research, excessive body hair could also mean higher intelligence. Is correlating body hair with intelligence a valid method for studying the genetics of intelligence? Why or why not? What other factors contribute to intelligence? Is it logical to assume that individuals with little or no body hair are consistently less intelligent than their hairy counterparts? What type of study could be done to prove or disprove this idea?

Internet Activities

The following activities use the resources of the World Wide Web to enhance the topics covered in this chapter. To investigate the topics described below, log on to the book's home page at:

http://www.brookscole.com/biology

1. The National Alliance for the Mentally Ill home page regularly posts articles updating research in the field of genetics and mental illness. Use this site to locate information concerning research on manic-depression or schizophrenia. You can find pertinent information by clicking on many of the headings in the column on the left. For instance, click on "Links" to find related Web sites for either illness. Read at least three articles or search through three different sites for your information.
   a. What recent updates did you find with information from genetics research? Are any promising gene therapies suggested?
   b. From what you have read, have your own views changed as to whether manic-depression or schizophrenia is strongly linked to an inherited genetic component versus environmental causes such as stress, abuse, etc.? Why or why not?

2. The Internet Mental Health homepage from Canada provides a wealth of information about various disorders and has over 1,000 other mental health Web sites linked to it. Click on "Mental Disorders" in the left column and choose three that are of interest to you.
   a. Which did you choose? Did you discover genetic research for any or all of the three through the topical links provided? What did you find? (Hint: for each disorder there is a link, "Research Re: Cause" that provides a summary of recent research articles).
   b. Alcohol dependence is a major problem in many societies today. Click on that disorder heading. What genetic links are described in the research articles? Read one or two magazine articles and try to discover any difference between the genetics of male and female alcoholism. (If not listed here, The Harvard Mental Health Letter of Nov. 1994 has an excellent article and it is at the Web site for that periodical) What did you discover? Is there a major difference between the sexes?

3. Schizophrenia is considered to have a genetic component, yet little is known about the genetic link. Examine the details regarding the probabilities of inheriting schizophrenia in the various groups listed at the Health-Center.com web site. Using this information, can you suggest an explanation for how this disease is inherited?

4. Go to the website for Great Ideas in Personality. Click on behavior genetics. The study of behavioral disorders and their linkage to specific genes can be difficult. It is imperative that the researcher recognizes and reconciles environmental influences on behavior and be able to separate those influences from the actions of genes. What types of studies are done to investigate behavior genetics? Why must these studies be confined to the groups listed?

For Further Reading

Aldhous, P. (1992). The promise and pitfalls of molecular genetics. Science 257: 164–165.

Aston, C., & Hill, S. (1990). Segregation analysis of alcoholism in families ascertained through a pair of male alcoholics. Am. J. Hum. Genet. 46: 879–887.

Blum, K., Noble, E., Sheridan, P., Montgomery, A., Ritchie, T., Jagadeeswaran, P., Nogami, H., Briggs, A., & Cohn, J. (1990). Allelic association of human dopamine D(2) receptor gene in alcoholism. JAMA 263: 2055–2060.

Bolos, A., Dean, M., Lucas-Derse, S., Ramsburg, M., Brown, G., & Goldman, D. (1990). Population and pedigree studies reveal a lack of association between the dopamine D(2) receptor gene and alcoholism. JAMA 264: 3156–3160.

Brunner, H., Nelen, M., vanZandvoort, P., Abeling, N., van Gennip, A., Walters, E., Kulper, M., Ropers, H., & van Dost, B. (1993). X-linked borderline mental retardation with prominent behavioral disturbance: Phenotype, genetic localization, and evidence for disturbed monoamine metabolism. *Am. J. Hum. Genet. 52:* 1032–1039.

Brunner, H., Nelen, M., Breakfield, X., Ropers, H., & van Oost, B. (1993). Abnormal behavior associated with a point mutation in the structural gene for monoamine oxidase A. *Science 262:* 578–580.

Coryell, W., Endicott, J., Keller, M., Andreasen, N., Grove, W., Hirschfield, R., & Scheftner, W. (1989). Bipolar affective disorder and high achievement: A familial association. *Am. J. Psychiatry 146:* 983–988.

Devore, E., & Cloninger, C. (1989). The genetics of alcoholism. *Ann. Rev. Genet. 23:* 19–36.

Duclos, F., Boschert, U., Sirugo, G., Mandel, J-L., Hen, R., & Koenig, M. (1993). Gene in the region of Friedreich ataxia locus encodes a putative transmembrane protein expressed in the nervous system. *Proc. Nat. Acad. Sci. USA 90:* 109–113.

Freimer, N. B., Rens, V. I., Escamilla, M. A., McInnes, L. A., Spesny, M., Leon, P., Service, S. K., Smith, L. B., et al. (1996). Genetic mapping using haplotype, association and linkage methods suggest a locus for severe bipolar disorder (BPI) at 18q22-q23. *Nat. Genet. 12:* 436–441.

Ginns, E. I., Ott, S., Egeland, J. A., Allen, C. R., Fann, C. S., Pauls, D. L., Weissenbachoff, J., Carulli, J. P., et al. (1996). A genome-wide search for chromosomal loci linked to bipolar affective disorder in the Old Order Amish. *Nat. Genet. 12:* 431–435.

Haines, J. (1991). The genetics of Alzheimer disease—a teasing problem. *Am. J. Hum. Genet. 48:* 1021–1025.

Hamer, D., Hu, S., Magnuson, V., Hu, N., & Pattatucci, A. (1993). A linkage between DNA markers on the X chromosome and male sexual orientation. *Science 261:* 321–327.

Hu, S., Pattatucci, A. M., Patterson, C., Li, L., Fulker, D. W., Cherney, S. S., Krugylak, L., & Hamer, D. H. (1995). Linkage between sexual orientation and chromosome Xq28 in males but not in females. *Nat. Genet. 11:* 248–256.

Jamison, K. (1993). *Touched With Fire: Manic Depressive Illness and the Artistic Temperament.* New York: Free Press.

Kendler, K., Heath, A., Neale, M., Kessler, R., & Eaves, L. (1992). A population-based twin study of alcoholism in women. *JAMA 268:* 1877–1882.

Lawrence, S., Keats, B., & Morton, N. (1992). The AD1 locus in familial Alzheimer disease. *Ann. Hum. Genet. 56:* 295–301.

Levay, S. (1991). A difference in hypothalamic structure between heterosexual and homosexual men. *Science 253:* 1034–1037.

Moizes, H. W., Yang, L. Kristbjarnarson, H., Wiese, C., Byerley, W., Macciardi, F., Arott, V., Blackwood, D., et al. (1995). An international two-stage genome-wide search for schizophrenia susceptibility genes. *Nat. Genet. 11:* 321–324.

Pericak-Vance, M., Bebout, J., Gaskell, P., Yamaoka, L., Hung, W-Y., Alberts, M., Walker, A., Bartlett, R., Haynes, C., Welsh, K., Earl, N., Heyman, A., Clark, C., & Roses, A. (1991). Linkage studies in familial Alzheimer disease: evidence for chromosome 19 linkage. *Am. J. Hum. Genet. 48:* 1034–1050.

Powledge, T. (1993). The genetic fabric of human behavior. *Bioscience 43:* 362–367.

Powledge, T. (1993). The inheritance of behavior in twins. *Bioscience 43:* 420–424.

Risch, N., & Botstein, D. (1996). A manic depressive history. *Nat. Genet. 12:* 351–353.

Schellenberg, G., Paijami, H., Wigsman, E., Orr, H., Goddard, K., Anderson, L. Nemens, E., White, J., Alonso, M., et al. (1993). Chromosome 14 and late-onset familial Alzheimer disease (FAD). *Am. J. Hum. Genet. 53:* 619–628.

Tivol, E. A., Shalish, C., Schuback, D. E., Hsu, Y. P., and Breakefield, V. O. (1996). Mutational analysis of the human MAOA gene. *Am. J. Med. Genet. 67:* 92–97.

van de Wetering, B., & Heutink, P. (1993). The genetics of the Gilles de la Tourette syndrome: A review. *J. Lab. Clin. Med. 121:* 638–645.

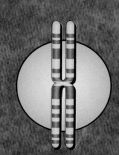

Genes in Populations

*A*rguments in which a father and son take opposite sides are neither new nor even unusual. All offspring occasionally disagree with their parents. It is rare, however, when such arguments result in ideas that are regarded as controversial some 175 years later. One such disagreement between Thomas Malthus and his father arose in the years following the French Revolution. This event, like the American Revolution that preceded it, was hailed by many (including the elder Malthus) as the beginning of a new era for humanity. According to his view, throwing off the repressions of monarchy presented an opportunity for unlimited progress in social, political, and economic areas. The younger Malthus argued that to the contrary, even if all social impediments were removed, there were immutable natural constraints that would limit progress. These limits would always result in the continuation of poverty and misery as part of the human condition. His essay on this subject was first published in 1798 as *The Essay on the Principle of Population as It Affects the Future Improvement of Society*. Later versions of this work were expanded and published as several editions of a book and also appeared in the 1824 supplement to the Encyclopedia Britannica.

In his essay, Malthus noted that populations grow geometrically, with the human population doubling in size every 25 years or so (▶ Figure 17.1). Resources such as living space and food supply are more limited and grow slowly, if at all. This means that population size will rapidly outstrip the ability of the environment to support a continual increase in the birth rate. When this point is reached, constraints such as war, disease, and starvation begin to limit population growth by increasing the death rate. The use of voluntary constraints on population growth such as delayed marriage, celibacy, and birth control can help limit population growth and bring about a reduction in human suffering and an improvement in living conditions. But the existence of sexual passion and human nature cause most people to ignore such voluntary constraints. According to the younger Malthus, the result is a continuation of

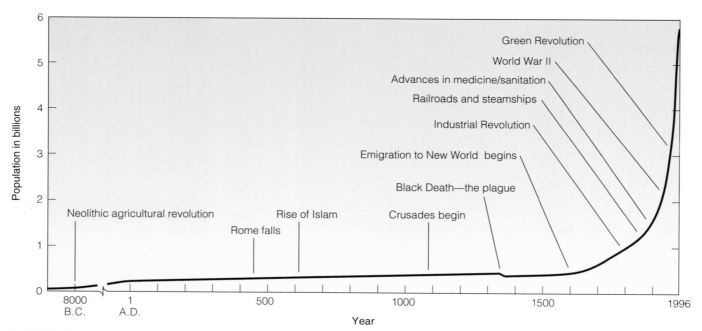

▶ **FIGURE 17.1** Exponential growth of the human population. It took 1800 years to reach 1 billion, but only another 130 years to reach 2 billion, and another 45 years to reach 4 billion. Before the year 2000, the population will be 6 billion.

unrelieved poverty and marginal living conditions, even in the most prosperous of nations.

In the 19th century, the writings of Malthus were used to argue that social reform and welfare were useless, because poverty was the result of natural law, not social inequities. In the 20th century, the debate over Malthus continues. Beginning in the 1960s, population growth was again regarded as a global threat that would override technological progress and, lead to a lowered standard of living and an increase in poverty and social problems. Movements calling for zero population growth and universal policies of birth control arose in the United States and other Western countries. More recently, China instituted a policy of one couple, one child. Many developing nations regard such policies as attempts to limit their potential for growth and economic expansion and, in the extreme, as policies designed to result in extinction for smaller, poorer countries.

Malthus correctly foresaw the implications and dangers of uncontrolled population growth, and he was one of the first to deal with the dynamics of populations. His work on the relationship between a population and its environment influenced both Wallace and Darwin, who incorporated Malthus' ideas into the theory of natural selection. In considering the effects of limited resources on populations, Wallace and Darwin concentrated their attention on those that lived rather than those that perished. Both observed that individuals who carry advantageous hereditary variations were more likely to survive and to leave the greatest number of offspring. Neither Wallace nor Darwin had any knowledge how this variation was generated, but they recognized its importance.

In the decades after Mendel's work became widely known, it became obvious that the phenotypic variation seen in populations of organisms had a genetic basis and that the genetic structure of populations is an essential part of the evolutionary process. Today the study of population genetics is closely tied to the study of evolution. Populations represent collections of alleles, and evolution is the result of changes in allele frequencies.

In this chapter we consider the population as a genetic unit and examine its organization, methods of measuring allele frequencies, and the ways in which the genetic structure of the population directly affects the incidence of human genetic disorders. We begin with a definition of populations, their subunits, and the concept of populations as reservoirs of genetic diversity. Then, we consider how allele frequencies can be measured in populations and how this information is used to answer practical questions about the frequency of disorders and heterozygotes in a population.

THE POPULATION AS A GENETIC RESERVOIR

Humans are distributed over geographic areas that include most of the land surfaces of the earth (❱ Figure 17.2). Because humans are clustered in many different regions, our species is subdivided into locally interbreeding units known as **populations**, or **demes**. Like individuals, populations are dynamic: They have life histories that include birth, growth, and response to the environment. Like individuals, populations can age and eventually die. Parameters such as age structure (❱ Figure 17.3), spatial distribution, birth and death rates, and allele frequencies can be used to describe populations.

Populations are more genetically diverse than individuals. For example, no single individual can have blood types A, B, and O. Only a group of individuals has the genetic capacity to carry all three blood types. The set of genetic information carried by a population is known as its **gene pool**. For a given gene, such as the ABO blood

■ **Population** A local group of organisms belonging to a single species, sharing a common gene pool; also called a deme.

■ **Gene pool** The set of genetic information carried by the members of a sexually reproducing population.

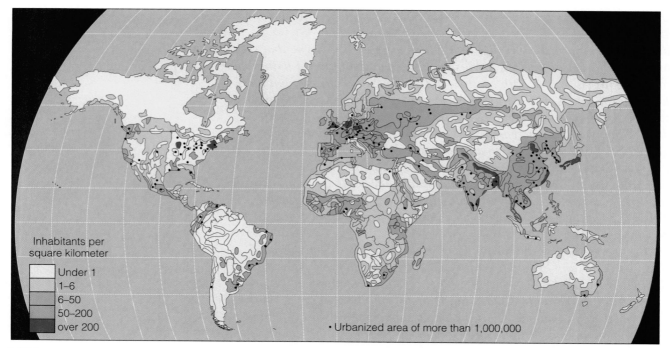

FIGURE 17.2 Human population density. Humans are not randomly distributed across the land areas of the world but are clustered into discrete populations.

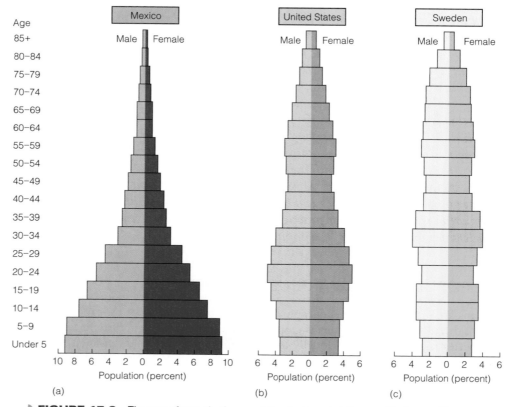

FIGURE 17.3 The rate of growth of a population is strongly influenced by its age. The population of Mexico (a) is pyramid-shaped, and has more individuals who have yet to reproduce. This type of population has the potential for explosive growth. (b) The population structure of the United States is tapered at the bottom and has more individuals who have already reproduced. (c) The population of Sweden is of roughly constant size across all age groups and is a stable population.

type, the pool includes all of the *A, B,* and *O* alleles in the population. Zygotes produced by one generation represent samples selected from the gene pool to form the next generation. The gene pool of a new generation is descended from the parental pool but for a variety of reasons, including chance, may have different allele frequencies than those in the parental pool. Over time, alterations in allele frequency can cause a population to undergo changes in phenotype frequency. The long-term effect of changes in the genetic structure of a population is evolutionary change.

MEASURING ALLELE FREQUENCIES IN POPULATIONS

Population size can remain stable or can undergo drastic changes if expansion occurs, or the population may decline and have only a few or no survivors. Factors that influence the size of a population include birth rate, disease, migration, and adaptation to environmental factors such as climate. As these factors change, the genetic structure of a population can also change (see Concepts and Controversies: The Thrifty Genotype).

In studying populations, geneticists analyze the genetic structure of a population by measuring **allele frequencies** and study the population over several generations to determine whether allele frequencies are stable.

In our discussion, the term *allele frequency* means the frequency with which alleles of a given genetic locus are present in the population. As several following examples will show, allele frequencies are not the same as genotype frequencies.

Allele frequencies cannot always be determined directly because we see phenotypes, not genotypes. However, if we consider codominant alleles, phenotypes are equivalent to genotypes, and we can determine allele frequencies directly. The MN blood group in humans is an example of a codominant system. In this case the gene *L* (located on chromosome 4) has two alleles, L^M and L^N, that are responsible for the M and N blood types, respectively. Each allele controls the synthesis and presence of a gene product on the surface of red blood cells, independently of the other allele. Thus individuals may be type M (L^M/L^M), type N (L^N/L^N), or type MN (L^M/L^N). The genotypes, blood types, and immunological reactions of the MN blood groups are shown in Table 17.1.

■ **Allele frequency** The percentage of all alleles of a given gene that are represented by a specific allele.

Codominant Allele Frequencies Can Be Measured Directly

In a codominant system, such as the MN blood group, allele frequency is determined simply by counting how many copies of each allele are present in a given population. For example, in a population of 100 individuals, suppose that blood typing shows that there are 54 MM homozygotes, 26 MN heterozygotes, and 20 NN homozygotes. The 54 MM individuals carry 108 copies of the *M* allele (54 individuals, each of whom carries 2 *M* alleles). The 26 MN heterozygotes carry an additional 26 *M* alleles, so the population has a total of 134 *M* alleles (108 + 26 = 134). Each member

Table 17.1 MN Blood Groups

Genotype	Blood Type	Antigens Present	Antibody Reactions
$L^M L^M$	M	M	Anti-M
$L^M L^N$	MN	M, N	Anti-M, Anti-N
$L^N L^N$	N	N	Anti-N

Table 17.2 **Determining Allele Frequencies for Codominant Genes by Counting Alleles**

Genotype	MM	MN	NN	Total
Number of individuals	54	26	20	100
Number of L^M alleles	108	26	0	134
Number of L^N alleles	0	26	40	66
Total	108	52	40	200

Frequency of L^M in population: 134/200 = 0.67 = 67%

Frequency of L^N in population: 66/200 = 0.33 = 33%

Table 17.3 **Frequencies of L^M and L^N Alleles in Various Populations**

Population	Genotype Frequency (%)			Allele Frequency	
	MM	MN	NN	L^M	L^N
U.S. Indians	60.00	35.12	4.88	0.776	0.224
U.S. Blacks	28.42	49.64	21.94	0.532	0.468
U.S. Whites	29.16	49.38	21.26	0.540	0.460
Eskimos (Greenland)	83.48	15.64	0.88	0.913	0.087

of our population of 100 individuals carries two copies of the *L* gene, so there is a total of 200 alleles (100 individuals, each of whom has 2 alleles = 200). So the frequency of the *M* allele is 134/200, or 0.67 (67%). The frequency of the *N* allele can be calculated similarly by counting 40 *N* alleles in the *NN* homozygotes (20 individuals, each of whom has 2 *N* alleles) and an additional 26 *N* alleles in the *MN* heterozygotes, a total of 66 copies of the *N* allele. The frequency of the *N* allele in the population is 66/200, or 0.33 (33%). This method of calculating gene frequencies in codominant populations is summarized in Table 17.2, and the frequencies of *M* and *N* alleles in several human populations are listed in Table 17.3.

Recessive Allele Frequencies Cannot Be Measured Directly

The preceding example used a codominant gene system for measuring allele frequencies. In codominant systems, there is a direct relationship between phenotype and genotype. But most human genes exhibit dominant or recessive phenotypes. If one allele is recessive, there is no direct relationship between phenotype and genotype because the heterozygote and the homozygous dominant individuals have identical phenotypes. In this situation, we cannot measure allele frequencies by counting alleles because we cannot determine how many individuals have the homozygous dominant genotype and how many are heterozygotes. However, a mathematical formula can be used to determine allele frequencies when one or more alleles are recessive and a number of conditions (described below) are met. This method, developed independently by Godfrey Hardy and Wilhelm Weinberg, is known as the **Hardy-Weinberg Law**.

■ **Hardy-Weinberg Law** The statement that allele and genotype frequencies remain constant from generation to generation when the population meets certain assumptions.

Concepts and Controversies

The Thrifty Genotype

*T*he Pima Indians of the American Southwest have one of the highest rates of adult diabetes of any population in the United States. Between 42 and 66% of the Pima over the age of 35 years are affected with diabetes. This disorder, accompanied by obesity, and developed in the Pima only in this century, became recognized as a serious health problem only after 1950. Diabetes is usually followed by blindness, kidney failure, and heart disease.

How is it that an inherited condition with so many deleterious effects can be present in a population at such a high frequency? James Neel of the University of Michigan speculated on this question. He observed that the frequency of diabetes is very low in hunter-gatherer societies, such as the Pima were before the 20th century. He also noted that females prone to diabetes become sexually mature at an earlier age. In addition, susceptible females give birth to larger-than-average babies (increased birth weight is linked to increased survival). He postulated that these characteristics might provide a reproductive advantage to diabetics and account for the high frequency of this disorder.

In addition, Neel proposed that in the feast-or-famine diet of hunter-gatherers, the diabetic represents a "thrifty" genotype that is more efficient in converting food into energy. In diabetics, more insulin is released after each meal, and more glucose is metabolized. According to Neel, as the age of the individual increases, these repeated cascades of insulin release produce a counterreaction that releases an antagonist to stop insulin action. In the hunter-gatherer culture, this overloading occurs infrequently, and adult-onset diabetes does not fully develop.

We now know that there is no antagonist as envisioned by Neel, but as he reviewed the concept in 1982, he noted that "although incorrect in detail, it may have been correct in principle." It appears that adult-onset diabetics release more insulin after a meal than nondiabetics. This quick response prevents loss of blood glucose through the kidneys and is an effective adaptation in hunter-gatherer societies. When the Pima diet converted to one that is high in refined carbohydrates, the release of extra insulin occurs over and over again, eventually causing a reduction in the number of cell-surface insulin receptors. As a result, although insulin is present in the blood, it is ineffective in mobilizing glucose, and the symptoms of diabetes appear.

The adult-onset diabetic genotype may represent one that was well adapted to the environment of the hunter-gatherer societies that prevailed for hundreds of thousands of years. In the case of the Pima Indians, environmental conditions changed dramatically in less than a century, and the genotype is now at a distinct disadvantage in an environment where carbohydrate-rich foods are freely available. This idea is interesting and if confirmed, would be an example of natural selection in action.

THE HARDY-WEINBERG LAW: MEASURING ALLELE AND GENOTYPE FREQUENCIES

After Mendel's work became widely known, there was a great deal of debate about whether the principles of Mendelian inheritance applies to humans. One of the first genetic traits identified in humans was a dominant mutation known as brachydactyly (MIM/OMIM 112500). Because the phenotypic ratio of dominant traits is 3:1 among progeny of a heterozygotes (1*AA*, 2*Aa*, 1*aa*), it was thought that over time, this would become the phenotypic ratio in the human population. However, because only a small fraction of the population has brachydactyly, the argument went, perhaps Mendelian inheritance is valid for plants and other animals but does not explain inheritance in humans.

An English mathematician, Godfrey Hardy, and a German physician, Wilhelm Weinberg, independently recognized that this reasoning was false because it failed to distinguish between the *mode of inheritance* (in this case a dominant trait with a 3:1 ratio) and the *frequency* of the dominant and recessive alleles in the population. Hardy and Weinberg each developed a simple mathematical model to estimate the frequency of alleles in a population and to describe how alleles combine to form genotypes.

SIDEBAR

Selective Breeding Gone Bad

Purebred dogs are the result of selective breeding over many generations, and worldwide, there are now more than 300 recognized breeds and varieties of dogs. Selective breeding to produce dogs that have desired traits, such as long noses and closely set eyes in collies, or the low, sloping hindlegs of German shepherds, can have unintended side effects. About 70% of all collies suffer from hereditary eye problems, and more than 60% of German shepherds are at risk for hip dysplasia. It is estimated that 25% of all purebred dogs have a serious genetic disorder.

The high level of genetic disorders in purebred dogs is a direct result of selective breeding. Over time, selective breeding has decreased genetic variability and increased animals homozygous for recessive genetic disorders.

Outbreeding is a simple genetic solution to this problem. In the world of dogs, this means mixed-breed dogs, or mutts. Crosses between collies and German shepherds or between Labrador retrievers and German shepherds combine the looks and temperaments of both breeds but reduce the risk of offspring that have genetic disorders.

Assumptions for the Hardy-Weinberg Law

The mathematical model developed by Hardy and Weinberg is based on a number of assumptions:

- The population is large. In practical terms, this means that the population is large enough so that there are no sampling errors in measuring allele frequencies.
- There is no selective advantage for any genotype; that is, all genotypes have equal ability to survive and reproduce.
- Mating between individuals in the population is random.
- Other factors that cause changes in allele frequency such as mutation and migration, are absent or rare events and can be ignored.

These assumptions make the Hardy-Weinberg method less exact than counting alleles directly, as in a codominant system. Because the assumptions of the Hardy-Weinberg method only rarely exist in natural situations, allele frequencies determined in this way are regarded as estimates.

Calculating Allele and Genotype Frequencies

Let us illustrate how the model works by considering a population that carries an autosomal gene with two alleles, *A* and *a*. The frequency of the dominant allele *A* in gametes is represented by p, and the frequency of the recessive allele *a* in gametes is represented by q. Because the sum of p and q represents 100% of the alleles for that gene in the population, then $p + q = 1$. A Punnett square can be used to derive the genotypes produced by the random combination of gametes that carry these alleles (Figure 17.4).

FIGURE 17.4 The frequency of the dominant and recessive alleles in the gametes of the parental generation determines the frequency of the alleles and the genotypes of the next generation.

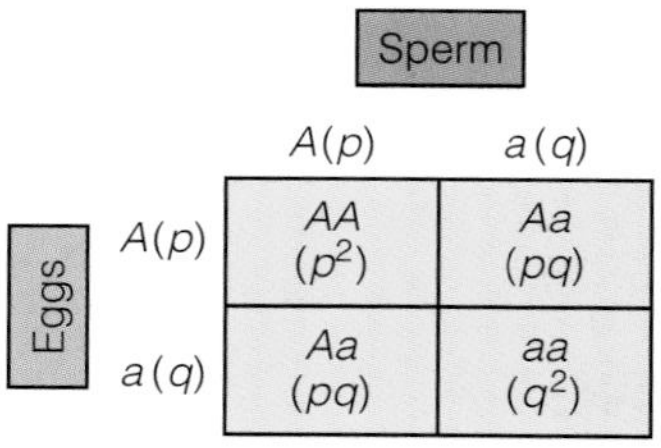

In the combination of gametes that produces the next generation, the chance that both the egg and sperm will carry the *A* allele is $p \times p = p^2$. The chance that the gametes will carry unlike alleles is $(p \times q) + (p \times q) = 2pq$. The chance that a homozygous recessive combination of alleles might result is $q \times q = q^2$. It is important to note that although p^2 is the chance that both gametes will carry an *A* allele, p^2 represents the frequency of the homozygous *AA* genotype. Similarly, $2pq$ is a measure of heterozygote (*Aa*) frequency, and q^2 represents the frequency of homozygous recessive (*aa*) individuals. In other words, the distribution of genotypes in the next generation can be expressed as:

$$p^2 + 2pq + q^2 = 1$$

where 1 represents 100% of the genotypes present. This formula expresses the Hardy-Weinberg Law, which states that both allele and genotype frequencies will remain constant from generation to generation in a large, interbreeding population where mating is random and there is no selection, migration, or mutation.

The formula can be used to calculate allele frequencies (*A* and *a* in our example) and the frequency of the various genotypes in a population. To show how the model works, let's begin with a population for which we already know the frequency of the alleles. Suppose that we have a large, randomly mating population in which the frequency of an autosomal dominant allele *A* is 60% and the frequency of the recessive allele *a* is 40%. This means that $p = 0.6$ and $q = 0.4$. Because *A* and *a* are the only two alleles, the sum of $p + q$ equals 100% of the alleles:

$$p\ (0.6) + q\ (0.4) = 1$$

In this population, 60% of the gametes carry the dominant allele A, and 40% carry the recessive allele a. The distribution of genotypes in the next generation is shown in ▶ Figure 17.5.

In the new generation, 36% ($p^2 = 0.6 \times 0.6$) of the offspring will have a homozygous dominant genotype AA, 48% ($2\ pq = 2[0.6 \times 0.4]$) will be heterozygous Aa, and 16% ($q^2 = 0.4 \times 0.4$) will have a homozygous recessive genotype, aa.

We can also use the Hardy-Weinberg equation to calculate the frequency of the A and a alleles in the new generation. The frequency of A is

$$p^2 + 1/2\ (2pq)$$

$$0.36 + 1/2\ (0.48)$$

$$0.36 + 0.24 = 0.60 = 60\%$$

For the recessive allele a, the frequency is

$$q^2 + 1/2\ (2pq)$$

$$0.16 + 1/2\ (0.48)$$

$$0.16 + 0.24 = 0.40 = 40\%$$

Because $p + q = 1$, we could have calculated the value for the recessive allele a by subtraction:

$$p + q = 1$$

$$q = 1 - p$$

$$q = 1 - 0.60$$

$$q = 0.40 = 40\%$$

▶ **FIGURE 17.5** The frequency of alleles and genotypes in a new generation where the alleles in the parental generation are present at 0.6 for the dominant allele (A) and 0.4 for the recessive allele (a).

Sperm

Eggs	A (p = 0.6)	a (q = 0.4)
A (p = 0.6)	AA ($p^2 = 0.36$)	Aa ($pq = 0.24$)
a (q = 0.4)	Aa ($pq = 0.24$)	aa ($q^2 = 0.16$)

Genetic equilibrium The situation when the allele frequency for a given gene remains constant from generation to generation.

Populations Can Be in Genetic Equilibrium

In the previous example, the frequencies of A and a in the new generation are the same as in the parental generation. Populations in which the allele frequency of a given gene remains constant from generation to generation are in a state of **genetic equilibrium** for that gene. This does not mean that the population is in a state of equilibrium for all alleles. On the contrary, if forces such as mutation, selection, or migration are operating on other genotypes, the frequency of other alleles may change from one generation to the next.

The presence of a Hardy-Weinberg equilibrium illustrates why dominant alleles do not increase in frequency as new generations are produced. In the case of brachydactyly, if conditions for the Hardy-Weinberg Law are met, a genetic equilibrium will be established. Because allele and genotype frequencies determine phenotype frequencies, brachydactyly will not increase in the population and reach a 3:1 frequency but instead will be maintained at an equilibrium frequency from generation to generation.

In addition, genetic equilibrium also helps to maintain genetic variability in the population. In the previous example, at equilibrium we can be assured that 60% of the alleles for gene A will be dominant (A) and 40% will be recessive (a) in generation after generation. The presence and maintenance of genetic variability is important to the process of evolution.

USING THE HARDY-WEINBERG LAW IN HUMAN GENETICS

The Hardy-Weinberg Law is one of the foundations of population genetics and has many applications in human genetics and human evolution. We consider only a few of its uses, primarily those that apply measuring allele and genotype frequencies.

Autosomal Dominant and Recessive Alleles

In most autosomal traits, the homozygous dominant and heterozygous genotypes have the same phenotype. If the trait being examined is inherited recessively, the frequency of the recessive allele can be calculated by first counting the number of homozygous recessive individuals in the population. For example, cystic fibrosis is an autosomal recessive trait, and homozygous recessive individuals can be identified by their distinctive phenotype. Suppose that 1 in 2500 individuals in a population is affected with cystic fibrosis. These individuals have the recessive genotype aa. According to the Hardy-Weinberg equation, the frequency of this genotype in the population is given by q^2. This means that the frequency of the a allele in this population is equal to the square root of q^2:

$$q^2 = 1/2500 = 0.0004$$
$$q = \sqrt{0.0004}$$
$$q = 0.02 = 1/50$$

For the dominant allele in this case, once we know that the frequency of the a allele is 0.02 (2%), we can calculate the frequency of the dominant allele A by subtraction:

$$p + q = 1$$
$$p = 1 - q$$
$$p = 1 - 0.02$$
$$p = 0.98 = 98\%$$

In this population, 98% of the alleles for gene A are dominant (A), and 2% are recessive (a). This method can be used to calculate the allele and genotype frequencies for any dominant or recessive trait.

Autosomal Codominant Alleles

As outlined earlier, the allele and genotype frequencies for autosomal codominant traits can be determined directly from the phenotype because each genotype gives rise to a distinctive phenotype, as in the MN blood groups. Allele and genotype frequencies can be determined for this mode of inheritance by counting individuals in the population. Further, there is no need to make assumptions about Hardy-Weinberg conditions because the genotype and phenotype frequencies are identical.

X-Linked Traits

One of our underlying assumptions in estimating the allele frequency for the autosomal recessive trait that controls cystic fibrosis was that the frequency of the dominant allele A or the recessive allele a is the same in sperm as it is in eggs. That is, 98% of the sperm and 98% of the eggs in this population should carry the dominant allele A, and 2% of the sperm and 2% of the eggs should carry the recessive allele a. This situation does not hold true for X-linked traits. Human females carry two X chromosomes and two copies of all genes on the X. Males, on the other hand, have only one X chromosome and are hemizygous for all loci on the X chromosome. This means that genes on the X chromosome are not distributed equally in the population: Females (and their gametes) carry two-thirds of the total number, and males (and their gametes) carry one-third of the total. As we will see below, the Hardy-Weinberg equation can be used to calculate genotype frequencies in females for recessive X-linked traits. However, because males are hemizygous for all traits on the X chromosome, the allele frequency for recessive X-linked traits can be calculated directly by counting the number of males in the population who carry the mutant phenotype. For example, in the United States, about 8% of males are color-blind. Therefore, the frequency of the allele for color blindness in this population is 0.08 ($q = 0.08$).

Because females have two copies of the X chromosome, genotypic frequencies for X-linked recessive traits in females can be calculated using the Hardy-Weinberg equation. If color blindness in males occurs with a frequency of 8% ($q = 0.08$), then we would expect color blindness in females to occur with a frequency of q^2, or 0.0064 (0.64%). Given the allelic frequency of 0.08, this means that in a population of 10,000 males, 800 would be color-blind, but in a population of 10,000 females, only 64 would be color-blind. This example reemphasizes the fact that males are at much higher risk for deleterious traits carried on the X chromosome. The relative values for the frequency of X-linked recessive traits in males and females are listed in Table 17.4.

Multiple Alleles

Until now we have discussed allele frequencies in genes that have only two alleles. For genes such as the ABO blood type, however, three alleles of the isoagglutinin locus (I) are present in the population. In this system, the alleles A and B are codominant, and both are dominant to O. This system has six possible genotypic combinations:

$$AA, AO, BB, BO, AB, OO$$

Homozygous AA individuals and heterozygous AO individuals are phenotypically identical, as are BB and BO individuals. This results in four phenotypic combinations, known as blood types A, B, AB, and O.

The Hardy-Weinberg Law can be used to calculate both the allele and genotype frequencies for this three-allele system by adding another term to the equation. For the three blood group alleles,

$$p(A) + q(B) + r(O) = 1$$

In other words, when you add up the frequency of the A, B, and O alleles, you have accounted for 100% of the alleles for this gene that are present in the population. The genotypic frequencies are given by the equation,

$$(p + q + r)^2 = 1$$

The allele frequency values for A, B, and O can be calculated from the distribution of phenotypes in a population, if random mating is assumed.

If we know the frequency of the A, B, and O alleles for a given population, we can then calculate the genotypic and phenotypic frequencies for all combinations of these alleles. The genotypic combinations can be calculated by an expansion of the Hardy-Weinberg equation:

$$p^2 (AA) + 2pq (AB) + 2pr (AO) + q^2 (BB) + 2qr (BO) + r^2 (OO) = 1$$

That is, the frequency of AA individuals is predicted to be p^2, AB individuals $2pq$, etc. The frequencies for the A, B, and O alleles in different populations in the world are listed in Table 17.5. The geographic distribution of ABO alleles is shown in ▶ Figure 17.6. By using the equations shown above and the values in Table 17.5, we can calculate the genotypic and phenotypic frequencies for the populations shown in Figure 17.6.

Estimating the Frequency of Heterozygotes in a Population

An important application of the Hardy-Weinberg Law is estimating heterozygote frequency in a population. The majority of deleterious recessive genes in human populations are carried in the heterozygous condition. To calculate the frequency of individuals who are heterozygous for recessive traits, we usually begin by counting the number of homozygous recessive individuals (all of whom show the recessive phenotype). For example, earlier we calculated the frequency of the allele for cystic fibrosis, an autosomal recessive trait with a frequency of 1 in 2500 among white Americans. (The disease is much rarer among American blacks and Asians). These

Table 17.4
Frequency of X-Linked Recessive Traits in Males and Females

Males	Females
1/10	1/100
1/100	1/10,000
1/1000	1/1,000,000
1/10,000	1/100,000,000

Table 17.5 — Frequency of ABO Alleles in Various Populations

Population	Frequency A(p)	B(q)	O(r)
Armenians	36.0	10.4	53.6
Basques	25.5	—	74.5
Eskimos	35.5	4.6	59.9
Belgium	27.0	5.9	67.1
Denmark	29.4	7.7	62.9
Greece	22.9	8.2	68.9
Poland	25.9	14.0	60.1
Russia (Urals)	29.5	19.5	51.0
Russia (Siberia)	13.0	25.1	61.9
Russia (Tadzhikistan)	21.1	37.1	41.7
Sri Lanka (Sinhalese)	14.0	15.2	70.8
India (Assum)	19.2	11.1	69.7
India (Madras)	16.5	20.5	63.0
China (Hong Kong)	19.1	19.1	61.8
Japan	26.2	18.3	55.5
Nigeria (Ibo)	13.2	9.5	77.3
Nigeria (Yoruba)	13.8	14.6	71.6
Upper Volta	14.8	18.2	67.0
Kenya	17.2	14.0	68.8

homozygous individuals can be distinguished from the rest of the population by clinical symptoms that indicate defects in the function of exocrine glands. (To review the symptoms, see Chapter 4.) The frequency of the homozygous recessive genotype is 1 in 2500, or 0.0004, and is represented by q^2. As before, the frequency of the recessive allele q in the population can be calculated as:

$$q = \sqrt{q^2}$$
$$q = \sqrt{0.0004}$$
$$q = 0.02 = 2\%$$

Because $p + q = 1$, we can calculate the frequency of the dominant allele p:

$$p = 1 - q$$
$$p = 1 - 0.02$$
$$p = 0.98 = 98\%$$

Knowing the allele frequencies, we can use the Hardy-Weinberg equation to calculate genotype frequencies. Recall that in the Hardy-Weinberg equation, $2pq$ gives the frequency of the heterozygous genotype. Using the values we have calculated for p and q, we can determine the frequency of heterozygotes:

$$2pq = 2(0.98 \times 0.02)$$
$$2pq = 2(0.0196)$$
$$\text{Heterozygote frequency} = 0.039 = 3.9\%$$

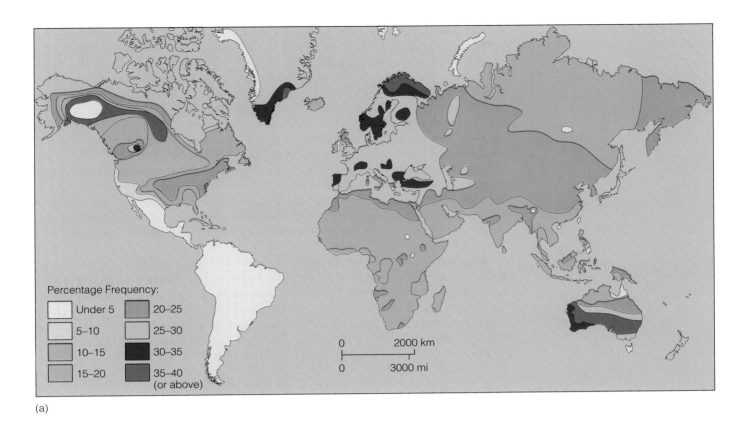

(a)

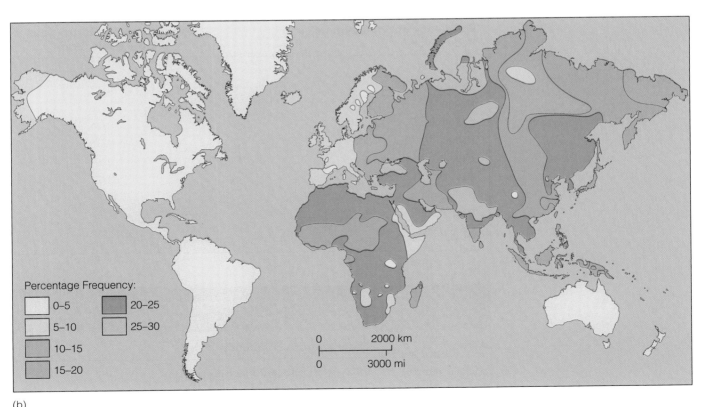

(b)

▶ **FIGURE 17.6** The distribution of alleles in the ABO system. (a) The distribution of the *A* allele in the indigenous population of the world (before 1600 A.D.). (b) Distribution of the *B* allele.

This means that 3.9%, (let's call it 4%), or about 1 in 25 white Americans carry the gene for cystic fibrosis.

Sickle cell anemia is an autosomal recessive trait that affects 1 in 500 black Americans. That means that the frequency of these homozygous recessive individuals (q^2) is 0.002. Using the Hardy-Weinberg equation, we can calculate that 8.5% of black Americans, or 1 in every 12, are heterozygous carriers for this trait. The frequencies of heterozygous carriers of recessive alleles whose frequencies range from 1 in 10 to 1 in 10,000,000 are listed in Table 17.6. The heterozygote frequencies for some common human autosomal recessive traits are listed in Table 17.7.

Many people are surprised to learn that heterozygotes for recessive traits are so common in the population. In most cases, because they assume that if a trait is relatively rare (say 1 in 10,000 individuals), that the number of heterozygotes must also be rather low. In fact, under these circumstances, 1 in 50 (2%) of the population is a heterozygote (Table 17.6), and there are about 200 times as many heterozygotes as there are homozygotes.

What are the chances that two heterozygotes will marry and produce an affected child? We can calculate how this occurs as follows: The chance that two heterozygotes will marry is $1/50 \times 1/50 = 1/2500$. Because they are heterozygotes, the chance that they will produce a homozygous recessive offspring is one in four. So, therefore, the chance that they will marry and produce a child affected by a recessive trait is $1/2500 \times 1/4 = 1/10,000$. If the trait is present in 1 out of every 10,000 individuals in a population, 1 in 50 individuals must be a heterozygous carrier of the recessive allele.

Table 17.6 Heterozygote Frequencies for Recessive Traits

Frequency of Homozygous Recessives (q^2)	Frequency of Heterozygous Individuals ($2pq$)
1/100	1/5.5
1/500	1/12
1/1000	1/16
1/2500	1/25
1/5000	1/36
1/10,000	1/50
1/20,000	1/71
1/100,000	1/158
1/1,000,000	1/500
1/10,000,000	1/1582

Table 17.7 Frequency of Heterozygotes for Some Recessive Traits (U.S. population)

Trait	Heterozygote Frequency
Cystic fibrosis	1/22 whites; much lower in blacks, Asians
Sickle cell anemia	1/12 blacks; much lower in most whites and in Asians
Tay-Sachs disease	1/30 among descendants of Eastern European Jews; 1/350 among others of European descent
Phenylketonuria	1/55 among whites; much lower in blacks and those of Asian descent
Albinism	1/10,000 in Northern Ireland; 1/67,800 in British Columbia

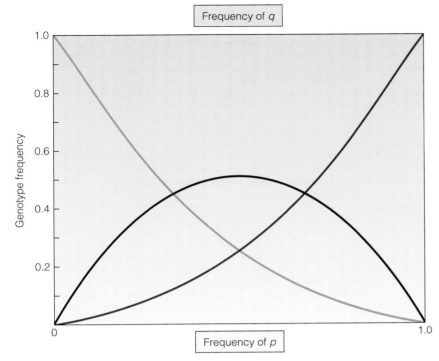

FIGURE 17.7 The relationship between allelic and genotypic frequency in a population that is in Hardy–Weinberg equilibrium. As the frequencies of the homozygous genotypes (p^2 and q^2) decline, the frequency of the heterozygote ($2pq$) rises.

Once the frequency of either allele is known, we can calculate the frequency of the homozygous dominant genotypes and the heterozygotes. Remember that the frequency of the genotypes depends on the allele frequency. The relationship between allele and genotype frequency is shown in ▶ Figure 17.7. As the frequencies of p and q move away from zero, the percentage of heterozygotes in the population increases rapidly. This again illustrates the point that in traits, such as cystic fibrosis or sickle cell anemia, the majority of the recessive alleles in a population are carried by heterozygotes, where the deleterious effects of the allele are not expressed.

ANTHROPOLOGY, POPULATION STRUCTURE, AND GENETICS

In many situations, anthropologists use properties such as language, dress, social customs, taboos, and food sources to determine how closely related cultures are and to work out the ancestral sources of these cultures. The analysis of allele frequencies and genotypes has proven to be a valuable tool in these studies, and the application of population genetics to anthropology has given rise to a discipline known as **anthropological genetics**.

The relationship between linguistic or cultural differences and biological differentiation is illustrated by the population structure on Bougainville Island in Papua, New Guinea (▶ Figure 17.8). Bougainville, which is roughly 130 miles long and 40 miles at its maximum width, is home to 17 different languages (▶ Figure 17.9). Most of these languages are spoken over a range of only about 10 miles, and many are subdivided into distinct dialects or even sublanguages that are found over much smaller ranges of only a mile or two. Traditionally, the populations on this island have little social or political contact with anyone outside their immediate area. The linguistic evidence and other aspects of their culture suggest that these local populations are quite isolated from one another.

■ **Anthropological genetics** The union of population genetics and anthropology to study the effects of culture on gene frequencies.

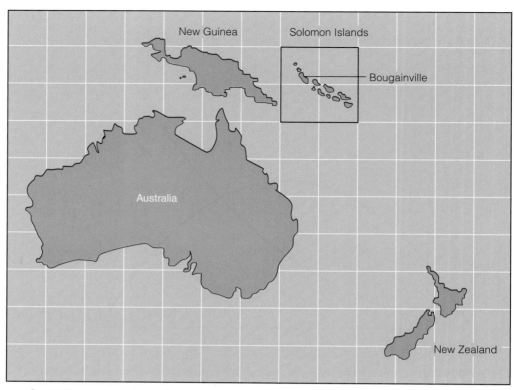

▶ **FIGURE 17.8**   Map of the southwest Pacific, showing the Solomon Islands and the location of Bougainville, which is part of Papua, New Guinea.

▶ **FIGURE 17.9**   Language divisions on Bougainville. Villages included in the genetic study are numbered 1 through 18. Blood-type data for these groups are given in Table 17.8.

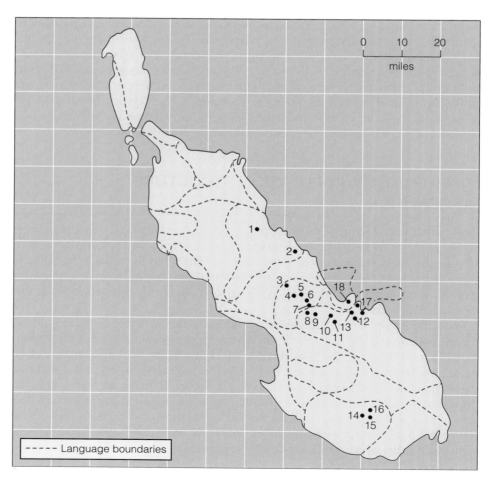

Table 17.8 Observed ABO Phenotypes in 18 Villages on Bougainville Island

GROUP	ABO PHENOTYPE (%)			
	A	O	B	AB
1	63.7	16.3	10.3	9.4
2	47.9	52.0	0.00	0.00
3	41.6	58.3	0.00	0.00
4	41.5	58.4	0.00	0.00
5	71.2	28.7	0.00	0.00
6	56.4	43.5	0.00	0.00
7	41.9	58.06	0.00	0.00
8	45.6	53.5	0.78	0.00
9	40.6	59.3	0.00	0.00
10	41.2	57.7	0.004	0.004
11	43.5	56.4	0.00	0.00
12	53.9	40.8	0.043	0.008
13	40.0	41.6	15.0	0.033
14	22.4	56.2	15.7	5.61
15	51.8	35.0	10.5	2.6
16	38.4	50.0	7.69	3.8
17	23.0	47.7	24.7	4.5
18	47.5	26.4	15.8	10.1

If these populations are in fact, isolated by culture, then little or no matings should take place among these cultures. If this is the case, then, the populations may also be genetically distinct from one another. To examine this possibility, blood-type phenotypes that represent eight of these language groups were determined in 18 villages. Some of these data are presented in Table 17.8. There is a wide range in the observed frequencies of blood types in the 18 villages. As we will see in Chapter 18, if there were interbreeding between or among these populations, the Hardy-Weinberg equation would predict that the blood-type frequencies would be equalized within a generation or two. Because the frequencies for ABO alleles are quite different among most villages, these results support the idea that these populations are genetically and culturally isolated. The notion that human activities such as culture can modify allele frequencies and form genetic barriers between populations is explored in more detail in Chapter 18.

Case Studies

CASE 1

Jane, a healthy woman, was referred for genetic counseling because she had a brother Matt and a sister Edna who had cystic fibrosis and died at the ages of 32, and 16, respectively. She married John, who has no family history of cystic fibrosis. Jane wants to know the probability that she and John will bear a child who has cystic fibrosis. The genetic counselor used the Hardy-Weinberg model to calculate the probability that this couple will have an affected child.

The counselor explains to Jane that there is a two in three chance that she is a carrier for a mutant CF gene by using a Punnet square to explain how she determined this. The probability that John is a carrier is equal to the population carrier frequency ($2pq$). The probability that John and Jane will have a child who has CF equals the probability that Jane is a carrier (2/3) multiplied by the probability that John is a carrier ($2pq$) multiplied by the probability that they will have an affected child if they are both carriers (1/4).

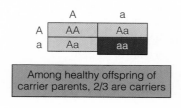

Among healthy offspring of carrier parents, 2/3 are carriers

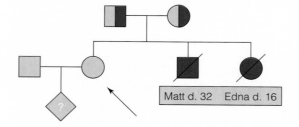

Matt d. 32   Edna d. 16

CASE 2

A population geneticist has sampled 200 people on a small tropical island and wants to determine the allelic frequency of the MN blood group. He knows that this single locus has two codominant alleles, M and N. Thus, three phenotypes, M, MN, N correspond to three genotypes MM, MN, and NN, respectively. Among the 200 individuals sampled, 72 are M, 104 are MN, and 24 are N. The frequencies of the respective genotypes are 0.36 (MM), 0.52 (MN), and 0.12 (NN). The allelic frequencies can be precisely determined as follows:

This sample consists of 200 (N) individuals and 400 (2N) alleles. Each M individual carries 2 M alleles ($2 \times 72 = 144$); each MN carrier 1 M allele ($1 \times 104 = 104$). This sample contains $144 + 104 = 248$ M alleles (out of 400 alleles).

Therefore, the frequency of the M allele is $248/400 = 0.62$ and the frequency of the N allele $= 1 - 0.62 = 0.38$.

Summary

1. The work of Charles Darwin and Alfred Russel Wallace on evolution and natural selection was influenced by Malthus' essay on population dynamics. Both Darwin and Wallace recognized that the struggle for existence favors those who carry variations that helped them adapt to the existing conditions, allowing them to leave more offspring.

2. In the early decades of this century, genes were recognized as the agents that cause such variations, giving rise to the field of population genetics. After the mathematical and theoretical basis of this field was established, experimentalists began to study allele frequencies in populations rather than in the offspring of a single mating. This work has produced the basis for our understanding of evolution.

3. The Hardy-Weinberg Law provides a means of measuring gene frequencies within populations and determining whether the population is in equilibrium.

4. The Hardy-Weinberg equation assumes that the population is large and randomly interbreeding and that factors such as mutation, migration, and selection are absent. The presence of equilibrium in a population explains why dominant alleles do not replace their recessive alleles.

5. The Hardy-Weinberg Law can also be used to estimate the frequency of heterozygotes in a population and to establish when allele frequencies are shifting in the population. The conditions that lead to changing allele frequencies in a population are those that produce evolutionary change.

Questions and Problems

1. Draw a graph showing the difference between a population that grows at a geometric rate and one that is only permitted to grow at a constant arithmetic rate. Be sure to label each axis.
2. What are some of the natural constraints referred to by Malthus that keep the size of human populations in check?
3. Define the following terms:
   a. population
   b. gene pool
   c. gene frequency
   d. genotype frequency
4. Explain the connection between changes in population allele frequencies and evolution, and relate this to the observations made by Darwin and Wallace concerning natural selection.
5. Do you think populations can evolve without changes in allele frequencies?
6. Design an experiment to determine if a population is evolving.
7. Suppose you are monitoring the allelic and genotypic frequencies of the MN blood group locus in a small human population. You find that for 1-year-old children the genotypic frequencies are MM = 0.25, MN = 0.5, and NN = 0.25, while the genotypic frequencies for adults are MM = 0.3, MN = 0.4, and NN = 0.3.
   a. Compute the *M* and *N* allele frequencies for the 1-year-olds and adults.
   b. Are the allele frequencies in equilibrium in this population?
   c. Are the genotypic frequencies in equilibrium?
8. Drawing on your newly acquired understanding of the Hardy-Weinberg equilibrium law, point out why the following statement is erroneous: "Since most of the people in Sweden have blond hair and blue eyes, the genes for blond hair and blue eyes must be dominant in that population."
9. In a population where the females have the allelic frequencies A = 0.35 and a = 0.65 and the frequencies for males are A = 0.1 and a = 0.9, how many generations will it take to reach Hardy-Weinberg equilibrium for both the allelic and genotypic frequencies? Assume random mating, and show the allelic and genotypic frequencies for each generation.
10. Explain why a population carries more genetic diversity than an individual.
11. If a trait determined by an autosomal recessive allele occurs at a frequency of 0.25 in a population, what are the allelic frequencies? Assume Hardy-Weinberg equilibrium, and use *A* and *a* to symbolize the dominant wild-type and recessive alleles, respectively.
12. Five percent of the males of a population express a sex-linked recessive trait.
    a. What are the frequencies of the dominant and recessive alleles?
    b. What are the genotypic frequencies for the males and females in the population?
13. In a given population, the frequencies of the 4 phenotypic classes of the ABO blood groups are found to be A = 0.33, B = 0.33, AB = 0.18, and O = 0.16. What is the frequency of the O allele?
14. In Table 17.3 which pairs of populations appear to be most closely related in terms of their allelic frequencies at the MN locus? What scenario would you propose to account for these relationships?
15. For each population in Table 17.3 use the given allelic frequencies to determine if the genotypic frequencies are in general agreement with the Hardy-Weinberg law.
16. Using Table 17.6, determine the frequencies of *p* and *q* that result in the greatest proportion of heterozygotes in a population.

Internet Activities

The following activities use the resources of the World Wide Web to enhance the topics covered in this chapter. To investigate the topics described below, log on to the book's home page at:

http://www.brookscole.com.biology

1. GenBank is the NIH's database of all known nucleotide and protein sequences coupled with supporting bibliographic and biological data. In the next two activities you will use two features of GenBank, the ENTREZ system to search for a DNA sequence and BLAST, to find similar sequences in GenBank.

   Begin this activity by searching the nucleotide database of ENTREZ.

   Enter the gene designation "glycosyltransferase" (an enzyme which places blood group antigens on red blood cell surfaces) in the query block. Select "retrieve" to find all data in GenBank on this particular enzyme. Select one of the documents related to this gene in humans and then retrieve the sequence in FASTA format. Highlight a portion of the sequence (300-500 bases) and copy it to the clipboard.
   In this activity you will search the GenBank database for sequence similarity between the sequence you highlighted and the sequence data of various species. Access BLAST.
   Enter the sequence you copied to the clipboard by pasting it into the query box. Submit the query. How

many different species have similar sequences to the one you selected? What percentage homology do these sequences have? Are there other human sequences which are identical or similar to the sequence you selected?

When used in the above manner, GenBank can be helpful in determining phylogenic relationships between species and also provide information about the evolution of a particular gene.

2. While the Hardy-Weinberg principle is easily demonstrated using mathematics, it may be difficult to conceptualize without some sort of visual aid or activity. The Access Excellence Activities exchange describes a Hardy-Weinberg demonstration using several decks of playing cards. The exercise should be carried out with a large group (greater than 20 students). In each subsequent round of the activity, allelic and genotype frequencies should be first be predicted and then calculated. After playing the game as suggested, modify it by incorporating one of the following variables: founder effect or mutation. How do these variables modify expected frequencies?

For Further Reading

Barrantes, R., Smouse, P. E., Mohrenweiser, H. W., Gershowitz, H., Azofeifa, J., Arias, T. D., & Neel, J. V. (1990). Microevolution in lower Central America: Genetic characterization of the Chibcha-speaking groups of Costa Rica and Panama, and a consensus taxonomy based on genetic and linguistic affinity. *Am. J. Hum. Genet.* 46: 63–84.

Bodmer, W. F., & Cavelli-Sforza, L. L. (1976). *Genetics, Evolution and Man.* San Francisco: Freeman.

Chakraborty, R., Smouse, P. E., & Neel, J. V. (1988). Population amalgamation and genetic variation: Observations on artificially agglomerated tribal populations of Central and South America. *Am. J. Hum. Genet.* 43: 709–725.

Constans, J. (1988). DNA and protein polymorphism: Application to anthropology and human genetics. *Anthropol. Anz.* 46: 97–117.

Deevey, E. S., Jr. (1960). The human population. *Sci. Am.* 203 (September): 194–205.

Feldman, M. W., & Christiansen, F. B. (1985). *Population Genetics.* Palo Alto, CA: Blackwell Scientific.

Friedlaender, J. S. (1975). *Patterns of Human Variation.* Cambridge, MA: Harvard University Press.

Mettler, L. E., Gregg, T. G., & Schaffer, H. E. (1988). *Population Genetics and Evolution,* 2nd ed. Englewood Cliffs, NJ: Prentice Hall.

Neel, J. V. (1978). The population structure of an Amerindian tribe, the Yanomama. *Ann. Rev. Genet.* 12: 365–413.

Relethford, J. H. (1985). Isolation by distance, linguistic similarity and the genetic structure on Bougainville Island. *Am. J. Physiol. Anthropology* 66: 317–326.

Romeo, G., Devoto, M., & Galietta, L. J. (1989). Why is the cystic fibrosis gene so frequent? *Hum. Genet.* 84: 1–5.

Spiess, E. B. (1989). *Genes in Populations,* 2nd ed. New York: Wiley.

Woo, S. L. (1989). Molecular basis and population genetics of phenylketonuria. *Biochemistry* 28: 1–7.

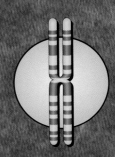

Human Diversity and Evolution

One day, about three and a half million years ago, two humanlike creatures, one larger, one smaller, walked across a muddy field of volcanic ash in a region of what is now Tanzania, Africa. Then as now, the region contained small lakes, patches of vegetation, and a volcano. The pair passed through the area during or just after a rain, and shortly after an eruption had covered the ground with a layer of volcanic ash. Along the way, the smaller one stopped, turned away to look at something, then resumed walking. This unremarkable event was later to provide important clues about several aspects of human evolution. Soon after these creatures passed, several new layers of volcanic ash covered their footprints.

After about three and a half million years, erosion by wind and rain partly exposed the fragile footprints, and they were discovered in 1978 by a team of fossil hunters led by Mary Leakey. Altogether, more than fifty prints covering a distance of about a hundred feet were recovered and preserved as casts.

These prints demonstrate that ancestral primates, known as hominids, had feet with well-developed arches, heels, and toes. More importantly, these prints reveal that one of the key events in human evolution, upright posture and walking, evolved more than three million years ago. Other evidence from fossil remains suggest that these small, primitive hominids had apelike faces with receding foreheads and small brains. However, their pelvis and leg anatomy allowed them to stand upright and walk much like modern humans, freeing their hands for other tasks, such as carrying food or young.

In this chapter, we consider the evidence how and where our species, Homo sapiens, arose from primate ancestors through a series of evolutionary adaptations over a period spanning millions of years. The broad outlines of this evolutionary process are now well established. New discoveries and interpretations of the evidence are constantly refining our knowledge of the nature and sequence of the historical events of our own species. Although much of the evidence consists of bits of fossilized bones dug out of rocks, there is a direct and thrilling connection between us and our ancestors when we view humanlike footprints that have survived for more than three million years.

MEASURING GENETIC DIVERSITY IN HUMAN POPULATIONS

The key to understanding the evolutionary history of the human species lies in identifying factors that generate genetic variation among members of a species and the way these variations are acted upon by natural selection. In the following sections, we explore how genetic variation is produced and the role of forces including natural selection and culture as a force in changing allele frequencies.

Blood types, such as the ABO locus, are ideal for studies of genetic variation in populations. Everyone has a blood type, and it is easily analyzed from a few drops of blood. Soon after their discovery, detailed maps of the geographic distribution of blood-type alleles became available (Figure 18.1).

The ABO blood types are genetic **polymorphisms** because two or more distinct alleles of the gene are present in the population at frequencies greater than 1%. The study of genetic polymorphisms provides information about the amount of genetic variability in a population. Some polymorphisms are responsible for genetic disorders, such as cystic fibrosis or sickle cell anemia, and others, such as the ABO blood types are of interest only as genetic markers. In this chapter we examine the forces that generate genetic diversity and those that change the frequency of alleles in a population.

New Alleles Are Generated by Mutation

The gene pool in each generation is reshuffled to produce the genotypes of the off-spring, and in such circumstances, genetic variation is produced by recombination

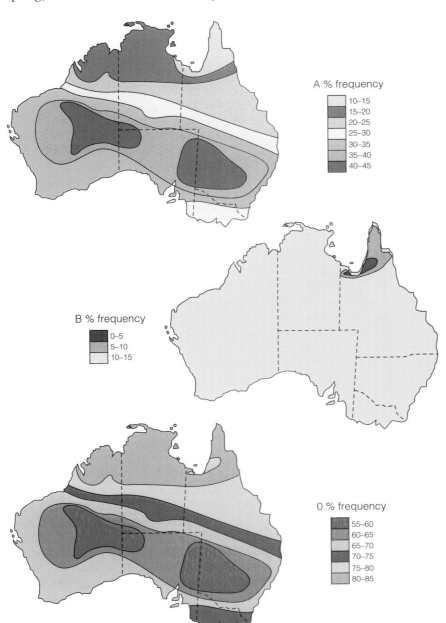

 FIGURE 18.1 A reconstruction of the geographic distribution of the ABO alleles in the prehistoric population of Australia.

and Mendelian assortment alone. These processes can generate new combinations of alleles, but they do not produce any new alleles. Mutation is the ultimate source of all new alleles and is the origin of all genetic variability.

The mechanisms and rates of mutation in humans were presented in Chapter 11. In this section, we consider the effect of mutation on allele frequencies in a population. Once the rate of mutation for a gene is known, we can use the Hardy-Weinberg Law to calculate the change in allele frequency that results from mutations in that gene in each generation. As an example, consider the dominant trait, achondroplasia. Statues and murals indicate that this form of dwarfism was known in Egypt more than 2500 years ago, or about 125 generations (allowing 20 years per generation) ago. If copies of the gene for achondroplasia were introduced into the gene pool only by mutation in each generation beginning 2500 years ago, how much has the frequency of achondroplasia changed over this period? Should we expect a higher frequency of achondroplasia among residents of an ancient city, such as Cairo, than among residents of a recently established city like Houston?

For this calculation we assume that initially, only homozygous recessive individuals who had the genotype *dd* (normal stature) were present in the population and that mutation has added new mutant (*D*) alleles to each generation at the rate of 1×10^{-5}. The change in allele frequency over time brought about by this rate of mutation is shown in ▶ Figure 18.2. To change the frequency of the recessive allele (*d*) from 1.0 (100%) to 0.5 (50%) at this rate of mutation will require 70,000 generations, or 1.4 million years. Thus the frequency of achondroplasia need not be any higher in Houston than in Cairo. Therefore, our conclusion is that although mutation is the ultimate source of all genetic variation, mutation alone has a minimal impact on the genetic variability of a population.

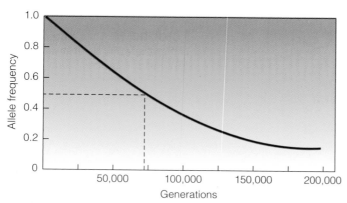

▶ **FIGURE 18.2** The rate of replacement of a recessive allele *a* by the dominant allele *A* by mutation alone. Even though the initial rate of replacement is high, it will take about 70,000 generations or 1.4 million years to drive the frequency of the allele *a* from 1.0 to 0.5.

Genetic Drift Can Change Allele Frequencies

Occasionally populations begin from a small group of individuals, known as founders. Alleles carried by these founders, whether they are advantageous or detrimental, become established in the new population. These events take place simply by chance and are known as **founder effects.**

Random changes in allele frequency that occur from generation to generation in small populations are examples of **genetic drift.** In addition to founder effects, genetic drift can occur in small populations and by temporary but drastic reductions in population size. These reductions, called population bottlenecks, are often caused by natural disasters. In extreme cases, drift can lead to the elimination of one allele from all members of the population. Small interbreeding groups on isolated islands often provide examples of genetic drift.

The population history for one such isolated island, Tristan da Cuhna (▶ Figure 18.3), is known in detail. Located in the southern Atlantic Ocean, it is one of the most remote locations on Earth. It is 2900 km (about 1800 mi) from Capetown, South Africa, and 3200 km (about 2000 mi) from Rio de Janeiro, Brazil. The island was occupied by the British in 1816 to prevent the rescue of Napoleon, who was in exile on the island of St. Helena. After Napoleon's death, Corporal William Glass, his wife, and two daughters received permission to remain after the British army withdrew. Others joined Glass at intervals, and the development of the isolated and highly inbred population that formed here can be traced with great accuracy.

Since 1961, the few hundred residents have been studied to determine the effects of isolation and inbreeding on the structure of the island's gene pool. As might be ex-

Founder effects Allele frequencies established by chance in a population that is started by a small number of individuals (perhaps only a fertilized female).

Genetic drift The random fluctuations of allele frequencies from generation to generation that take place in small populations.

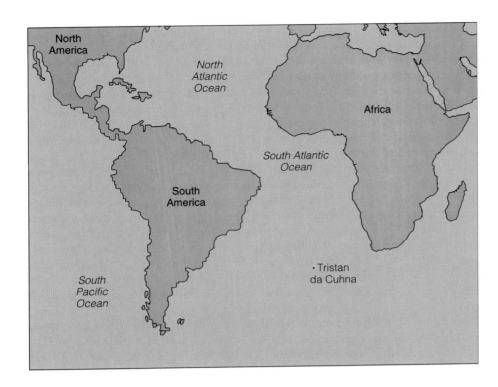

▶ **FIGURE 18.3** Location of the island of Tristan de Cuhna, first discovered by a Portuguese admiral in 1506.

■ **Clinodactyly** An autosomal dominant trait that produces a bent finger.

pected, one effect of inbreeding is an increase in homozygosity for recessive traits. Table 18.1 lists genetic markers for which all the islanders tested are homozygous.

Because of a founder effect, traits carried by one or a small number of early settlers are often found in a large fraction of the present population. On Tristan, a deformity of the fifth finger, known as **clinodactyly** (MIM/OMIM 112700), is present at a high frequency. This autosomal dominant trait is especially prominent in members of the Glass family, the first permanent residents of the island. The high frequency of this trait in the present island population can be explained by its presence in one of the original colonists.

This brief example illustrates how genetic drift can be responsible for changing allele frequencies in populations that are isolated, inbred, and stable for long periods of time. Most human populations, however, do not live on remote islands and are not subject to prolonged isolation and inbreeding. Yet there are differences in the distribution and frequency of alleles among populations, indicating that founder effects and drift are not the only factors that can change allele frequencies.

Natural Selection Acts on Variation in Populations

In formulating their thoughts on evolution, Wallace and Darwin recognized that not all members of a given population are equally viable or fertile, given the competition for limited resources such as food and mates. As a result, some individuals are better adapted to the environment than others. These better adapted individuals have an increased chance of leaving more offspring than those with other genotypes. The ability of a given genotype to survive and to reproduce is known as its **fitness**. Fitter genotypes are better able to survive and reproduce, and as a result, make a larger contribution to the gene pool of the next generation than other, less fit genotypes.

In time this differential reproduction of better adapted, or fitter, individuals leads to changes in allele frequencies within the population. The process of differential reproduction of fitter genotypes is known as **natural selection.** The relationship between the sickle cell allele and malaria is an example of the way natural selection changes allele frequency. Sickle cell anemia is an autosomal recessive condition associated with a mutant form of hemoglobin. Affected individuals have a wide range

Table 18.1
Homozygous Markers among Tristan Residents

Transferrins

Phosphoglucomutase

6-Phosphogluconate
  dehydrogenase

Adenylate kinase

Hemoglobin A variants

Carbonic anhydrase (2 forms)

Isocitrate dehydrogenase

Glutathione peroxide

Peptidase A, B, C, D

SOURCE: *Data from T. Jenkins, P. Beighton, & A. G. Steinberg, (1985) Ann. Hum. Biol., 12, 363–371,* **Table 2.**

■ **Fitness** A measure of the relative survival and reproductive success of a given individual or genotype.

■ **Natural selection** The differential reproduction shown by some members of a population that is the result of differences in fitness.

of clinical symptoms (see Chapter 4 for a review of the symptoms). In spite of the fact that untreated homozygotes have a high rate of childhood death, the mutant allele is present in very high frequencies in certain populations. In some West African countries, 20% of the population may be heterozygous for this trait, and in regions along rivers, such as the Gambia, almost 40% of the population is heterozygous. If homozygous recessive individuals die before they reproduce, why hasn't this mutant allele been eliminated from the population?

The reason that the mutant allele is present at a high frequency in certain West African countries and in certain regions of Europe, the Middle East and Asia is the presence of an infectious disease, malaria. In the tropics, malaria is caused by a protozoan parasite, *Plasmodium falciparum*. The disease is transmitted to humans by infected mosquitoes, and affected individuals suffer recurring episodes of illness throughout life. Victims of malaria are more likely to contract other diseases, often with fatal results. Thus, malaria victims have a reduced fitness. Malaria may seem like an exotic and rare disease to those of us in more developed countries, but more than two million people die from malaria each year, and more than 300 million individuals worldwide are infected with this disease. Because of population growth in the developing world, and the spread of drug-resistant strains of *Plasmodium*, malaria is actually increasing rather than decreasing.

The geographic distributions of malaria and sickle cell are shown in ▶ Figure 18.4. Research has shown that the mutant allele for sickle cell anemia confers resistance to malaria, and experiments on human volunteers have confirmed this. In heterozygotes and in recessive homozygotes, the mutant hemoglobin (Hb S) alters the properties of the plasma membrane of red blood cells and makes them resistant to infection by the malarial parasite. As a result, heterozygotes are fitter than those who have the homozygous normal genotype. Those who have the homozygous recessive genotype are also resistant to malaria but are less fit than those who have the heterozygous or homozygous normal genotypes because they suffer from sickle cell anemia. In this case, selection favors the survival and differential reproduction of heterozygotes.

▶ **FIGURE 18.4** (a) The distribution of sickle cell anemia in the Old World. (b) The distribution of malaria in the same region.

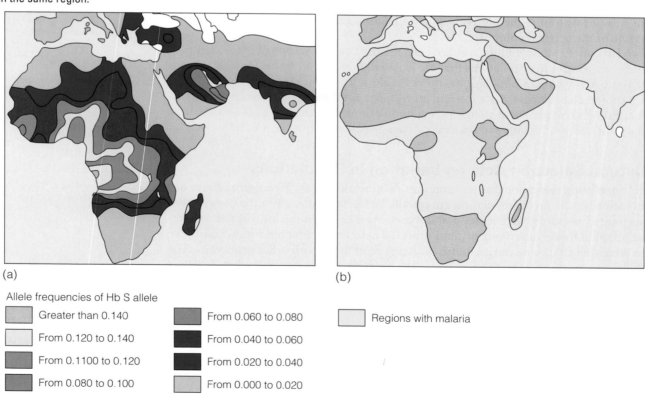

(a)

(b)

Allele frequencies of Hb S allele

Greater than 0.140	From 0.060 to 0.080
From 0.120 to 0.140	From 0.040 to 0.060
From 0.1100 to 0.120	From 0.020 to 0.040
From 0.080 to 0.100	From 0.000 to 0.020

Regions with malaria

As a result of increased fitness, heterozygotes make a larger contribution to the gene pool of the next generation than the other genotypes, and the HbS allele is spread through the population and maintained at a high frequency.

Selection and the Genetic History of a Population

The effects of selection on allele frequency can also be seen in subpopulations geographically separated from one another but derived from a single ancestral population. There have been several emigrations from Israel during the last 2500 years. These dispersed subpopulations moved to Europe, North Africa, the Middle East, and Asia. The frequency of one allele of the enzyme glucose-6-phosphate dehydrogenase (*G6PD*) (MIM/OMIM 305900) deficiency is now very different among these subpopulations (▶ Figure 18.5). The frequency ranges from almost zero in populations from Central Europe (Ashkenazi) to around 70% in Kurdish Jews.

There are two likely explanations for such differences in allele frequency: founder effects and selection. The allele for *G6PD* deficiency has a geographic distribution similar to that of malaria. This distribution and the fact that dramatic differences in allele frequency arose in the short span of 100 to 125 generations indicate a role for selection. In this case the selective force once again is malaria. Malarial parasites reproduce well only in cells that contain the enzyme G6PD. As a result, homozygous females and hemizygous males who are G6PD deficient are resistant to malaria. If we assume that the ancestral population had a low frequency of this allele, selection has caused a dramatic change in frequency in a relatively short time, illustrating the power of natural selection on allele frequency.

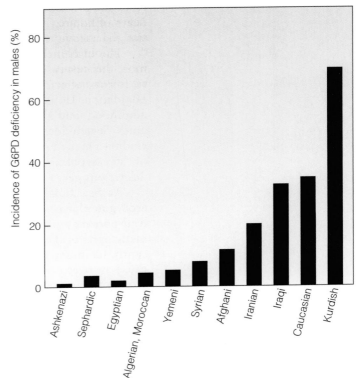

▶ FIGURE 18.5 Distribution of G6PD deficiency in various Jewish populations. Because intermarriage with native populations is rare, the differences in frequency are attributed to selection. G6PD deficiency confers resistance to malaria, and the frequency of the allele is highest in regions where malaria is endemic.

NATURAL SELECTION AND THE FREQUENCY OF GENETIC DISORDERS

Many genetic disorders are disabling or fatal, so why are they so common? In other words, what keeps natural selection from eliminating the deleterious alleles responsible for these disorders? In analyzing the frequency and population distribution of human genetic disorders, it is clear that there is no single answer. One conclusion, drawn from the Hardy-Weinberg Law, is that rare lethal or deleterious recessive alleles survive because the vast majority of them are carried in the heterozygous condition and thus are hidden in the gene pool. Other factors, however, can cause the differential distribution of alleles in human populations, and several of these are discussed below.

Almost all affected individuals who have the X-linked disorder, Duchenne muscular dystrophy (*DMD*) (MIM/OMIM 310200), die without reproducing. If this is the case, eventually the mutant allele should be eliminated from the population. However, because the mutation rate for *DMD* is high (perhaps as high as 1×10^{-4}), mutation replaces those *DMD* alleles lost when affected individuals die without reproducing. Thus, the frequency of the *DMD* allele in a population represents a balance between alleles introduced by mutation and those removed by the death of affected individuals.

Selection can cause spreading of detrimental alleles through large, historically well-established populations. We have already examined the classic case of heterozygote advantage in the relationship between malaria and sickle cell anemia. For some other genetic diseases, if there is a selective advantage for heterozygotes, it is less obvious or perhaps no longer observable.

Tay-Sachs disease (MIM/OMIM 272800) is inherited as an autosomal recessive disorder that is fatal in early childhood. Although this is a rare disease in most populations, there is a tenfold higher incidence of Tay-Sachs in populations of Ashkenazi Jews (those who live in or have ancestors from Eastern Europe). In these populations, the frequency of heterozygotes can be as high as 11%. There is indirect evidence that Tay-Sachs heterozygotes are more resistant to tuberculosis, a disease endemic to cities and towns, where most of the European Jews lived. As in sickle cell anemia, the death of homozygous Tay-Sachs individuals is the genetic price paid by the population that allows the higher fitness and survival of the more numerous heterozygotes.

The high incidence of cystic fibrosis (CF) (MIM/OMIM 219700) has been a more difficult case to explain. In European and European-derived populations, cystic fibrosis affects about 1 in 2000 births, and heterozygotes make up 4–5% of the population. Until recently, this autosomal recessive disease has been lethal in early adulthood, and almost all cases have been the result of matings between heterozygotes. Several ideas have been advanced to explain the frequency of this disease, including a high mutation rate, higher fertility in heterozygotes, and genetic drift. None of these hypotheses has made a convincing case to explain the high frequency of this deleterious gene.

At the cellular level, CF impairs chloride ion transport in secretory cells, and heterozygotes have a reduced level of chloride transport. This reduced level of chloride transport may make heterozygotes more resistant to certain infectious diseases. In many parts of the world, today as in the past, bacterial diarrhea contributes significantly to infant mortality. Because certain toxins produced in bacterial diarrhea cause an oversecretion of chloride ions, it has been postulated that infant CF heterozygotes are more likely to survive such illnesses that kill by electrolyte depletion and dehydration. More recent evidence indicates that CF heterozygotes may have high resistance to one of these diseases, typhoid fever.

These examples illustrate that many factors contribute to the frequency of disease alleles in human populations and that each disease must be analyzed individually. In some cases the frequency of mutant alleles is maintained by mutation. In other cases migration and founder effects can greatly increase the frequency of deleterious alleles. In still other cases, natural selection favors heterozygous carriers, whereas affected homozygotes bear the burden of conferring advantages on other genotypes.

HUMAN ACTIVITY AS A FORCE IN CHANGING ALLELE FREQUENCIES

In addition to natural selection, human activities can also influence the frequency and distribution of genes and their alleles. Collectively, these activities are known as culture. Social customs and technology are powerful forces that dictate and shape human mating patterns and as a result, can change allele frequencies. For example, the invention of technology for long-distance travel in the 16th century led to worldwide movement of populations, causing a redistribution of alleles. Social constraints that limit mate selection to those who have some common bonds (language, religion, and economic status) prevent random mating and can change allele frequencies. In this section we consider how two human activities, migration and mate selection, bring about such changes.

Migration Reduces Genetic Variation between Populations

Mapping the distribution of the B allele of the ABO blood type (▶ Figure 18.6) reveals a large-scale genetic effect of migration. The highest frequency of the B allele is

FIGURE 18.6 The gradient of the B allele of the ABO locus is due to waves of migration from Asia into Europe.

in indigenous populations of Central Asia. Several waves of migration from Central Asia into Europe have produced a gradient (called a cline) that runs from east to west. The highest frequencies in this gradient are in Eastern Europe, and the lowest in Southwest Europe.

Other studies have uncovered genetic relics of previous patterns of human migration. A large-scale study of 19 loci and 63 alleles carried out at more than 3000 locations in Europe has resulted in the identification of 33 boundaries that represent regions of sharp genetic changes (Figure 18.7). There is an abrupt shift in allele frequencies across each boundary, suggesting that little mixing of the neighboring populations has occurred. Of the 33 boundaries, 22 are physical boundaries (mountains or ocean) across which there is little genetic exchange. But the other 11 genetic boundaries are not associated with physical barriers; instead, they represent linguistic barriers that separate populations. (Some of the 22 physical boundaries also have linguistic barriers.)

The results of these studies suggest that the genetic structure of the populations in Europe is caused by adaptation to the local environment and reflects the diverse origins of populations that came into contact through migrations. The results also indicate that language is an effective barrier to gene flow and acts as a selective force that maintains genetic differences between populations. Although European populations

▶ **FIGURE 18.7** Genetic/linguistic barriers in Europe. Thirty-three genetic barriers have been discovered, each of which represents an abrupt shift in allele frequencies. Of these, 31 correspond to modern linguistic boundaries. The two that do not match are #1 and #32. The correlation between language and allele frequency suggests that culture in the form of language may play a major role in establishing and maintaining genetic boundaries.

have been in contact for hundreds or thousands of years, there has been little breakdown of allelic differences in some regions.

Mate Selection Is Usually Nonrandom

In a population at genetic equilibrium, one assumption of the Hardy-Weinberg Law is that mating is random. This ideal condition rarely occurs in human populations, partly as a result of geographic proximity and partly as a result of social structure and cultural limitations. One form of nonrandom mating is called **assortative mating.** Among humans, cultural factors such as common language, physical characteristics, economic status, and religion are often important in mate selection. These factors can influence allele frequency.

Inbreeding is another form of nonrandom mating that involves mating between related individuals who share common genes. The most extreme form of inbreeding in humans is incest, mating between parents and children or between siblings. In almost all societies, **incest** is culturally unacceptable. In some societies, however, incest or mating with close relatives was common among royalty. Such matings were frequent in the Ptolemaic dynasties of Egypt (▶ Figure 18.8). The genetic effect of such mating is an increase in homozygosity and a decrease in heterozygosity. The fre-

■ **Assortative mating** Reproduction in which mate selection is not random but instead is based on physical, cultural, or religious grounds.

■ **Inbreeding** Production of offspring by related parents.

■ **Incest** Sexual relations between parents and children or between brothers and sisters.

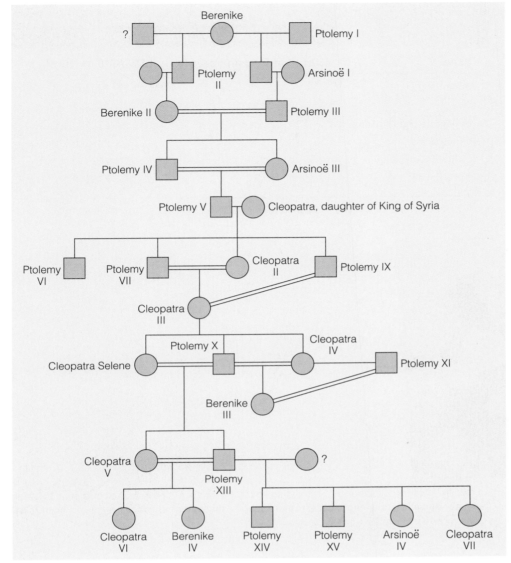

▶ **FIGURE 18.8** Pedigree of the Ptolemaic dynasty of Egypt. Incest and consanguineous matings (indicated by the double horizontal lines) were common. Cleopatra VII, the great Cleopatra, is known for her romantic involvement with Julius Caesar and Marc Antony. She had one son by Caesar and three children by Antony.

quencies of some autosomal recessive conditions among the offspring of **consanguineous matings** are shown in Table 18.2.

During the 19th and early 20th centuries, consanguineous matings involving cousins were common in many cultures, including those of Western Europe (▶ Figure 18.9a). But as population mobility has increased, the incidence of such matings has decreased, contributing to a more random system of matings (Figure 18.9b). Many states in the United States have laws regulating the degree of consanguinity that is permitted in marriages, but these laws are based on social or religious customs rather than on genetics.

■ **Consanguineous mating** Matings between two individuals who share a common ancestor in the preceding two or three generations.

Culture Is a Force in Altering Allele Frequencies

Collective human activity in the form of cultural practices can act as a selective force to alter the frequency of genotypes and alleles. The interaction of culture, environment,

Table 18.2	Percentage of First-Cousin Parents in Children Who Have Recessive Genetic Diseases	
Disease	**% of First-Cousin Parents**	**MIM/OMIM Number**
Albinism	10	203100
PKU	10	261600
Xeroderma pigmentosum	26	278700
Alkaptonuria	33	203500
Ichthyosis congenita	40	242300
Microcephaly	54	215200

SOURCE: S. Reed (1980) Counseling in Medical Genetics (3rd ed.). New York: Alan Liss, p. 77, Table 10.1.

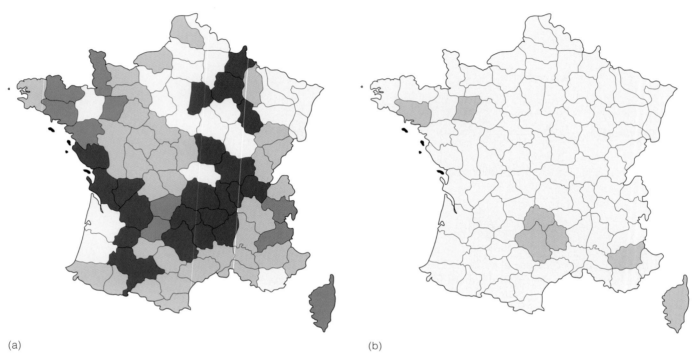

(a) (b)

▶ **FIGURE 18.9** (a) Relative levels of consanguinity in France between 1926 and 1930. Darker shades represent higher levels of inbreeding. (b) Relative levels of consanguinity in France between 1956 and 1958.

and allele frequency associated with herding of dairy animals and lactose digestion is an example.

Lactose is the principal sugar in milk (human milk is 7% lactose) and is a ready energy source. In the small intestine of infants, the enzyme lactase converts lactose into glucose and galactose, sugars that are easily absorbed by the intestine. The production of lactase in most humans slows at the time of weaning, and lactase production stops as children grow into adults. Adults with low lactase levels are unable to metabolize lactose (MIM/OMIM 223100). Such adults have bad reactions to milk or dairy products characterized by gas, cramps, diarrhea, and nausea. In these individuals, undigested lactose passes from the small intestine into the colon, where it is digested by bacteria, producing gas and the resulting diarrhea.

In some human populations, lactase is produced throughout adulthood. These individuals are called lactose absorbers (LA). Population surveys of lactose absorp-

Table 18.3	Lactose Absorption in Various Populations
Population	**Percent Lactose Absorbers (LA)**
Eskimos—Greenland	15
!Kung—Africa	2.5
Tuareg—Africa	85
Bantu—Africa	0
Arabs—Saudi Arabia	86
Sephardic Jews—North Africa	38
Danes—Europe	98
Czechs—Europe	100
U.S.—Asians	3
Aborigines	15
U.S.—Blacks	25

tion show that the frequency of the allele for lactose absorption varies from 0.0 to 100% (Table 18.3). Genetic evidence indicates that adult lactose utilization (and the adult production of lactase) is inherited as an autosomal dominant trait.

Why does the frequency of this allele vary so widely across populations? The answer appears to be that cultural practices are a selective force. According to this hypothesis, human populations originally resembled other land mammals and were lactose-intolerant as adults. As dairy herding developed in some populations, adult lactose absorbers had the selective advantage of being able to derive nutrition from milk. This improved their chances of survival and success in leaving offspring. As a result, the cultural practice of maintaining dairy herds provided the selective factor that conferred an increased advantage on the LA genotype.

Population surveys provide a strong correlation between a cultural history of dairying and the frequency of lactose absorbers. The Tuareg of North Africa are nomadic herders who have been in the Central Sahara for 2000 to 3000 years (▶ Figure 18.10). Other food is often unavailable, and adult consumption of several liters of milk per day is common. There is a high frequency of the LA allele in this population. In fact,

▶ **FIGURE 18.10** The Tuareg people of Africa inhabit the central region and the southern edge of the Sahara desert.

all populations that have high LA frequencies (60% to 100%) have a history of dairy herding that can be traced back for over 1000 years. In contrast, in populations in the tropical forest belt of Africa, where sleeping sickness carried by the tsetse fly prevents dairy herding, the frequency of lactose absorption is low (0% to 20%). Taken together, the evidence from genetics, anthropology, and geography supports the idea that the variation in the frequency of lactose absorption found in present human populations is derived from cultural practices that act as a selective force on allele frequencies. The role of culture as a selective force makes humans unique among land mammals in having populations with a high frequency of adult lactose absorbers.

GENETIC VARIATION IN HUMAN POPULATIONS

Evidence indicates that a great deal of genetic variation is present in the human genome. The process of mutation has introduced all of this variation. To be detectable as a polymorphism, the mutant allele must somehow spread through the population. Natural selection and drift are the primary mechanisms by which alleles spread through local population groups.

The Spread of Polymorphisms

One of the ways to monitor the spread of alleles through a population is to study polymorphisms that exhibit a geographical gradient. This distribution may result from migratory patterns, selection, or a combination of factors. An interesting polymorphism is the gradient of blond or tawny hair among Australian aborigines (▶ Figure 18.11). This allele apparently originated in the west-central desert region (▶ Figure 18.12) and spread from there to much of the west coast. The trait is inherited codominantly.

The reason for this gradient in hair color isn't known. Possibly, tawny hair is only a phenotypic by-product of a gene that controls a biochemical trait with more direct selective significance. The aborigines live in a harsh desert environment, and

▶ **FIGURE 18.11** An Australian aborigine near Kolomburu in Western Australia who has a polymorphic hair color.

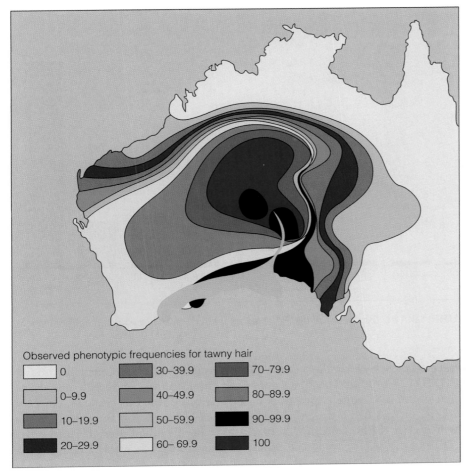

FIGURE 18.12 The frequency of tawny hair color in Australian aborigines forms a gradient from a center in the west-central desert region.

this allele may contribute to an increase in fitness or may interact with other genes to increase fitness. On the other hand, perhaps those with tawny hair are more often selected as mates.

Measuring Gene Flow among Populations

There has been a long-standing interest in both anthropology and human genetics in estimating gene flow among divergent populations to study and reconstruct the origin and history of hybrid populations formed when European and non-European populations come into contact. The best documented case is gene flow into the American black population from Europeans, but other populations have also been studied.

Most of the black population in the United States originated in West Africa, and the majority of the white population arrived from Europe. The frequency of the Duffy blood group allele Fy^o (MIM/OMIM 110700) is close to 100% in African populations, whereas this allele in Europeans has a frequency close to zero. Europeans are almost all Fy^a or Fy^b, and these alleles have very low frequencies in African populations. By measuring the frequencies of the Fy^a and Fy^b alleles in the U.S. black population, we can obtain an estimate of how much genetic mixing has occurred over the last 300 years. The frequency of the Fy^a allele among black populations in West Africa and in several locations in the United States is shown in ▶ Figure 18.13. Using this as an average gene, we can calculate that about 20% of the genes in the black population in some Northern cities are derived from Europeans.

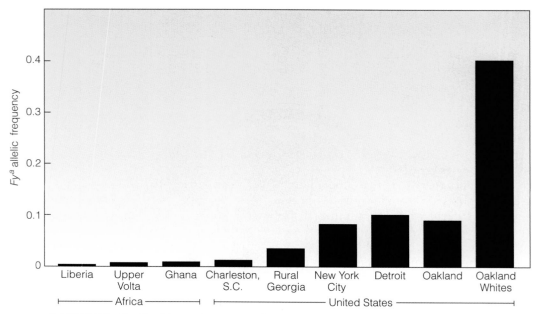

FIGURE 18.13 Distribution of the Fy^a allele among African and American Black populations.

Other studies give similar results. The frequency of PKU (phenylketonuria) (MIM/OMIM 261600) in U.S. blacks has been estimated as 1 in 50,000, about one-third that of the U.S. white population. Using analysis of the PKU gene and molecular markers, such as RFLP sites, the origins of the PKU mutation and the phenylalanine hydroxylase (PAH) gene in U.S. blacks have been studied. Results suggest that about 20% of the PAH genes in U.S. blacks originated from a Caucasian population, whereas the rest are likely to be of West African origin.

A more comprehensive study using 15 genes looked at 52 alleles in U.S. blacks from the Pittsburgh region, including 18 unique alleles of African origin, and found that the proportion of European genes in this black population is about 25%. Another approach to the study of gene flow between European and U.S. black populations employed both nuclear genes (ABO, MN, Rh, etc.) and mitochondrial markers to examine gene flow. Because mitochondrial markers are maternally inherited, they can be used to estimate the maternal contribution to the population under study. In this work, the results from nuclear alleles also indicate that about 25% of the genes in the black population studied (60 U.S. blacks) originated from Europeans.

However, not all contact between genetically distinct populations leads to a reduction in genetic differences. Using a combination of genetic and historical methods, researchers investigated the origin and extent of European-American admixture in the Gila River American Indian community of central Arizona. Results from the genetic study indicate a European admixture of 5.4%, whereas the demographic data indicate an admixture of 5.9%. The first contact between Europeans and the Gila River community occurred in 1694. Using 20 years as the estimate for one generation, about 15 generations of contact have occurred. On this basis, because gene flow is about 0.4% year, the Gila River American Indian community has retained almost 95% of its native gene pool even after close contact with other gene pools for 300 years.

THE QUESTION OF RACE

Race A genotypically distinct group within a species.

In the 19th century, biologists used the term **race** to describe groups of individuals within a species that could be distinguished in some phenotypic way from other groups within that species. For example, two-spotted beetles and four-spotted beetles found in the same population might be separate races, even though some two-spotted and four-spotted beetles might be siblings.

Races as Subspecies

As population genetics emerged as a separate field of study, the phenotypic concept of race was replaced by a genetic definition. Geneticists regard races as populations that have significant differences in allele frequencies when compared to other populations. To the biologist studying nonhuman organisms, races represent groups that are genetically differentiated to the point of being subspecies but have not yet formed separate species.

How does this happen? If a population splits into two isolated groups, the single gene pool is also split into two. Over time, these isolated groups can become genetically different from one another. If they continue to diverge, they accumulate enough genetic differences to be classified as geographic races or subspecies.

The concept of geographic race does not clarify the difficulty of defining what we mean by the term race as it applies to humans. Clearly, there are phenotypic differences among humans. Residents of the Kalahari Desert are rarely mistaken for close relatives of Aleutian seal hunters, for example. In spite of these visible phenotypic differences, from a genetic point of view, is there a need or a value in classifying humans into racial groups?

Are There Human Races?

In one sense, it is easy and perhaps logical to assume that the physical differences that exist among groups of humans are evidence for underlying genetic differences. However, a decision to divide the human species into races depends on showing significant genetic differences, not phenotypic differences.

Beginning in the early 1970s, enough information was available to study allele frequencies in various populations and to relate these differences to the concept of human races. In one study, Richard Lewontin used protein variants to analyze allele frequencies at 15 loci. Data was gathered from populations all over the world. After measuring the amount of variation, he determined how much of this variation exists within a population and how much is present between populations. He also asked how much genetic variation exists among what are considered conventional racial groups. His results indicate that 85% of the detected variation is present within populations and that less than 7% of the variation is present between racial groups. Studies of other protein markers confirmed these findings. These results, however, had little impact on the idea that the human species is divided into racial groups distinguished by large biological differences.

More recently, geneticists have used DNA markers to study genetic variation in populations. One study analyzed the distribution of variation within populations and among populations using 109 DNA markers (microsatellite alleles and RFLPs). The differences were analyzed for members of the same population, between and among populations on the same continent, and among four or five different geographic groups. Again, as in the protein studies, most of the genetic variation was within groups (about 85%), and only about 10% of the variation was among groups on different continents (equivalent to racial groups).

Taken together, the work on proteins and DNA indicates that most of the genetic variation in the species is present within human populations and that there is little variation among populations, including those classified as different racial groups. In keeping with the genetic definition of race, if humans are to be divided into racial groups, large-scale genetic differences should occur along sharp boundaries. If such genetic differences exist, they have not yet been observed. On the other hand, is it possible that the small amount of genetic variation observed among groups is enough to warrant the classification of humans into racial groups? Almost all geneticists would answer no to that question and agree that presently, there is no genetic basis for subdividing our species into racial groups.

Human Genetic Variation: Summing Up

In earlier sections of the chapter we considered genetic differences over geographic distances in the form of clines (ABO blood types, HbS, G6PD) and genetic borders that coincide with linguistic borders. How can these findings be reconciled with the idea that as a species, we have a highly homogeneous gene pool?

The patterns found in clines or gradients of allele frequencies represent responses to forces such as migration or natural selection. These patterns can be detected only when allele frequencies for one or a small number of genes are analyzed. When a large number of alleles is examined (using either protein markers or DNA markers), it is clear that natural selection does not have a substantial effect on the rest of the genome. As a result, there may be large differences in the frequency of one or two alleles among populations, but when many loci are considered, few differences are found.

The same is true for the pattern of geographic and linguistic boundaries. Although sharp boundaries in allele frequency can be detected, only a small number of genes is involved. On a larger scale, only a small amount of genetic variation among populations can be detected. In fact, clines and genetic/linguistic boundaries represent some of the genetic differences that exist among populations.

One other conclusion can be drawn from studying genetic diversity across human populations. If diversity is analyzed by the size of populations, ranging from small villages to large nations and continents, a large fraction of the genetic variation in our species is present in small populations. This is consistent with the idea that our species has undergone a recent expansion from a small, common population. This insight plays an important part in our current understanding of how, when, and where our species originated. The following sections explore the evidence that surrounds the evolution of our species.

PRIMATE EVOLUTION AND HUMAN ORIGINS

The process of species formation itself is difficult to observe, mainly because of the timescale over which it operates. In the case of human evolution, the events have been reconstructed from the study of fossils and DNA variations.

Primate evolution from the early Miocene Epoch (20 to 25 million years ago) to the appearance of modern man some 100,000 years ago has been reconstructed mainly from the fossil record. Primates that are ancestors to both humans and apes are found as fossils from the Miocene (▶ Figure 18.14). These **hominoids** originated in Africa between four and eight million years ago and underwent a series of rapid evolutionary steps, giving rise to a diverse array of species. This diversity, coupled with the fragmentary fossil record, makes it difficult to draw conclusions about phylogenetic relationships among the early hominoids and to identify the ancestral line leading to modern **hominids**.

Humanlike Hominids Appeared about Four Million Years Ago

The earliest human fossils (hominids) are from a species called *Australopithecus afarensis* that first appeared 3.6–4.0 million years ago. This species, which lived in East Africa, had a combination of apelike and humanlike characteristics. Members of this species walked upright on two legs (they were the creatures described in the opening vignette of the chapter), but the body proportions were apelike, with short legs and relatively long arms (▶ Figure 18.15). The arm bones of australopithecines are long and curved like those of chimpanzees but have humanlike elbows that could not support body weight while knuckle walking (which is how

Hominoid A member of the primate superfamily Hominoidea, including the gibbons, great apes, and humans.

Hominid A member of the family Hominidae, which includes bipedal primates such as *Homo sapiens*.

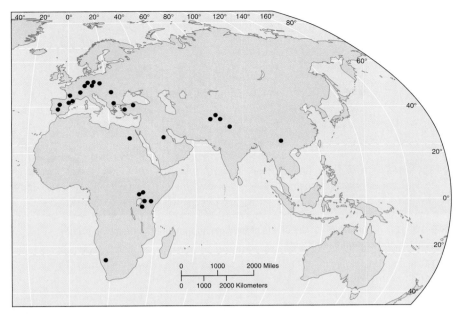

FIGURE 18.14　Hominid fossil sites from the Miocene (22.5 to 5 million years ago).

FIGURE 18.15　A reconstruction of a group of *Australopithecus afarensis* gathering and eating fruits and seeds.

chimpanzees and gorillas move around). This combination of features is well suited for a species that spends time climbing in trees but walks on two legs across the ground. The skulls of australopithecines (▶ Figure 18.16) are apelike and have small brains, receding faces, and large, canine teeth. Perhaps the most famous specimen of *Australopithecus* is a 40% complete skeleton recovered in the 1970s

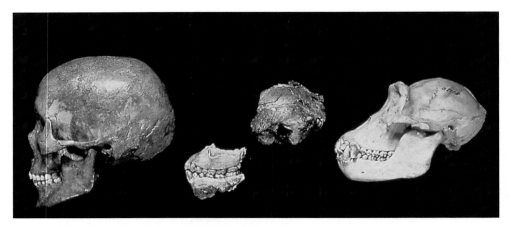

▶ **FIGURE 18.16** Comparison of the skulls of modern humans (left), *Australopithecus* fragments (center), and chimpanzee (right).

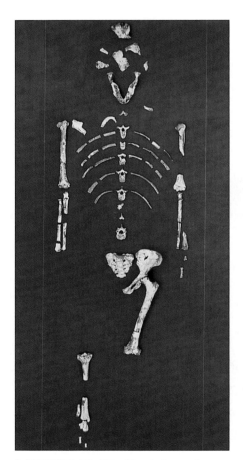

▶ **FIGURE 18.17** The skeleton of Lucy, a 3.5 million-year-old female *Australopithecus afarensis*.

and known as Lucy—named after the Beatles song, "Lucy in the Sky with Diamonds" (▶ Figure 18.17). A new set of fossils, older and anatomically more apelike than Lucy, has recently been discovered in East African rock layers dating to 4.4 million years ago. This species has been named *Australopithecus ramidus*, and its relationship to *A. afarensis* remains to be established.

There is general, though not universal, agreement that *A. afarensis* is the ancestral stock from which all other known hominid species are derived (▶ Figure 18.18). This species was remarkably stable for more than a million years after it appeared in the fossil record. About 2 million years ago, however, there was a burst of evolutionary change—probably related to climatic changes that accompanied the advancing glaciers in Europe. The fossil record indicates that at least three and perhaps as many as six or more species of hominids occupied different or overlapping habitats in East Africa at this time. These hominid species can be placed into two groups. One was composed of several species of more graceful (gracile), small-brained forms, including *A. africanus*, and heavier, more robust species, including *A. robustus*. This group gradually died out, and the last species survived until about one million years ago.

The other group was characterized by relatively large brains, a smaller facial structure, and reduced teeth. These relatively large-brained, less robust species are grouped in the genus *Homo*, the genus to which our species belongs. Although the fossil record is incomplete, it is generally agreed that the australopithecines are ancestral to the members of our genus. From the fossil record, it is clear that evolutionary adaptations in primates developed at different rates, showing a pattern called mosaic evolution. Skeletal changes that led to walking and dental changes that led to diet diversification were early adaptations. Later changes include an increase in brain size accompanied by the development of tools, the use of fire, and the origins of language.

The Genus *Homo* Appeared about Two Million Years Ago

The earliest humans classified as members of the genus *Homo* probably appeared in a cluster of species that developed about 2–2.5 million years ago. This group contains several species. Members of this species cluster, which arose about 2 million years ago, had a brain size about 20% larger than that of the australopithecines. They also had differently shaped skulls and teeth distinct from those of australopithecines. At least one and probably several species of the genus *Homo* lived along-

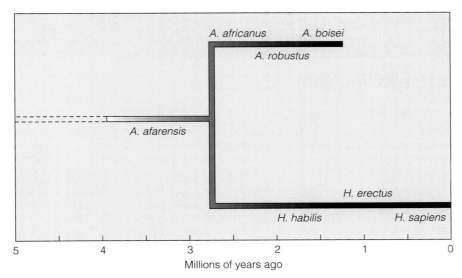

FIGURE 18.18 About three million years ago, the *A. afarensis* line split. One line led to several other species of australopithecines, and the other line led to *Homo sapiens*.

side the robust australopithecines in East Africa for about a million years, and other members of the genus *Homo* lived in South Africa at the same time. After this period, the australopithecines became extinct. Sometime during this period, a species that became our immediate ancestor arose within the *Homo* line. This hominid, called *Homo erectus* (▶ Figure 18.19), represents an important turning point in human evolution (see Concepts and Controversies: Tool Time: Did *Homo erectus* Use Killer Frisbees?)

Homo erectus Originated in Africa

African fossils of *H. erectus* date to about 1.8 million years ago, but the species may actually be much older. Although *H. erectus* originated in Africa, members migrated into Asia and Europe. Fossils of *H. erectus* have been recovered from Indonesia (Java Man), China (Peking Man), and sites in North Africa. The recent dating of Indonesian fossils of *H. erectus* to about 1.8 million years ago indicates that as a species, *H. erectus* is probably older than 2 million years.

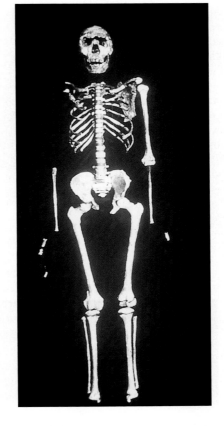

FIGURE 18.19 A skeleton of *Homo erectus.* This is the most complete skeleton of *H. erectus* found to date and is of a boy who lived some 1.6 million years ago.

Physically *H. erectus* is different from earlier members of the genus *Homo* in several respects: increased brain size, a flatter face, and prominent brow ridges. There are physical similarities between *H. erectus* and modern humans, including height and walking patterns, but there are also some significant differences. In some populations of *H. erectus,* the skull is pointed in the back, not dome-shaped as in *H. sapiens* (▶ Figure 18.20). In addition, *H. erectus* had a receding chin, a prominent brow ridge, and some differences in teeth.

Concepts and Controversies

Tool Time: Did *Homo erectus* Use Killer Frisbees?

*T*he appearance of *Homo erectus* about two million years ago represents a turning point in human evolution. Studies of tooth wear patterns indicate that meat became an important component in the diet of *H. erectus*. This species was also the first hominid to move out of Africa into Europe and Asia, exploiting new habitats as resources. Evidence indicates that these and other changes were accompanied by technological innovation in the form of new tools. The use of tools predates *H. erectus* by about a million years, but these early tools (called Oldowan tools) were small, were used mainly for chopping, and remained unchanged over a span covering a million years.

The tools of *H. erectus* (called Acheulian tools) were large, had two cutting faces, and included hand axes, cleavers, and picks (see art). The typical tool kit also included scrapers and trimmers for producing and sharpening new tools. Over time, Acheulian tools were gradually refined but changed little in the million years they were used. Few new tools were added to the basic tool kit. These tools disappeared about 200,000 years ago, at a time when toolmaking entered a period of technological innovation and refinement that marks the Middle Paleolithic (Middle Stone Age).

Of all the tools in the kit of *H. erectus,* the possible uses of the hand ax have remained a subject of speculation. Modern-day anthropologists have learned how to make such tools and have used them as small axes for chopping, or as heavy-duty knives for slicing animal hides or skinning carcasses. Examination of fossil hand axes by electron microscopy indicates that these tools may have had a range of uses on many materials including hide, meat, bone, and even wood. One form of

the hand ax is ovoid and has a pointed end (art). Their size and shape has led to the suggestion that they may have been thrown like frisbees into a herd of small animals, stunning or wounding one, which could then be overtaken and killed. Flight tests of fossils and replicas support the idea of killer frisbees, but this use is not widely accepted by anthropologists and remains among the most speculative proposals for the function of hand axes. Whatever their uses, the tools of *H. erectus* were associated with dramatic changes in behavior, including diet, migration, systematic hunting, the use of fire, and the establishment of home bases or camps. The role of new technology in promoting or supporting these new behaviors remains an area of intense investigation in paleoanthropology.

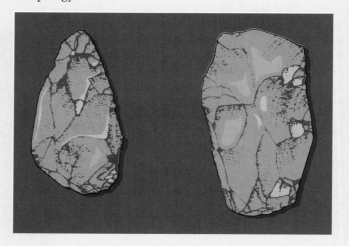

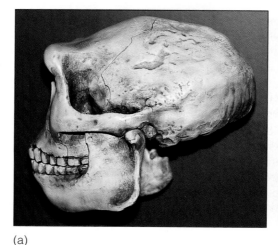

▶ **FIGURE 18.20** Skulls of (a) *Homo erectus* and (b) *Homo sapiens.* The skull of *H. erectus* is low and pointed in the rear, whereas the *H. sapiens* skull is high and dome-shaped.

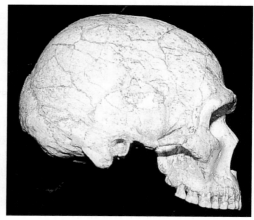

(a)

(b)

THE APPEARANCE AND SPREAD OF *HOMO SAPIENS*

Tracing the origins of our species has become a multidisciplinary task, using the tools and methods of anthropology, paleontology, archeology the techniques of genetics and recombinant DNA technology—and more recently, satellite mapping from space. These methods are being used to reconstruct the origins and ancestry of populations of *H. sapiens* and to determine how and when our species originated and became dispersed across the globe.

Two Theories Differ on How and Where *Homo sapiens* Originated

From the evidence provided by fossils and artifacts, there is agreement that populations of *H. erectus* moved out of Africa one to two million years ago and spread through parts of the Middle East and Asia. What currently divides paleoanthropologists is the question of how and where *H. sapiens* originated. In a general sense, there are two opposing views about the origin of modern humans. One idea (often called the out-of-Africa hypothesis) argues that after *H. erectus* moved out of Africa, populations that remained behind continued to evolve and gave rise to *H. sapiens* about 200,000 years ago. From this single source somewhere in Africa, modern humans migrated to all parts of the world, displacing the populations descended from *H. erectus,* which then became extinct. According to this model, modern human populations are all derived from a single speciation that took place in a restricted region within Africa and should show a relatively high degree of genetic relatedness.

This model of human evolution was first proposed from evidence that members of African populations show the greatest amount of genetic diversity as measured by differences in mitochondrial DNA nucleotide sequence. Members of non-African populations show much less diversity. The underlying assumption in these studies is that mutational changes in mitochondrial DNA accumulate at a constant rate, providing a "molecular clock" that can be calibrated by studying the fossil record.

Studies of mitochondrial DNA reveal a single ancestral mitochondrial lineage for our species that originated in Africa. Calculations using the molecular clock indicate that our species originated about 200,000 years ago from an African population that might have been made up of about 10,000 individuals.

The second idea about the origin of our species is called the multiregional hypothesis. According to this idea, after populations of *H. erectus* spread from Africa over the Middle East and Asia, *H. sapiens* developed as the result of an interbreeding network descended from the original colonizing populations of *H. erectus.* The evidence to support this model is derived from a combination of genetic and fossil evidence. The fossil record shows a gradual transition from archaic to modern humans that took place at multiple sites outside of Africa. In this model, *H. erectus* became gradually transformed into *H. sapiens,* instead of being replaced by *H. sapiens.*

The two opposing ideas are hotly debated and have received a great deal of attention in the press and other media. These alternative explanations for the appearance of modern humans show that scientists can reach different conclusions about the same problem.

Although the accuracy of the molecular clock and the method used to construct mitochondrial phylogenetic trees has been called into question, studies of genetic variation in nuclear genes supports a single point of origin and a timescale consistent with the out-of-Africa hypothesis. Recent work using RFLPs on the Y chromosome (which is passed from father to sons) indicates that our species developed at a single site about 200,000 to 270,000 years ago. Two genetic polymorphisms at a locus on chromosome 12 studied in 42 diverse populations also support an African origin for *H. sapiens* about 100,000 to 300,000 years ago.

The issue over the origins of *H. sapiens* has not been resolved. Although we have considered only two possible origins for our species, further work may produce additional ideas. Each will have to be evaluated by using the available information and new information, and perhaps by applying new techniques.

The Spread of Humans Across the World

Genetic evidence from the study of mitochondrial DNA, nuclear genes and markers, and markers on the Y chromosome furnish us with two lines of evidence about the origins of modern forms of *H. sapiens* and the amount of present-day variation in our genome. Taken together, these two lines provide strong support for the idea that *H. sapiens* originated in Africa and spread from there to other parts of the world (❱ Figure 18.21). Based on the molecular evidence and some fossil evidence, it appears that modern *H. sapiens* originated in Africa some 130,000 to 170,000 years ago and that a small subset of this population emigrated from Africa about 137,000 years ago. There may have been one primary migration or several from a base in northeast Africa. The emigrants carried a subset of the variation present in the African population, consistent with the finding that present-day non-African populations have a small amount of genetic variation.

Modern forms of *H. sapiens* spread through Central Asia some 50,000 to 70,000 years ago and into Southeast Asia and Australia about 40,000 to 60,000 years ago. *H. sapiens* moved into Europe some 40,000 to 50,000 years ago, displacing the Neandertals, who lived there from about 100,000 years ago to about 30,000 years ago. Recent analysis of DNA from the bones of a Neandertal skeleton indicates that they were not the ancestors to European populations of *H. sapiens,* further supporting the out-of-Africa hypothesis. North and South America were populated by three or four waves of migration that occurred 15,000 to 30,000 years ago.

Although this model explains most of the data available, there are other issues that remain to be resolved. Nonetheless, it is clear that genetic analysis of present-day populations coupled with anthropology, archaeology, and linguistics can provide a powerful tool for reconstructing the history of our species.

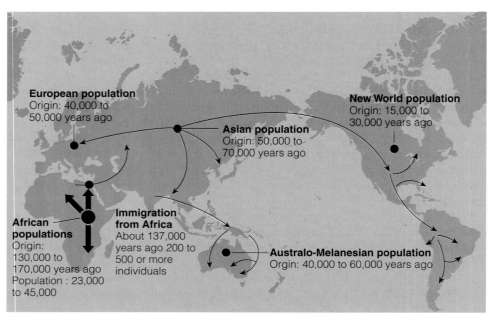

❱ **FIGURE 18.21** The origin and spread of modern *H. sapiens* reconstructed from genetic and fossil evidence.

CASE 1

Natural selection may alter genotypic proportions by increasing or decreasing fitness (i.e., differential fertility or mortality). There are several examples of selection associated with human genetic disorders. Sickle cell anemia and other abnormal hemoglobins are the best examples of selection in humans. Carriers of the sickle and other hemoglobin mutations are more resistant to malaria than either homozygous class. Therefore, in areas where malaria is endemic, carriers are less likely to die of malaria and will pass on proportionally more genes than homozygotes. Carrier frequencies for more "common" recessive disease, such as cystic fibrosis in Europeans and Tay-Sachs in the Ashkenazi Jewish population, may also have been influenced by balancing selection, but the selective agent is not known for certain.

Selection may favor homozygotes over heterozygotes, which results in an unstable polymorphism. One example of this is selection against heterozygous fetuses if an Rh-negative mother carries an Rh-positive (heterozygous) fetus. This should result in a gradual elimination of the Rh-negative allele. However, the high frequency of the Rh-negative allele in so many populations suggests that other unknown factors may contribute to the maintenance of the Rh-negative allele in human populations.

Summary

1. Studies indicate that human populations carry a large amount of genetic diversity. Natural selection acts on genetic diversity in populations to drive the process of evolution.

2. All genetic variants originate by mutation, but mutation is an insignificant force in bringing about changes in allele frequency. Other forces, including drift and selection, act on the genetic variation in the gene pool and are primarily responsible for changing the frequency of alleles in the population. Drift is a random process that acts in small, isolated populations to change allele frequency from generation to generation. Examples include island populations or those separated from general population by socioreligious practices.

3. Selection increases the reproductive success of fitter genotypes. As these individuals make a disproportionate contribution to the gene pool of succeeding generations, genotypes change. The differential reproduction of fitter genotypes is known as natural selection. Darwin and Wallace identified selection as the primary force in evolution that leads to evolutionary divergence and the formation of new species. The high frequency of genetic disorders in some populations is the result of selection that often confers increased fitness on the heterozygote.

4. Humans have developed a set of adaptive mechanisms collectively known as culture. Culture has a profound effect on the interaction of the gene pool with the environment and can also act as a force that brings about changes in allele frequencies. Although both plant and animal species employ strategies of dispersal and migration, human technology makes it possible for humans to migrate to any place on earth within a short time. The net effect of this short migratory time is the reduction of allele frequency differences among populations. In some instances, migration into small, relatively isolated populations can rapidly generate changes in allele frequencies.

5. Patterns of culture also influence mating patterns, tending to make them nonrandom. This process of assortative mating can also change allele frequencies, although slowly. It is thought that food choices, coupled with selection during thousands of years, have influenced the frequency of a gene that enables adults to use milk as a food source.

6. The fossil record from the Miocene Epoch can be used to trace the evolution of hominoids and hominids, primates that gave rise to the present human species. The incomplete nature of the fossil record makes it difficult to construct a phylogeny, but now techniques of molecular biology are being used to clarify the origin of humans.

7. A combination of linguistics, archaeology, anthropology, and genetics is being used to reconstruct the dispersal of human populations across the globe. The evidence available suggests that North and South America were populated by four waves of migration sometime during the last 15,000 to 30,000 years. In Europe, the alignment of genetic and linguistic barriers suggests that culture, in the form of language, can be a force in establishing and maintaining genetic differences among populations.

Questions and Problems

1. Distinguish between mutations and polymorphisms. How are polymorphisms used in the study of genetic variation?
2. Why is it that mutation, acting alone, has little effect on gene frequency?
3. What is the relationship between founder effects and genetic drift?
4. The theory of natural selection has been popularly summarized as "survival of the fittest." Is this an accurate description of natural selection? Why or why not?
5. What are examples of nonrandom mating, and how does this behavior affect genetic equilibrium?
6. How do you think the development of culture affected the process of human evolution? Is our present culture affecting selection? Can you give specific examples?
7. Do the differences we see in the ABO blood group polymorphisms represent adaptive changes, or do they reflect some other process, not important to fitness of populations in which they occur?
8. How would a drastic reduction in a population's size affect the population?
9. How are genetic polymorphisms maintained in a population?
10. Will a recessive allele that is lethal in the homozygous condition ever be removed from a large population as a result of natural selection?
11. A specific mutation in the *BRCA1* gene has been estimated to be present in about 1% of Ashkenazi Jewish women of Eastern European descent. This specific alteration, 185delAG, is found about three times more often in this ethnic group than the combined frequency of the other 125 mutations found to date. It is believed that the mutation is the result of a founder effect from many centuries ago. Explain the founder principle.
12. Successful adaptation is defined by:
    a. evolving new traits
    b. producing many offspring
    c. an increase in fitness
    d. moving to a new location
13. The major factor causing deviations from Hardy-Weinberg equilibrium is:
    a. selection
    b. nonrandom mating
    c. mutation
    d. migration
    e. early death
14. Do you think that our species is still evolving, or are we shielded from natural selection by civilization? Is it possible that misapplications of technology will end up exposing our species to more rather than less natural selection (consider the history of antibiotics)?
15. Is gene transfer a form of eugenics? Is it advantageous to use gene transfer to eliminate some genetic disorders? Can this and other technology be used to influence the evolution of our species? Should there be guidelines for the use of genetic technology to control its application to human evolution? Who should create and enforce these guidelines?

Internet Activities

The following activities use the resources of the World Wide Web to enhance the topics covered in this chapter. To investigate the topics described below, log on to the book's home page at:

http://www.brookscole.com/biology

1. For a good overview of hominid evolution, descriptions and photographs of the important fossil finds which currently shape our ideas on hominid evolution, and rebuttals of creationists attacks on evolution theory, study the Web site by Jim Foley at the Talk Origins page.
   a. Compare and contrast the scientific evidence and hypotheses with the creationist's views for the australopithecines, *Homo erectus* and the Neanderthals. You should read the Overview article of Creationists Arguments, also. For more sites on either side of the issues, click on Links. Finally, look at the Hominid Fossils page for each of these. Briefly summarize the differences between the creationists and evolutionists. Which arguments do you find more persuasive, the evolutionists or the creationists? Do you feel the "debate" has merit?
2. The Leakey foundation was established to support research into human origins. Browse the foundation website and investigate recent discoveries in the world of anthropology.
3. Polymorphisms in mitochondrial DNA can provide us with information about genetic relatedness as well as diversity and species origins. Solve the mystery of the Romanovs at the Cold Spring Harbor DNA Learning Center by comparing single nucleotide polymorphisms (SNPs) in the mitochondrial DNA sequence.

For Further Reading

Ayala, F. (1984). Molecular polymorphism: How much is there, and why is there so much? *Dev. Genet. 4*: 379–391.

Bahn, P. (1993). 50,000 year old Americans of Pedra Furada. *Nature 362*: 114–115.

Barbujani, G., & Sokol, R. (1990). Zones of sharp genetic change in Europe are also linguistic boundaries. *Proc. Nat. Acad. Sci. USA 87*: 1816–1819.

Bodmer, W. F., & Cavelli-Sforza, L. L. (1976). *Genetics, Evolution and Man*. San Francisco: Freeman.

Cavalli-Sforza, L., Menzonni, P., & Piazza, A. (1993). Demic expansion and human evolution. *Science 259*: 639–646.

Durham, W. (1991). *Coevolution: Genes, Culture and Human Diversity*. Stanford: Stanford University Press.

Gould, S. J. (1982). Darwinism and the expansion of evolutionary theory. *Science 216*: 380–387.

Gyllensten, U. B., & Erlich, H. A. (1989). Ancient roots for polymorphism at the HLA-DQ alpha locus in primates. *Proc. Nat. Acad. Sci. USA 86*: 9986–9990.

Kennedy, K. (1976). *Human Variation in Space and Time*. Dubuque, Iowa: Brown.

Lewontin, R. (1974). *The Genetic Basis of Evolutionary Change*. New York: Columbia University Press.

Little, B. B., & Malina, R. M. (1989). Genetic drift and natural selection in an isolated Zapotec-speaking community in the valley of Oaxaca, Southern Mexico. *Hum. Hered. 39*: 99–106.

Mayr, E. (1963). *Animal Species and Evolution*. Cambridge, MA: Harvard University Press.

Molnar, S. (1983). *Human Variation*. Englewood Cliffs, NJ: Prentice Hall.

Relethford, J. H. (1988). Heterogeneity of long-distance migration in studies of genetic structure. *Ann. Hum. Biol. 15*: 55–63.

Roberts, D. F. (1988). Migration and genetic change. Raymond Pearl lecture 1987. *Hum. Biol. 60*: 521–539.

Romero, G., Devoto, M., & Galietta, L. J. (1989). Why is the cystic fibrosis gene so frequent? *Hum. Genet. 84*: 1–5.

Schanfield, M. (1992). Immunoglobulin allotypes (GM and KM) indicate multiple founding populations of native Americans: Evidence of at least four migrations to the New World. *Hum. Biol. 64*: 381–402.

Semino, O., Torrino, A., Scozzari, R., Brega, A., De Benedictis, G., & Santachiara-Benerecetti, A. S. (1989). Mitochondrial DNA polymorphisms in Italy: III. Population data from Sicily: A possible quantitation of maternal African ancestry. *Ann. Hum. Genet. 53*: 193–202.

Shields, G., Schmiechen, A., Frazier, B., Redd, A., Voevoda, M., Reed, J., & Ward, R. (1993). mtDNA sequences suggest a recent evolutionary divergence for Beringian and northern North American populations. *Am. J. Hum. Genet. 53*: 549–562.

Smith, J. M. (ed.). (1982). *Evolution Now: A Century after Darwin*. San Francisco: Freeman.

Stanton, W. (1960). *The Leopard's Spots*. Chicago: University of Chicago Press.

Torroni, A., Schurr, T., Cabell, M., Brown, M., Neel, J., Larsen, M., Smith, D., Vullo, C., & Wallace, D. (1993). Asian affinities and continental radiation of the four founding native American mtDNAs. *Am. J. Hum. Genet. 53*: 563–590.

Torroni, A., Sukernik, R., Schurr, T., Starikovskaya, Y., Cabell, M., Crawford, M., Comuzzie, A., & Wallace, D. (1993). mtDNA variation of aboriginal Siberians reveals distinct genetic affinities with native Americans. *Am. J. Hum. Genet. 53*: 591–608.

Towne, B., & Hulse, F. S. (1990). Generational changes in skin color variation among Habbani Yemeni Jews. *Hum. Biol. 62*: 85–100.

Wallace, D., & Torroni, A. (1992). American Indian prehistory as written in the mitochondrial DNA: A review. *Hum. Biol. 64*: 403–416.

Yunis, J. J., & Prakash, O. (1982). The origin of man: A chromosomal pictoral legacy. *Science 215*: 1525–1530.

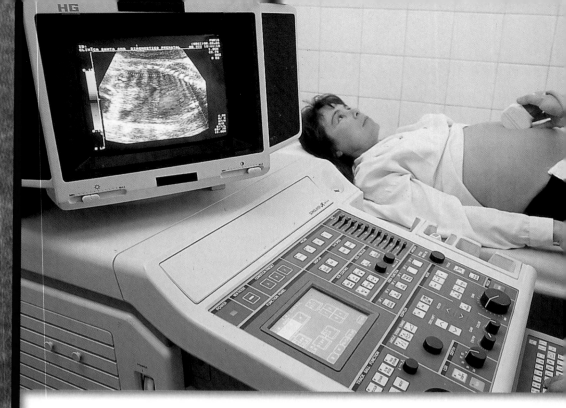

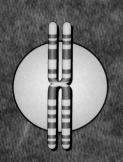

Genetic Screening and Genetic Counseling

Huntington disease is inherited as an autosomal dominant trait. This disorder is characterized by gradual loss of motor coordination, degenerative personality changes, and progressive dementia. The disease develops slowly and leads to death within 15 years after its onset. Most affected individuals develop symptoms of the disease around the age of 40, beginning with involuntary limb tremors and behavioral changes. Through a combination of pedigree analysis and molecular biology, the gene for Huntington disease has been mapped to chromosome 4. The gene has been cloned, and genetic markers within and near the gene are used in genetic testing to identify those at risk for developing this disorder.

Suppose that your uncle, age 49, has recently been diagnosed with Huntington disease. His physician calls you to inform you of the diagnosis and indicates that you have a 25% chance of developing the disease. You are 25 years old, married, and have one child. Should you undergo testing to determine whether you are at risk? Can you live with the outcome if the results indicate that you will develop the disease? Do you want your children to be informed of your condition? Do you want them to be tested? Do you want them to know if they will develop the disease? Should you have any more children? What if you were not tested, but your child was, and found to have the Huntington disease gene? You are the only parent who could have transmitted the mutant allele. You now have information about your genotype that you may not have wanted. Does this mean that if you do not wish to be tested, you can prevent your child from being tested?

This scenario and the questions it raises are not hypothetical. Although Huntington disease affects only a small fraction of the population (3 or 4 out of 100,000), it is now possible to identify those who will develop this devastating disease. Helping individuals to understand the implications of genetic testing and to make informed decisions about actions to be taken are goals of genetic counseling.

In this chapter we survey the field of genetic screening and explore the rationale, the methods of testing, and the potential for its use and misuse. We also consider the role of the genetic counselor in informing and educating those who undergo genetic screening. With the growth in genetic technology, these fields will have a great impact on our own lives and personal decisions and on those of our family members and friends.

GENETIC SCREENING

Genetic screening is the systematic and organized search for individuals who have certain genotypes. The term includes testing for those who have or may carry a genetic disease, those at risk of producing a genetically defective child, and those who may have a genetic susceptibility to environmental agents. Genetic screening is conducted for a variety of reasons. Prenatal screening is often used as the basis for selective abortion. Newborn screening is conducted to diagnose and treat a range of metabolic disorders. Adult screening to identify heterozygotes is used in genetic counseling and family planning. Occupational screening is used to identify those who may have an inherited susceptibility to materials or conditions in the workplace.

Several outcomes of genetic screening need to be considered. First, identification of an individual who has or is at risk for a genetic disorder, often leads to the discovery of other affected or at-risk individuals within the same family. Second, screening often identifies individuals who will develop genetic disorders later in adult life.

Genetic screening The systematic search for individuals who have certain genotypes.

When the genetic defect diagnosed is traumatic and fatal, such as in Huntington disease, this knowledge often has serious personal and social effects. Third, the results of genetic screening often have a direct impact on the offspring of the screened individual.

Newborn and Carrier Screening

Phenylketonuria (PKU), an autosomal recessive disorder was the first genetic disease to be systematically screened in newborns. Screening for PKU is now mandatory in all states and in 20 foreign countries. Over 100 million children have been tested for PKU, and more than 10,000 affected individuals have been identified and treated (▶ Figure 19.1). Although PKU is relatively uncommon (1 in 12,000), it is disabling, imposes a large personal and financial burden, and can be easily uncovered by using an inexpensive test. In addition, the discovery that the effects of the disease can be controlled by dietary treatment was instrumental in establishing mandatory screening programs. PKU is the model disease in newborn screening programs. Most of the other diseases currently screened in newborns are metabolic deficiencies (Table 19.1).

Carrier screening is the identification of phenotypically normal individuals who are heterozygous for an autosomal recessive or X-linked recessive disease. Some genetic disorders, including Tay-Sachs disease and sickle cell anemia, have been screened for on a large scale. The development of screening programs for these two diseases has been made possible by three factors:

- The diseases occur mainly in defined populations. Tay-Sachs carriers are found most frequently among Jews of East European origin, and sickle cell carriers are most common in U.S. blacks of West African origin.
- Carrier detection for these disorders is inexpensive and rapid.
- The existence of prenatal testing gives couples at risk the option of having only unaffected children.

Tay-Sachs Disease   Tay-Sachs disease is inherited as an autosomal recessive trait whose incidence in the general population is 1 in 360,000. In Ashkenazi

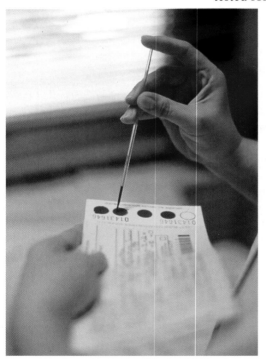

▶ **FIGURE 19.1**   In a Guthrie test for PKU, paper disks that contain a drop of blood are placed on the surface of a bacterial culture that will grow only in the presence of phenylalanine. If an infant has PKU, rapid growth occurs around the disk and is visible as a halo.

Table 19.1 Genetic Screening in Newborns (as of 1994)	
Genetic Disorder	**Number of States (Plus District Of Columbia) Screening For**
Phenylketonuria	51
Congenital hypothyroidism	51
Galactosemia	44
Hemoglobinopathies (sickle cell, thalassemias)	42
Maple syrup urine disease	25
Homocystinuria	22
Biotinidase deficiency	16
Congenital adrenal hyperplasia	13
Tyrosinemia	6
Cystic fibrosis	1

Jews, the rate is almost 100 times higher (1 in 4800 births). In the 1970s, carrier-screening programs were set up to identify heterozygotes. More than 300,000 individuals were tested in the first 10 years of screening. Of these, 268 couples were identified in which both members were carriers and had not yet had an affected child. The programs were coupled with counseling sessions that provided education about the risks of having an affected child, the availability of prenatal screening, and reproductive options. None of these screening programs is mandatory, although some states have laws that require informing couples that screening for Tay-Sachs disease is available. In 1970 there were 50 to 100 Tay-Sachs births annually in the United States. Because of screening programs, there are now fewer than 10 such births each year.

Sickle Cell Trait Sickle cell anemia is an autosomal recessive condition that differentially affects Americans whose family origins are in West Africa, and in the lowlands of the Mediterranean Sea, including Sicily, Italy, Greece, Lebanon, and Israel. Blood tests and recombinant DNA-based tests can identify heterozygous carriers and affected homozygotes. In 1971 Connecticut instituted a program of screening black schoolchildren in grades 7-12, with parental consent, for **sickle cell trait,** a term used to designate heterozygous carriers. In 1972 Congress passed the National Sickle Cell Anemia Control Act, part of which was designed to establish carrier-screening programs. Some of the programs established as a result of this legislation were compulsory, requiring that black children be screened before attending school; others required screening before obtaining a marriage license. Professional football players were screened, as were cadets at the Air Force Academy in Colorado Springs, where heterozygotes were excluded from enrollment. The exclusion was based on the assumption that heterozygotes might undergo sickling of red blood cells at high altitudes under reduced oxygen concentrations. This policy was reversed in 1981 under the threat of a lawsuit. Other individuals who tested as positive heterozygotes were reportedly turned down for insurance and employment, even though carriers do not have any inherent health problems.

Some of these screening programs were criticized for laxity in confidentiality of records and the failure to provide counseling to those identified as heterozygotes. In the late 1970s, many of the sickle cell screening programs were cut back or reorganized, and currently only 10 states offer sickle cell screening.

Cystic Fibrosis Against the background of an uneven record of previous carrier screening programs, a new program of universal screening to detect heterozygous carriers of cystic fibrosis (CF) has been proposed. Cystic fibrosis (see Chapter 4 for a detailed discussion) is an autosomal recessive condition that occurs in all ethnic groups, although at different rates (Table 19.2). Between 25,000 and 30,000 individuals in the United States are affected by CF, and eight million others are heterozygous carriers.

The issues surrounding CF screening are legal, ethical, and economic. Testing for CF heterozygotes is one of the first recombinant-DNA-based screening tests to become available, and decisions about its implementation and availability will set precedents for the dozens of DNA-based tests that will be available in the next few years. The concerns that surround CF screening and CF screening as a model for other forms of genetic testing resolve into a number of issues:

- Standards of care. Should screening be offered to everyone, or only those with a family history?
- Confidentiality and discrimination. Who will have access to test results? Will identification as a carrier result in exclusion from health insurance coverage or employment?
- Quality and reliability of tests. How accurate are the tests? How often do false positive or false negative results occur?
- Cost effectiveness. What proportion of the population must participate in screening for it to be cost effective? How can participation be encouraged?

■ **Sickle cell trait** The symptoms shown by those heterozygous for sickle cell anemia.

Table 19.2
Cystic Fibrosis among Live Births in the United States

Population	Incidence at Birth
Caucasian	1 in 2,500
Hispanic	1 in 9,600
African American	1 in 18,000
Asian American	1 in 90,000

Note: From Office of Technology Assessment, 1992.

Still to be resolved are questions about who is qualified to do genetic testing and screening, how the costs of screening will be recovered, and how the public can be educated about the procedures and evaluation of risk.

In 1997, NIH issued a consensus statement about CF screening. This statement, put together by an independent panel of experts, recommends offering testing for CF to adults who have a positive family history of CF, to partners of those with CF, to couples currently planning a pregnancy, and to couples seeking prenatal testing. The panel does not recommend testing newborn infants or offering CF testing to the public at large. The panel also recommends that CF screening programs be phased in over a period of time to ensure that adequate education, appropriate testing, and counseling services are available to all persons being tested.

In addition, the panel emphasizes that every attempt be made to protect the rights of those being tested and that testing information not be used to discriminate in insurance, employment, and educational opportunities.

Occupational Screening

Two classes of tests are used for genetic screening in the workplace: tests to screen workers for inherited susceptibility to specific environmental agents and tests to monitor the amount of genetic damage actually produced in susceptible workers. We will consider only the first class of tests. Genetically determined sensitivity to certain environmental agents is well documented; in other cases the relationship between genetic factors and environmental agents is uncertain or is based on inadequate information.

Approximately 50 different genetic traits are thought to be related to susceptibility to environmental agents. Some of these are listed in Table 19.3. Many of these traits are rare, testing for carriers is difficult, and in some cases, the relationship to a specific environmental agent is based on inadequate data. To justify use in occupational screening, the traits must be present in a sufficient fraction of the workforce, and the trait should be associated with a clear risk to carriers. Most often the trait must be present in at least 1% of the workforce, and exposure should result in serious illness or risk of death for inclusion in a program of screening. A list of traits that fulfill many of these requirements is contained in Table 19.4.

The current status of genetic testing in the workplace is difficult to evaluate. Originally, five genetic conditions were identified as suitable for occupational screening: serum alpha-1-antitrypsin deficiency, glucose-6-phosphate dehydrogenase deficiency, carbon disulfide sensitivity, hypersensitivity to organic isocyanates, and sickle cell trait (heterozygotes). Recommendations for these as candidates for testing were

Table 19.3	Some Genetic Factors Affecting Susceptibility to Environmental Agents
Trait	**Agents**
G6PD deficiency	Oxidants, such as ozone and nitrogen dioxide
Sickle cell trait	Carbon monoxide, cyanide
Thalassemias	Lead, benzene
Erythrocyte porphyria	Lead, drugs including sulfanilomide, and barbiturates
Gout	Lead
Sulfite oxidase deficiency	Sulfite, bisulfate, sulfur dioxide
Wilson disease	Copper, vanadium
Pseudocholinesterase variants	Carbamate insecticides, muscle relaxants
Cystinuria	Heavy metals

Table 19.4	Genetic Traits Associated with Workplace Hazards	
Trait	**Environmental Agent**	**Status of Interaction**
Glucose-6-phosphate dehydrogenase (G6PD) deficiency	Primaquine, fava beans	Definite
Methemoglobin reductase deficiency	Nitrates, acetanilide, amines, sulfanomides	Definite
N-Acetyl transferase deficiency	Isoniazid hydrochloride, dapsone, hydralazine	Definite
PTC nontaster	Thiouria, related compounds	Possible
Slow alcohol metabolism	Ethanol toxicity	Possible
Sickle cell trait (heterozygotes)	Low oxygen concentration	Inconclusive

made because these conditions met "the prerequisites for industrial applications of bettering job assignment, improving coverage of industrial air limits, and hence reducing risk to worker health." Unfortunately, the recommendations for screening were not entirely based on a firm scientific foundation.

In 1983 the federal Office of Technology Assessment (OTA) conducted a thorough review of genetic screening in the workplace. The report indicates that there are clear-cut and well-known relationships between some genetic traits and agents present in the workplace. For example, males who are hemizygous for G6PD deficiency, or females homozygous for the deficiency may develop severe anemia when exposed to chemical oxidizing agents. This conclusion is based on the *in vitro* exposure of cells from G6PD-deficient individuals to these chemicals. Although many of these same chemicals can be encountered in industrial settings, the report calls for studies that assess directly whether exposure to these chemicals in the workplace actually poses any hazards to G6PD-deficient individuals. The report concludes that in the absence of direct evidence of harm to workers actually in contact with such agents, there is no justification for occupational screening programs of any kind. Critics of this report argue that sufficient information exists from laboratory studies on cells and tissues to show direct harm to certain genotypes upon exposure and that little or no research is being done on workers in the workplace. In effect, they charge, workers are being used as guinea pigs, and testing will start only after workers who have susceptible genotypes are seriously affected. As a result of information generated by the Human Genome Project and the development of new tests, the issue of occupational screening will be increasingly important in the coming years.

Reproductive Screening

Artificial insemination by a donor (AID) is often used when male fertility is low or, more rarely, to avoid genetic risks to the offspring by the transmitting a genetic defect. Some 172,000 women are artificially inseminated each year, resulting in 65,000 births. Of these, 30,000 births are the result of insemination with anonymously donated sperm obtained through physicians or sperm banks. In all cases, care must be taken to prevent genetic defects from being transmitted through the donor sperm. Because everyone carries some deleterious mutations, genetic defects in the donors cannot be eliminated. In one study of more than 600 potential semen donors, 6% were excluded as donors based on a detailed screening procedure. Of these, 2.6% were excluded for cytogenetic reasons and 3.4% for genetic reasons. The chromosomal abnormalities detected included breaks, translocations, partial aneuploidy, and the

Table 19.5
Some Traits Uncovered In Screening 676 Potential Sperm Donors

Trait	Number
Polyposis coli (predisposes to cancer)	3
Ankylosing spondylitis	2
Epilepsy	3
Manic-depressive psychosis	2
Dominant renal disease (not defined)	1
Severe hip dislocation	3

Note: From J. Selva, C. Leonard, M. Albert, J. Auger, & D. David, (1986) Genetic screening for artificial insemination by donor (AID). Clin. Genet. 29: 389–396.

presence of fragile sites. Table 19.5 lists some of the genetic traits uncovered in family histories of the prospective donors. They include single-gene traits and polygenic or familial traits, such as epilepsy.

The Office of Technology Assessment, a branch of Congress, surveyed 15 sperm banks and 367 physicians to determine whether sperm donors had been screened for infectious diseases, genetic defects, or both. According to the report, published in 1987, 14 sperm banks tested all donors for the human immunodeficiency virus (HIV), and the other tested only men from high-risk groups. Twelve of the sperm banks screened for transmissible diseases; 13 screened for genetic diseases. Interestingly, only 44% of the physicians tested for HIV, and fewer than 30% tested for transmissible diseases, such as syphilis or hepatitis. Moreover, only 48% screened for any genetic defects.

The most disturbing aspect of the report indicates that physician screening of sperm donors for genetic diseases is unreliable. Twenty-five percent of the physicians said they would accept sperm from a healthy donor who has a family history of Huntington disease. Huntington disease is an autosomal dominant disorder that does not appear until the individual is more than 40 years of age, whereas most sperm donors are under 30 years of age. On the other hand, 49% of physicians would reject a healthy sperm donor who has a family history of hemophilia. Recall that hemophilia is an X-linked trait expressed from birth. Unaffected males do not carry the trait and are incapable of passing it on to their offspring. The report has sparked calls for the Food and Drug Administration (FDA) to require physicians and sperm banks to screen sperm samples for HIV and to use fresh sperm only when the donor is known to the recipient. In the meantime, the results from these surveys indicate some degree of genetic risk from artificial insemination.

In a 1996 survey of commercial sperm banks, 100% of the banks screened prospective donors by medical history, family history and physical examination. Only 56% of the banks screened prospective donors by karyotype, and only 19% screen all prospective donors for Tay-Sachs disease, sickle cell anemia, and thalassemia. The report concluded that family history, reported by the prospective donor is the major source of information used by commercial sperm banks as the basis for acceptance or rejection, and that genetic professionals have a minimal role in the evaluation of semen donors.

PRENATAL TESTING

Genetic testing uses specific assays to determine the genetic status of individuals suspected to be at high risk for a particular inherited condition. Prenatal testing is used to detect genetic diseases and birth defects in the fetus. More than 200 single-gene disorders can be diagnosed prenatally (Table 19.6). In most cases the disorders are rare, and genetic testing is usually done when there is a family history or an indication that warrants testing. For conditions, such as Tay-Sachs disease or sickle cell anemia, the parents can be tested biochemically to determine if either is a carrier. If the tests for both parents are positive, the fetus has a 25% chance of being affected. In such cases prenatal testing can determine whether the fetus is a recessive homozygote afflicted with the disease. Similarly, if the mother is a carrier for certain deleterious, X-linked genetic disorders, testing can be offered.

For other genetic conditions, such as Down syndrome (trisomy 21), analysis of the fetal chromosomes is the most direct way to detect an affected fetus. In this situation testing is not carried out because of a familial history of genetic disease or detection of heterozygotes in the parents but usually because of advanced maternal age. Because the risk of Down syndrome increases dramatically with maternal age (see Chapter 6), cytogenetic testing is recommended for all pregnant females older than 35 years of age.

Table 19.6	Some Metabolic Diseases and Birth Defects That Can Be Diagnosed by Prenatal Testing	
Acatalasemia	Mannosidosis
Adrenogenital syndrome	Maple syrup urine disease
Chédiak-Higashi syndrome	Marfan syndrome
Citrullinemia	Muscular dystrophy, X-linked
Cystathioninuria	Niemann-Pick disease
Cystic fibrosis	Oroticaciduria
Fabry disease	Progeria
Fucosidosis	Sandhoff disease
Galactosemia	Spina bifida
Gaucher disease	Tay-Sachs disease
G6PD deficiency	Thalassemia
Homocystinuria	Werner syndrome
I-cell disease	Xeroderma pigmentosum
Lesch-Nyhan syndrome	

In addition to genetic disorders, some birth defects associated with abnormal embryonic development can be diagnosed prenatally. Among these are defects in the formation of the neural tube, a structure that arises in the first two months of development. One such defect, spina bifida, is a condition in which the spinal column is open or partially open. Neural tube defects can be diagnosed accurately by testing the amniotic fluid for elevated levels of alpha-fetoprotein. In about 80% of cases, alpha-fetoprotein levels in the maternal blood serum are also elevated. A test using maternal blood can identify mothers for whom further tests, such as amniocentesis, are recommended.

Samples or images for prenatal testing can be obtained in several ways, including amniocentesis, chorionic villus biopsy, ultrasonography, and fetoscopy. The fluids and cells obtained for testing can be analyzed by several techniques, including cytogenetics, biochemistry, and recombinant DNA technology. Images obtained by ultrasonography can be used to diagnose conditions, such as neural tube defects, cardiac abnormalities, and malformations of the limbs associated with chromosomal aberrations.

Because recombinant DNA technology analyzes the genome directly, it is the most specific and sensitive method currently available. The accuracy, sensitivity, and ease with which recombinant DNA technology can be used to assemble a profile of the genetic diseases and susceptibilities carried by an individual have raised a number of legal and ethical issues that have yet to be resolved.

We previously discussed two methods of obtaining samples for genetic testing (amniocentesis and chorionic villus sampling; see Chapter 6). In this section, we examine some other methods of obtaining genetic information about the fetus, including ultrasonography, fetoscopy, and blastomeric isolation. We also examine the consequences and potential problems associated with such tests. The process of genetic counseling, discussed later in this chapter, often uses these methods to determine whether a fetus is affected by a genetic disorder.

Ultrasonography

Ultrasonography is a technique based on sonar technology, which was developed for military use during World War II. For prenatal diagnosis, a transducer is placed on the abdomen over the enlarged uterus (Figure 19.2). The probe emits pulses

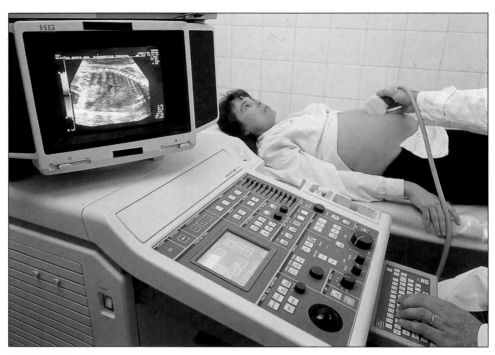

▶ **FIGURE 19.2**   A pregnant woman is undergoing an ultrasound examination.

of ultrasonic energy, and as the ultrasound strikes the surface of the fetus, some waves are reflected from the fetus and return to the transducer. These reflected waves are electronically converted to images and displayed on a screen (▶ Figure 19.3).

Ultrasound can be used to diagnose multiple pregnancy; determine fetal sex; and identify neural tube defects and skeletal disorders, limb malformations, other central nervous system defects, and congenital heart defects.

Fetoscopy

Fetoscopy is the direct visualization of the fetus by a fiber-optic device known as an endoscope. In this procedure a hollow needle is inserted through the abdominal wall into the amniotic cavity, and the fiber-optic cable is threaded through the needle. The image is transmitted to a video screen and can be viewed and recorded from the screen (▶ Figure 19.4). The technique is most useful when a disorder cannot be diagnosed by cytogenetic or biochemical methods. Fetoscopy can also be used to obtain samples of fetal blood, allowing diagnosis of some genetic diseases, such as hemophilia and certain forms of thalassemia. Fetoscopy poses a danger to the fetus, however, and there is a 2% to 5% chance of spontaneous abortion in using this method.

Testing Embryonic Blastomeres

A new method of prenatal genetic screening, combining microsurgery and recombinant DNA technology, is being used for genetic testing of human preimplantation embryos at the 6- to 10-cell stage of development (see the discussion of this topic in Chapter 13). In this procedure embryos derived from *in vitro* fertilization are incubated until they reach a multicellular stage of development. Then, using a micromanipulator, a hole is made in the material surrounding the embryo, and a single cell is removed for analysis. The DNA from this single cell, called a blastomere, is screened using the polymerase chain reaction (PCR) to detect the presence of mutant genes, such as those for muscular dystrophy or hemophilia. Operated embryos continue to

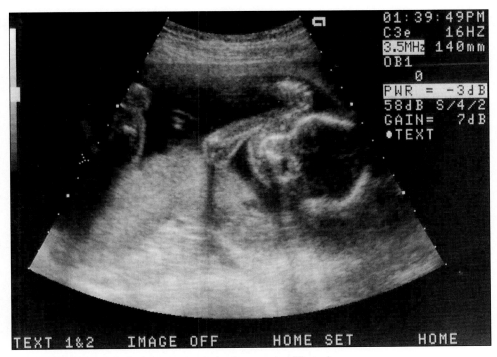

FIGURE 19.3 Baby Lavery as seen in ultrasound at 20 weeks.

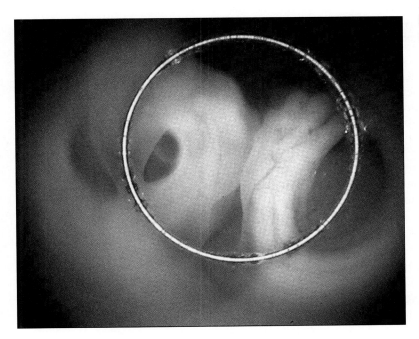

FIGURE 19.4 The hand and face of a nine-week-old fetus seen by fetoscopy.

develop *in vitro* and after uterine transfer, develop to full term. (see Concepts and Controversies: The Business of Making Babies).

Risks and Problems Associated with Prenatal Testing

Although many genetic disorders and birth defects can be detected by prenatal testing, the technique has some limitations. These include measurable risks to the mother and fetus, including infection, hemorrhage, fetal injury, and spontaneous abortion. Keep in mind, however, that overall, there is a 3.5% risk of bearing a child who has a birth defect, genetic condition, or mental retardation.

Conventional strategies for prenatal testing will not always detect the majority of certain defects. In Down syndrome, for example, amniocentesis is recommended for all mothers over the age of 35 years. (In reality, only a small percentage of pregnant women over 35 are tested by amniocentesis.) However, some 65% of all Down syndrome births occur to mothers under the age of 35. The differential distribution of Down syndrome births reflects differences in the number of pregnancies in women under and over the age of 35. Younger mothers may have 65% of the Down syndrome children, but they also have 93% of all births. Older women have about 7% of all children but 20% of the Down syndrome births, emphasizing once again the relationship between maternal age and increased risk of Down syndrome.

In the case of neural tube defects, some 90% of all affected infants are born to parents who have no family history of such conditions. Thus, testing couples who have had an affected child will have little effect on the overall rate of prenatal detection for this birth defect. On the other hand, screening of all pregnant females is not cost-effective or possible, given the limited number of prenatal screening clinics.

GENETIC COUNSELING

Genetic counseling A process of communication that deals with the occurrence or risk that a genetic disorder will occur in a family.

Genetic counseling is a process of communication that deals with the occurrence or risk that a genetic disorder will occur in a family. Counseling involves one or more appropriately trained persons who help an individual or family to understand

- the medical facts, including the diagnosis, probable course of the disorder, and the available treatment and management
- the way heredity contributes to the disorder and the risk of recurrence
- the alternatives for dealing with the risk of recurrence
- how to adjust to the disorder in an affected family member or to the risk of recurrence

The genetic counseling community attempts to achieve these goals in a nondirective way. Genetic counselors provide all of the information available to and desired by an individual or family, so that the person or family can make the decisions most suitable to them based on their own cultural, religious, and moral beliefs. The National Society of Genetic Counselors has developed guidelines used by counselors in this process. These guidelines include respect for the autonomy and privacy of the individual, observing the need for confidentiality and informed consent, and providing information to the patient in a nondirective way. Genetic counselors also provide supportive counseling for families, serve as patient advocates, and refer individuals and families to community and state support services. They serve as educators and resource people for other health care professionals and for the general public. Some genetic counselors work in administrative positions, and others engage in research activities related to the field of medical genetics and counseling.

Who Are Genetic Counselors?

Genetic counselors are health care professionals who have specialized graduate training and experience in the areas of medical genetics, psychology, and counseling. They usually work as members of a multidisciplinary health care team and offer information and support to families that have relatives with genetic conditions and to families who may be at risk for a variety of inherited conditions. Genetic counselors identify families at risk, investigate the problem in the family, interpret information about the disorder, analyze inheritance patterns and risk of recurrence, and review available options with the family (◗ Figure 19.5).

Typically, genetic counselors are graduates of a two-year master of science (MS) degree program. There are currently 22 genetic counseling programs in the United States. Approximately from four to seven students are accepted into each program each year, and an average of 130 new counselors enter the work force

The Business of Making Babies

*R*ecombinant DNA methods are revolutionizing the fields of genetic testing and genetic counseling. New technology has also made the business of human fertilization a part of private enterprise. One in six couples in the United States (over three million couples of childbearing age) is classified as infertile, and most of these couples want to have children. The first successful in vitro fertilization (IVF) was accomplished in 1981 at the Medical College of Virginia at Norfolk. Since then, more than 150 hospitals and clinics that use these and other techniques have opened. Many of these clinics are associated with university medical centers, but others are operated as free-standing businesses. Some are public companies that have sold stock to raise start-up money or to cover operating costs. It is estimated that a capitalization of about $1 million is required to start an IVF clinic and that 50–60 fertilization attempts per month are necessary for the venture to be profitable. Each IVF attempt costs between $5,000 and $10,000, and several attempts (four to six) are usually required for success. Because these costs are not usually covered by insurance, IVF is a major expense for couples who want children.

In IVF, egg maturation is induced with drugs, and the mature eggs are recovered by laparascopy (a small incision is made in the abdominal wall, and a fiber-optic device is used to recover the eggs). In an alternative procedure, an ultrasonically guided needle is inserted into the vagina and moved up to the ovary to remove the eggs. The eggs are fertilized in a dish (in vitro means, literally, "in glass"). After the fertilized egg begins to develop, it is implanted into the woman's uterus.

If extra eggs are recovered, they are fertilized, and the resulting embryos are frozen in liquid nitrogen. This eliminates the need to retrieve eggs every month for fertilization. If fertilization is successful, the extra embryos can remain in storage for implantation at a later time, or they can be donated to another couple.

Several companies, such as IVF Australia, are open in several locations in the United States. Because of the high start-up costs and expertise required, it is possible that the field will be dominated by a small number of companies through franchising agreements. Some investment analysts predict that IVF will grow into a $6 billion annual business. In parallel, a genetic testing and screening industry is beginning to emerge that offers tests in the areas of prenatal diagnosis, newborn screening, carrier screening, adult-onset screening, and forensic testing.

❱ **FIGURE 19.5**　In a genetic counseling session, the counselor uses the information from pedigree construction, medical records, and genetic testing to educate and inform a couple about their risks for genetic disorders.

each year. Students in these programs are trained in biology, genetics, molecular biology, biochemistry, psychology, clinical work, and laboratory methods. They also receive training in ethical, social, and legal issues related to genetic disorders. The American Board of Genetic Counseling offers a certification examination for counselors every three years. Most counselors work in university medical centers or at large hospitals in metropolitan areas. However, many smaller community hospitals and physicians in private practice are becoming aware of how valuable genetic counseling can be for their patients. As genetic testing and screening methods proliferate, it is hoped that genetic counseling services will become available to an increasing number of those who request genetic testing.

Reasons to Seek Genetic Counseling

There are many reasons that someone should seek genetic counseling services. The most typical case is an individual/family that has questions about his or her reproductive future. For example, individuals who have a family history of a genetic disorder, family risk of cancer, birth defect, developmental disability, or an affected child may have concerns about the risk of recurrence in future offspring. Women more than 35 years of age and individuals from specific ethnic groups in which particular genetic conditions are seen more frequently will learn of their increased risk for genetic or chromosomal disorders and the diagnostic testing that is available. Other reasons for referral include multiple miscarriages, maternal diseases, such as diabetes or lupus, known carrier status for a genetic disorder, parental anxiety, and environmental exposures, such as drugs or infections. Fetal anomalies suspected as a result of a maternal screening test or an ultrasonogram may also lead a couple to genetic counseling. Anyone who has unanswered questions about diseases or traits in their family should consider genetic counseling. People who might be especially interested include

- women who are pregnant or are planning to become pregnant after age 35
- couples who already have a child with mental retardation, an inherited disorder, or a birth defect
- couples who would like testing or more information about genetic defects that occur more frequently in their ethnic group
- couples who are first cousins or other close blood relatives
- people concerned that their jobs, lifestyle, or medical history may pose a risk to a pregnancy, including exposure to radiation, medications, chemicals, infection, or drugs
- women who have had two or more miscarriages or babies who died in infancy
- couples whose infant has a genetic disease diagnosed by routine newborn screening
- those who have, or are concerned that they might have, an inherited disorder or birth defect
- pregnant women who, based on ultrasound tests or blood tests for alpha-fetoprotein, have been told their pregnancy may be at increased risk for complications or birth defects

How Does Genetic Counseling Work?

Most individuals are referred to genetic counseling services after a prenatal test or after the birth of a child who has a genetic condition. In either case, the couple is concerned about the risks to the fetus in the current pregnancy or to future pregnancies based on this abnormal result. The counselor usually begins by constructing a detailed family and medical history, or pedigree. For prenatal counseling, all birth defects, causes of deaths, ages at death, and other health conditions are noted for each person in the family for at least three generations. It is also important to note the occurrence of any pregnancy losses and the stage (first, second, or third trimester) when the pregnancy was lost. This information gives the counselors clues about certain genetic conditions that appear only in one sex or conditions that are associated with repeated miscarriages.

Prenatal screening, cytogenetic, or biochemical tests performed on the expectant couple or on the developing fetus can be used along with the pedigree to help determine risk of occurrence or recurrence. The counselor uses as much information as possible to establish whether the trait is genetically determined.

Before pregnancy, genetic counseling can only offer probabilities that a specific birth defect may occur. The general population's risk of bearing a child who has a serious genetic problem is about 2–5%. This is considered the background risk for every couple, regardless of family or medical history. This risk can increase if a genetic condition is in the family but can never decrease. During pregnancy, chromosomal errors, many genetic disorders, and other conditions that may not have a genetic basis (such as heart defects) can be ruled out through prenatal tests.

If it is found that a condition in a family is genetically determined, the counselor constructs a risk assessment for the couple. In this process, the counselor uses all of the information available to explain the risk of having another child affected with the condition or to explain the risk that the individual who is being counseled will be affected with the condition. Conditions that are considered high risk include dominant conditions (50% risk if one parent is heterozygous), simple autosomal recessive (25% when both parents are heterozygotes), and certain chromosomal translocations. Often conditions are difficult to assess because they involve polygenic traits or conditions that have high mutation rates (like neurofibromatosis).

Effectively communicating risk estimates so those being counseled can clearly understand can be a difficult task for the genetic counselor. Telling a couple that they have a one in four chance of having a child affected with a genetic condition may sound straightforward, but several barriers may exist that prevent accurate understanding of this risk. Many individuals lack a fundamental understanding of elementary probability. Some couples believe that if they have had one child affected with the condition, then the next three children will not be affected, so their risk of recurrence drops to zero. This is not the case. The one in four, or 25%, risk, is independent of the number of pregnancies that a couple has. To put it another way, this is the risk for each pregnancy (Table 19.7).

Learning about abnormal test results or a genetic disorder in the family can be devastating news for individuals. A genetic counselor offers emotional support and understanding during what can be a very difficult time. Intense emotions expressed by an individual or a couple during a counseling session can inhibit effective understanding of the risks of recurrence or the details of the condition. Emotional strife can influence decisions that may need to be made about a pregnancy, the care of a child, having more children, or the ability of the family to cope with ongoing problems.

Table 19.7 **Some Risk Factors in Families That Have One Affected Child**	
Trait	**Risk of More Affected Children**
Autosomal recessive	25%
Autosomal dominant	50%
Rare, sex-linked recessive	0% females, 50% males
Chromosome abnormality	<1% to 100%
Genetic anomaly; not a simple mode of inheritance	Generally <10%
Nongenetic malformation	2%

When necessary, counselors refer patients to parent organizations that deal with a specific genetic condition or to medical specialists, education specialists, or family support groups.

Genetic counselors explain basic concepts of biology and inheritance to all couples. This helps them understand how genes, proteins, or cell-surface antigens are related to the defects seen in their child or family. The counselor provides information that allows informed decision making about future reproductive choices. Reproductive alternatives, such as adoption, artificial insemination, *in vitro* fertilization, egg donation, and surrogate motherhood are options that the counselor presents to the couple.

Future Directions

As the Human Genome Project accelerates the number of genetic disorders that can be detected by heterozygote and prenatal screening and as these techniques become more available, the role of the genetic counselor will become more important. The Human Genome Project is changing the focus of genetic counseling from reproductive risks to adult-onset conditions, such as cancer, polycystic kidney disease, and Huntington disease. Although counseling sessions address reproductive risks for these conditions, the primary focus is on the individual being counseled. The areas addressed include the risk of inheriting the gene, the potential severity of the condition, and the age of onset.

Advances in recombinant DNA technology are elucidating the genetic basis for conditions such as coronary artery disease, diabetes, and cancer. DNA tests for susceptibility to these adult-onset, common conditions are now being developed. Ultimately, it may be possible to treat these conditions following presymptomatic diagnosis. Presymptomatic genetic tests are already available for some forms of cancer, including breast and ovarian cancer. Genetic counseling has expanded its focus from reproductive risks to include common adult disease.

THE IMPACT OF GENETIC TESTING AND GENETIC SCREENING

The development of genetic screening and counseling programs has provided many benefits to individuals and society at large. But it has also created a number of associated problems and raised serious questions whether screening should be mandatory, who is to have access to the results of screening, and whether individuals identified as carriers of genetic defects are socially stigmatized. In this section, we briefly examine several aspects of these problems.

Personal Consequences

The information that one is a carrier of a genetic disease often has a devastating psychological effect. Many identified carriers suffer a loss of self-image and regard themselves as worthless. This feeling is often reinforced by the feelings of family members toward carriers. For example, in some parts of rural Greece, marriages are arranged by parents and relatives. In one village in which screening for sickle cell was conducted, carriers were regarded as unsuitable marriage partners for anyone, not just other carriers.

To counter these effects, screening programs must be coupled with effective counseling programs for carriers, their families, and the general public. The education process must stress that carriers are not at risk for the disease, nor should they be prevented from marrying other carriers. Options for matings between heterozygotes should be carefully distinguished, including adoption, artificial insemination, and prenatal diagnosis coupled with selective abortion. The effect of genetic testing on childbearing decisions has been documented in a number of studies. In one such study, couples at risk for having children afflicted with a severe form of beta-thalassemia were counseled about the availability of prenatal diagnosis. Before such services were available, couples known to be at risk (through birth of an affected child or carrier screening) had stopped having children altogether, and almost all pregnancies that occurred were reported as accidental. Of these pregnancies, 70% were terminated for fear of bearing a child who has beta-thalassemia. The availability of prenatal diagnosis brought about a significant change in childbearing decisions in such couples. In fact, reproductive patterns returned to almost normal levels, and less than 30% of all pregnancies were terminated because of thalassemia. Other surveys have reported similar findings and emphasized the impact of genetic testing on individual lives.

Predictive genetic testing for autosomal dominant fatal genetic disorders that first appear in middle age (Huntington disease, polycystic kidney disease) has been evaluated to determine its psychological and social impact. Recently, the Canadian Collaborative Study of Predictive Testing reported on the psychological consequences of predictive testing for Huntington disease (HD). This form differs from other genetic testing because it requires testing other family members to produce informative results. Consequently, a request for testing affects the other members and forces them to consider whether they wish to be tested.

In the study reported by the Canadian group, 200 individuals, each of whom had an affected parent, were followed after HD testing. They were separated into three groups: those who had increased risk, those who had no change in risk (mostly from uninformative test results), and those who had decreased risk. The results suggest that testing has positive benefits for many participants. Clearly, those in the low-risk group showed an increase in well-being and psychological health. But those in the increased-risk group did not show a negative response to their conditions. In fact, they reported less depression and an increased sense of well-being 12 months after testing. It appears that knowledge of status, whether for increased or decreased risk, has psychological value, whereas those whose status is uncertain remain susceptible to depression and have a lowered sense of well-being. Further tests and follow-up will be required to determine whether positive psychological effects are a hallmark of predictive genetic testing.

Social Consequences

At the broader level, we can compare the response of the Jewish community to screening for Tay-Sachs disease with some of the responses in the Black community to sickle cell screening. In Tay-Sachs disease screening, the program was voluntary and welcomed by the community, which was involved in its planning and implementation. An adequate educational and counseling program accompanied the screening. In contrast, screening for sickle cell disease (homozygous recessive individuals afflicted with the disease) and sickle cell trait (heterozygous individuals unaffected by the disease) was largely mandated by law, and prominent members of the Black community were not involved in the planning and initiation of screening. The origin of the screening program from outside the community coupled with sporadic problems (lack of confidentiality and inadequate education and counseling) generated suspicion and resentment about it.

Better planning and implementation of this large-scale screening program could have avoided many of these problems. Voluntary rather than mandatory participation, coupled with adequate education and counseling and community involvement, would undoubtedly have eased many fears and suspicions. Perhaps it would be better to offer testing and counseling to those who request it rather than screening large groups to identify and label individuals as "carriers." Others argue that only mandatory screening programs can be effective. If, say, only 10% of those at risk take advantage of screening programs, the program is ineffective, and the cost-benefit ratio would not justify the existence of the program.

The lessons from earlier attempts at sickle cell screening are particularly important in the light of new discoveries about sickle cell disease. A recent study has revealed that children who have sickle cell disease and are less than 3 years of age have poor resistance to bacterial infections, particularly those caused by *Streptococcus*. Children who have sickle cell anemia have a 15% chance of dying from a bacterial infection during the first three years of life. The study also demonstrated that doses of the antibiotic penicillin are highly effective in preventing illness and death.

These findings have prompted a National Institutes of Health (NIH) panel to recommend that sickle cell screening be made available to all newborns, whether or not they are members of high-risk ethnic groups. This recommendation has been made because it would avoid errors in classifying individuals as members of certain ethnic groups and because screening is easy and inexpensive. Sickle cell screening can be done using a small blood sample and materials costing $0.22 per test. According to the panel, the expenditure of approximately $880,000 per year (4 million births × $0.22 per test) would produce a 15% reduction in the mortality rate among small children who have sickle cell anemia. Recently, a panel convened by the U.S. Public Health Service recommended implementation of a program to screen all newborns in the U.S. for sickle cell anemia. If this recommendation is accepted, newborn screening will begin in the next few years.

It remains to be seen whether sickle cell screening programs will be adopted in any of the 40 states that do not currently test for this condition. This recommendation by the NIH panel also raises the questions whether such screening should be mandatory or voluntary and whether the states should screen all newborns or only those in high-risk groups. These issues are certain to be debated once again by community groups and state legislatures in the near future.

Legal Implications

Genetic screening has raised a number of legal issues, many of which have not yet been resolved. Here, we consider several questions about genetic screening to illustrate that genetic methodology and practice are several steps ahead of legislation, legal decisions, and social consensus.

If a child is born who has a genetic defect that can be diagnosed prenatally, can the physician be held responsible for not informing the parents that prenatal screening for this defect is available? If an insurance company pays for a genetic screening test, does it have the right to know the results of the test? Can health or life insurance companies require genetic testing as a condition for obtaining insurance? Should individuals who test positive for Huntington disease or other genetic disorders be denied health or life insurance?

The Occupational Safety and Health Administration has categorized 24 chemicals that may be associated with reproductive hazards and excludes all fertile women from jobs that involve exposure to these chemicals. Is this protection or a form of sexual discrimination? Does knowing that these rules may apply to 20 million jobs change your answer?

These questions illustrate that many problems that involve the development and use of genetic screening need to be resolved. These issues involve science, sociology, law, and ethics. In decisions about genetic screening, the rights of individuals must be considered and balanced against the rights of employers and society. Health policy is an area in which all citizens need to be educated and informed. As the constellation of genetic screening tests grows, these problems must be faced and solved.

Case Studies

CASE 1

Trudy was a 33-year-old woman who went with her husband, Jeremy, for genetic counseling. Trudy has had three miscarriages. The couple has a 2-year-old daughter who is in good health and is developing normally. Chromosomal analysis was done on the products of conception of the last miscarriage and was found to be 46, XY. The last miscarriage occurred in January 1999. Peripheral blood samples for both parents were taken at the time and sent to the laboratory. Trudy's chromosomes were 46, XX and Jeremy's were 46, XY, t(6;18) (q21;q23). Jeremy appears to have a balanced translocation between chromosome number 6 and 18. There is no family history of stillbirths, neonatal death, infertility, mental retardation, or birth defects. Jeremy's parents both died in their 70s from heart disease, and his is unaware of any pregnancy losses experienced by his parents or siblings.

The recurrence risks associated with a balanced translocation between chromosome 6 and 18 were discussed in detail. The counselor used illustrations to demonstrate the approximate 50% risk of unbalanced gametes; the other 50% of the gametes result in either normal or balanced karyotypes. The family was informed that empirical risk for unbalanced conceptions is significantly less than the 50% relative risk. Prenatal diagnostic procedures were described, including amniocentesis and chroionic villus sampling. The benefits, risks, and limitations of each were described.

The couple indicated a desire to pursue another pregnancy and were interested in proceeding with an amniocentesis.

CASE 2

Genetic counselors frequently use empirical risk estimates for multifactorial conditions or in situations were Mendelian principles do not apply. Let's assume that a couple comes for genetic counseling with no other positive family history except for one child affected with a disorder that could be either multifactorial or chromosomal. The couple wants to know the risk of having another affected child. If the child is affected with spina bifida, the empirical risk to a subsequent child is approximately 4%. If the child has Down syndrome, the empirical risk of recurrence would be approximately 1% if the karyotype were trisomy 21, but it might be substantially higher if one of the parents were a carrier of a Robertsonian translocation involving chromosome 21.

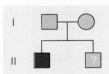

Summary

1. Genetic screening is the search for individuals who have a particular genotype. In prenatal and newborn screening, several considerations are important. Treatable diseases are favorable for screening even when they are rare in the population. These include PKU, galactosemia, and maple syrup urine disease.

2. Carrier screening is the search for heterozygotes who may be at risk of producing a defective child. An increasing number of autosomal recessive diseases can be screened by molecular probes, including sickle cell anemia, Huntington disease, cystic fibrosis, and Duchenne muscular dystrophy. Large-scale carrier screening has been conducted for two autosomal recessive diseases that affect discrete population segments: Tay-Sachs disease and sickle cell anemia. The programs have been technically successful but were accepted somewhat differently in the affected segments of the population. A study of their implementation and the community reaction to them will be valuable in designing and planning other carrier screening programs.

3. Occupational screening is used to detect individuals who are genetically susceptible to agents in the workplace that can cause the development of disease. Although a number of agents that can cause adverse reactions have been identified and a number of diseases can be screened, no large-scale studies have been conducted to establish that such chemicals are harmful to sensitive individuals in the workplace. Because occupational screening can be used to exclude individuals from employment, its implementation should be restricted to those genetic conditions in which a danger has been clearly demonstrated.

4. The rapid development of methodology for genetic screening has generated a number of problems, including whether screening should be voluntary or mandatory and whether the results of screening tests can be used to deny services, such as insurance or health care coverage. These issues will undoubtedly be the subject of much debate and a fair amount of legislation in the next few years.

5. Prenatal screening can also detect chromosome abnormalities, such as Down syndrome, and birth defects, such as spina bifida, that may have a genetic component. One of the considerations in prenatal screening is identifying the risk group to be screened. The frequency of Down syndrome increases rapidly as maternal age increases over 35 years, and it is easy to see that screening should be made available to all pregnant females over the age of 35. Yet most Down syndrome births occur to younger mothers because they have many more children than older mothers. Should screening be made available to all mothers for Down syndrome? Given the limited resources available now for such screening, is this cost-effective? Several techniques are used in prenatal diagnosis, and each carries a risk to the mother and the fetus. Less invasive methods, such as the detection of fetal cells in the maternal circulation, are under development, making prenatal screening for some diseases safer and more economical.

6. Genetic counseling is a service that undertakes the accurate assessment of a family history to determine the risk of genetic disease in subsequent children. In most cases this is done after the birth of a child affected with a genetic disorder, but in other cases counseling is entirely retrospective. Decisions about whether to have additional children or to undergo abortion or even to marry are always left to those being counseled.

Questions and Problems

1. List the types of genetic screening covered in this chapter, and briefly summarize the unique characteristics of each type.

2. The measurement of alpha-fetoprotein levels is used to diagnose neural tube defects. For every 1000 such tests, approximately 50 positive cases will be detected. However, up to 20 (40%) of these cases may be false positives. In a false positive, the alpha-fetoprotein level is elevated, but the child has no neural tube defect. Your patient has undergone testing of the maternal blood for alpha-fetoprotein, and the results are positive. She wants to abort a defective child but not a normal one. What are your recommendations?

3. The reaction to screening for Tay-Sachs disease and sickle cell anemia offers an interesting contrast in the institution and administration of genetic screening programs. Cystic fibrosis is an autosomal disease that mainly affects the white population, and 1 in 20 whites are heterozygotes. Now that the gene has been mapped to chromosome 7, assume that RFLP markers are available to diagnose heterozygotes. Should a genetic screening program for cystic fibrosis be instituted? Should the federal government fund this? Should the program be voluntary or mandatory, and why?

4. As a genetic counselor, a couple who wishes to have children visits you. There is a history of a deleterious, recessive trait in males in the woman's family but not in the man's family. The couple is convinced that because his family shows no history of this genetic disease, they are at no risk of having affected children. What steps would you take to assess this situation and educate this couple?

5. A couple has had a child born with neurofibromatosis. They come to your genetic counseling office for help. After taking an extensive family history, you determine that there is no history of this disease on either side of the family. The couple wants to have another child and want to be advised about risks of another child with neurofibromatosis. What advice do you give them?

6. The initial step in the process of genetic counseling is to construct a family pedigree and analyze it for inheritance patterns. In pedigree analysis, what are the major differences between autosomal recessive and autosomal dominant conditions?

7. An interesting polymorphism in the human population has to do with the ability to roll one's tongue (curl up the sides of the tongue to make a trough). Some people can do this trick, and others cannot. Hence it is an example of a dimorphism. Its significance is a mystery. In one family, a boy was unable to roll his tongue, but to his great dismay his sister could. Furthermore, both his parents were rollers, and so were both grandfathers and one paternal uncle and one paternal aunt. One paternal aunt, one paternal uncle, and one maternal uncle could not. Draw the pedigree for this family, clearly defining your symbols and deduce the genotypes of as many individuals as possible.

8. Phenylketonuria (PKU) is a human hereditary disease that prevents the body from processing the amino acid phenylalanine, which is contained in dietary protein. Symptoms of PKU present in early infancy and, if it remains untreated leads to severe mental retardation. PKU is caused by a recessive allele with simple Mendelian inheritance. A couple intends to have children but seeks genetic counseling because the woman has a sister with PKU and the man has a brother with PKU. There are no other known cases in their families. They ask you, the genetic counselor, to determine the probabilities that their first child will have PKU. What is this probability?

9. You are a genetic counselor and have been asked to review a proposal to screen local high school students

for cystic fibrosis (CF) carrier status. The investigator's protocol states that the test will be done by extracting cells from inside the mouth with a cotton swab from all Caucasian students. Prior to testing, students will be given a booklet about CF and about the test and asked to sign a consent form. Results will be distributed to students in sealed envelopes with a toll-free number to call if they have questions about the test results. Of the various objections that you might raise with this screening protocol, the LEAST compelling of the following is that. . .

a. non-Caucasian students will not be able to be tested, if they wish to be

b. the method of informing the students of results does not adequately protect privacy or provide for appropriate follow-up counseling

c. offering the screening in the classroom may unduly pressure students into being tested or stigmatize them

d. the students' parents must be involved in providing consent

e. the students are too young to benefit from the information

10. You are a genetic counselor and your patient has asked to be tested to determine if she carries a gene that predisposes her to early onset cancer. If your patient has this gene there is a 50:50 chance that all of her siblings inherited this gene and there is also a 50:50 chance that it will be passed on to their offspring. Your patient is very concerned about confidentiality and does not want anyone in her family to know she is being tested, including her identical twin sister. Your patient is tested and found to carry an altered gene that gives her an

85% lifetime risk of developing breast cancer and a 60% lifetime risk of developing ovarian cancer. At the result disclosure session, she once again reiterates that she does not want anyone in her family to know her test results.

a. Knowing that a familial mutation is occurring in this family, what would be your next course of action in this case?

b. Is it your duty to contact members of this family despite the request of you patient? Where do your obligations lie . . . with your patient or with the patient's family?

c. Would it be inappropriate to try and convince the patient to share her results with her family members? What other problems can you foresee with this case?

11. A young woman (proband) and her partner are referred for prenatal genetic counseling because the woman has a family history of sickle cell anemia. The proband has sickle cell trait (*Ss*) and her partner is not a carrier nor does he have sickle cell anemia (*SS*). Prenatal testing indicates that the fetus is affected with sickle cell anemia (*ss*). The results of this and other tests indicate that the only way the fetus could have sickle cell disease is if the woman's partner is not the father of the fetus. The couple is at the appointment seeking their test results.

a. How would you handle this scenario? Should you have contacted the proband beforehand to explain the results and the implications of these results?

b. Is it appropriate to keep this information from the partner since he believes he is the father of the baby? What other problems do you see with this case?

Internet Activities

The following activities use the resources of the World Wide Web to enhance the topics covered in this chapter. To investigate the topics described below, log on to the book's home page at:

http://www.brookscole.com/biology

1. Human Genome Project Information for the U.S. Dept. of Energy is a very valuable web site. Access it and then click Ethical, Legal and Social Issues. Once there, scroll down to The Genetic Privacy Act and Commentary under the heading Products of ELSI Research. Click and read through Parts A and B under the heading Commentary.

a. Do you think adequate safeguards have been established to protect individuals' genetic information?

b. We are often curious about dead celebrities or major political figures. If a person, such as Abraham Lincoln, has been dead over one hundred years, should the public have the right to test a sample of hair or other body substance to determine if he

carried any of the genetic markers for Marfan syndrome or manic-depression? Why or why not? What rules would you propose to protect their descendants?

c. If you were counseling a patient whose family history had evidence that a genetic disease like Huntington disease, a genetic condition that is damaging to the nervous system and can lead to an early death, would you advise that patient to have the genetic test done? If you were the patient, would you want to know if you had the genetic marker for Huntington disease? Give some pros and cons for both positions.

2. On the ELSI home page there is an entry for the 1993 Insurance Task Force. Click on that title and read the brief report and its recommendations.

a. Should insurance companies be denied all genetic information about applicants who want health or life insurance? Why or why not?

b. If research has shown that women with specific identifiable genes have a 95% chance of developing a fatal type of breast cancer, should the insurance companies be able to have that information? Give three reasons why they should have that information available and three other reasons why they shouldn't.

3. NOAH (The New York Online Access to Health) has an excellent home page for genetic counseling. Access it through the address provided. There should be brief scenarios provided as an introduction.

   a. Click and read the entry: Is Genetic Counseling for You? Think about yourself or your family members. Are there any possible genetic conditions you may be concerned about? If so, there is useful information here about what genetic counselors can do for you and information on how to contact a counselor in your geographic area.

4. The Human Genome Project web site has many resources which address the issues raised by genetic research including two new publications, "To Know Ourselves" and "Your Genes Your Choices." Request or download both publications. Use them as a starting point for class discussion regarding issues raised by genetic research.

5. While the promise of genetic testing and gene therapy seem to bode well for our future health, the moral, social, and legal implications of our new-found knowledge will raise particular questions we may not be ready to answer. In the "new" human genetics, it seems that technology has outpaced the practical applications of the science. The legal community is attempting to assess the situation, so that it will be better able to deal with cases involving the application, interpretation, and confidentiality of genetic tests. Read the article in *The Judges Journal* on genetic testing. Are there any issues raised in the article that you disagree with? Is there an issue associated with testing for any of the diseases mentioned that the article fails to raise? Do you think the legal profession has or will have the scientific expertise to effectively defend or prosecute cases involving the use or misuse of genetic testing?

For Further Reading

Agency for Health Care Policy and Research. Sickle Cell Disease Panel. Sickle cell disease: comprehensive screening and management in newborns and infants. Rockville, MD, Public Health Service, Department of Health and Human Services, April, 1993.

Billings, P., & Beckwith, J. (1992). Genetic testing in the workplace: A view from the USA. *Trends Genet.* 8: 198–202.

Calabrese, E. J. (1986). Ecogenetics: Historical foundation and current status. *J. Occup. Med.* 28: 1096–1102.

Dagenais, D., Courville, L., & Dagenais, M. (1985). A cost-benefit analysis for the Quebec Network of Genetic Medicine. *Soc. Sci. Med.* 20: 601–607.

Draper, E. (1991). *Risky Business: Genetic Testing and Exclusionary Practices in the Hazardous Workplace.* New York: Cambridge University Press.

Emery, A. E. H., & Pullen, I. (1984). *Psychological Aspects of Genetic Counseling.* New York: Academic.

Fuhrmann, W., & Vogel, F. (1983). *Genetic Counseling,* 3rd ed. New York: Springer-Verlag.

Gibbs, R. A., & Caskey, C. T. (1989). The application of recombinant DNA technology for genetic probing in epidemiology. *Ann. Rev. Pub. Health* 10: 27–48.

Hodgson, S. V., & Bobrow, M. (1989). Carrier detection and prenatal diagnosis in Duchenne and Becker muscular dystrophy. *Br. Med. Bull.* 45: 719–744.

Jinks, D. C., Minter, M., Tarver, D. A., Vanderford, M., Hejtmancik, J. F., & McCabe, E. R. B. (1989). Molecular genetic diagnosis of sickle cell disease using dried blood specimens on blotters used for newborn screening. *Hum. Genet.* 81: 363–366.

Kolata, G. (1986). Genetic screening raises questions for employers and insurers. *Science* 232: 317–319.

Modell, B., & Kuliev, A. (1993). A scientific basis for cost-benefit analysis of genetics services. *Trends Genet.* 9: 46–52.

Murray, R. F. (1986). Tests of so-called genetic susceptibility. *J. Occup. Med.* 28: 1103–1107.

Rowley, P. (1984). Genetic screening: Marvel or menace? *Science* 225: 138–144.

Sandovnick, A., & Baird, P. (1985). Reproductive counseling for sclerosis patients. *Am. J. Med. Genet.* 20: 349–354.

Selva, J., Leonard, C., Albert, M., Auger, J., & David, G. (1986). Genetic screening for artificial insemination by donor (AID). *Clin. Genet.* 29: 389–396.

Sommer, S. S., Cassady, J. D., Sobell, J. L., & Bottema, C. D. (1989). A novel method for detecting point mutations or polymorphisms and its application to population screening for carriers of phenylketonuria. *Mayo Clin. Proc.* 64: 1361–1372.

U.S. Congress, Office of Technology Assessment, Cystic Fibrosis and DNA Tests: Implications of Carrier Screening, OTA-BA-532. Washington, DC: U.S. Government Printing Office, August, 1992.

Uzych, L. (1986). Genetic testing and exclusionary practices in the workplace. *J. Public Health Policy Spring* 1986: 37–57.

Wapner, R. J., & Jackson, L. (1988). Chorionic villus sampling. *Clin. Obstet. Gynecol.* 31: 328–344.

Williams, C., Weber, L., Williamson, R., & Hjelm, M. (1988). Guthrie spots for DNA-based carrier testing in cystic fibrosis. *Lancet ii:* 693.

Probability

Mendel's use of mathematics to analyze the results of his experiments is frequently over-looked as an important contribution to biology. His application of mathematical reasoning to the analysis of data helped transform an observational and descriptive science into a quantitative and experimental one. When Mendel carried out his experiments, statistics and statistical methods were not highly developed or widely used. In analyzing the results of his crosses, Mendel converted the numbers of individuals with particular genotypes or phenotypes into ratios. From these ratios, he was able to deduce the mechanisms of inheritance.

As we now know, the ratios Mendel observed are the result of random segregation and assortment of genes into gametes during meiosis and their union at fertilization in random combinations. This randomness provides an element of chance in the outcome, and prevents us from making exact predictions. In counting pea seeds in the F2, we may expect three fourths of the seeds to be yellow, but we cannot be absolutely certain that the first seed in an unopened pod will be yellow. The rules of probability can, however, help us guess how often such an event will take place.

DEFINING PROBABILITY

Most people have an innate sense of probability that seems part of common sense. For example, almost everyone would agree with the idea that a January snowfall is more probable in Minneapolis than in Miami. Other aspects of probability also seem obvious. When a coin is flipped, the probable outcomes are heads or tails. In the birth of a child, we expect the outcome to be a boy or a girl.

Unfortunately, the use of intuition alone in matters of probability is not always reliable. For example, what would you say is the probability that in a crowd of 20 people, two individuals share the same birth date? Considering that there are 365 days in the year (excluding leap year), intuition may say that it is not very likely. In fact, in a group of 20, there is almost an even chance that two people share the same birthday. The probability of a shared birthday for groups of various sizes is as follows: for a group of 23, the probability is 51%; for a group of 30, it is 71%; for 40, it is 89%; and for 50, there is a 97% chance that two people will share the same birthday. We will not explore the mathematical reasoning behind this probability, but it is based on the fact that if one person can have any of the 365 days for his or her birthday, the second person can have any of the remaining 364 days, the third person can have any of the remaining 363 days, and so on.

From the preceding example, it should be clear that in order to be useful in genetics and in science, probability needs to be expressed in more quantitative terms. The use of a quantitative approach to probability allows us to assign a numerical value to the probability that a given event will occur, and prevents us from leaping to conclusions about the possible outcome of genetic crosses.

QUANTIFYING PROBABILITY

In quantifying probability, let us begin at the limits. If an event is certain to occur, it has a probability of 1; if the event is certain not to occur, then the probability is 0. In genetics as in most other areas, we usually deal with events that are a mixture of degrees of certainty. While it is certain that we will all die (a probability of 1), when we die is less certain and therefore must be assigned a probability somewhere between 0 and 1. Insurance companies spend a great deal of time and effort in attempting to determine such probabilities, although we may prefer not to think about them.

In general terms, we can express the probability (p) of an event as the proportion of times that such an event occurs (r) out of the number of times that the event can occur (n):

$$p = r/n$$

In other words, if an event occurs r times in n trials, the probability that the event will take place is r/n. This probability is somewhere between the limits of 0 and 1. If we toss a coin, it may land with heads up or tails up. The probability that it will land with heads up is

$$p = r/n = 1/2$$

Likewise, the probability of a child being a boy or a girl is 1/2. Other events have different probabilities. In a pair of dice, each die has six faces. When a die is thrown, the probability of any of the faces being up is

$$p = r/n = 1/6$$

In a deck of 52 cards, the probability of drawing any given card (the ace of spades for example) is

$$p = r/n = 1/52$$

In roulette, the wheel contains the numbers 1–36 plus 0 and 00. The probability of the ball landing on any number is therefore

$$p = r/n = 1/38$$

In a monohybrid cross, the probability that an offspring of the self-fertilized F1 pea plant will have a dominant phenotype is

$$p = 3/4$$

In considering probability, we must consider not only the probability of one type of outcome but also the probability of other outcomes. If an event has a probability of p, the probability of an alternative outcome is $q = 1-p$. In other words, the sum of the probability of p and q equals 1. In the preceding examples, the probability of drawing an ace of spades is 1/52; the probability of drawing another card is 51/52, which when added to 1/52 equals 1. In the monohybrid cross, the probability of a dominant phenotype is 3/4, and the probability of a recessive phenotype is 1/4. Since we are certain that the F2 will have either a dominant or recessive phenotype, adding the probabilities of both phenotypes (3/4 + 1/4 = 1) covers all the possible phenotypic combinations.

COMBINING PROBABILITIES

Two rules of probability are useful in analyzing genetics problems. The first, called the product rule, is used when we wish to calculate the probability of two or more independent events occurring at the same time. The second, called the sum rule, is used when two or more events are mutually exclusive, or are alternative events.

In the product rule, we are asking what is the probability that event A and event B will occur together. This rule can be summarized as follows: The probability that independent events will occur together is the product of their independent probabilities.

When a coin is tossed, the probability that it will be heads is 1/2, and the chance that it will be tails is 1/2. If we toss a coin four times, and it turns up heads each time, the probability that it will turn up tails on the fifth try is still 1/2. In other words, chance has no memory. The probability that two heterozygotes will have a child with cystic fibrosis is 1/4. This does not mean that if their first child has cystic fibrosis, they can be assured of having three unaffected children. It means that for each child, there is a 1 in 4 chance that it will have cystic fibrosis, no matter whether they have 1 child or 20 children. If they have four unaffected children, the chance that their fifth child will have cystic fibrosis is still 1/4.

If, on the other hand, we want to ask what is the probability that we can toss a coin four times and get heads each time, we use the product rule. Since each coin toss is an independent event, the probability of getting heads four times out of four is $1/2 \times 1/2 \times 1/2 \times 1/2 = (1/2)^4 = 1/16$. Similarly, the probability that heterozygous parents will have four children affected with cystic fibrosis is $1/4 \times 1/4 \times 1/4 \times 1/4 = 1/256$.

In using the sum rule, we are asking how often one or the other of two mutually exclusive events can occur. For example, in rolling a die, what is the probability of a three or a five coming up? Since on a single throw only one number can come up, it is impossible to get both numbers on a single throw. The probability of a three is 1/6, and the probability of a five is 1/6. If we want to know what is the probability of either a three **or** a five coming up on a single throw, we add the individual probabilities:

$$1/6 + 1/6 = 2/6 = 1/3$$

In considering the possible genotypic combinations in children of parents heterozygous for cystic fibrosis, what is the probability that a child will have either one or two copies of the dominant allele? The probability of being homozygous dominant (two copies) is 1/4, and the probability of being heterozygous (one copy) is 1/2. Therefore, the probability of being either heterozygous or homozygous dominant is 1/2 + 1/4 = 3/4. This is in fact the proportion of individuals with the dominant phenotype seen in the F2 of a monohybrid cross.

In applying probability to the analysis of genetic problems, first determine whether you want to know the probabilities of event A and event B, or the probability of event A or event B. If you want A and B, use the product rule and multiply the probabilities of A and B. If you want the probability of A or B, use the sum rule, and add the probability of event A to the probability of event B. For example, the frequency of albinism is about 1/10,000, and the frequency of cystic fibrosis is about 1/2000. If we want to know the probability of having both albinism and cystic fibrosis, we multiply the probabilities:

$$1/10,000 \times 1/2000 = 1/20,000,000$$

If we want to know the probability of having either albinism or cystic fibrosis, we add the probabilities:

$$1/10,000 + 1/2000 = 6/10,000 = 1/1666$$

As you can see, the probabilities are very different and reflect whether we are asking that both events occur, or that one or the other event will occur.

Answers to Selected Questions and Problems

Chapter 2

1. e
5. d
6. Homologous chromosomes are similar in that they contain alleles of the same gene at the same position along the length of the chromosome. One member of the pair is maternally derived and the other is paternally derived. The homologues may or may not carry the same allele of a gene.
8.

	Mitosis	Meiosis
Number of daughter cells produced	2	4
Number of chromosomes per daughter cell	$2n$	n
Number of cell divisions	1	2
Do chromosomes pair? (Y/N)	N	Y
Does crossing over occur? (Y/N)	N	Y
Can the daughter cells divide again? (Y/N)	Y	N
Do the chromosomes replicate before division? (Y/N)	Y	Y
Type of cell produced	SOMATIC	GAMETE

10. Cell furrowing involves the constriction of the cell membrane that causes the cell to eventually divide. It is associated with the process of cytokinesis, cytoplasmic division of the cell. If cytokinesis does not occur in mitosis, the cell will be left with a $2 \times 2n$ the number of chromosomes or $4n$ (tetraploid).
12. left to right: e, c, b, d

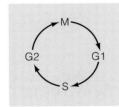

16. Cells undergo a cycle of events involving growth, DNA replication and division. Daughter cells undergo the same series of events. During S phase, DNA synthesis and chromosome replication occurs. During M, mitosis takes place.
20. **a.** mitosis **b.** meiosis I **c.** meiosis II
23. Chiasma (chiasmata)
25. Cells undergo a finite number of divisions before they die, this is the Hayflick limit. If a mutation occurs in a gene or genes that control this limit, the Hayflick limit may be reduced so that cells age and die earlier than normal.

Chapter 3

1. **a.** A gene is the fundamental unit of heredity. The gene encodes a specific gene product (i.e. a pigment involved in determining eye color). Alleles are alternate forms of a gene that may cause various phenotypic effects. For example, there may be a blue eye color allele, a brown eye color allele, and a green eye color

allele of a gene. The brown eye color allele may be the dominant allele where the blue and green alleles may be recessive. In many cases, such as with disease causing genes, the term "normal" allele (which gives the normal phenotype) is used in contrast to "mutant" allele (the allele that causes the disease). There may be several types of mutant alleles causing variable phenotypic effects associated with the disease.

   b. Genotype refers to the genetic constitution of the individual (*AaBb* or *aabb*). Notice that the genotype always includes at least two letters, each representing one allele of a gene pair in a diploid organism. A gamete would contain only one allele of each gene due to its haploid state (*Ab* or *ab*). Phenotype refers to an observable trait. For example, *Aa* (the genotype) will cause a normal pigmentation (the phenotype) in an individual where *aa* will cause albinism. Similarly *Bb* or *bb* (the genotypes) would give rise to brown hair versus blonde hair (the phenotypes) respectively.

   Phenotypes can refer to properties that are not observable to the naked eye. For example, in sickle cell anemia, heterozygotes do not "look" sick however, their red blood cells are abnormal in shape (slightly sickled) causing an abnormal cellular phenotype. This cellular phenotype is seen only with the use of a microscope.

   c. Dominance refers to a trait that is expressed in the heterozygous condition. Therefore, only one copy of a dominant allele needs to be present to express the phenotype. Recessiveness refers to a trait that is not expressed in the heterozygous condition. It is masked by the dominant allele. To express a recessive trait, two copies of the recessive allele must be present in the individual.

   d. Complete dominance occurs when a dominant allele completely masks the expression of a recessive allele. In pea plants, yellow seed color is dominant to green. In heterozygotes, the phenotype of the seeds is the same as in seeds homozygous for the yellow allele.

   Incomplete dominance occurs when the phenotype of the heterozygote is intermediate to the two homozygotes. In *Mirabilis*, a red flower crossed with a white flower will give a pink flower.

   In codominance, both alleles are fully expressed, as in heterozygotes (*MN*) for the MN blood group.

7. a. Homologous chromosomes are segregating, or gene pairs are segregating (*a* from *a*, *D* from *d*).
   b. Members of the *aa* gene pair are segregating independently from members of the *Dd* gene pair.
   c. 8
   d. 2
   e. *aaDd*
   f. albino, but not deaf
   g. metaphase of meiosis I
   h. 2 chromatids, 2 chromosomes

8. Punnett Square:

	AB	Ab	aB	ab
AB	AABB	AABb	AaBB	AaBb
Ab	AABb	AAbb	AaBb	Aabb
aB	AaBB	AaBb	aaBB	aaBb
ab	AaBb	Aabb	aaBb	aabb

phenotypes: A_B_ 9/16   *genotypes:* AABB 1/16
              A_bb 3/16   AABb 2/16
              aaB_ 3/16   AAbb 1/16
              aabb 1/16   AaBB 2/16
                                           AaBb 4/16
                                           Aabb 2/16
                                           aaBB 1/16
                                           aaBb 2/16
                                           aabb 1/16

Fork-line:
  phenotypes:

$$\begin{array}{l} \tfrac{3}{4}\,\text{A} \Big\langle \begin{array}{l} \tfrac{3}{4}\,\text{B} \rightarrow \tfrac{9}{16}\,\text{A\_B\_} \\ \tfrac{1}{4}\,\text{b} \rightarrow \tfrac{3}{16}\,\text{A\_bb} \end{array} \\ \tfrac{1}{4}\,\text{a} \Big\langle \begin{array}{l} \tfrac{3}{4}\,\text{B} \rightarrow \tfrac{3}{16}\,\text{aaB\_} \\ \tfrac{1}{4}\,\text{b} \rightarrow \tfrac{1}{16}\,\text{aabb} \end{array} \end{array}$$

  genotypes:

$$\begin{array}{l} \tfrac{1}{4}\,\text{AA} \Big\langle \begin{array}{l} \tfrac{1}{4}\,\text{BB} \rightarrow \tfrac{1}{16}\,\text{AABB} \\ \tfrac{2}{4}\,\text{Bb} \rightarrow \tfrac{2}{16}\,\text{AABb} \\ \tfrac{1}{4}\,\text{bb} \rightarrow \tfrac{1}{16}\,\text{AAbb} \end{array} \\[4pt] \tfrac{2}{4}\,\text{Aa} \Big\langle \begin{array}{l} \tfrac{1}{4}\,\text{BB} \rightarrow \tfrac{2}{16}\,\text{AaBB} \\ \tfrac{2}{4}\,\text{Bb} \rightarrow \tfrac{4}{16}\,\text{AaBb} \\ \tfrac{1}{4}\,\text{bb} \rightarrow \tfrac{2}{16}\,\text{Aabb} \end{array} \\[4pt] \tfrac{1}{4}\,\text{aa} \Big\langle \begin{array}{l} \tfrac{1}{4}\,\text{BB} \rightarrow \tfrac{1}{16}\,\text{aaBB} \\ \tfrac{2}{4}\,\text{Bb} \rightarrow \tfrac{2}{16}\,\text{aaBb} \\ \tfrac{1}{4}\,\text{bb} \rightarrow \tfrac{1}{16}\,\text{aabb} \end{array} \end{array}$$

11. Let *S* = smooth, and *s* = wrinkled and *Y* = yellow and *y* = green. The parents are: *SSYY* x *ssyy*. The F1 offspring are: *SsYy*.

12. a. Using the symbols from the problem above, the parents are: *SsYy* x *SsYy*.
    b. The genotypes of the F1 are: *SsYy*, *Ssyy*, *ssYy* and *ssyy*.

13. a. All F1 plants will be long-stemmed.
    b. Let *S* = long-stemmed and *s* = short-stemmed. The long-stemmed P1 genotype is *SS*, the short-stemmed P1 genotype is *ss*. The long-stemmed F1 genotype is: *Ss*.
    c. Approximately 225 long-stemmed and 75 short-stemmed.
    d. The expected genotypic ratio is: 1 *SS*:2 *Ss*: 1 *ss*

17. The P1 generation is: *FF* × *ff*. The F1 generation is: *Ff*. The mode of inheritance is incomplete dominance.

20. Possible genotypes of parents:
    a. *BbHH* × *BbHH* or *BbHH* × *BbHh*
    b. *BbHh* × *bbHh*
    c. *BbRr* × *bbrr*

22. ¾ for *A* × ½ for *b* × 1 for *C* = ⅜ for *A*, *b*, *C*

24. Since neither species produces progeny resembling a parent, simple dominance is ruled out. The species producing pink-flowered progeny from red and white (or very pale yellow) suggests incomplete dominance as a mode of inheritance. However, in the second species, the production of orange-colored progeny cannot be explained in this fashion. Orange would result from an equal production of red and yellow; instead in this case, codominance is suggested, with one parent producing bright red flowers and the other producing pale yellow flowers.

26. During meiotic prophase I, the replicated chromosomes synapse or pair with their homologues. These paired chromosomes align themselves on the equatorial plate during metaphase I. During anaphase I it is the homologues (each containing two chromatids) that separate from each other. There is no preordained orientation for this process–it is equally likely that a maternal or a paternal homologue will migrate to a given pole. This provides the basis for the law of random segregation. Independent assortment results from the fact that the polarity of one set of homologues has absolutely no influence on the orientation of a second set of homologues. For example, if the maternal homologue of chromosome 1 migrated to a certain pole, it will have no bearing on whether the maternal or paternal homologue of chromosome 2 migrates to that same pole.

28. a. unaffected female
    b. affected male
    c. affected female
    d. heterozygous female
    e. dizygotic (fraternal) twin boys

Chapter 4

2. d.
3. Autosomal dominant
5. a.

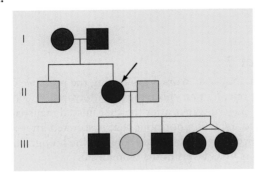

   b. The mode of inheritance is consistent with an autosomal dominant trait. Both of the proband's parents are affected. If this trait were recessive, all of their children would have to be affected (*aa* × *aa*

can only produce *aa* offspring). As we see in this pedigree, the brother of the proband is not affected indicating that this is a dominant trait. His genotype is *aa*, the proband's genotype is *AA* or *Aa* and both parents' genotype is most likely *Aa*.

   c. Since the proband's husband is unaffected, he is *aa*

8. a. This pedigree is consistent with autosomal recessive inheritance.
   b. If inheritance is autosomal recessive, the individual in question is heterozygous.

11. a. 50% chance for sons
    b. 50% for daughters

13. c. X-linked dominant
14. d. X-linked recessive
18. 20% of 90 = 18
20.

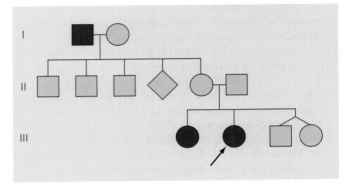

24. a. Seven phenotypes are possible: A, B, C, D, AB, AC, BC
    b. Ten genotypes are possible: *a/a, b/b, c/c, d/d, a/d, b/d, c/d, a/b, a/c, b/c*

27.

```
          4 map units      8 map units
     C--------------------A---------------------B
```

29. Due to the rarity of the disease, we assume the paternal grandfather is heterozygous for the gene responsible for Huntington disease. His son has a 1/2 chance of possessing the deleterious allele. In turn, should the father carry the HD allele, his son would have a 1/2 chance of inheriting it. Therefore, at present, the child has a 1/2 × 1/2 = 1/4 chance of having inherited the HD allele.

Chapter 5

2. a. Height in pea plants is determined by a single pair of genes with dominant and recessive alleles. Height in humans is determined by polygenes.
   b. For traits determined by polygenes, the offspring of matings between extreme phenotypes show a tendency to regress toward the mean phenotype in the population.

5. a. F1 genotype = $A'AB'B$, phenotype = height of 6 ft.

   b. $A'AB'B \times A'AB'B$

   ↓

Genotypes	Phenotypes
$A'A'B'B'$	7 ft
$A'A'B'B$	6 ft 6 in.
$A'A'BB$	6 ft
$A'AB'B'$	6 ft 6 in.
$A'AB'B$	6 ft
$A'ABB$	5 ft 6 in.
$AAB'B'$	6 ft
$AAB'B$	5 ft 6 in.
$AABB$	5 ft

6. Liability is caused by a number of genes acting in an additive fashion to produce the defect. If exposed to certain environmental conditions, the person above the threshold will most likely develop the disorder. The person below the threshold is not predisposed to the disorder and will most likely remain normal.

11. Relatives are used because the proportion of genes held in common by relatives is known.

13. b.

14. This suggests that environmental factors play a major role in the expression of the trait. However, since there is a significant concordance difference between MZ and DZ twins, this trait is also shown to be genetic.

19. a. The study included only men who were able to pass a physical exam which eliminated markedly obese indiviudals, and so the conclusions cannot be generalized beyond the group of men inducted into the armed forces.

    b. To design a better study, include MZ and DZ twin men and women, maybe even children. Include a cross section of various populations (ethnic groups, socioeconomic groups, weight classifications, etc. Control the diet so that it remains a constant. Another approach is to study MZ and DZ twins that were reared apart (and presumably in different environments), or adopted and natural children who where raised in the same household (same environment). There are other possible answers.

21. Heritability (H), would be zero.

24. The heritability difference observed between the ethnic groups for this trait cannot be compared because heritability measures variation within one population at the time of the study. Heritability cannot be used to estimate genetic variation between populations.

Chapter 6

1. a. Chemical treatment of chromosomes resulting in unique banding patterns

   b. Q banding with quinacrine and G banding with Giemsa

5. Condition 2 is most likely lethal. This condition involves a chromosomal aberration, trisomy. This has the potential for interfering with the action of all genes on the trisomic chromosome. Condition 1 involves an autosomal dominant lesion to a single gene, which is more likely to be tolerated by the organism.

7. Triploidy

11. Two or three possibilities should be considered. The child could be monosomic for the relevant chromosome. The child has the paternal copy carrying the allele for albinism (father is heterozygous) and a nondisjunction event resulted in failure to receive a chromosomal copy from the homozygous mother. The second possibility is that the maternal chromosome carries a small deletion, allowing the albinism to be expressed. The third possibility is that the child represents a new mutation, inheriting the albino allele and having the other by mutation. Since monosomy is lethal, either the second or third possibility seems likely.

12. In theory, the chances are 1/3.

15. The embryo will be tetraploid. Inhibition of centromere division results in nondisjunction of an entire chromosome set. After cytoplasmic division, some cytoplasm is lost in an inviable product lacking genetic material and the embryo develops from the tetraploid product.

19. Primary oocytes begin meiosis I before birth and do not complete meiosis until fertilization. Thus, these germ cells are exposed to years of environmental mutagens. In contrast, meiosis in males takes 48 days. Spermatogonial cells that suffer a mutation may never undergo meiosis.

23.

	Answers:
loss of a chromosome segment	deletion
extra copies of a chromosome segment	duplication
reversal in the order of a chromosome segment	inversion
segment moved to another chromosome	translocation

26. A fragile site appears as a gap or break near the tip of the X chromosome long arm. Affected males have long, narrow faces with protruding chins and large ears.

Chapter 7

4. Pregnancy tests work by detecting the presence of a hormone, human chorionic gonadotropin (hCG) as early as two weeks after the first missed menstrual period. hCG is a hormone that is secreted after a fertilized egg implants into the uterine lining. The test detects hCG levels in the woman's urine.

6. There are significant economic and social consequences associated with FAS, including the costs of surgery for facial reconstruction, treatment of learning disorders and mental retardation, and caring for institutionalized individuals.

   Prevention depends on the education of pregnant women and the early treatment of pregnant women with alcohol dependencies. Other answers are possible.

8. d.
9. A mutation causing the loss of the SRY, testosterone, or testosterone receptor gene function. Also a defect in the conversion from testosterone to DHT can cause the female external phenotype until puberty.
13. Female
14. Calico cats are heterozygous females where one allele encodes black fur color and the other allele encodes orange/yellow fur color. These alleles are X-linked. In these heterozygous females, one of the X chromosomes gets inactivated early in development leaving only one X-linked allele active in any patch of cells. Therefore, some patches of the cat are black and some are orange/yellow. This illustrates the fact that female mammals are mosaics. X-inactivation occurs in the homozygous females also.
16. Random inactivation in females, so the genes from both X chromosomes are active in the body as a whole

Chapter 8

2. Proteins are found in the nucleus. Proteins are complex molecules composed of 20 different amino acids, nucleic acids are composed of only four different nucleotides. Cells contain hundreds or thousands of different proteins, only two main types of nucleic acids.
4. Protease destroyed any small amounts of protein contaminants in the transforming extract. Similarly, treatment with RNAse destroyed any RNA present in the mixture.
6. The process is transformation, discovered by Frederick Griffith. The P bacteria contain genetic information that is still functional even though the cell has been heat-killed. However, it needs a live recipient host cell to accept its genetic information. When heat killed P and live D bacteria are injected together, the dead P bacteria can transfer its genetic information into the live D bacteria. The D bacteria are then "transformed" into P bacteria and can now cause polkadots.
8. Chargaff's Rule: A = T and C = G
   If A = 27%, then T must equal 27%
   If G = 23%, then C must equal 23%
   Base composition:
   A = 27%
   T = 27%
   C = 23%
   G = 23%
        100%
11. b.
13. c.
14. a. number of chains — 2
    b. polarity — opposite
    c. bases interior or exterior — interior
    d. sugar/phosphate interior or exterior — exterior
    e. which bases pair — A:T, G:C
    f. right or left handed helix — right

18. c.
20. One G − C base pair

Chapter 9

2. b.
4. a. rRNA — structural and functional part of a ribosome
   b. tRNA — transports amino acids to translation machinery. Contains an anticodon (nucleotide triplet) that recognizes the codon (its complement) on the mRNA
   c. mRNA — intermediate between DNA and proteins. Carries genetic information from the nucleus to the cytoplasm where translation occurs
8. 1. removal of introns: to generate a contiguous coding sequence that can make an amino acid chain
   2. addition of the 5′ cap: ribosome binding
   3. addition of the 3′ polyA tail: mRNA stability
11. Answer: 25%   Total length: 10kb
                    Coding region: 2.5kb
18. a. UAC
    b. ACC
    c. UC(A/G) or AG(A/C/U/G)
    d. GA(U/C/A/G) or AA(U/C)
19. Transcription: DNA, TATA box, RNA polymerase, nucleotides, pre-mRNA, enhancer. Translation: ribosomes, tRNA, amino acyl synthetase, A site, anticodon, amino acids
21. a. No
    b. Yes

Chapter 10

2. c.
5. Accumulation of one or more precursors may be detrimental. Overuse of an alternative minor pathway may result in the accumulation of toxic intermediates. Deficiency of an important product may occur. Other reactions may be blocked.
7. Let *H* = the mutant allele for hypercholesterolemia
           Let *h* = the normal allele
   Answer: *HH* heart attack as early as the age of 2, definite heart disease by age 20, death in most cases by 30. No functional LDL receptors produced.
           Hh heart attack in early 30's. Half the number of functional receptors are present and twice the normal levels of LDL.
           hh normal. Both copies of the gene are normal and can produce functional LDL receptors.
10. The disease: phenylketonuria (PKU), a deficiency of the enzyme phenylalanine hydroxylase. This enzyme is used to convert the substrate phenylalanine to the product tyrosine.
    phenotype: enhanced reflexes, convulsive seizures and mental retardation; also, they have lighter hair and skin color than siblings and other family members

15. Normal. Phenylketonuria and alkaptonuria are caused by mutations in different genes affecting different enzymes. The children will be normal in phenotype because they will carry one normal copy of the PKU gene, the phenylalanine hydroxylase gene, from the AKU parent. Similarly, the children will have one normal copy of the AKU gene, the homogentisic acid oxidase gene, from the PKU parent.

Let the mutant PKU gene be *p* and the normal gene be *P*. Let the mutant AKU gene be *a* and the normal gene be *A*. Both diseases are autosomal recessive.

parental genotypes:	*ppAA*	×	*PPaa*
	(PKU parent)		(AKU parent)
The genotypes of the children will be:			*PpAa*
			(normal)

19.

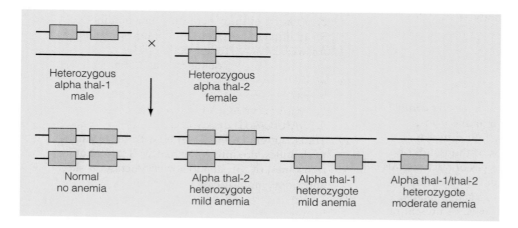

20. **a.** Buildup of substance A, no substance C
    b. Buildup of substance B, no substance C
    c. Buildup of substance B, as long as A is not limiting factor
    d. 1/2 the amount of C
21. **a.** yes. each will carry the normal gene for the other enzyme (individual 1 will be mutant for enzyme 1 but normal for enzyme 2. This is because enzyme 1 and 2 are encoded by two different genes)
    b. Let *D* = dominant mutation in enzyme 1, let normal allele = *d*
    Let *A* = dominant mutation in enzyme 2, let normal allele = *a*

    Ddaa × *ddAa*

offspring:	*DdAa*	mutation in enzyme 1 and 2, A buildup, no C
	Ddaa	mutation in enzyme 1, A buildup, no C
	ddAa	mutation in enzyme 2, B buildup, no C
	ddaa	no mutation, normal

    Ratio would be 1:2:1 for substance B buildup, no C: substance A buildup, no C: normal

22. Alleles for enzyme 1: *A* (dominant, 50% activity); *a* (recessive, 0% activity). Alleles for enzyme 2: *B* (dominant, 50% activity); *b* (recessive, 0% activity).

	Enzyme 1	Enzyme 2	Compound		
			A	B	C
1*AABB*	100	100	N	N	N
2*AaBB*	50	100	N	N	N
4*AaBb*	50	50	N	N	N
2*AABb*	100	50	N	N	N
1*AAbb*	100	0	N	B	L
2*Aabb*	50	0	N	B	L
1*aaBB*	0	100	B	L	L
2*aaBb*	0	50	B	L	L
1*aabb*	0	50	B	L	L

N, normal; B, buildup; L, less.

23. Drugs usually act on proteins. Different people have different forms of proteins. Different proteins are inherited as different alleles of a gene.
29. People have different abilities to smell and taste chemical compounds such as phenylthiocarbamide (PTC); some people are unable to smell skunk odors; different reactions to succinylcholine, a muscle

relaxant, and to primaquine, an antimalarial drug. Others are sensitive to the pesticide parathion.

Chapter 11

2. 245,000 births represent 490,000 copies of the achondroplasia gene, since each child carries two copies of the gene. The mutation rate is therefore 10/490,000 or 2×10^{-5} per generation.

4. a. A mutation caused by the addition or deletion of nucleotide residues from a coding portion of a gene destroys the triplet reading frame (any additions or deletions that alter the frame except multiples of 3 bp).
   b. A measure of the occurrence of mutations per individual per generation.
   c. Nucleotide sequences that increase in number through generations and are associated with disease phenotypes.

7. Missense-same
   Nonsense-shorter
   Sense-longer

12. a. See pedigree below
    b. No. Mutation occured in germ cell, not retinal cell
    c. Assuming that affected individuals are heterozygotes, the chance is 50%.

17. X-rays can cause mutations. Medical professionals need to be aware of the dose of X-rays that a patient receives. The benefits of the X-ray method need to outweigh the risks.

18. Muscular dystrophy is an X-linked disorder. A son receives an X chromosome from his mother and a Y chromosome from his father. In this case, the mother was a heterozygous carrier of muscular dystrophy, and passed the mutant gene to her son. The father's exposure to chemicals in the work place is unrelated to this condition in his son.

Chapter 12

1. d.
3. b.
4. b.
7. DNA may be cloned by inserting it into a plasmid or phage vector that can replicate in a host cell. DNA is prepared for cloning by using a restriction enzyme to cleave it into fragments with sticky ends. DNA can be joined to the cloning vector by using DNA ligase.

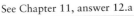
See Chapter 11, answer 12.a

9. A linear DNA segment, such as a human gene, cannot be replicated in a host cell and passed to progeny cells. Vectors are self replicating molecules which means that they contain specific DNA sequences that will cause them to be duplicated before cell division and then passed to progeny cells. Therefore, any DNA segment ligated to the vector will also be duplicated and many copies of the DNA segment can be generated in this manner.

10. *Eco*R1: 2 kb, 11.5 kb, 10 kb
    *Hin*dIII/*Pst*1: 7 kb, 3 kb, 7.5 kb, 6 kb
    *Eco*R1/*Hin*dIII/*Pst*1: 2 kb, 5 kb, 3 kb, 3.5 kb, 4 kb, 6 kb

15. a.
17. DNA derived from individuals with sickle cell anemia will lack one fragment contained in the DNA from normal individuals, and in addition, there will be a large (uncleaved) fragment not seen in normal DNA.

Chapter 13

2. c.
3. c.
8. Restriction fragment length polymorphism. A restriction fragment is a DNA segment that has been cut with a restriction enzyme. The length of the DNA refers to the number of nucleotides it contains, as measured by Southern blot analysis. Polymorphism means that alleles that differ in the length of the DNA fragment are common in the human population.

12. Potential dangers could include the escape of genetically altered organisms, and undesirable side effects of gene therapy. Gene therapy, as one form of recombinant DNA technology raises a number of ethical questions about germ-line therapy, in which altered genomes are transmitted to offspring, and use of gene therapy for cosmetic reasons, such as altering height.
    Some of these potential dangers can be reduced or eliminated by research, while others require public policy decisions.

13. Conclusion: Evidence excludes suspect #1 and includes suspect #2. If Suspect #1 has a different DNA pattern than the evidence, the suspect is excluded from having committed the crime. But that evidence sample of DNA cannot possibly be Suspect #1's. Suspect #2's DNA corresponds perfectly at each of the four places of variation on the human chromosomes examined.

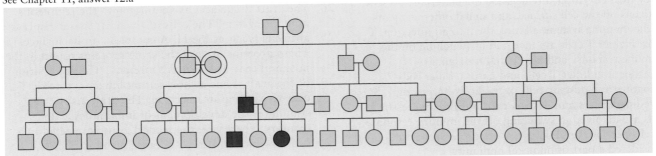

14. The probe could be any DNA fragment on chromosome 21. The fluorescent probe should bind to three chromosomes in a Down syndrome individual rather than the normal two.

Chapter 14

1. a.
3. d.
6. The inheritance is dominant because only one mutant allele causes the predisposition to retinoblastoma. However, the second allele must also be mutated in at least one eye cell to produce the disease. Therefore, the expression of retinoblastoma is recessive.
8. a. The father did not acquire a second mutation in the *Rb* gene in any of his eye cells. He remained heterozygous and therefore no cancer resulted.
   b. 50%
   c. Their son can be tested frequently for eye tumors. Early detection and treatment means that their son may be able to retain normal eyesight.
10. c.
13. A proto-oncogene is a gene normally found in the human genome that promotes cell division. An oncogene is a mutated proto-oncogene that promotes uncontrolled cell division that leads to cancer.
17. Cigarette smoking, dietary fat vs. dietary fiber, asbestos, vinyl chloride, etc.
18. Conditions a and d would produce cancer. The loss of function of a tumor suppressor gene would allow cell growth to go unchecked. The overexpression of a proto-oncogene would promote more cell division than normal.
25. Diet is suspected as the cause in both cases. When Japanese move to the U.S. and adopt an American diet, the rate of breast cancer goes up, but the rate of colon cancer goes down. The reverse is also the case, where the Japanese diet in Japan predisposes to colon cancer, but not to breast cancer.

Chapter 15

1. Phagocytes (macrophages and neutrophils): recognize nonself organisms and cellular debris and engulf it. They then enzymatically degrade the materials, present antigens on the cell surface, and signal other elements of the immune system, alerting them to an infection. T cells: Helper T cells are involved in switching on the immune response and are crucial for signal transduction: killer T cells and natural killer cells recognize and destroy cells of the body harboring invading viruses and destroy cancer cells; suppressor T cells are involved in switching off the immune response. B cells produce antibodies but are more properly considered a part of humoral immunity.

3. helper T cells–activate B cells to produce antibodies
   supressor T cells–stops the immune response of B and T cells
   cytotoxic T cells–targets and destroys infected cells
5. Vaccines aim to inject antigens to induce a primary immune response and the production of memory cells. A second injection (booster) of antigens then creates a secondary response to increase the number of memory cells. The antigen is often a killed or weakened strain of the microorganism. Vaccination using a weakened strain may cause some symptoms of illness but the antigen will not cause a life threatening disease.
9. a. The mother is Rh⁻. She will produce antibodies against the Rh antigen if her fetus is Rh^+. This happens when blood from the fetus enters the maternal circulation.
   b. the mother already has circulating antibodies against the Rh protein from her first Rh^+ child. She can mount a greater immune response against the second Rh^+ child by generating a large number of antibodies.
12. For any antibody class, the genes encoding the single heavy chain and the two light chains are complex. There are several hundred V or variable region gene segments, a few (5-10) J or joining region gene segments, and a single segment encoding the C or constant region. Recombination results in the juxtaposition of V, J, and C gene segments to form a functional gene. This process of recombination can therefore result in a large variety of functional antibody genes within the body's population of B cells.
14. Autosomal codominant inheritance
17. The infant can only be type O and the genotype is therefore I^O/I^O. The man on trial can only be type AB and his genotype must therefore be I^A/I^B. The mother must be type A since she only possesses antibodies to type B blood. Since there is no question of maternity involved here, she contributed an I^O allele to the infant. Therefore the mother's genotype is I^A/I^O. The court can dismiss the suit because the father of the infant had to possess an I^O allele, and the man on trial doesn't carry one.
18. The son would be able to receive blood from both his parents. His genotype would be I^A/I^B and therefore he would express both the A and the B antigens. Therefore, he would not have the antibodies against the A or the B antigen.
20. The graft tissue acts as an antigen to the recipient's immune system. The cells of the graft tissue displays antigens (such as the HLA complex) on its surface. These proteins are not the same proteins found in the host and they are seen as "foreign". The recognition of the graft tissue as non-self mobilizes cytotoxic T cells to cause graft rejection. The graft also stimulates the development of immunological memory. Therefore, the second graft is rejected in a shorter amount of time because memory T cells, memory B cells, and antibodies specific to the graft tissue are already circulating in the blood.

24. Blood type B individuals do not make the A antigen. Therefore, the patient will generate antibodies against the A antigen because it is "foreign". The antibodies will bind to the transfused red blood cells that contain the A antigen causing them to clump and burst. Thus, hemoglobin, the major protein in the red blood cells, is released into the blood which crystallizes in the kidney. This can cause kidney failure.

26. They do not produce an immune response in most people.

28. He has X-linked agammaglobulinemia (XLA). Affected individuals tend to be boys who are highly susceptible to bacterial infections. The antibody-mediated immunity is not functioning. B cells may be absent or immature B cells may be unable to mature and produce antibodies. The cell-mediated immunity is normal.

Chapter 16

1. If the case involves a single gene, pedigree analysis and linkage and segregation studies, including the use of RFLP markers and other recombinant DNA technologies are the most appropriate methods. In polygenic cases, twin studies to determine concordance and heritability are common. Recently, geneticists have studied the offspring of twins to overcome certain problems such as the limitations of small ample sizes. Also, recombinant DNA studies have begun with twins to locate polygenic genes.

3. *Drosophila* has many advantages for the study of behavior. Mutagenesis and screening for behavior mutants allows the recovery of mutations that affect many forms of behavior. The ability to perform genetic crosses and recover large numbers of progeny over a short period of time also enhances the genetic analysis of behavior. This organism can serve as a model for human behavior, because cells of the nervous system in both *Drosophila* and humans use similar mechanisms to transmit impulses and store information.

7. From a single study, perfect pitch (MIM/OMIM 159300) can be classified as a familial trait; its pattern of inheritance strongly suggests that it is an autosomal dominant trait. To confirm this, more studies covering several generations in families with perfect pitch would be desirable.

9. The heritability of Alzheimer disease, a multifactorial disorder, cannot be established due to interactions between genetic and environmental factors. Less than 50% of Alzheimer cases can be attributed to genetic causes, indicating that the environment plays a large role in the development of this disease. Other non-genetic factors may involve aluminum and prions.

11. His monozygotic twin is at high risk for the disease. Studies show that there is a 46% concordance where identical twins both develop schizophrenia. Studies also indicate that MZ twins raised apart display the same level of concordance as those raised together.

15. c.

Chapter 17

1.
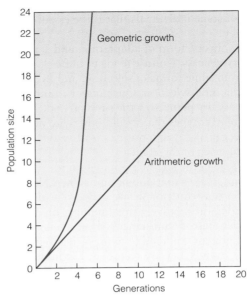

3. a. population: Local groups of individuals occupying a given space at a given time.
   b. gene pool: The set of genetic information carried by a population.
   c. allele frequency: The frequency of occurrence of particular alleles in the gene pool of a population.
   d. genotype frequency: The frequency of occurrence of particular genotypes among the individuals of a population.

7. a. Children: $M = 0.5$, $N = 0.5$. Adults: $M = 0.5$, $N = 0.5$.
   b. Yes, allelic frequencies are unchanged.
   c. No. The genotypic frequencies are changing within each generation.

10. In the case of multiple alleles, such as the ABO blood system, the greatest number of alleles that can be carried by an individual is two. A population, however, has the capacity to carry many different alleles for any given locus.

13. O = 0.04

Chapter 18

1. All polymorphisms are mutations. When a mutation resulting in a particular allele is present in at least 1% of the population, it is termed a polymorphism. This percentage is chosen because it is clearly higher than can be accounted for by mutation alone. Because such genetic variations cannot be accounted for by mutation alone, other forces such as natural selection must be active. The study of such variation is important in understanding the process of evolution.

4. No it is not an accurate description. Natural selection depends on fitness, the ability of a given genotype to survive and reproduce. It is the differential reproduction of some individuals that is the essence of natural selection.

6. Culture represents one form of adaptation, and is itself an evolutionary process. Culture tends to buffer populations from some forms of selection, and so alters the process of natural selection. Our present culture is affecting selection by permitting the survival and reproduction of genotypes that would otherwise not reproduce. Over the long run, this can lead to an increase in frequency of these alleles.

8. A drastic decrease in the size of a population for whatever reason, makes that population vulnerable to genetic drift. Such a reduction in size can also lead to inbreeding among the relatively small number of survivors. Under these circumstances, those alleles which are still present will tend to express themselves phenotypically, even if they are recessive. The limited variability that is characteristic of small populations plays a role in species extinction.

10. No, it is difficult to eliminate deleterious recessive alleles from the gene pool of a large population by natural selection. These alleles survive in populations because the majority of them are carried in the heterozygous condition and are not expressed phenotypically. As a result, natural selection cannot act on these alleles, and they remain hidden in the gene pool.

12. c.

Chapter 19

1. Prenatal screening: used to diagnose genetic defects and birth defects in the unborn.
 Newborn screening: used to diagnose genetic defects in neonatal and early infants stages of life.
 Carrier screening: used to identify phenotypically normal individuals who carry deleterious recessive traits.
 Occupational screening: used to identify those who are genetically susceptible to agents found in the workplace.
 Reproductive screening: used for identification of gamete donors carrying genetic or chromosomal defects.

2. The first recommendation would be for another α-fetoprotein test to determine whether the first test may have been a false positive. If the second test is positive, amniocentesis is recommended. If the results from this procedure are ambiguous, fetoscopy may be considered, balancing the risks of this procedure against the benefits.

5. Since NF is a dominant trait, and there is no evidence of this trait on either side of the family, the birth of the NF child is probably the result of a spontaneous mutation. However, because the phenotype of NF is so variable and some cases may escape detection, members of the immediate family should be examined by a specialist to rule out familial transmission. If it is confirmed that there is no family history, the recurrence risk is low.

8. If we let the allele causing PKU be p and the normal allele be P, then the sister and the brother of the woman and the man, respectively must have been pp. In order to produce these affected individuals, all four grandparents must have been heterozygous normal. The pedigree can be summarized as:

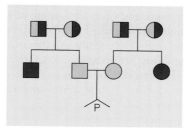

The only way the man and woman can have a PKU child is if both of them are heterozygotes, since they do not have the disease. Both the grandparental matings are simple Mendelian monohybrid crosses expected to produce progeny in the following proportions:

¼ – PP	–	Normal
½ – Pp	–	Normal
¼ – pp	–	PKU

We know that the man and the woman are normal, so the probability of either being a heterozygote is ⅔, because within the P_ classes, ⅔ are Pp and ⅓ are PP. The probability of both the man and the woman being heterozygotes is ⅔ × ⅔ = 4/9. If they are both heterozygous, then one-quarter of their children would have PKU. This means that the probability that their first child will have PKU is ¼, and the total probability of their being heterozygous and of their first child having PKU is 4/9 × ¼ = 4/36 or 1/9.

Glossary

5′ flanking region A nucleotide region adjacent to the 5′ end of a gene that contains regulatory sequences.

ABO groups Three alleles of a gene on human chromosome 9 that specify the presence and/or identity of certain molecules on the surface of red blood cells.

Achondroplasia An autosomal dominant trait associated with dwarfism.

Acquired immunodeficiency syndrome (AIDS) An infectious disease characterized by infection with human immunodeficiency virus (HIV) and the loss of T4 helper lymphocytes, causing an inability to mount an immune response.

Acrocentric A chromosome with the centromere placed very close to, but not at, one end.

Adenine and guanine Purine nitrogenous bases found in nucleic acids.

Affect Pertaining to mood or feelings.

Alkaptonuria A relatively benign autosomal recessive genetic disorder associated with the excretion of high levels of homogentisic acid.

Allele One of the possible alternative forms of a gene, usually distinguished from other alleles by its phenotypic effects.

Allele expansion Increase in gene size caused by an increase in the number of trinucleotide sequences.

Allele frequency The percentage of all alleles of a given gene that are represented by a specific allele.

Allergens Antigens that provoke an immune response.

Alpha thalassemia Genetic disorder associated with an imbalance in the ratio of alpha and beta globin caused by reduced or absent synthesis of alpha globin.

Alzheimer disease A heterogeneous condition associated with the development of brain lesions, personality changes, and degeneration of intellect. Genetic forms are associated with loci on chromosomes 14, 19, and 21.

Ames test A bioassay developed by Bruce Ames and his colleagues for identifying mutagenic compounds.

Amino group A chemical group (NH_2) found in amino acids and at one end of a polypeptide chain.

Amniocentesis A method of sampling the fluid surrounding the developing fetus by inserting a hollow needle and withdrawing suspended fetal cells and fluid. Used in diagnosing fetal genetic and developmental disorders; usually performed in the 16th week of pregnancy.

Anaphase A stage in mitosis during which the centromeres split and the daughter chromosomes begin to separate.

Androgen insensitivity An X-linked genetic trait that causes XY individuals to develop into phenotypic females.

Anaphylaxis A severe allergic response in which histamine is released into the circulatory system.

Aneuploidy A chromosome number that is not an exact multiple of the haploid set.

Ankylosing spondylitis An autoimmune disease that produces an arthritic condition of the spine; associated with the HLA allele B27.

Anthropological genetics The union of population genetics and anthropology to study the effects of culture on gene frequencies.

Antibody A class of proteins produced by plasma cells that couple or bind specifically to the class of proteins that stimulate the immune response.

Antibody-mediated immunity Immune reaction that protects primarily against invading viruses and bacteria by means of antibodies produced by plasma cells.

Anticipation Onset of a genetic disorder at earlier ages and with increasing severity in successive generations.

Anticodon loop The region of a tRNA molecule that contains the three-base sequence (known as the anticodon) that pairs with a complementary sequence (known as a codon) in an mRNA molecule.

Antigen A foreign molecule or cell that stimulates the production of antibodies.

Antigenic determinant The site on an antigen to which an antibody binds, forming an antigen-antibody complex.

Assortative mating Reproduction in which mate selection is not at random but instead is based on physical, cultural, or religious grounds.

Assortment The random distribution of members of homologous chromosome pairs during meiosis.

Atherosclerosis Arterial disease associated with deposition of plaques on the inner surface of blood vessels.

Autosomes Chromosomes other than the sex chromosomes. In humans, chromosomes 1-22 are autosomes.

Background radiation Radiation in the environment that contributes to radiation exposure.

B cells White blood cells that originate in bone marrow and mature in the bone marrow.

Bacteriophage A virus that infects bacterial cells. Genetically modified bacteriophages are used as cloning vectors.

Beta thalassemia Genetic disorder associated with an imbalance in the ratio of alpha and beta globin caused by reduced or absent synthesis of beta globin.

Bipolar disorder An emotional disorder characterized by mood swings that vary between manic activity and depression.

Blastocyst The developmental stage at which the embryo implants into the uterine wall.

Blood type One of the classes into which blood can be separated based on the presence or absence of certain antigens.

B-memory cells   Long-lived B cells produced after exposure to an antigen that play an important role in secondary immunity.

Camptodactyly   A dominant human genetic trait that is expressed as immobile, bent little fingers.

Cap   A modified base (guanine nucleotide) that is attached to the 5′ end of eukaryotic mRNA molecules.

Carboxyl group   A chemical group (COOH) found in amino acids and at one end of a polypeptide chain.

Caretaker gene   A gene that helps maintain the integrity of the genome, for example, DNA repair genes.

Cell cycle   The sequence of events that takes place between successive mitotic divisions.

Cell furrow   A constriction of the cell membrane that forms at the point of cytoplasmic cleavage during cell division.

Cell-mediated immunity   Immune reaction mediated by T cells that is directed against body cells that have been infected by viruses or bacteria.

Centromere   A region of a chromosome to which fibers attach during cell division. Location of a centromere gives a chromosome its characteristic shape.

Charcot-Marie-Tooth disease   A heritable form of progressive muscle weakness and atrophy. One form, CMT-1, can be produced by mutation at any of three loci.

Chiasmata   (singular: chiasma) The crossing of nonsister chromatid strands seen in the first meiotic prophase. Chiasmata represent the structural evidence for crossing over.

Chorion   A membrane outside and surrounding the amnion, from which projections known as villi extend to the uterine wall, forming the placenta.

Chorionic villus sampling (CVS)   A method of sampling fetal chorionic cells by inserting a catheter through the vagina or abdominal wall into the uterus. Used in diagnosing biochemical and cytogenetic defects in the embryo. Usually performed in the 8th or 9th week of pregnancy.

Chromatid   One of the strands of a longitudinally divided chromosome, joined by a single centromere to its sister chromatid.

Chromatin   The component material of chromosomes, visible as clumps or threads in the nuclei examined under a microscope.

Chromosome theory of inheritance   The theory that genes are carried on chromosomes and that the behavior of chromosomes during meiosis is the physical explanation for Mendel's observations on the segregation and independent assortment of genes.

Chromosomes   The thread-like structures in the nucleus that carry genetic information.

Clinodactyly   An autosomal dominant trait that produces a bent finger.

Clones   Genetically identical organisms, cells, or molecules all derived from a single ancestor. Cloning is the method used to produce such clones.

Codominance   Full phenotypic expression of both members of a gene pair in the heterozygous condition.

Codon   A triplet of bases in messenger RNA that encodes the information for the insertion of a specific amino acid in a protein.

Color blindness   Defective color vision caused by reduction or absence of visual pigments. There are three forms: red, green, and blue blindness.

Complement system   A chemical defense system that kills microorganisms directly, supplements the inflammatory response, and works with (complements) the immune system.

Concordance   Agreement between traits exhibited by both members of a twin pair.

Consanguineous matings   Matings between two individuals who share a common ancestor in the preceding two or three generations.

Continuous variation   A distribution of phenotypic characters from one extreme to another in an overlapping, or continuous, fashion.

Correlation coefficient   A measure of the degree to which variables vary together.

Covalent bond   A chemical bond that results from electron sharing between atoms. Covalent bonds are formed and broken during chemical reactions.

Cri-du-chat syndrome   A deletion of the short arm of chromosome 5 associated with an array of congenital malformations, the most characteristic of which is an infant cry that resembles a mewing cat.

Crossing over   The process of exchanging parts between homologous chromosomes during meiosis; produces new combinations of genetic information.

C-terminus   The end of a polypeptide or protein that has a free carboxyl group.

Cystic fibrosis   A fatal recessive genetic disorder, common in the U.S. white population, associated with abnormal secretions of the exocrine glands.

Cytokinesis   The process of cytoplasmic division that accompanies cell division.

Cytosine, thymine, and uracil   Pyrimidine nitrogenous bases found in nucleic acids.

Cytoskeleton   A system of protein microfilaments and microtubules that allows a cell to have a characteristic shape.

Deletion   A chromosomal aberration in which a segment of a chromosome is deleted or missing.

Deoxyribonucleic acid (DNA)   A molecule consisting of antiparallel strands of polynucleotides that is the primary carrier of genetic information.

Deoxyribose and ribose   Pentose sugars found in nucleic acids. Deoxyribose is found in DNA, ribose in RNA.

Dermatoglyphics   The study of the skin ridges on the fingers, palms, toes, and soles.

Dihybrid cross   A mating between two individuals who are heterozygous at two loci (e.g., $BbCc \times BbCc$).

Diploid   The condition in which each chromosome is represented twice, as a member of a homologous pair.

Discontinuous variation   Phenotypes that fall into two or more distinct, nonoverlapping classes.

Dispermy   Fertilization of a haploid egg by two haploid sperm, forming a triploid zygote.

Dizygotic (DZ) twins   Twins derived from two separate and nearly simultaneous fertilization events, each involving one egg and one sperm. Such twins share, on average, 50% of their genes.

DNA fingerprint   A pattern of restriction fragments that is unique to an individual.

DNA polymerase   An enzyme that catalyzes the synthesis of DNA using a template DNA strand and nucleotides.

DNA replication   The process of DNA synthesis.

DNA restriction fragment   A segment of a longer DNA molecule produced by the action of a restriction endonuclease.

DNA sequencing   A technique for determining the nucleotide sequence of a fragment of DNA.

Dominant   The trait expressed in the F1 (or heterozygous) condition.

Dosage compensation   A mechanism that regulates the expression of sex-linked gene products.

Dosimetry   The process of measuring radiation.

Duplication   A chromosomal aberration in which a segment of a chromosome is repeated and therefore is present in more than one copy within the chromosome.

Ecogenetics   A branch of genetics that studies genetic traits related to response to environmental agents.

Endoplasmic reticulum (ER) A system of cytoplasmic membranes arranged into sheets and channels that functions in transport of gene products.

Environmental variance The phenotypic variance of a trait in a population that is attributed to exposure to differences in the environment by members of the population.

Epidemiology The study of the factors that control the presence, absence, or frequency of a disease.

Essential amino acids Amino acids that cannot be synthesized in the body and must be supplied in the diet.

Eugenics The attempt to improve the human species by selective breeding.

Eukaryote An organism composed of one or more cells that contain membrane-bound nuclei and that undergo mitosis and meiosis.

Exons DNA sequences that are transcribed and joined to other exons during mRNA processing, and are translated into the amino acid sequence of a protein.

Expressivity The range of phenotypes resulting from a given genotype.

Familial adenomatus polyposis (FAP) An autosomal dominant condition associated with the development of growths known as polyps in the colon. These polyps often develop into malignant growths, causing cancer of the colon and/or rectum.

Familial fructose intolerance An autosomal recessive trait caused by a mutation in a gene encoding an enzyme of fructose metabolism. If left untreated by dietary restrictions, it can result in malnutrition, growth retardation, and death.

Familial hypercholesterolemia An autosomal dominant genetic condition associated with defective or absent LDL receptors, which function in cholesterol metabolism. Affected individuals are at increased risk for cardiovascular disease.

Fertilization The fusion of two gametes to produce a zygote.

Fetal alcohol syndrome A constellation of birth defects caused by maternal drinking during pregnancy.

Fitness A measure of the relative survival and reproductive success of a given individual or genotype.

Fluorescent in situ hybridization (FISH) A method of mapping DNA sequences to metaphase chromosomes using probes labeled with fluorescent dyes.

Founder effects Gene frequencies established by chance in a population that is started by a small number of individuals (perhaps only a fertilized female).

Fragile X An X chromosome that carries a nonstaining gap, or break, at band q27. Associated with mental retardation in hemizygous males.

Frameshift mutations Mutational events in which one to three bases are added to or removed from DNA, causing a shift in the codon reading frame.

Fructosuria An autosomal recessive condition associated with the inability to metabolize the sugar fructose, which accumulates in the blood and urine.

Galactosemia An autosomal recessive trait associated with the inability to metabolize the sugar galactose. Left untreated, high levels of galactose-1-phosphate accumulate, causing cataracts and mental retardation.

Gamete A haploid reproductive cell, such as the egg or sperm.

Gatekeeper gene Genes that regulate cell growth and passage through the cell cycle, for example, a tumor suppressor gene.

Gene The fundamental unit of heredity.

Gene pool The set of genetic information carried by the members of a sexually reproducing organism.

Genetic counseling Process by which genetic counselors analyze the genetic risk within families and present the family with available options to avoid or reduce risks.

Genetic drift The random fluctuations of gene frequencies from generation to generation that take place in small populations.

Genetic equilibrium The situation when the frequency of alleles for a given gene remain constant from generation to generation.

Genetic goitrous cretinism A hereditary disorder in which the failure to synthesize a needed hormone produces physical and mental abnormalities.

Genetic library In recombinant DNA technology, a collection of clones that contains all the genetic information in an individual. Also known as a gene bank.

Genetic map The arrangement and distance between genes on a chromosome, deduced from studies of genetic recombination.

Genetics The scientific study of heredity.

Genetic screening The systematic search for individuals of certain genotypes.

Genetic variance The phenotypic variance of a trait in a population that is attributed to genotypic differences.

Genome The set of genes carried by an individual.

Genomic imprinting Phenomenon in which the expression of a gene depends on whether it is inherited from the mother or the father. Also known as genetic or parental imprinting.

Genotype The specific genetic constitution of an organism.

Golgi apparatus Membranous organelles composed of a series of flattened sacs. They sort, modify, and package proteins synthesized in the ER.

Haploid The condition in which each chromosome is represented once, in an unpaired condition.

Haplotype A set of closely linked genes that tend to be inherited together, as with the HLA complex.

Hardy-Weinberg Law The statement that allele frequencies and genotype frequencies will remain constant from generation to generation when the population meets certain assumptions.

Helper T cells A type of lymphocyte that stimulates the production of antibodies by B cells when an antigen is present.

Hemizygous A gene present in a single dose on the X chromosome that is expressed in males in both the recessive and dominant condition.

Hemoglobinopathies Disorders of hemoglobin synthesis and function.

Hemoglobin variants Alpha and beta globins with variant amino acid sequences.

Hemolytic disease of the newborn (HDN) A condition that results from Rh incompatibility and is characterized by jaundice, anemia, and an enlarged liver and spleen. Also known as erythroblastosis fetalis.

Hereditary nonpolyposis colon cancer (HNPCC) A form of colon cancer associated with genetic instability of microsatellites.

Heritability An expression of how much of the observed variation in a phenotype is due to differences in genotype.

Heterogametic The production of gametes that contain different kinds of sex chromosomes. In humans, males produce gametes that contain X or Y chromosomes.

Heterozygous Carrying two different alleles for one or more genes.

Histones Small DNA-binding proteins that function in the coiling of DNA to produce the structure of chromosomes.

Hominid A member of the family Hominidae, which includes bipedal primates such as *Homo sapiens*.

Hominoid A member of the primate superfamily Hominoidea, including the gibbons, great apes, and humans.

Homogametic The production of gametes that contain only one kind of sex chromosome. In humans, all gametes produced by females contain only an X chromosome.

Homologues Members of a chromosome pair.

Homozygous Having identical alleles for one or more genes.

Huntington disease   An autosomal dominant genetic disorder characterized by involuntary movements of the limbs, mental deterioration, and death within 15 years of onset. Symptoms appear between 30 and 50 years of age.

Hydrogen bond   A weak chemical bonding force that holds polynucleotide chains together in DNA.

Hypertension   Elevated blood pressure, consistently above 140/90 mm.

Hypophosphatemia   An X-linked dominant disorder. Those affected have low phosphate levels in blood, and skeletal deformities.

Immunoglobulins   The five classes of proteins to which antibodies belong.

Inborn error of metabolism   The concept advanced by Archibald Garrod that many genetic traits are the result of alterations in biochemical pathways.

Inbreeding   Production of offspring by related parents.

Incest   Sexual relations between parents and children or between brothers and sisters.

Incomplete dominance   Failure of a dominant phenotype to be expressed in the heterozygous condition. Such heterozygotes have a phenotype that is intermediate between those of the homozygous forms.

Independent assortment   The random distribution of genes into gametes during meiosis.

Inflammatory response   The body's reaction to invading microorganisms.

Inner cell mass   A cluster of cells in the blastocyst that gives rise to the body of the embryo.

Intelligence quotient (IQ)   A score derived from standardized tests that is calculated by dividing the individual's mental age (determined by the test) by his or her chronological age, and multiplying the quotient by 100.

Interphase   The period of time in the cell cycle between mitotic divisions.

Introns   Sequences present in some genes that are transcribed, but removed during processing, and therefore are not present in mRNA.

Inversion   A chromosomal aberration in which the order of a chromosome segment has been rotated 180 degrees from its usual orientation.

***In vitro* fertilization (IVF)**   A procedure in which gametes are fertilized in a dish in the laboratory, and the resulting zygote is implanted in the uterus for development.

Ionizing radiation   Radiation that produces ions during interaction with other matter, including molecules in cells.

Karyotype   The chromosome complement of a cell line or a person, photographed at metaphase and arranged in a standard sequence.

Killer T cells   T cells that destroy body cells infected by viruses or bacteria. Can also directly attack viruses, bacteria, cancer cells, and cells of transplanted organs.

Klinefelter syndrome   Aneuploidy of the sex chromosomes resulting in a male with an XXY chromosome constitution.

Leptin   A hormone produced by fat cells that signals the brain and ovary.

Lesch-Nyhan syndrome   An X-linked recessive condition associated with a defect in purine metabolism that causes an overproduction of uric acid.

Leukemia   A form of cancer associated with uncontrolled growth of leukocytes (white blood cells) or their precursors.

Linkage   A condition in which two or more genes do not show independent assortment; rather, they tend to be inherited together. Such genes are located on the same chromosome. By measuring the degree of recombination between such genes, the distance between them can be determined.

Lipoproteins   Particles with protein and phospholipid coats that transport cholesterol and other lipids in the bloodstream.

Locus   The position occupied by a gene on a chromosome.

Lod (log of the odds) method   A probability technique used to determine whether genes are linked.

Lod score   The ratio of the probability that two loci are linked to the probability that they are not linked, expressed as the $\log_{10}$. Scores of three or more are taken as establishing linkage.

Lymphocytes   White blood cells that arise in bone marrow and mediate the immune response.

Lyon hypothesis   The proposal that dosage compensation in mammalian females is accomplished by the random inactivation of one of the two X chromosomes.

Lysosomes   Membrane-enclosed organelles containing digestive enzymes.

Macrophages   Large, white blood cells that are phagocytic and involved in mounting an immune response.

Marfan syndrome   An autosomal dominant genetic disorder that affects the skeletal system, the cardiovascular system, and the eyes.

Mast cells   Cells that synthesize and release histamine during an allergic response, or during an inflammatory response.

Meiosis   The process of cell division during which one cycle of chromosome replication is followed by two successive cell divisions to produce four haploid cells.

Messenger RNA (mRNA)   A single-stranded complementary copy of the base sequence in a DNA molecule that constitutes a gene.

Metabolism   The sum of all biochemical reactions by which living organisms generate and use energy.

Metacentric   A chromosome with a centrally placed centromere.

Metaphase   A stage in mitosis during which the chromosomes move and arrange themselves near the middle of the cell.

Metaphase plate   The cluster of chromosomes aligned at the equator of the cell during mitosis.

Millirem   Each rem is equal to 1000 millirems.

Missense mutation   A mutation that causes the substitution of one amino acid for another in a protein.

Mitochondria   (singular: mitochondrion) Membrane-bound organelles present in the cytoplasm of all eukaryotic cells. They are the sites of energy production within cells.

Mitosis   Form of cell division that produces two cells, each with the same complement of chromosomes as the parent cell.

Molecule   A structure composed of two or more atoms held together by chemical bonds.

Monocytes   White blood cells that clean up viruses, bacteria, and fungi and dispose of dead cells and debris at the end of the inflammatory response.

Monohybrid cross   A mating between individuals who are each heterozygous at a given locus (e.g., $Bb \times Bb$).

Monosomy   A condition in which one member of a chromosome pair is missing; having one less than the diploid number $(2n-1)$.

Monozygotic (MZ) twins   Twins derived from a single fertilization event involving one egg and one sperm; such twins are genetically identical.

Mood   A sustained emotion that influences perception of the world.

Mood disorders   A group of behavior disorders associated with manic and/or depressive syndromes.

Mosaic   An individual composed of two or more cell types of different genetic or chromosomal constitution. In this case, both cell lines originate from the same zygote.

mRNA   A single stranded complementary copy of the nucleotide sequence in a DNA molecule that constitutes a gene.

Müllerian inhibiting hormone (MIH)   A hormone produced by the developing testis that causes the breakdown of the Müllerian ducts in the embryo.

Multifactorial traits   Traits that result from the interaction of one or more environmental factors and two or more genes.

Multiple alleles   Genes with more than two alleles have multiple alleles.

Muscular dystrophy   A group of genetic diseases associated with progressive degeneration of muscles. Two of these, Duchenne and Becker muscular dystrophy, are inherited as X-linked, allelic, recessive traits.

Mutation rate   The number of events producing mutated alleles per locus/per generation.

Natural selection   The differential reproduction shown by some members of a population that is the result of differences in fitness.

Neurofibromatosis   An autosomal dominant condition characterized by pigmentation spots and tumors of the skin and nervous system.

Nitrogen-containing base   A purine or pyrimidine that is a component of nucleotides.

Nondisjunction   The failure of homologous chromosomes to properly separate during meiosis or mitosis.

Nonhistone proteins   The array of proteins other than histones that are complexed with DNA in chromosomes.

Nonsense mutation   A mutation that changes an amino-acid-specifying codon to one of the three termination codons.

N-terminus   The end of a polypeptide or protein that has a free amino group.

Nucleolus   (plural: nucleoli) A nuclear region that functions in the synthesis and assembly of ribosomes.

Nucleosome   A bead-like structure composed of histones wrapped by DNA.

Nucleotides   The basic building blocks of DNA and RNA. Each nucleotide consists of a base, a phosphate, and a sugar.

Nucleotide substitutions   Mutations that involve substitutions, insertions, or deletions of one or more nucleotides in a DNA molecule.

Nucleus   The membrane-bounded organelle present in most eukaryotic cells; contains the chromosomes.

Oncogene   A gene that induces or continues uncontrolled cell proliferation.

One gene–one enzyme hypothesis   The idea that individual genes control the synthesis and therefore the activity of a single enzyme. This idea provides the link between the gene and the phenotype.

One gene–one polypeptide hypothesis   A refinement of the one gene-one enzyme hypothesis made necessary by the discovery that some proteins are composed of subunits encoded by different genes.

Oogonia   Mitotically active cells that produce primary oocytes.

Ootid   The haploid cell produced by meiosis that will become the functional gamete.

Organelle   A cytoplasmic structure having a specialized function.

Palindrome   A word, phrase, or sentence that reads the same in both directions. Applied to a sequence of base pairs in DNA that reads the same in the 5′ to 3′ direction on complementary strands of DNA. Many recognition sites for restriction enzymes are palindromic sequences.

Pattern baldness   A sex-influenced trait that acts like an autosomal dominant trait in males and an autosomal recessive trait in females.

Pedigree analysis   Use of family history to determine how a trait is inherited, and to determine risk factors for family members.

Pedigree chart   A diagram listing the members and ancestral relationships in a family; used in the study of human heredity.

Penetrance   The probability that a disease phenotype will appear when a disease-related genotype is present.

Pentose sugar   A five-carbon sugar molecule found in nucleic acids.

Pentosuria   A relatively benign genetic disorder of sugar metabolism characterized by the accumulation of xylulose in the blood and urine.

Peptide bond   A chemical link between the carboxyl group of one amino acid and the amino group of another amino acid.

Pharmacogenetics   A branch of genetics that is concerned with the inheritance of differences in the response to drugs.

Phenotype   The genetically controlled, observable properties of an organism.

Phenylketonuria (PKU)   An autosomal recessive disorder of amino acid metabolism that results in mental retardation if untreated.

Philadelphia chromosome   An abnormal chromosome produced by an exchange of portions of the long arms of chromosome 9 and 22.

Plasma cells   Cells produced by mitotic division of B cells that synthesize and secrete antibodies.

Plasmids   Extrachromosomal DNA molecules found naturally in bacterial cells. Modified plasmids are used as cloning vectors or vehicles.

Pleiotropy   The appearance of several apparently unrelated phenotypic effects caused by a single gene.

Polar body   A cell produced in the first or second division in female meiosis that contains little cytoplasm and will not function as a gamete.

Polygenic trait   A phenotype resulting from the action of two or more genes.

Polymerase chain reaction (PCR)   A method for amplifying DNA segments that uses cycles of denaturation, annealing to primers, and DNA-polymerase directed DNA synthesis.

Polymorphism   The occurrence of two or more genotypes in a population in frequencies that cannot be accounted for by mutation alone.

Polypeptide   A polymer made of amino acids joined together by peptide bonds.

Polyploidy   A chromosome number that is a multiple of the normal diploid chromosome set.

Polyps   Growths attached to the substrate by small stalks. Commonly found in the nose, rectum, and uterus.

Population   A local group of organisms belonging to a single species, sharing a common gene pool; also called a deme.

Porphyria   A genetic disorder inherited as a dominant trait that leads to intermittent attacks of pain and dementia, with symptoms first appearing in adulthood.

Positional cloning   Identification and cloning of a gene responsible for a genetic disorder that begins with no information about the gene or the function of the gene product.

Prader-Willi syndrome   A deletion of a small segment of the long arm of chromosome 15 that produces a syndrome characterized by uncontrolled eating and obesity.

Precocious puberty   An autosomal dominant trait expressed in a sex-limited fashion. Heterozygous males are affected, but heterozygous females are not.

Preimplantation testing   Testing for a genetic disorder in an early embryo; testing is done by removing a single cell from the embryo.

Pre-mRNA   The original transcript from a DNA strand, converted to messenger RNA (mRNA) molecules by removal of certain sequences, and addition of others.

Prenatal testing   Testing to determine the presence of a genetic disorder in an embryo or fetus; commonly done by amniocentesis or chorionic villus sampling.

Presymptomatic testing   Genetic testing for adult-onset disorders; testing can be done at any time before symptoms appear.

Primary oocytes   Cells in the ovary that undergo meiosis.

Primary spermatocytes   Cells in the testis that undergo meiosis.

Primary structure   The amino acid sequence in a polypeptide chain.

Prion   An infectious protein that is the cause of several disorders, including Creutzfeldt-Jakob syndrome and mad-cow disease.

Proband   First affected family member seeking medical attention.

Probe   A labeled nucleic acid used to identify a complementary region in a clone or genome.

Product   The specific chemical compound that is the result of enzyme action. In biochemical pathways, a compound can serve as the product of one reaction and the substrate for the next reaction.

Progeria   A rare autosomal recessive genetic trait in humans, associated with premature aging and early death.

Prokaryote   An organism whose cells lack membrane-bound nuclei with true chromosomes. Cell division is usually by binary fission.

Promoter   A region of a DNA molecule to which RNA polymerase binds and initiates transcription.

Promutagen   A nonmutagenic compound that is a metabolic precursor to a mutagen.

Prophase   A stage in mitosis during which the chromosomes become visible and split longitudinally except at the centromere.

Proto-oncogene   A gene that normally controls cell division and may become a cancer gene (oncogene) by mutation.

Pseudogene   A gene that closely resembles a gene at another locus, but is nonfunctional because of changes in its base sequences that prevent transcription or translation.

Pseudohermaphroditism   An autosomal recessive genetic condition that causes XY individuals to develop the phenotypic sex of females, but change to a male phenotype at puberty.

Purine   A class of double-ringed organic bases found in nucleic acids.

Pyrimidines   A class of single-ringed organic bases found in nucleic acids.

Quantitative trait   A trait controlled by two or more genes.

Quaternary structure   Structure formed by the interaction of two or more polypeptide chains in a protein.

R group   A term used to indicate the position of an unspecified group in a chemical structure.

Radiation   The process by which electromagnetic energy travels through space or a medium such as air.

Recessive   The trait unexpressed in the F1 but reexpressed in some members of the F2 generation.

Reciprocal translocation   A chromosomal aberration resulting in a positional change of a chromosome segment. This changes the arrangement of genes, but not the number of genes.

Recombinant DNA technology   Technique for joining DNA from two or more different organisms to produce hybrid, or recombined, DNA molecules.

Recombination   The process of exchanging chromosome parts between homologous chromosomes during meiosis; produces new combinations of genetic material.

Recombination theory   The idea that functional antibody genes are created by the recombination of DNA segments during B-cell maturation.

Regression to the mean   In a polygenic system, the tendency of offspring of parents with extreme differences in phenotype to exhibit a phenotype that is the average of the two parental phenotypes.

Rem   The unit used to measure radiation exposure and damage in humans. It is the amount of ionizing radiation that has the same effect as a standard amount of X rays.

Restriction enzymes   Enzymes that recognize a specific base sequence in a DNA molecule and cleave or nick the DNA at that site.

Restriction fragment length polymorphism (RFLP)   Variations in the length of DNA fragments generated by a restriction endonuclease. Inherited in a codominant fashion, RFLPs are used as markers for specific chromosomes or genes.

Retinoblastoma   A malignant tumor of the eye that arises in the retinal cells, usually occurring in children, with a frequency of 1 in 20,000. Associated with a deletion on the long arm of chromosome 13.

Retrovirus   Viruses that have RNA as their genetic material. During the virus life cycle, the RNA is transcribed into DNA. The name retrovirus symbolizes this backward order of transcription.

Ribonucleic acid (RNA)   A nucleic acid molecule that contains the pyrimidine uracil and the sugar ribose. The several forms of RNA function in gene expression.

Ribosomal RNA (rRNA)   A component of the cellular organelles known as ribosomes.

Ribosomes   Cytoplasmic particles composed of two subunits that are the site of gene product synthesis.

RNA polymerase   An enzyme that catalyzes the formation of an RNA polynucleotide chain using a template DNA strand and ribonucleotides.

Robertsonian translocation   Breakage in the short arms of acrocentric chromosomes followed by fusion of the long parts into a single chromosome.

Sarcoma   A cancer of connective tissue. One type of sarcoma in chickens is associated with the retrovirus known as the Rous sarcoma virus.

Schizophrenia   A behavioral disorder characterized by disordered thought processes and withdrawal from reality. Genetic and environmental factors are involved in this disease.

Secondary immunity   Resistance to an antigen the second time it appears, due to T and B memory cells. The second response is faster, larger, and lasts longer than the first.

Secondary oocyte   The haploid cell produced by meiosis that will become a functional gamete.

Secondary structure   The pleated or helical structure in a protein molecule that is brought about by the formation of bonds between amino acids.

Segregation   The separation of members of a gene pair from each other during gamete formation.

Semiconservative replication   A model of DNA replication that results in each daughter molecule containing one old strand and one newly synthesized strand. DNA replicates in this fashion.

Sense mutation   A mutation that changes a termination codon into one that codes for an amino acid. Such mutations produce elongated proteins.

Severe combined immunodeficiency disease (SCID)   A disease characterized by the complete lack of ability to mount an immune response; inherited as an X-linked recessive and in another form as an autosomal recessive trait.

Sex chromosomes   The chromosomes involved in sex determination. In humans, the X and Y chromosomes are the sex chromosomes.

Sex-influenced genes   Loci that produce a phenotype that is conditioned by the sex of the individual.

Sex-limited genes   Loci that produce a phenotype that is produced in only one sex.

Sex ratio   The relative proportion of males and females belonging to a specific age group in a population.

Sickle cell anemia   A recessive genetic disorder associated with an abnormal type of hemoglobin, a blood transport protein.

Sickle cell trait   The symptoms shown by those heterozygous for sickle cell anemia.

Sister chromatids   Two chromatids joined by a common centromere. Each chromatid carries identical genetic information.

Southern blot   A method for transferring DNA fragments from a gel to a membrane filter, developed by Edward Southern for use in hybridization experiments.

Sperm   Male haploid gametes produced by morphological transformation of spermatids.

Spermatids   The four haploid cells produced by meiotic division of a primary spermatocyte.

Spermatogenesis   The process of sperm production, including meiosis and the cellular events of sperm formation.

Spermatogonia   Mitotically active cells in the gonads of males that give rise to primary spermatocytes.

SRY   A gene called the sex-determining region of the Y, located near the end of the short arm of the Y chromosome, that plays a major role in causing the undifferentiated gonad to develop into a testis.

Start codon or initiator codon   A codon present in mRNA that signals the location for translation to begin. The codon AUG functions as an initiator codon.

Stem cells   Cells in bone marrow that produce lymphocytes by mitotic division.

Stop codons   Codons present in mRNA that signal the end of a growing polypeptide chain. UAA, UGA, and UAG function as stop codons.

Submetacentric   A chromosome with a centromere placed closer to one end than the other.

Substrate   The specific chemical compound that is acted upon by an enzyme.

Suppressor T cells   T cells that slow or stop the immune response of B cells and other T cells.

Synapsis   The pairing of homologous chromosomes during prophase I of meiosis.

T cells   White blood cells that originate in bone marrow and undergo maturation in the thymus gland.

Telocentric   A chromosome with the centromere located at one end.

Telophase   The last stage of mitosis, during which division of the cytoplasm occurs, the chromosomes of the daughter cells disperse, and the nucleus reforms.

Template   The single-stranded DNA that serves to specify the nucleotide sequence of a newly synthesized polynucleotide strand.

Teratogen   Any physical or chemical agent that brings about an increase in congenital malformations.

Tertiary structure   The three-dimensional structure of a protein molecule brought about by folding on itself.

Testicular feminization   An X-linked genetic trait that causes XY individuals to develop into phenotypic females.

Testosterone   A steroid hormone produced by the testis; the male sex hormone.

Tetraploidy   A chromosome number that is four times the haploid number, having four copies of all autosomes and four sex chromosomes.

Thymine dimer   A molecular lesion in which chemical bonds form between a pair of adjacent thymine bases in a DNA molecule.

Tourette syndrome   An autosomal dominant behavioral disorder characterized by motor and vocal tics and inappropriate language. Genetic components are suggested by family studies showing increased risk for relatives of affected individuals.

Transcription   Transfer of genetic information from the base sequence of DNA to the base sequence of RNA brought about by RNA synthesis.

Transfer RNA (tRNA)   A small RNA molecule that contains a binding site for a specific type of amino acid and a three-base segment known as an anticodon that recognizes a specific base sequence in messenger RNA.

Transformation   The process of transferring genetic information between cells by means of DNA molecules.

Transforming factor   The molecular agent of transformation: DNA.

Translation   The process of converting the base sequence in an RNA molecule into the linear sequence of amino acids in a protein.

Translocation   A chromosomal aberration in which a chromosome segment is transferred to another, nonhomologous chromosome.

Trinucleotide repeats   A form of mutation associated with the expansion in copy number of a nucleotide triplet in or near a gene.

Triploidy   A chromosome number that is three times the haploid number, having three copies of all autosomes and three sex chromosomes.

Trisomy   A condition in which one chromosome is present in three copies while all others are diploid; having one more than the diploid number ($2n + 1$).

Trisomy 13   The presence of an extra copy of chromosome 13 that produces a distinct set of congenital abnormalities resulting in Patau syndrome.

Trisomy 18   The presence of an extra copy of chromosome 18 that results in a clinically distinct set of invariably lethal abnormalities known as Edwards syndrome.

Trisomy 21   An aneuploid condition involving the presence of an extra copy of chromosome 21, resulting in Down syndrome.

Trophoblast   The outer layer of cells in the blastocyst that gives rise to the membranes surrounding the embryo.

Tubal ligation   A contraceptive procedure in women in which the oviducts are cut, preventing ova from reaching the uterus.

Tumor suppressor gene   A gene that normally functions to suppress cell division.

Turner syndrome   A monosomy of the X chromosome (45, X) that results in female sterility.

Uniparental disomy   A condition in which both copies of a chromosome are inherited from a single parent.

Unipolar disorder   An emotional disorder characterized by prolonged periods of deep depression.

Vaccine   A preparation containing dead or weakened pathogens that elicit an immune response when injected into the body.

Variable-number tandem repeats (VNTRs)   Short nucleotide sequences, repeated in tandem, that are clustered at many sites in the genome. The number of repeats at homologous loci is variable.

Vasectomy   A contraceptive procedure in men in which the vas deferens is cut and sealed to prevent the transport of sperm.

Vector   A self-replicating DNA molecule used to transfer foreign DNA segments between host cells.

Werner syndrome   A genetic trait in humans that causes aging to accelerate in adolescence, leading to death by about age 50.

X-linkage   The pattern of inheritance that results from genes located on the X chromosome.

X-linked agammaglobulinemia (XLA)   A rare, sex-linked, recessive trait characterized by the total absence of immunoglobulins and B cells.

XYY karyotype   Aneuploidy of the sex chromosomes resulting in a male with an XYY chromosome constitution. Such males are disproportionately represented in penal institutions.

Yeast artificial chromosome (YAC)   A cloning vector with telomeres and a centromere that can accommodate large DNA inserts, and that uses the eukaryote, yeast, as a host cell.

Y-linked   Genes located only on the Y chromosome.

Zygote   The diploid cell resulting from the union of a male haploid gamete and a female haploid gamete.

Credits

Figure 6.20a   Photo courtesy of Dr. Irene Uchida, Genetic Services, Oshawa General Hospital, Hamilton, Ontario, Canada

Figure 6.20b   ©Martin M. Rotker

Figure 6.21   Courtesy of Cytogenetics Laboratory, Loyola University Medical Center, Maywood, Illinois

Figure 6.30   ©Grant R. Sutherland/Visuals Unlimited

CHAPTER 7

Opener   ©David Phillips/Photo Researchers

Figure 7.1   ©David M. Phillips/Visuals Unlimited

Figure 7.6   From Lennart Nilsson, A Child Is Born, ©1966, 1977 Dell Publishing Company, Inc.

Figure 7.10   Photo courtesy of Dr. Marilyn Miller, University of Illinois-Chicago

Figure 7.16   From: Zourlas, P. et al. 1965. Clinical histologic and cytogenetic findings in male hermaphroditism. Obstetrics and Gynecology 25: 768–778, Figure 7

Figure 7.18a, b   ©M. Abbey/Photo Researchers

Figure 7.19   ©R. McNerling/Taurus Photos

Figure 7.20a   Photo courtesy of Dr. George Brewer, University of Michigan

Figure 7.20b   Photo courtesy of Dr. George Brewer, University of Michigan

CHAPTER 8

Opener   ©Lawrence Berkeley Laboratory/SPL/Photo Researchers

Figure 8.5   From: Franklin, R. and Gosling, R. G. 1953. Molecular configuration in sodium thymonucleate. Nature 171: 740-741.

Figure 8.10a   ©H. Swift and D. W. Fawcett/Visuals Unlimited

Figure 8.10b   ©D. R. Wolstenholme, I. B. Dawid, and D. W. Fawcett/Visuals Unlimited

Figure 8.11   Reprinted by permission. American Scientist. Nucleosomes: The structural quantum in chromosomes, by Olin D. and Olin A. 66: 704-711

Figure 8.12   From: Harrison, C. et al., 1982. High resolution scanning electron microscopy of human metaphase chromosomes. J. Cell Sci. 56: 409–422, Figure 3. The Company of Biologists, Limited

CHAPTER 9

Opener   ©Professor Oscar L. Miller/SPL/Photo Researchers

Figure 9.2   ©Arthur M. Siegelman/Visuals Unlimited

Figure 9.12   ©S. L. McKnight and O. L. Miller, Jr./SPL/Photo Researchers

CHAPTER 10

Opener   ©Walter Reinhart/CNRI/Photo Researchers

Figure 10.1   ©Museo del Prado, Madrid, Spain/Giraudon, Paris/Superstock

Figure 10.13a   B. Carragher, D. Bluemke, R. Josephs. From the Electron Microscope and Image Processing Laboratory, University of Chicago

Figure 10.13b   B. Carragher, D. Bluemke, R. Josephs. From the Electron Microscope and Image Processing Laboratory, University of Chicago

Figure 10.17a   ©Ray Coleman/Photo Researchers

Figure 10.17b   ©Ken Brate/Photo Researchers

CHAPTER 11

Opener   NASA/Science Source/Photo Researchers

Figure 11.11   ©Kenneth E. Greer/Visuals Unlimited

CHAPTER 12

Opener   ©P. A. McTurk, University of Leicester and David Parker/SPL/Photo Researchers

Figure 2.2   ©E. Webber/Visuals Unlimited

Figure 12.4   ©Le Corre-Ribeiro/Liaison Agency

Figure 12.9   ©1993 SPL/Custom Medical Stock Photo

Figure 12.13   ©1993 Michael Gabridge/Custom Medical Stock Photo

Figure 12.17   ©Alfred Pasieka/SPL/Photo Researchers

Figure 12.18   ©1992 NIH. All rights reserved.

Figure 12.20   Courtesy of BioRad Corporation, Richmond, California

CHAPTER 13

Opener   ©Louise Lockley/CSIRO/SPL/Photo Researchers

Figure 13.4   Drs. T. Reid and D. Ward/Peter Arnold, Inc.

Figure 13.6   ©Dr. Yorgos Nikas/SPL/Photo Researchers

Figure 13.7   Image courtesy of Affymetrix, Inc., Santa Clara, California

Figure 13.9   ©Leonard Lessin/Peter Arnold, Inc.

Unnumbered photo page 326 UPI/Corbis

Figure 13.11   ©Phillipe Plailly/Eurelios/SPL/Photo Researchers

Figure 13.13a   Courtesy of BioRad Corporation, Santa Clara, California

Figure 13.14   Photo courtesy of Calgene

CHAPTER 14

Opener   ©Nancy Kedersha/Immunogen/SPL/Photo Researchers

Figure 14.3   ©1991 Custom Medical Stock Photo

Figure 14.7   ©K. G. Murti/Visuals Unlimited

Figure 14.10   ©I. Associates/Custom Medical Stock Photo

CHAPTER 15

Opener   ©J. L. Carson/Custom Medical Stock Photo

Figure 15.6a, b   Photo courtesy of Dorothea Zucker-Franklin, New York University School of Medicine

Figure 15.15   ©Baylor College of Medicine/Peter Arnold, Inc.

Figure 15.16   ©David M. Phillips/Visuals Unlimited

Figure 15.17   ©J. L. Carson/Custom Medical Stock Photo

CHAPTER 16

Opener   ©Tim Beddow/SPL/Photo Researchers

Figure 16.2   ©Columbus Instruments/Visuals Unlimited

Figure 16.5c   ©C. Raines/Visuals Unlimited

Figure 16.6   ©Biophoto Associates/Science Source/Photo Researchers

Figure 16.8a   ©Science VU/E. R. Lewis, T. E. Everhard, and Y. Y. Zeevi/University of California/Visuals Unlimited

Figure 16.8c   ©T. Reese and D. W. Fawcett/Visuals Unlimited

Figure 16.9   ©Corbis

Figure 16.10   Dr. Monte Buschbaum, Mt. Sinai Medical Center, New York, New York

Figure 16.12   Dr. Dennis Selkoe, Center for Neurologic Diseases, Harvard Medical School, Boston, Massachusetts

CHAPTER 17

Opener   ©Tim Davis/Tony Stones Images

CHAPTER 18

Opener   ©Paul Hanny/Liaison Agency

Figure 18.10   ©Olivier Martel/Photo Researchers

Figure 18.11   ©Photo courtesy of Peter Weigand.

Figure 18.16   Institute of Human Origins

Figure 18.17   Photo courtesy of Cleveland Museum of Natural History

Figure 18.19   National Museum of Kenya/Visuals Unlimited

Figure 18.20a   ©John D. Cunningham/Visuals Unlimited

Figure 18.20b   Dr. David Frayer

CHAPTER 19

Opener   ©Carlos Goldin/SPL/Photo Researchers

Figure 19.1   ©1991 Custom Medical Stock Photo

Figure 19.2   ©Carlos Goldin/SPL/Photo Researchers, Inc.

Figure 19.3   ©Photo courtesy of Terry and Kerry Lavery.

Figure 19.4   ©Alexander Tsiarus/Science Source/Photo Researchers

Figure 19.5   Photo courtesy of Shelly Cummings

Index